SOLAR MATERIALS SCIENCE

Academic Press Rapid Manuscript Reproduction

Based on the Distinguished Lecture Series
sponsored by the New Mexico Joint Center
for Materials Science, September–December 1979.

SOLAR MATERIALS SCIENCE

edited by

Lawrence E. Murr

Department of Metallurgical and Materials Engineering
New Mexico Institute of Mining and Technology
Socorro, New Mexico

ACADEMIC PRESS   **1980**

A Subsidiary of Harcourt Brace Jovanovich, Publishers

New York   London   Toronto   Sydney   San Francisco

ACADEMIC PRESS, INC.
111 Fifth Avenue, New York, New York 10003

United Kingdom Edition published by
ACADEMIC PRESS, INC. (LONDON) LTD.
24/28 Oval Road, London NW1 7DX

Library of Congress Cataloging in Publication Data

Main entry under title:

Solar materials science.

Based on a lecture series sponsored by the New Mexico
Joint Center for Materials Science, held Sept.–Dec.,
1979.
Includes bibliographical references and index.
1. Solar energy—Congresses. 2. Materials—Con-
gresses. I. Murr, Lawrence Eugene.
TJ810.S67   621.47′028   80-18959
ISBN 0-12-511160-6

PRINTED IN THE UNITED STATES OF AMERICA

80 81 82 83   9 8 7 6 5 4 3 2 1

CONTENTS

Contributors   ix
Preface   xi

INTRODUCTION   1

Chapter 1 Introduction to Solar Materials Science   3
            Richard S. Claassen and Barry L. Butler

Chapter 2 Introduction to the Role of Crystal Defects in Solar   53
            Materials
            L. E. Murr and O. T. Inal

Chapter 3 Surface and Interface Studies and the Stability of Solid   93
            Solar Energy Materials
            A. W. Czanderna

SECTION 1   SOLAR COLLECTOR (PHOTOTHERMAL) MATERIALS

Chapter 4 The Optical Properties-Microstructure Relationship in   151
            Particulate Media: Optical Tailoring of Solar
            Absorbers
            Alex Ignatiev

Chapter 5 Solar Mirror Materials: Their Properties and Uses in   171
            Solar Concentrating Collectors
            R. B. Pettit and E. P. Roth

Chapter 6 The Effect of Soiling on Solar Mirrors and Techniques   199
            Used to Maintain High Reflectivity
            E. P. Roth and R. B. Pettit

Chapter 7 The Emissivity of Metals   229
            A. J. Sievers

Chapter 8   Fundamental Limits to the Spectral Selectivity of   255
             Composite Materials
             A. J. Sievers

Chapter 9   Composite Film Selective-Absorbers   277
             R. A. Buhrman and H. G. Craighead

Chapter 10   Corrosion Science and Its Application to Solar   319
             Thermal Energy Material Problems
             Steven L. Pohlman

SECTION 2   SOLAR STORAGE AND THERMOCHEMICAL MATERIALS

Chapter 11   Thermal Storage in Salt-Hydrates   377
             Maria Telkes

Chapter 12   Thermodynamic Basis for Selecting Heat Storage   405
             Materials
             Maria Telkes

Chapter 13   The Application of Reversible Chemical Reactions to   439
             Solar Thermal Energy Systems
             Raymond Mar

Chapter 14   Materials Science Issues Encountered During the   459
             Development of Thermochemical Concepts
             Raymond Mar

SECTION 3   SOLAR CONVERSION (PHOTOVOLTAIC) MATERIALS

Chapter 15   Introduction to Photovoltaics: Physics, Materials, and   489
             Technology
             Lawrence Kazmerski

Chapter 16   Research and Device Problems in Photovoltaics   551
             Lawrence Kazmerski

Chapter 17   Heterojunctions for Thin Film Solar Cells   585
             Richard H. Bube

Chapter 18   The Optimization of Solar Conversion Devices   619
              D. S. Ginley, M. A. Butler, and C. H. Seager

Chapter 19   Introduction to Basic Aspects of Plasma-Deposited   665
              Amorphous Semiconductor Alloys in Photovoltaic
              Conversion
              Richard W. Griffith

PROBLEMS AND REFERENCES

Chapter 20   Problems and References   735
              L. E. Murr

PROBLEM SOLUTIONS AND DISCUSSION

Chapter 21   Problem Solutions and Discussion   761
              L. E. Murr

Index   *781*

CONTRIBUTORS

Numbers in parentheses indicate the pages on which authors' contributions begin.

Richard H. Bube (585), *Department of Material Science and Engineering, Stanford University, Stanford, California*

Robert A. Buhrman (277), *School of Applied and Engineering Physics, Cornell University, Ithaca, New York*

Barry L. Butler (3), *Materials Branch, Solar Energy Research Institute, 1536 Cole Boulevard, Golden, Colorado*

M. A. Butler (619), *Solid State Materials, Division 5154, Sandia Laboratories, Albuquerque, New Mexico*

Richard S. Claassen (3), *Materials and Process Sciences, Organization 5800, Sandia Laboratories, Albuquerque, New Mexico*

H. G. Craighead (277), *Bell Telephone Laboratories, Holmdel, New Jersey*

A. W. Czanderna (93), *Materials Branch, Solar Energy Research Institute, 1536 Cole Boulevard, Golden, Colorado*

David Ginley (619), *Solid State Materials, Division 5154, Sandia Laboratories, Albuquerque, New Mexico*

Richard W. Griffith (665), *Department of Energy and Environment, 480, Brookhaven National Laboratory, Upton, New York*

Alex Ignatiev (151), *Department of Physics, University of Houston, Central Campus, Houston, Texas*

Osman T. Inal (53), *Department of Metallurgical and Materials Engineering, New Mexico Institute of Mining and Technology, Socorro, New Mexico*

Lawrence Kazmerski (489, 551), *Photovoltaics Branch, Solar Energy Research Institute, 1536 Cole Boulevard, Golden, Colorado*

Raymond W. Mar (439, 459), *Exploratory Chemistry, Divison 1-8313, Sandia Laboratories, Livermore, California*

L. E. Murr (53, 735, 761), *Department of Metallurgical and Materials Engineering, New Mexico Institute of Mining and Technology, Socorro, New Mexico*

Richard B. Pettit (171, 199), *Thermophysical Properties, Division 5842, Sandia Laboratories, Albuquerque, New Mexico*

Steve Pohlman (319), *Materials Branch, Solar Energy Research Institute, 1536 Cole Boulevard, Golden, Colorado*

E. P. Roth (171, 199), *Thermophysical Properties, Division 5842, Sandia Laboratories, Albuquerque, New Mexico*

C. H. Seager (619), *Solid State Materials, Division 5154, Sandia Laboratories, Albuquerque, New Mexico*

A. J. Sievers (229, 255), *Laboratory of Atomic and Solid-State Physics, Cornell University, Ithaca, New York*

Maria Telkes (377, 405), *Solar Thermal Storage Development, The American Educational Complex, American Technological University, P.O. Box 1416, Killeen, Texas*

PREFACE

Solar energy has been described as a limitless, cheap, and clean energy source of the future rivaling fusion power. However, like many other technologies, it is today undeveloped and inefficient in large measure. Development requires the use of specialized materials and materials that are abundant, cheap, and reliable enough to render the process cost effective or profitable. In a large sense, solar efficiencies and solar technology applications are currently materials limited. However, the science of materials has made large strides in the past two decades. Properties of materials in particular are becoming understandable and predictable on the basis of structure and microstructure, and the application of materials science has led in some cases to materials design criteria based on fundamental microstructural manipulations that include defect chemistry.

One of the principal drawbacks in the development of contemporary solar and other related energy technologies has been the lack of professionals, trained or expert, in these specialized areas or able to immediately apply their particular skills or professional training and experience. This can be rectified in the long term by alterations in curricula and the establishment of new curricula to address a particular energy technology area, while in the short term continuing education approaches are the more expedient routes. The lecture series upon which this book is based, sponsored by the New Mexico Joint Center for Materials Science, sought to serve both these approaches and at the same time provide for an overview of the state-of-the-art in specific solar technology areas. This book is therefore intended as a timely exposition of the materials aspects of contemporary and future solar energy development. In general, the major objectives are to provide advanced undergraduate and beginning graduate students with a formal vehicle introducing basic concepts of materials science, demonstrating their utilization in solar technology, and providing contemporary and state-of-the-art examples of this utilization and the technology; and to provide a basis for continuing education in the general area of solar materials science, and a concept of the application of materials science and related skills of technical or professional people to the development of a viable solar technology.

The subject matter addressed by the post-introductory chapters is divided into three categories: (1) solar collector (photothermal) materials; (2) solar storage and photochemical materials; and (3) solar conversion (photovoltaic) materials. Lectures originally composing the series upon which this book is based were organized to present an introduction to a topic, or an overview of a solar materials area, and then to present a state-of-the-art description of a research topic in this area, or

xi

encompassing the particular phenomena discussed in the overview. Thus, while this book is not specifically intended as an introduction to solar materials science, it does, in many areas, contain a significant, if not sufficient, level of introductory or background information, and can indeed serve as an introduction to solar materials science. To a large extent, this book depicts the theme of application of materials science and engineering in solar energy and solar technology development. It is, as was the original lecture series, intended as a single semester course for upper division undergraduates and graduate students in a wide range of curricula and programs, including metallurgy, metallurgical and materials engineering, materials science, electrical and mechanical engineering, engineering science, solid-state physics and chemistry, and other related areas and disciplines. Its organization is also intended to serve as a self-study guide and reference for a wide range of researchers and other professionals and for continuing education use. An extensive list of problems keyed to the chapters, and each containing a list of reference readings, along with a special section on problem solutions and discussion are intended to extend the topic coverage, and provide a mechanism for emphasizing and reinforcing the applications of materials science principles in solar energy research and development. The presentation assumes some background in the materials sciences as well as mathematics (through differential equations), chemistry, and physics (through modern or introductory solid-state physics). Introductory chapters provide some review of the necessary background or suggest adequate reference sources that may be consulted. In addition, the original lectures were video-taped in order to provide additional circulation of the program, and to aid in self-paced or independent study programs. These tapes are available through the New Mexico Joint Center for Materials Science by contacting the editor.

The New Mexico Joint Center for Materials Science sponsored the Distinguished Lecture Series, which composes this volume as a continuing education and engineering curriculum enrichment program for both upper division and graduate credit in the Fall, 1979 academic semester. It is a pleasure to acknowledge the financial support of the federal laboratory members: Sandia Laboratories, the Air Force Weapons Laboratory, and Los Alamos Scientific Laboratory. The facilities and participation made available through the University of New Mexico and the New Mexico Institute of Mining and Technology are also gratefully acknowledged. Finally, the typing of front matter and other sections of the complete camera-ready manuscript and the related editorial assistance of Lorraine Valencia were an important and very much appreciated contribution to the final composition of this book.

LAWRENCE E. MURR
Socorro, New Mexico

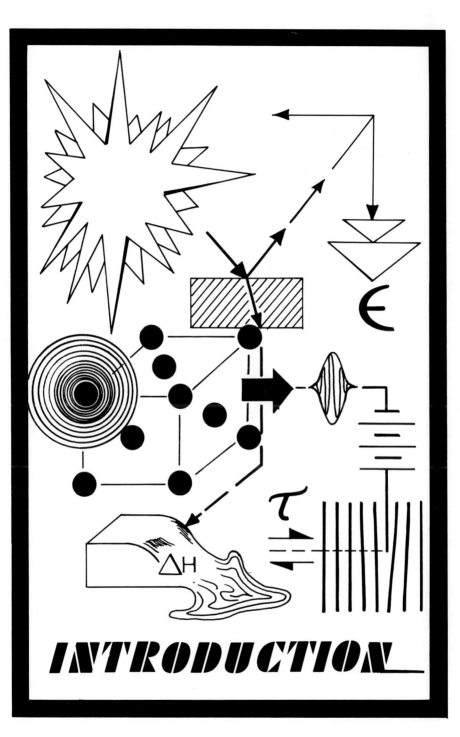

"As for the supplies of mechanical power, it is axiomatic that the exhaustion of our coal and oil-fields is a matter of centuries only. As it has often been assumed that their exhaustion would lead to the collapse of industrial civilization. I may perhaps be pardoned if I give some of the reasons which lead me to doubt this proposition.

Water-power is not, I think, a probable substitute on account of its small quantity, seasonal fluctuation, and sporadic distribution. It may perhaps, however, shift the centre of industrial gravity to well-watered mountainous tracts such as the Himalayan foothills, British Columbia, and Armenia. Ultimately we shall have to tap those intermittant but inexhaustible sources of power, the wind and the sunlight. The problem is simply one of storing their energy in a form as convenient as coal or petrol. If a windmill is one's back garden could produce a hundred weight of coal daily (and it can produce its equivalent in energy), our coalmines would shut down. Even tomorrow a cheap, foolproof, and durable storage battery may be invented, which will enable us to transform the intermittent energy of the wind into continuous electric power." [From J.B.S. HALDANE, "Daedalus or Science and The Future" (A paper read to the Heretics, Cambridge, on February 4, 1923)]

In the first of three introductory chapters, the enormous distortions in the predictions relating to world energy posture noted above by Haldane are made self evident, but the conclusions he arrives at some fifty-five years ago are unalterably reinforced. More importantly, the dependence of solar energy implementation, and its success as a viable energy alternative upon the science of materials is demonstrated. The range of solar energy systems is outlined and the materials problems as well as materials utilization in these systems is illustrated with specific examples. The overwhelming theme of this chapter is that materials along with design criteria pervade all aspects of the solar materials problems in the United States, and that there is a need for confidence in reliable performance and system lifetime projections, which can only be obtained through a knowledge of materials and materials problems. The second chapter provides for a review of crystal structure and crystal defects as well as an overview of the structure-property relationships which can be considered to figure prominently in the selection and performance of materials for specific solar energy applications. This chapter also provides for a fundamental understanding of solar materials microstructures as well as examples of principal techniques for their observation and analysis. The third and final chapter in this introductory section outlines the analytical approaches for the elucidation and study of surface and interfacial phenomena crucial to all solar materials systems to be described in three principal sections of this book. Some selected examples of current solar materials problems are also presented to emphasize the application of the techniques described.

CHAPTER 1

INTRODUCTION TO SOLAR MATERIALS SCIENCE[1]

Richard S. Claassen

Sandia Laboratories[2]
Albuquerque, New Mexico

Barry L. Butler

Solar Energy Research Institute[2]
Golden, Colorado

I. INTRODUCTION

Solar radiation can be converted to forms of energy which man
finds useful through a number of natural phenomena or by the use
of technology. A broad program is underway to understand and
exploit the natural phenomena as well as to develop and improve
the full range of technologies needed to make solar energy a significant contributor to our national energy supply. One important aspect of the technology development is materials. The purpose of this series is to present an up-to-date report on the
status of materials research and development which will play a
major role in the more promising approaches to converting solar

[1]*This work is supported by the Division of Solar Technology,
U.S. Department of Energy (DOE), under contract DE-AC04-76-DP00
789 and contract EG-77-C-01-4042.*
[2]*U.S. DOE facility*

3

radiation to other useful energy forms.

It is convenient to view the entire solar energy program from two broad frames of reference; the first programmatic, the second technical. The national solar program in the United States has been organized according to the conversion paths indicated in Figure 1 (1). This series of lectures will focus on the technological side of Figure 1 where materials science and engineering have the greatest opportunity to influence program success. From a conceptual, or technical viewpoint, it is appropriate to divide all solar systems into two broad classes as illustrated in Figure 2 (2); 1) those in which the primary interaction is a discrete quantum process, and 2) those in which the primary process is production of heat.

This series has necessarily been limited to three principle areas, namely, solar collector materials, photochemical conversion and storage, and solar conversion materials. Although many important elements of the solar program have been left out, the examples covered will illustrate the central importance of materials development and will indicate the challenges to the materials community. The next section provides a general framework in which materials development can be considered starting with emphasis on performance, life, and cost and ending with a strong recommendation for systems studies prior to expending effort on materials development. The third section discusses optical elements which are common to both thermal and quantum conversion processes. The fourth section identifies examples of problems in conversion processes. The final section is a brief summary of our viewpoint.

II. FRAMEWORK FOR MATERIALS DEVELOPMENT IN SOLAR ENERGY

A. *Principle Issues*

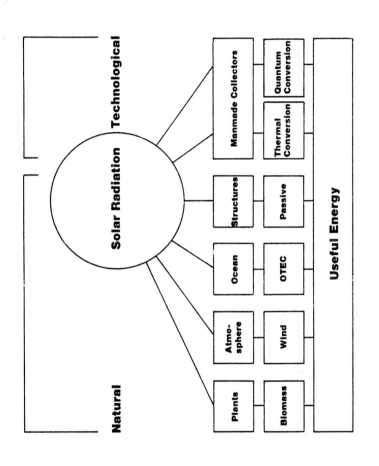

Figure 1.   Solar Energy Conversion Paths

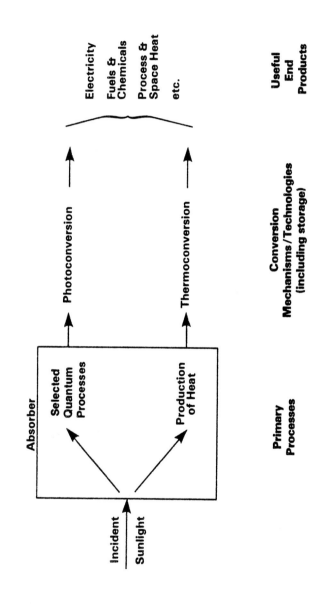

FIGURE 2. Solar Energy Conversion Processes (2)

It must be understood clearly that the principle issues in the development of solar energy are neither scientific proofs nor engineering feasibility demonstrations. We know that solar energy will work. What remains to be established is acceptable performance, adequate life, and low enough cost to enable widespread use.

At first glance it might seem that materials performance in solar devices would be modest, but in many cases, the requirements on materials are quite demanding. A concentrator, for example, which focuses the sun on a small receiver must cover large areas and have accurate optics, high specular reflectance, and long-term durability in a hostile environment. Photovoltaic devices (solar cells) must have the maximum possible efficiency to minimize overall systems costs while maintaining long-term durability in the terrestrial environment.

Although solar energy systems require no fuel, they are characterized by high initial cost. Service life measured in decades is required to amortize that initial cost at acceptable rates. Elements which are exposed to the atmosphere such as mirrors or surface covers (glazings) will be subject to degradation mechanisms such as deposition of dust, corrosion by water vapor and chemicals in the atmosphere and photochemical degradation. Because of the daily variation of solar insolation and because of short term fluctuations resulting from cloud passage some elements of the solar system will be subject to cyclic fatigue. Problems such as these must be resolved in advance so that the buying public can have confidence in the expected lifetimes of the solar devices.

Cost reduction is an essential element of the solar energy program. Concern for cost is reflected in considerations of materials availability, in the development and demonstration of mass production processes which are inherently inexpensive and in

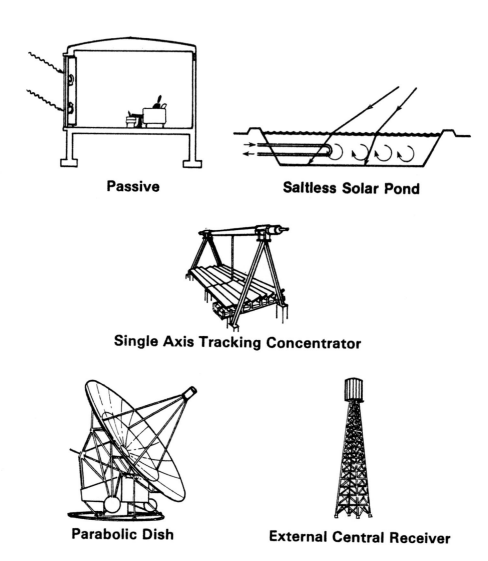

FIGURE 3.   Solar Energy Collector

the design of solar energy systems of minimum complexity.

Anyone working on materials development for solar energy systems should frequently check his progress against the dominating
parameters of performance, life, and cost.

B. Generic Systems Referred to in this Series

Figures 3 and 4 provide pictorial illustrations of the principle solar energy conversion technologies. This series will
emphasize the lower three optical concentrators shown in Figure
3: namely, the single-axis tracking system and the two-axis
tracking parabolic dish and the central receiver systems. Secondary solar systems, such as ocean thermal energy conversion and
wind which use the earth as the collector, will not be considered
in this report. In central receiver systems the elevated focal
point is surrounded by a large number of tracking heliostats.
Solar cells or photochemical cells may be deployed in an array as
illustrated in Figure 4 or may be placed at the focal point of
one of the concentrating systems.

C. Components, Functions, and Disciplines

The energy conversion concepts just outlined above require a
broad range of system components shown in Figure 5. These system
components require specific materials functions. The materials
functions can be loosely defined as: optical functions including
reflection, transmission and absorption; energy storage and transfer functions which include the use of energy transfer media and
phase change materials; and finally, the structural functions
which provide the strength and stiffness required. Research
disciplines can be identified with each of these required materials functions. Specific examples will be cited later in this
lecture series. But to give an example, a transmitting material

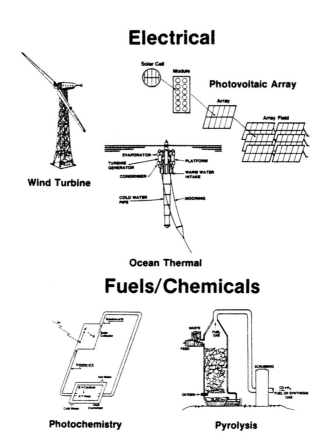

FIGURE 4. Solar and Renewable Energy Collectors

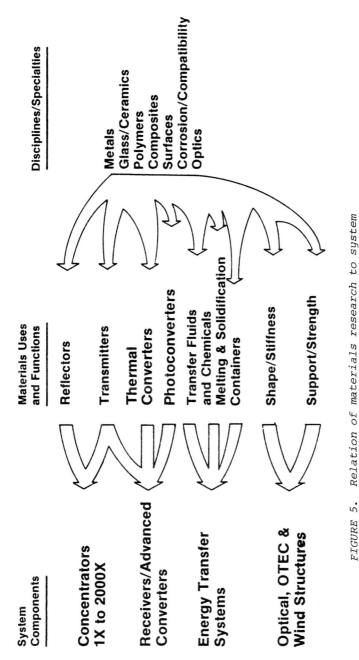

FIGURE 5. Relation of materials research to system components.

would require the disciplinary studies of optics, solid state
chemistry, glass and polymers. These studies would be aimed at
understanding the optical transmittance of these materials as a
function of material chemistry, radiation environment, and terrestrial environment. The output of this materials research
would be a basic understanding which could lead to improved
transmittance, longer life, or a materials processing breakthrough which could yield lower capital cost. Thus, the materials
research to support the solar conversion technologies will provide the basis for improving the elements of life cycle costs.

D. *Discipline Relationships*

The problems posed by using materials in solar energy conversion systems are approached from a disciplinary point of view
because they are similar to problems found with the application
of materials to other areas of technology. Studying the basic
materials properties and structures property relationships of
certain materials such as metals, ceramics, polymers, composites,
etc., allows the understanding generated in one technology to be
applied to another. Disciplinary research will aid in the development and tailoring of materials for solar energy systems.

In general materials in solar applications must perform in
the terrestrial environment for a long time without degradation.
In all systems which either use a glazing or a reflecting surface
or refracting optics, dust and dirt attachment and adhesion to
the mirror, glazing or lens will decrease the transmitted or reflected light which the converter can utilize. Refurbishing or
cleaning can be a major expense in a large area solar array.
Thus, materials research in the area of dust adhesion to glasses
and plastics is an area which is of generic importance to the
optical collecting system. Once the solar optical materials
needs are identified the heat transfer and energy transfer needs

come next. The major concern in the area of energy transfer is
the fluid and container compatibility which will limit system
life. Compatibility of materials used in photovoltaic device
electrical contacts and optical encapsulating materials is another
concern. Containment contamination or poisoning of quantum processes which would hinder the conversion of light to chemical energy is also important in photochemical conversion systems. In
the area of non-optical collecting systems such as wind and OTEC,
the major issues respectively are fatigue life of the turbine
blades as well as their costs and in OTEC systems it is the performance degradation due to corrosion and biofouling which will
take place over the life of the plant.

The materials program that must be developed in support of
solar energy requires both the ability to select the materials
with the right properties and also to develop basic supporting
understandings of these materials. The material selection requires the applications of solid state physics and chemistry and
chemical engineering. For example, the selection of the right
material for a photovoltaic device would require an understanding
of solid state physics. The understanding of a material for a
photochemical process would require a basic understanding of the
interaction of light and chemicals. For energy transfer systems,
one would need to know which chemical reactions to select and
under what types of conditions they would operate. Physics and
chemical engineering support the materials selection for the conversion process and materials engineering would support materials
selection for the more classical stress, strain and corrosion
and compatibility conditions.

Once the materials are identified for each of the areas of
technology, then the supporting disciplines of metallurgy,
ceramics, polymers, and surface science come into play to understand the material properties and their interactions with each

other in the system environment. By way of example, Table I has
been included to indicate where costs, performance, and life improvements could result from research in various materials
disciplines.

E. Identifying Critical Materials Issues

A critical materials issue is one that limits the cost, performance, life or availability of a material or a particular solar energy conversion system. Before expending development
effort on particular material problems, it is necessary to develop a perspective of the importance of that problem in the overall solar energy program. A system study can provide that perspective. Because the materials requirements are system specific
a series of examples will be used to illustrate the systems
approach to identify the critical materials areas. The first example is a broad study of materials availability; the second example traces all the important parameters in a specific collector
design; and the third which appears in the subsequent section
treats the absorber material.

Ray Watts, et. al., at Battelle Pacific Northwest Laboratory,
have developed a methodology for identifying material constraints
in the photovoltaic program (3). Their analysis places emphasis
on material supply and cost. They then apply the general concept
to specific designs for silicon and gallium arsenide cells. One
output of this study is the identification of aluminum, antimony,
copper, and gallium as significant material availability problems.
Ways are suggested to mitigate material supply problems using the
framework of the study.

George Treadwell of Sandia Laboratories has made a systems
analysis of a parabolic-cylindrical solar collector (4). The
study is a design tradeoff of the many parameters involved;
namely, the rim angle, the receiver tube size, asymetries in the

TABLE I. *Impacts of Materials Research on Life Cycle Cost Elements*

	Performance Changes	Cost	Life
Metallurgy			
Containment Cyclic Loading	Lower Corrosion Rates Higher Temperatures	Thinner Walls Lower Alloy Additions	Predictable Lives More Cycles to Failure
Reflectors	Ag—Alloys (Al) Increase Reflectance	Same as Ag	More Atmospheric Resistance
Polymers			
Encapsulants Transmitters	Lower Permeability Higher Solar Transmittance	Thin Films Thin Films	UV Resistance UV Resistance
Ceramics/Glass			
Transmitters Absorbers	Lower Fe^{+2} Ion Content Higher Temperatures	Thin Sections Cheap Raw Materials	Very Stable Thermal Shock Resistance
Surface Science			
Optical Degradation of Mirrors	Higher Reflectivity of Ag	Processing	Atmospheric Resistance
Absorbers	Higher α and Lower ε	Processing	Collector Environment Resistance
Catalysis	Higher Reaction Rates	Thin Films	Low Poisoning

collector, reflection and transmission properties of the mirror
and receiver tube envelope respectively, and the optical absorption and infrared emittance of the coating on the receiver tube.
In addition, the questions of whether to have a glass jacket a-
round the receiver tube is addressed as is the gap between the
jacket and receiver. Energy balance calculations are used to
test the sensitivity of the overall system performance to each
of the above parameters. A number of conclusions can be drawn.
For example, rim angles beyond 90° do not add much to system
performance. On the other hand, considerable gain can be a-
chieved by increasing the transmission of the glass jacket beyond that of conventional glass. Anti-reflection coatings on the
glass surface make an important improvement in system performance.

III. OPTICAL ELEMENTS

Our definition for this topic can be stated: optical elements
interact with incoming light by transmission, reflection, and
absorption but do not in themselves produce useful output energy.
For convenience we include support structures in this category.
This section will highlight important materials problems in
trough structures, selective absorbers, transmitters, and reflectors. Tables II, III, and IV give nominal optical properties
and durability data for a wide range of materials (5).

A. *Collector Types*

The solar conversion systems that we will be dealing with are
nonconcentrating systems, single axis concentrating systems and
two axis concentrating systems. The nonconcentrating systems include some photovoltaic devices, shallow solar ponds, passive
applications and thermal flat plate applications. The nonconcen-

TABLE II. Properties of Selected Solar Absorber Surfaces (6)

Material	Technique	Supplier (S)/ Developer (D)	α_s	$\varepsilon_t(T)$	T Stability** (°C)
Black Chrome	electro-deposited	Many	0.94-0.96	0.05-0.10(100) 0.20-0.25(300)	300
Pyromark	paint	Tempil	0.95	0.85(500)	<750
S-31 (nonselective)	paint	Rockwell International	0.8-0.85	0.8-0.85	>550
SOLARTEX	electro-deposited	Dornier (W. Germany)	0.93-0.96	0.14-0.18(310)	700
SOLAROX (proprietary)	"	"	0.92	0.20	200
Black Epoxy	paint	Amicon Corp.	NA	NA	NA
436-3-8	"	Bostik (U.S.M. Corp.)	0.90	0.92	NA
Enersorb	"	Desoto	0.96	0.92	NA
7729	"	C. H. Hare	0.96	0.90-0.92	NA
R-412	"	Rusto-leum Co.	0.95	0.87	NA
5779	"	"	0.95	0.90	NA
Nextel (nonselective)	"	3-M	0.97-0.98	>0.90	150
NOVAMET 150 (proprietary)	"	Ergenics	0.96	0.84	800 (1 hr)
MAXORB	(Proprietary)	Ergenics	0.97($\pm$.01)	0.10($\pm$.03)	150(20 wks) <400(1 hr)
Tabor Black (NiS/ZnS)	electrodeposited + overcoat	Miromit	0.91	0.14	

TABLE III. Thermal and Optical Properties of Cover Plate Materials (7)

Material	Index of Refraction	Normal Incident Short-wave Transmittance (λ=0.4-2.5μ)	Normal Incident Long-wave Transmittance (λ=2.5-40μ)	Thickness* (m)	Density (kg/m^3)	Specific Heat (J/°K-kg)	Thermal** Capacity (W-hr/°K-m^2)	References
Glass	1.518	0.840	0.020	3.175×10^{-3}	2.489×10^3	0.754×10^3	1.659	(8)
Fiberglass Reinforced Polyester (Sunlite)	1.540	0.870	0.076	6.350×10^{-4}	1.399×10^3	1.465×10^3	0.361	(8)
Acrylic (Plexiglas)	1.490	0.900	0.020	3.175×10^{-3}	1.189×10^3	1.465×10^3	1.534	(8)
Polycarbonate (Lexan)	1.586	0.840	0.020	3.175×10^{-3}	1.199×10^3	1.193×10^3	1.260	(8)
Polytetrafluoroethylene (Teflon)	1.343	0.960	0.256	5.080×10^{-5}	2.148×10^3	1.172×10^3	0.036	(8,9)
Polyvinyl Fluoride (Tedlar)	1.460	0.920	0.207	1.016×10^{-4}	1.379×10^3	1.256×10^3	0.049	(8)
Polyester (Mylar)	1.640	0.870	0.178	1.270×10^{-4}	1.394×10^3	1.046×10^3	0.051	(8)
Polyvinylidene Fluoride (Kynar)	1.413	0.930	0.230	1.016×10^{-4}	1.770×10^3	1.255×10^3	0.063	(10)
Polyethylene (Marlex)	1.500	0.920	0.810	1.016×10^{-4}	0.910×10^3	2.302×10^3	0.059	(8, 11)

*These values correspond to the thickness associated with the stated transmittances. They were used in the simulations to compute thermal capacity and are representative of commercially available film thicknesses.

**Thermal capacity = (Thickness) (Density) (Specific heat)

Material	Supplier	Estimates of Solar Weighted Reflectance[b] at Receiver Acceptance Angle τ			
		τ=4mr	10mr	18mr	$R_s(2\pi)$
I. Second-Surface Glass					
(a) Laminated Float Glass - 2.7mm thick -silvered	Carolina Mirror Co.	0.83	0.83	0.83	0.83
(b) Laminated Low-Iron Sheet Glass - 3.35mm thick - silvered	Gardner Mirror Co.	0.90	0.90	0.90	0.90
(c) Corning Silvered Microsheet Co.-0.114 mm thick- Mounted on optically flat plate	Corning Glass	0.76	0.87	0.92	0.95
(d) Corning 0317 Glass - 1.5 mm thick - Evaporated silver	Corning Glass	0.95	0.95	0.95	0.95
II. Metallized Plastic Films					
(a) 3M Scotchcal 5400 Laminated to backing sheet	3M Company	0.60	0.84	0.85	0.85
(b) 3M FEK-163 Laminated to backing sheet	3M Company	0.83	0.85	0.85	0.85
(c) Aluminized 2 mil FEP Teflon (G405600) Laminated to backing sheet	Sheldahl	0.70	0.81	0.82	0.87
(d) Silvered 2 mil FEP Teflon (G400300) Mounted on Optically Flat Plate	Sheldahl[a]	0.73	0.82	0.90	0.96
(e) Silvered 5 mil FEP Teflon (G401500) Mounted on Optically Flat Plate	Sheldahl[a]	0.77	0.83	0.89	0.95
(f) Front Surface Aluminized Mylar (200XM648A) stretched membrane	Boeing	0.88	0.88	0.88	0.88
III. Polished, Bulk Aluminum					
(a) Alzak Type I Specular	Alcoa				0.85
Perpendicular to rolling marks		0.61	0.68	0.76	
Parallel to rolling marks		0.68	0.76	0.83	
(b) Kinglux No. C4	Kingston Ind.				0.85
Perpendicular to rolling marks		0.67	0.71	0.75	
Parallel to rolling marks		0.69	0.71	0.75	
(c) Type 3002 High Purity Al - Buffed and Bright Anodized	Metal Fabrications, Inc.[a]	0.44	0.60	0.71	0.84

a) Experimental materials not produced in high production, so cost information is lacking.
b) Estimated from ~500 nm specularity data (Ref. 12) and solar weighted total hemispherical reflectance data. Standard deviation of the estimates is about 2%.

trating systems usually require a glazing which allows the light to reach the absorber/converter element and may be called upon to minimize heat loss from that element. In most cases, the transmitting glazing protects the components from the harsh terrestrial environment. Nonconcentrating systems also require a large area of an absorber/converter which could be a photoconversion device or a thermal conversion device.

Single axis concentrating systems such as parabolic troughs or linear fresnel lenses concentrate light onto a line where the photo or thermal conversion functions take place. Figure 6 shows a typical trough. These concentrating systems require either a high performance optical reflector which has both high lens focal quality and also high reflectance or a high quality transmitting lens. The converter system is now much reduced in size to as

FIGURE 6. A parabolic-cylindrical collector system.

little as 1/20th to 1/100th of what would have been needed in a
flat plate system of the same aperture area. The operating temperatures of such systems can approach 300°C. The materials problems generated by this type of optical concentrator include the
design of low cost stable structures to form the focusing reflecting or refracting optical elements, corrosion resistant mirror materials to reflect the solar energy, stable lens materials, receiver heat transfer materials which can stand high temperatures
and high rate of change of temperature produced when the sun is
obscured by a cloud. These systems also offer the potential for
quantum optical reactions to take place at the focus which may
require the transfer of solar energy directly to a photochemical
or photobiological process or to a photovoltaic device which under
these circumstances would probably need to be cooled.

   As is true for all solar energy technology, there is great
pressure to reduce the cost of the structure for a trough. Since
cost correlates with weight, we desire a lightweight structure
fabricated from low-cost raw material. The performance requirements for the trough are: 1) to provide the correct optical
shape to the reflective surface; 2) to maintain this shape within specified tolerances during operation; 3) to survive and protect the reflective surface under extreme weather conditions; 4)
to withstand long-term exposure to the environment. The first
item is conveniently treated in terms of the slope error at each
point on the mirror. Biggs and Vittitoe have developed a
stochastic methodology for deducing system performance from experimental measurements of geometrical errors (13). Accelerated
aging tests have shown that materials have a strong influence on
the degradation of the focal image quality with time.

   An engineering estimate suggests that three pounds per
square foot is a practical target for a lightweight structure of
minimum cost which will minimize structural loads on the base

support. The structure must have adequate stiffness to retain
its optical shape. Of particular importance is adequate flexural rigidity to withstand wind loading. An engineering estimate for this value is 0.5×10^{6} pound-inch. Using the concepts
of low aerial density and flexural rigidity a figure of merit
may be developed to compare various material candidates (14).

$$\Lambda = \frac{E}{\rho^{3}(1 - \nu^{2})}$$

where E is Young's modulus, ρ is density and ν is Poisson's ratio. For the above values $\Lambda_{min} = 231\frac{Nm^{7}}{kg^{3}}$. Table V shows that
for simple slab construction some of the obvious material can-
didates fall short of the minimum figure of merit. By going to
more sophisticated structures, at least two promising candidates
have been identified. They are a trough structure section fabricated from sheet molding compound with an integral rib structure
(Figure 7) and steel bonded to aluminum honeycomb (15).

TABLE V. Figure of Merit for Slab Type Mirror Structures (14)

Material	Density gm/cm^{3}	Poisson's ratio ν	Modulus E GPa	Figure of Merit Λ $\frac{Nm^{7}}{kg^{3}}$
Steel	7.80	0.30	199.9	.5
Aluminum	2.77	0.33	68.9	3.6
Sheet Molding	1.80	0.32	12.4	2.4
Plywood (fir)	.55	0.30	7.6	50.2

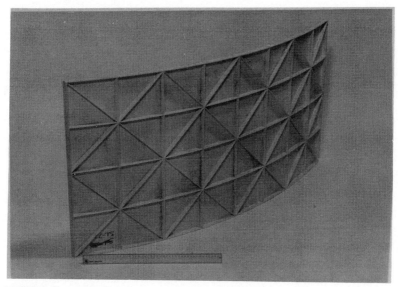

FIGURE 7. Parabolic reflector molded with ribs from sheet moulding compound.

Double axis concentrators such as parabolic dishes, circular Fresnel lenses, or central receiver systems provide concentration ratios much higher than can be achieved with   single-axis con- centrators, sometimes with ratios exceeding 2000. As the concen- tration ratio increases for both single and two-axis focusing systems the accuracy of the lens system must be improved because the equivalent system F number keeps increasing. In addition, the receiver temperatures can go as high as 600 to 1000°C causing many high temperature materials problems and in particular, a severe thermal shock problem when a cloud passes over. As the concentration ratio and power density increase so does the a- mount of area covered by the optical concentrator. This requires very large areas of high quality, high accuracy mirror or lens to be deployed. The problems associated with maintaining and ser- vicing a large mirror field require that very high reliability long life components be used.

B. *Selective Absorbers*

For thermal conversion systems it would be desirable to have
all of the incident light absorbed by the heat transfer surface
while a minimum of energy is reradiated. Properties of typical
commercial materials are listed in Table II. Figure 8 illustrates
why it may be possible to approach this goal while Figure 9,
drawn to the same scale, illustrates that strong selectivity in
the right wavelength range is provided by black chrome on nickel
(16). Note that a material, Pyromark® (Tempil Corporation) paint,
designed to be highly absorptive in the visible region also has
high absorption and therefore high emittance in the infrared.
The improvement which can be achieved by selective absorbers is
strongly dependent on specific design. Figure 10 shows the
difference in efficiency for a flat-plate, double-glazed collector
covered with a selective absorber and one with a non-selective

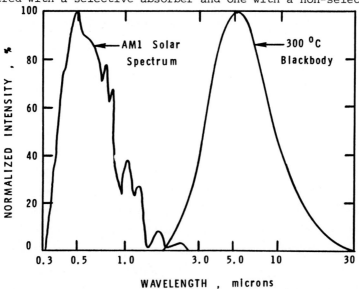

FIGURE 8. *Normalized intensity versus wavelength for solar
radiation which has passed through one atmosphere and for a black
body radiating at 300°C. The almost complete separation of spectra makes selective absorber effective.*

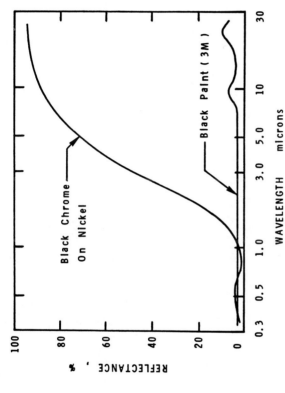

FIGURE 9. *Reflectance versus wavelength for two materials of widely different optical characteristics. The solar absorption coefficient, α_s, for the black chrome is 0.95; the emittance, ε, at $300^\circ C$ is 0.25 (16).*

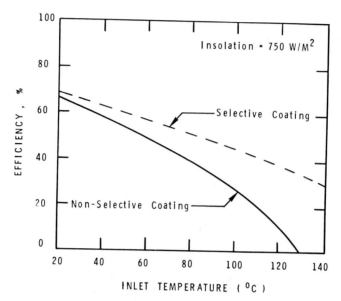

*FIGURE 10. Double glazed flat plate collector efficiency as a
function of fluid inlet temperature for a specific set of conditions given in the reference (17). The nonselective coating has
an emissivity ε of 0.95, the selective coating an ε of 0.10.*
absorber (17). Note that for low temperatures appropriate to

space heating there is little gain from the selective absorber

but at the higher temperatures needed for air conditioning ma-

chines the advantage is substantial. (A. Siever treats the sub-

ject of absorption, emission, and selectivity in a later chapter).

With linear concentrating collectors the influence of selec-

tive absorbers is significant. Figure 11 illustrates this in-

fluence of selective absorption on collector efficiency for a

system typical of a trough collector. Note that for operating

temperatures of 300°C efficiency is improved from 16 to 42% by

use of a selective coating (18).

For a central receiver, higher operating temperatures can be

used to provide greater thermodynamic efficiency in electric

generation. The higher flux on the receiver tends to mask the

importance of the power being reemitted. On the other hand, the

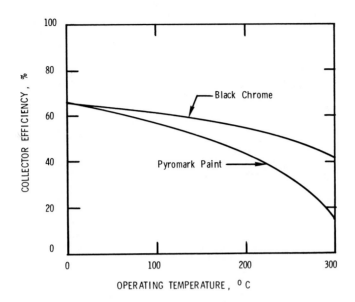

FIGURE 11.   Collector efficiency as a function of fluid operating temperature for a parabolic-cylindrical collector (18).

higher receiver temperatures increase the emittance proportional
to T^4. The relative importance of these two trends is highly
specific to the design and to the operating parameters. Some
feeling for these tradeoffs may be gained from Figure 12 which
indicates the improvement in efficiency which might be gained by
going from a nonselective absorber to an ideal selective coating.
Note that a plant would be operated to provide a constant temperature working fluid.   That means that a vertical trajec-
tory on the graph will be followed as the sun goes through its
daily cycle. A selective absorber which contributes little at
noon time may still be significant in the morning and afternoon.
We do not presently have selective absorbers which are stable at
temperatures above 400°C.   (R. A. Buhrman discusses cermet films
for high temperature selective absorbers in a later chapter).

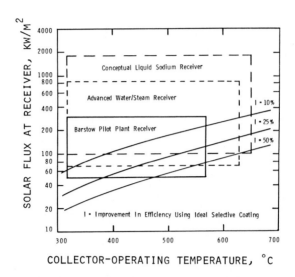

COLLECTOR-OPERATING TEMPERATURE, °C

FIGURE 12. A parametric display of improvement in solar flux to thermal energy efficiency in changing from nonselective to ideally selective absorbers. A practical system will probably operate at constant temperature so that the greater improvement at lower fluxes will be important during early and late daylight hours (19).

C. Transmitters

Transmitting materials are required to separate sensitive components from ambient atmosphere and to block re-radiation of long wavelength thermal energy. The chosen materials should have the highest possible transmission in the solar spectrum, but that desirable attribute is degraded by reflection at the front and back surfaces and absorption in the body of the transmitting material. Typical transmitting materials are listed in Table III. In most mirror designs the light passes through the transmitter twice, thereby doubling the losses by absorbtion.

Reflection at the surface is caused by the difference in index of refraction of the two media. The fraction reflected is given by the following formula:

$$\frac{I}{I_o} = (\frac{\eta_1 - \eta_2}{\eta_1 + \eta_2})^2$$

Where I and I_o are the incident and reflected intensity, and the indexes of refraction of air and glass are η_1 and η_2. For ordinary glass and air, the total reflection loss from the front and back surfaces is about 8%. It has been shown that a thin layer on the surface with an intermediate index of refraction can reduce reflection losses to about ½%. Beauchamp has shown that a silica film produced by precipitation can affect this improvement (20). The precipitation process, however, is expensive and is difficult to control in production. A practical, inexpensive process for producing the intermediate index of refraction layer would make an important contribution to the solar program.

Absorbtion in glass has been viewed as a potential problem. The best mirrors, from a geometric point of view, are made from float glass. In the final stage of production of float glass, the glass floats on a pool of tin which is stabilized by a reducing atmosphere. The iron impurity present in most common glass is reduced to Fe^{++} in the reducing atmosphere. Fe^{++} has a broad absorbtion band around 1000 nm. The marked effect can be seen in the solid curve of Figure 13. The process for glass production can be changed so that the iron impurity is either eliminated or oxidized to the Fe^{+++} state in the glass. The remaining absorbtion is then almost negligible as indicated by the dashed curve of Figure 13. Other transmitters such as polymers or other glasses will need measurement to be sure that absorbtion is not a problem.

Probably the most important questions about transmitters concern life and cleaning procedures. Dust accumulation on the first surface causes absorbtion and scattering. How quickly this degrades system performance depends on system design. Scattering is particularly important in collectors which depend on accurate

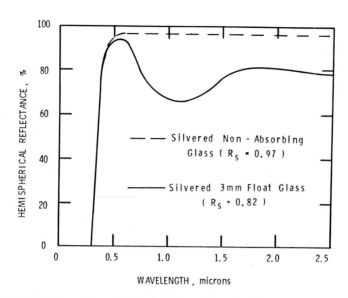

FIGURE 13. The hemispherical reflectance of glass silvered on the back surface. The principal absorbtion in the float glass is due to an iron impurity in the Fe⁺⁺ state (21).

focusing. Cleaning procedures are being studied but those investigated to date are expensive and may scratch the front surface which in turn will degrade performance. Degradation due to accumulation of dirt is an important problem. Local environment tests at Albuquerque have indicated that 5% reduction in reflection is a typical loss in a two-month exposure (22). The actual degradation is highly variable depending on local rain or snow precipitation. This great variability adds to the difficulty of estimating system performance. The problem of dust accumulation is discussed in later chapters by P. Roth and R. B. Pettit.

D. Reflectors

The principle use of reflectors is in concentrators which multiply the solar flux on a receiver. Typical reflector material systems are given in Table IV. Diffuse reflection does not contribute significantly so it is the specular reflectance which must be maximized. Figure 14 illustrates specular reflectance as

SPECULAR BEAM PROFILE

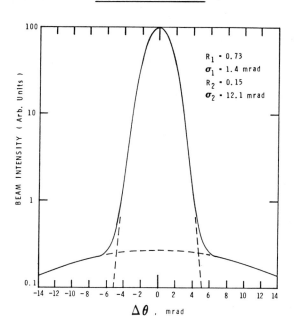

FIGURE 14. *Bidirectional reflection intensity for aluminized teflon plastic film illustrating two components of specular reflection. Note logarithmic scale (23).*

measured by a bidirectional reflectometer (23). The material for
this illustration is aluminized teflon, chosen because it has
two components of the reflected beam, namely a narrow specular
reflectance and a diffuse component scattered hemispherically.
Note that the beam intensity is displayed on a logarithmic
scale which exaggerates the width of the beam profiles.
Specular reflectance can be characterized by two parameters,
R_s the fraction of energy reflected in the specular beam, and
σ the standard deviation in the beam profile. A high-quality
reflector for concentrator use should have a high R_s and a
low σ. An example of system sensitivity to specular reflectance is provided by a reference design for a central re-

ceiver in the 25 to 300 MWe range utilizing an advanced sodium
receiver (24). The system study uses the technique of determining the break-even cost for a given component as the quality of
that component is reduced. It computes the required reduction in
component cost to maintain the same overall expense of producing
electric energy. Figure 15 displays the break even cost (value)
for various values of the specular reflectance of the heliostat

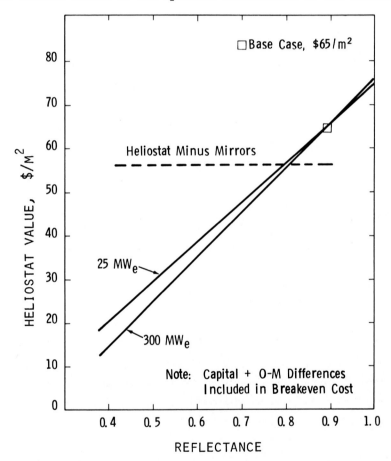

FIGURE 15. Sensitivity of heliostat value to mirror reflectance. The value is based on a full economic analysis of a
central receiver solar power plant (24).

mirrors. Note that the cost of the mirrors themselves would have
to drop drastically if reflectance falls a few percentage points
below the reference design of 0.90.

The importance of angular distribution may be considered in
the same way although the calculation of sensitivity to angular
error has actually been done for tracking errors in the heliostat.
Figure 16 indicates than an upper limit for σ is about 2 millirads.

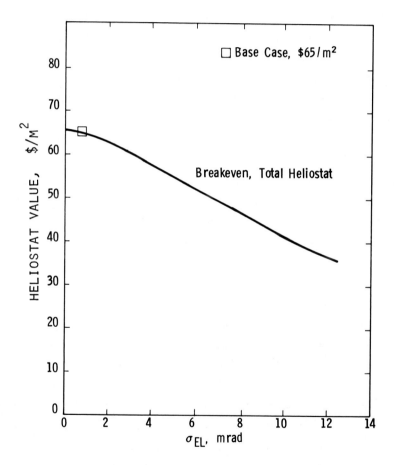

FIGURE 16. Heliostat value vs. standard deviation in angular accuracy for a central tower receiver (24).

Fortunately silvered glass mirrors can be held well within this
limit. Wren, however, points out that the importance of specularity is highly design specific (25).

IV. CONVERSION PROCESSES

A. *Thermal Processes*

A detailed framework of thermoconversion paths is presented
in Figure 17 (2). Selective absorbtion, an important step in the
conversion of insolation to heat, has already been discussed in
the preceding section. Additional areas of technical interest
and technological importance are containment and thermochemical
storage materials.

1. *Heat Transfer Materials*. Virtually all solar systems for
direct conversion to heat require a heat transfer material to
separate the working fluid from the atmosphere or other incompatible environments. A particularly difficult example is posed by
the Barstow Central Receiver Pilot Plant. The heat transfer fluid
is water/steam operating at high internal pressure which means
that the receiver tubes must be thick-walled. Incoloy 800 has been
selected as the base line material. Problems include creep at the
$550°C$ operating temperature, and low-cycle fatigue induced by the
diurnal cycle and by random clouds. Departure from nucleate boiling will cause high cycle fatigue of the nickel alloy. The combination of these degradation mechanisms over the 30-year life
of the system creates a creep-fatigue interaction mode with
which we have little experience. This application for alloys is
subject to the boiler code which presently qualifies only Incoloy
800 and 2-1/4 chrome 1 moly alloy. If these two alloys prove inadequate for the severe stresses of the central receiver application, then an extensive characterization program extending over
several years will be required to qualify a new alloy.

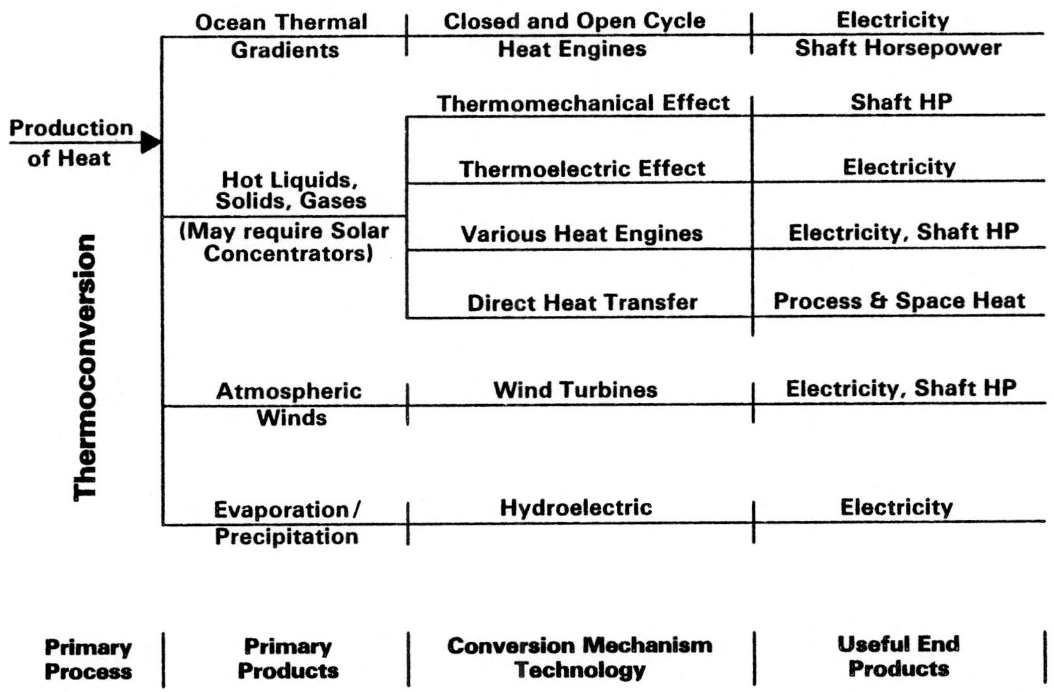

Primary Process	Primary Products	Conversion Mechanism Technology	Useful End Products
Thermoconversion	Ocean Thermal Gradients	Closed and Open Cycle Heat Engines	Electricity, Shaft Horsepower
	Hot Liquids, Solids, Gases (May require Solar Concentrators)	Thermomechanical Effect	Shaft HP
		Thermoelectric Effect	Electricity
		Various Heat Engines	Electricity, Shaft HP
		Direct Heat Transfer	Process & Space Heat
	Atmospheric Winds	Wind Turbines	Electricity, Shaft HP
	Evaporation/Precipitation	Hydroelectric	Electricity

Production of Heat

FIGURE 17. Detailed framework for thermal conversion paths (2).

2.   *Thermochemical Storage.* Thermochemical storage refers to
the use of reversible chemical reactions with high heat of reaction to store thermal energy. It offers a number of advantages
such as high energy density in the storage medium, and long-term
storage at ambient temperatures. Thermochemical storage systems
offer a wealth of material problems because the most attractive
materials in terms of specific energy density are also very
corrosive or chemically aggressive. The highest temperatures
and therefore the most corrosive environment are generally found
in the most complex and expensive components of the system,
namely, the reactor/heat exchangers. Side reactions, even at
very low rates, can destroy the reversibility of the chemical
reactions over many cycles. Many of the reactions of interest
for thermochemical storage require a catalyst in one or both
directions so that detailed knowledge must be obtained on catalyst
lifetime, degradation, and poisoning.

Ray Mar will discuss the broad subject of thermochemical
storage in a later chapter. To illustrate the challenge for material development, we provide one example now. The reaction

$$2SO_3 = 2SO_2 + O_2$$

is attractive because of the high heat of reaction of 98.3 kilo-
joules/mole. For practical rates of reaction in either direction,
catalysts are required although an inexpensive V_2O_5 based catalyst
is available for the lower temperature exothermic reaction (from
right to left in the above equation). The endothermic reaction
takes place at temperatures ($>1000^{\circ}K$) exceeding the melting point
of V_2O_5. Other transition metal oxides are candidates for development.

B.   *Quantum Processes*

The detailed framework for photoconversion paths is shown in
Figure 18 (2). Note that direct conversion of sunlight to a

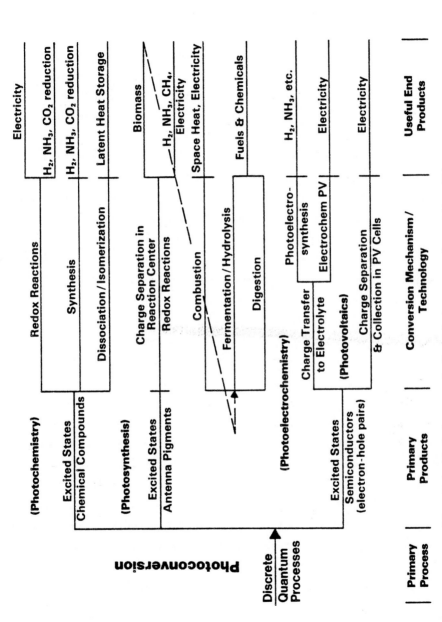

FIGURE 18. Detailed framework for photoconversion paths (2).

chemical or electrical product is a newer technology than thermal
conversion, it can be seen that the conversion technology is not
nearly as well developed. Photoconversion processes are materials specific. Photovoltaics and photochemistry require semiconductors or compounds with appropriate band gaps and configurations to produce charge separation of an electron and a hole that
have been generated by the interaction of a solar photon with the
material. It should be noted that not all photochemistry involves charge-separation, for example, charge separation does not
take place in unimolecular storage and homolytic and hetrolytic
dissociations (see Figure 4). In the area of photosynthesis, the
materials which take part in the reactions, and also the material
used to modify or digest primary photosynthetic products are of
critical importance to this conversion technology. The materials
challenges posed by these conversion technologies require a definitive understanding of the interaction of light with materials in
the solid and liquid states in both inorganic and organic materials
and possibly in the living biological environment.

   1. *Efficiency by Concentration.* Efficiency is of obvious
importance in all quantum processes. Because solar systems are
subject to cost pressure, system efficiency may be achieved with
concentrating collectors. For such concentrator systems, cost
may be less important than efficiency. Burgess has provided an
example by calculating system costs for photovoltaics used in
simple arrays compared to photovoltaics used in combination with
focusing collectors (26). Figure 19 is a display of the array
cost for the cells, structure, and directly associated equipment
as a function of the cell cost per unit area. The sharply curved
line is for a simple array fixed in orientation. The other two
curves for sun tracking collectors indicate that the array cost
is nearly independent of cell cost out to approximately $1000 per
square meter. The important point to observe in this figure

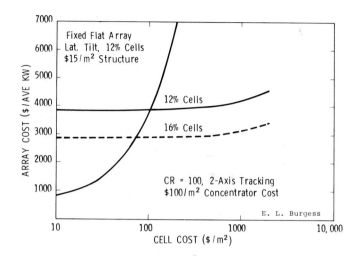

FIGURE 19.   The total array cost per average power vs. photo-voltaic cell cost for a fixed flat array and for a concentrator system of 100X for two cell efficiencies.   For concentrators, the cost is relatively insensitive to cell cost even at $200/m^2, the minimum present cost (26).

is that an increase of cell efficiency is directly reflected in lower array cost. We will consider ways to increase efficiency in photovoltaics in the next section.

2. *Photovoltaic Systems.* Silicon solar cells have performed admirably in our space program. High fabrication cost has prevented the transfer of this capability to widescale terrestrial use. Materials development has focused on high leverage problems identified by systems analysis.

The cost of silicon solar cells is being reduced in a step-by-step approach which addresses every element of cost. This program which lacks a single dramatic breakthrough is appealing because of the tremendous advantage of silicon technology. Extensive experience in working with silicon has been gained by the semiconductor industry. The basic physics and chemistry of silicon have been explored in minute detail. Successful large-

scale manufacturing takes place in many locations. The dramatic
cost reductions for solar cells do appear possible. Such reduction may be achieved by the economy of scale up to large production coupled with new technologies. Today silicon solar arrays
cost in the neighborhood of $20 per peak watt. The Department
of Energy target for 1986 is $0.50 per peak watt in 1978 dollars.

Table VI displays the cost goals for the several steps in producing silicon solar cell arrays (27). Note that in 1976 every
step was equal or greater in cost than the entire goal for 1986.
Each step or element of the manufacturing process must be reduced significantly in order to achieve the ultimate goal. Materials technology will play a central role in every step. Metallization provides a typical example. The 1976 technology
used silver for the electrical conductors deposited on the front
surface of the cell. The cost of the metal alone was more than
the total cost goal for solar cells. Alternative metallization
materials must be developed which contain little or no precious
metal. A nickel palladium alloy and screened aluminum conductors
are being investigated for this program.

TABLE VI. *Department of Energy Goals for Manufacturing Costs*
for Silicon Solar Cell Arrays in 1978 Dollars (27)

		Calendar Year					
		76	78	80	82	84	86
Silicon Material	¢/W	190	110	65	9	5	3
Ingot/Sheet Growth Value added	¢/W	480	210	160	190	52	14
Ingot Slicing Value added	¢/W	370	60	50	1	0	0
Cell Manufacture Materials	¢/W	80	40	35	20	8	7
Value added	¢/W	430	120	85	25	9	6
Array Fabrication Materials	¢/W	50	50	30	10	8	8
Value added	¢/W	400	110	80	45	18	12
Array Price Goals	$/W	20	7	5	3	1	0.50

Unresolved problems identified by the Jet Propulsion Laboratory are in the area of electrical contacts are: 1) improved understanding of process limitations and required controls and the development of low-cost processes that will repeatedly yield quality contacts (there are a large number of chemical and physical interactions that today are not predictably controlled); 2) verification of long-life contacts. This requires definition and understanding of the interactions of contacts with encapsulants; such as, moisture resistant contacts versus hermetic seals, and encapsulated module life testing and field demonstrations.

Solar cells can be made from III-V compounds for which gallium arsenide is the model. These more complex cells are of interest because of the limitations on silicon cells. Silicon has a band gap of about 1.1 eV. Photons with energy below that are not absorbed and photons with energy appreciably above that lose the excess energy in heat which does not contribute to cell output. A more general perspective is presented in Figure 20 which shows the various contributions to inefficiency in graphical form (28). For air mass 0, that is, insolation above the atmosphere, about 56% of the loss is due to wavelength mismatch. The remainder is due to practical limits within the cell. The cell inefficiency is more complicated than indicated by the diagram because the factors are influenced by cell temperature and by concentration of sunlight.

Inefficiency due to wavelength mismatch can be reduced by multiple or cascade cells as illustrated in Figure 21 (29). The best match to the solar spectrum is achieved for the lower band gap E_{G2} of about 1 eV and the upper band gap E_{G1} of about 1.7 eV. A dichroic mirror scheme is easier to construct. E. W. James, et. al., at Varian have constructed an experimental cell using a dichroic mirror and silicon for the lower band gap material (30). The higher band gap is provided by $(Al_{0.93}Ga_{0.07})As/(Al_{0.17}Ga_{0.83})As$

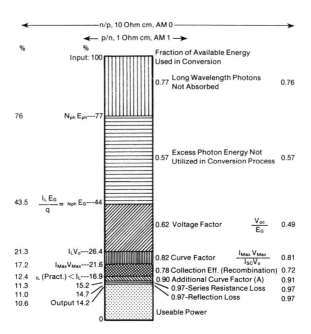

FIGURE 20. Bar chart of distribution of energy losses in the 1-Ω-cmρ-on-η and the 10-Ω-cmη-on-ρ silicon solar cell under air mass one and air mass zero sunlight, respectively (28).

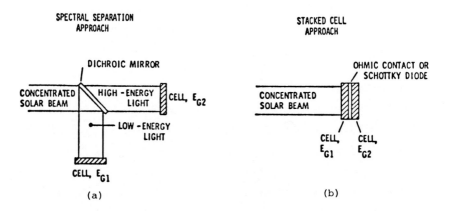

FIGURE 21. Two ways to implement the concept of spectral splitting to achieve higher cell efficiency. The bandgap E_{G1} is lower than E_{G2} (29).

which has a value of 1.65 eV.   Figure 22 shows the high
efficiency of this experimental cell for a range of solar flux.
The figure also displays the contributions to efficiency made by
each of the cells.

   One way in which materials technology might dramatically reduce the cost of solar cells is by providing mass production
techniques which are inherently inexpensive.   One approach under
study is the direct deposition of silicon onto a substrate by
vapor deposition or evaporation which produces a thin film of
polycrystalline silicon.   If successful this would bypass a very
expensive part of silicon cell production.   Note that in Table
VI the cost goals for growing and slicing single crystal silicon
are more than half of the total cost.   The problem with poly-

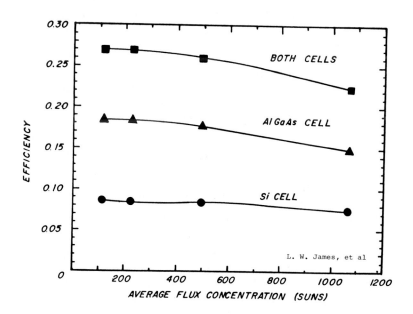

FIGURE 22.   The efficiency of a spectral splitting cell
constructed of silicon and $Al_{0.99}Ga_{0.07}As/Al_{0.17}Ga_{0.83}As$
for a range of solar concentrations.   Contributions from cell
are indicated (30).

crystalline silicon cells is a severe performance penalty because of carrier trapping at the grain boundaries. Chemical vapor deposition of silicon p-n layer on graphite substrate gave
an average size crystallite of 20-30 microns which resulted in a
cell efficiency of about 1.5% (31). Techniques are now being
developed for passivating grain boundaries so that charge mobility in polycrystalline silicon will be nearly equal to that in
single crystal silicon which in turn should enable high cell
efficiencies. Grain boundary passivation will be discussed by
D. Ginley in his chapter and amorphous seminconductors by R. W.
Griffith in his chapter.

3. *Photoconversion.* Photoconversion includes chemical and
biological processes in which radiant energy is converted directly
via quantum processes into other forms of energy (such as chemical feedstocks, liquid and gaseous fuels, and electricity). The
solar program includes basic research in photochemical, photoelectrochemical, and photobiological conversion of solar energy.
Photochemistry deals with the effects of light in producing
chemical change. The synthesis of photochemical model systems
that convert light energy into chemical potential (e.g., charge
separation) for processes such as CO_2-reduction or N_2-fixation
is an example of a solar energy application. Photoelectrochemistry is the interface between chemistry and solid-state physics.
Its solar applications include the production of chemicals or
electricity at a semiconductor-electrolyte interface. Photo-
biology encompasses the effects of light either in living systems
or in high molecular weight complexes isolated from living systems. Understanding the primary processes of photosynthesis holds
the key for potential applications in this area (22).

Although photoconversion mechanisms are capable of relatively
high theoretical conversion efficiencies (20% to 30%), photoconversion technology is in its infancy. Basic research is needed

in order to design and characterize new molecular systems and to
identify limitations of known systems. In all cases, a fundamental understanding of the relevant molecular processes is required to devise more efficient and controllable routes for direct production of fuels, chemicals and electricity from available sunlight.

Photochemical processes are attractive for conversion and
storage of solar energy because the sun's energy corresponds to
about one to three electron volts per photon, which is in the
same range as the chemical bonds in a wide variety of materials
(32). Photochemical processes for solar energy conversion are
presently the furthest from commercial application because there
does not exist a purely synthetic, operable photochemical system
However, there does exist a great deal of flexibility with chemical systems which would allow greater opportunity to assemble
and "tune" the components to specific systems requirements (33).

For example, three potentially useful solar-driven reactions
are:

$$2H_2O \rightarrow 2H_2 + O_2$$
$$N_2 + 3H_2O \rightarrow 2NH_3 + 3/2\ O_2$$
$$CO_2 + 2H_2O \rightarrow CH_4 + 2O_2$$

As written, these are 4-, 6-, and 8-electron processes, for
which the respective Gibbs free-energy requirements are, on the
average, 1.23, 1.17, and 1.06 eV/electron. Assuming that each
electron transfer is initiated by sequential absorption of one
photon, and taking the energy loss to be 0.8 eV/photon (34), we
see that these processes can, in principle, be driven by light
in the 600-660 nm range. If the energy loss were, say, 0.5 eV/
photon, light of 710-790 nm would be sufficiently energetic to
initiate the reactions.

It is unlikely, however, that any two electrons in such a sequential process will be transferred at the same potential. For

example, in the case of Co_2 reduction, it appears that the first step would occur closer to the redox potential of CO_2/CO_2^- (2.2 eV) (35) than to the average value of 1.06 eV. Thus, a spectrum of reduction potentials will probably be required. If specific, stepwise electron transfer from excited states can be effected, then for each step, it would be possible to determine the optimum wavelengths of light needed to initiate the processes. In this way, the complete solar spectrum might be utilized for photochemical generation of ammonia or methane from renewable resources (33).

Photoelectrochemical energy conversion is based on photoactive semiconductor electrodes that absorb visible light, thereby creating electron/hole pairs which separate in the space-charge layer produced at a semiconductor/liquid electrolyte interface. These separated electrons and holes are subsequently injected into the electrolyte to drive chemical reduction and oxidation reactions.

Photoelectrochemical cells can be configured to produce electricity or to drive chemical reactions (36). The emphasis of the program is on the latter systems. Chemical reactions driven uphill in energy (endoergic) produce fuels (for example, splitting of H_2O into H_2 and O_2); chemical reactions can also be driven downhill in energy (exoergic) and produced useful chemicals (for example, reduction of H_2 to NH_3). For the endoergic case, solar energy is converted into chemical energy, while for the exoergic case, solar energy provides the activation energy for the chemical reaction. The production of hydrogen by photoelectrochemical water splitting (photoelectrolysis) is a primary objective of the program.

Photoelectrochemical systems have high theoretical conversion efficiency (about 25%); they can be operated with inexpensive polycrystalline and/or amorphous electrodes without drastic loss

of efficiency; and simple and inexpensive photochemical reactor
systems based on photochemical diodes in a slurry or colloidal-
type system can be utilized. However, the major problem pre-
venting the implementation of these advantages is the lack of
sufficient chemical stability in the semiconductor electrode
materials that have band gaps in the optimum range (1.0 eV to
2.0 eV) in the electrolyte solutions. D. Ginley discusses
photoelectrochemistry in a later chapter.

The main emphasis of the photobiological efforts are in the
production of hydrogen from the interaction of light with biological materials (37). A number of biological systems will produce hydrogen when irradiated with solar radiation. These include
a number of green algaes and a number of photosynthetic micro-
organisms and photosynthetic bacteria. The goal of the researchers in this area is the production of photobiological solar cells. Because of the complexity of the reactions involved
in the absorption of light and long chain molecules, it is suggested that the reader study a recent review of this field for further details (38).

V. SUMMARY

In summary we would like to emphasize that materials along
with design pervade all aspects of the solar energy program. We
have provided examples where materials development can have a
large leverage on system costs. For solar energy to have a large
impact not only must acceptable cost be achieved but the public
must have confidence in the product and its performance over
the design lifetime. That means that we in the technical community need confidence in reliable performance and confidence in
our life projections. Confidence must be based on a thorough
knowledge of the materials and their properties in the intended

use environment. The materials community is responsible for that knowledge.

ACKNOWLEDGMENTS

The authors acknowledge the considerable help of staffs of both Sandia Laboratories and SERI. The help of Dr. J. Connolly in preparing the photochemistry section is also acknowledged. Thanks are given to Dr. J.C. Grosskreutz for providing the photo and thermoconversion path development schematic diagrams.

REFERENCES

1. *Solar Thermal Power Systems Program.* January 1978, DOE/ET-0018/1.
2. Grosskreutz, J.C., *Private Communication.*
3. Lichfield, J.W., et al., "A Method for Identifying Material Constraints to Implementation of Solar Energy Technologies", PNL-2711, July 1978.
4. Treadwell, G.W., "Design Considerations for Parabolic-Cylindrical Solar Collectors", SAND76-0082, July 1976.
5. Butler, B.L., Call, P.J., Jorgensen, G.L., and Pettit, R.B., "Solar Reflectance, Transmittance and Absorptance of Common Materials", October 1979, Proceedings of the Solar Industrial Process Heat Conference, Oakland, California, November 1979.
6. Call, P.J., "Applications of Passive Thin Films", Chapter 9, *Properties of Polycrystalline and Amorphous Thin Films and Devices,* L.L. Kazmerski, Editor; Academic Press (1979).
7. Jorgensen, G.L., "Long-Term Glazing Performance", Proceedings of Solar Glazing: 1979 Topical Conference, Mid-Atlantic Solar Energy Association, June 1979. (SERI/TP-31-193)
8. Ratzel, A.C., and Bannerot, R.B., "Commercially Available Materials for Use in Flat-Plate Solar Collectors",

Proceedings of 1977 Flat-Plate Solar Collector Conference, CONF-770253, 387 (1978).

9. Eidin, F.E., and Whillauer, D.E., "Plastic Films for Solar Energy Applications", *Proceedings of the United Nations Conference on New Sources of Energy,* Vol. 4, Rome, August 21-31, 519 (1961).

10. *Kynar 500. Polyvinylidene Fluoride for Architectural Finishes.* Pennwalt Corporation, Plastics Department, Three Parkway, Philadelphia, Pennsylvania. Brochure PL138-677-5M-B.

11. Hummel, D.O., *Infrared Analysis of Polymers, Resins and Additives, an Atlas,* Vol. 2, Part 2, Wiley-Interscience, New York (1969).

12. Butler, B.L., and Pettit, R.B., "Optical Evaluation Techniques for Reflecting Solar Concentrators", *SPIE 114,* 43 (1977).

13. Biggs, F., and Vittitoe, C., "The Helios Model for the Optical Behavior of Reflecting Solar Concentrators", SAND76-0347, March 1979.

14. Reuter, R.C., Jr., and Allred, R.E., "Structural and Material Optimization for Solar Collectors", Proceedings of 1979 International Congress of the International Solar Energy Society, May 28 - June 1, 1979.

15. Reuter, R.C., Jr., "Weight Minimization of Sandwich Type Solar Collector Panels", Proceedings of the 14th Intersociety Energy Conversion Engineering Conference, Vol. I, p. 1, August 5-10, 1979.

16. McDonald, G.E., "Spectral Reflectance Properties of Black Chrome for Use as a Solar Selective Coating", NASA TMX-71596, August 1974.

17. Duffie, J.A., and Beckman, W.A., *Solar Energy Thermal Processes,* John Wiley & Sons, Inc., 1974.

18. Treadwell, G.W., McCulloch, W.H., and Rusk, R.S., "Test Results from a Parabolic-Cylindrical Solar Collector",

SAND75-5333, July 1975.

19. Abrams, M., "The Effectiveness of Spectrally Selective Surfaces for Exposed, High-Temperature Solar Absorbers", SAND-75-8300, January 1978.

20. Beauchamp, E.K., "Low Reflection Film for Solar Collector Cover Plates", SAND75-0035, 1975.

21. Vitko, J., "Optical Studies of Second Surface Mirrors Proposed for Use in Solar Heliostats", SAND78-8228, April 1978.

22. Freese, J.M., "Effects of Outdoor Exposure on the Solar Reflectance Properties of Silvered Glass Mirrors", SAND78-1649, September 1978.

23. Pettit, R.B., "Characterization of the Reflected Beam Profile of Solar Mirror Material", *Solar Energy 19,* 733, 1977.

24. Fish, J.J., and Dellin, T.A., "Heliostat Design Cost/Performance Tradeoff", SAND79-8248, November 1979.

25. Wren, L., *J. Energy 3,* 82, 1979.

26. Burgess, E.L., Proceedings of the Society of Photo-Optical Instrumentation Engineers, August 24-25, 1976.

27. Excerpted from First Annual Report, Low-Cost Silicon Solar Array, ERDA/JPL 1012-76/5.

28. "Solar Cells: Outlook for Improved Efficiency", Published by an Ad Hoc Panel on Solar Cell Efficiencies, National Research Council, National Academy of Sciences, Washington, D.C., 1972.

29. Hovel, H.J., *IBM J. Res. Devel. 22,* 112-121, 1978.

30. James, L.W., VanderPlas, H.A., and Moon, R.L., "Novel Solar Cell Concentrator Photovoltaic Converter System", Varian Associates, SAND79-7048.

31. Chu, T.L., et al., *Solar Energy 17,* 229, 1975.

32. Milne, T.A., Connolly, J.S., Inman, R.E., Read, T.B., and Seibert, M., "Research Overview of Biological and Chemical Conversion Methods and Identification of Key Research Areas for SERI", SERI/TR-33-067, 1978.

33. Connolly, J.S., "Solar Photochemistry", Solar Energy Symposia of the 1978 Annual Meeting, American Section of the International Solar Energy Society, H.W. Boër and A.F. Jenkins, Eds., Part I, p. 2-5, 1978.

34. Bolton, J.R., *Science 202,* 705-711, 1978.

35. Jordan, J., and Smith, P.T., *Proc. Chem. Soc.,* 246-247, 1960.

36. Nozik, J., "Photoelectrochemistry: Applications to Solar Energy Conversion", *Annual Review of Physical Chemistry, Vol. 29:* 1978, p. 189.

37. Weaver, P., Lien, S., and Seibert, M., "Photobiological Production of Hydrogen--A Solar Energy Conversion Option", Solar Energy Research Institute, SERI/TR-33-122, 1979.

38. Seibert, N., Connolly, J.S., Milne, T.A., and Reed, T.B., "Biological and Chemical Conversion of Solar Energy at SERI", *AICHE Symposium Series, Vol. 74,* No. 181, p. 42, 1978.

CHAPTER 2

INTRODUCTION TO THE ROLE OF CRYSTAL
DEFECTS IN SOLAR MATERIALS[1]

L.E. Murr
O.T. Inal

Department of Metallurgical and Materials Engineering
New Mexico Institute of Mining and Technology
Socorro, New Mexico

I. INTRODUCTION

To a large extent, the science of materials is dominated by
the role of imperfections in determining or controlling the
properties of materials - physical, chemical, mechanical, electrical, etc. Conversely, the characterization of materials imperfections has led to a basic understanding of many properties,
the selective control of imperfections, and as a consequence,
many instances of selective control of residual properties of
materials, or the "tailor making" of materials to accommodate
specialized property requirements, including particular operating
environments. Solar materials are, in the broadest sense, any
materials selected to meet specific requirements involved in
solar applications, which can include collection, conversion,
storage, and transmission or distribution of energy from the sun.
Specific properties of interest can therefore involve optical and
thermal response, corrosion, erosion, and related degradation
resistance, electrical or electronic properties, thermal and

[1]Based upon research supported by the Division of Materials
Science, Office of Basic Energy Sciences, U.S. Department of
Energy, under grants DOE-ER-78-04-4266 and DE-FG04-79AL10887.

electrical conduction, as well as mechanical properties including
hardness, strength, ductility, etc. In certain cases, one or
many of these properties may be required, while many systems will
consist of an integrated system or composite of many materials
and properties to facilitate a specified function. In such cases,
properties desirable in one portion of the system, which result
by the presence of selective imperfections, may be incompatible
with those required in another portion. That is, imperfections
required to achieve one property may be noticeably detrimental to
another. Consequently, some sort of compromise or optimization
may be required. This can mean a very delicate variation or
control of internal structure, or the number and kind of imperfections which must be achieved through changes in crystal
structure, chemical composition, or both.

Crystal defects, or imperfections, in crystalline materials
are now well documented (1-4). In general, they are grouped into
regimes of zero-dimensional (or point) defects, one-dimensional
(or line) defects, two-dimensional (or planar) defects, and
three-dimensional (or volume) defects. Vacancies, interstitials,
and substitutional impurities constitute the more common point
defects (5) while charge balance requirements in ionic solids
require pairs of such defects to be formed, or some other
charge-compensating mechanism, such as the formation of a color
center where an electron is trapped by an anion vacancy (6).
Dislocations constitute the more common line defects (2), while
interfaces such as grain and phase boundaries, and free surfaces
(the solid-vapor or liquid-vapor interfaces) are characteristic
of planar imperfections (7,8). Point defect aggregates or other
larger voids account for the major types of volume defects (9).
The way in which such defects are formed or interact is an important part of understanding how they will effect residual materials properties (3,10). In addition, the amorphous state, characterized by a general lack of structural order, is also of some
practical importance because residual properties are also very

much dependent upon the degree of disorder, or the number and
kind of crystal defects.

II.   CRYSTAL STRUCTURES AND ORDER-DISORDER PHENOMENA

   Crystal defects are characterized as regular types of deviations from (or alterations of) perfect, periodic arrangements of
atoms in a crystal lattice array.   In describing such defects, it
is therefore necessary at the outset to define perfect lattice
arrays.   Such arrays can be characterized by the fourteen different Bravais lattices illustrated in Fig. 1.   Solid materials
which are composed of periodic arrays of such lattices (having
unit cell dimensions a, b, c as shown) are called crystalline
materials while solids which do not possess any of the identifi-
able unit cell structures shown in Fig. 1 are called non-crystalline or amorphous materials.   In some respects, amorphous solids
might be considered to be characterized by a defect continuum
because there is no periodic reference regime.   However, even
amorphous solids can contain geometric units such as silicate
tetrahedra composing common glass, which is perhaps the most
common example of an amorphous solid.   Crystalline materials can
be described by either long-range or short-range order (or atomic
lattice periodicity) while an amorphous solid exhibits complete
disorder.   Since lattice disorder is a state of high entropy, it
is not a minimum energy configuration and as a consequence all
materials will, under certain conditions, crystallize; forming a
periodic lattice array of either short-range or long-range order;
which can include a superlattice (4).   Many solids which are
non-crystalline (disordered) will therefore slowly crystallize
(or become ordered) when energy (especially thermal energy) is
supplied to drive such a process.
   While all truly non-crystalline solids are disordered, not
all crystalline solids are ordered in the sense that an ordered,
solid solution superstructure is created (4).   An order-disorder

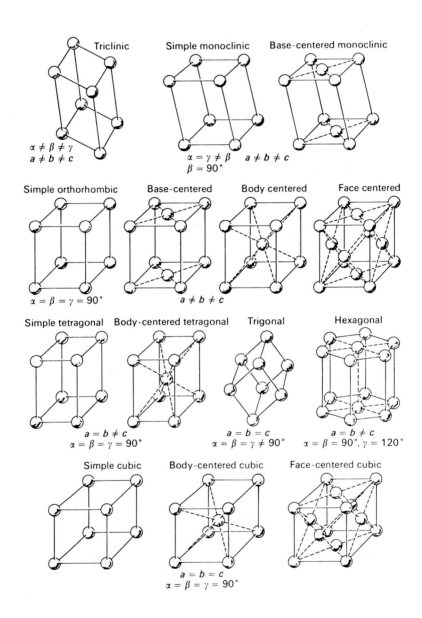

FIGURE 1.   Fourteen Bravais lattices composing seven different crystal systems.   Each lattice represents a particular unit cell
in a periodic atomic arrangement.   α, β, and γ are the vertex
angles; a, b, c are the unit cell dimensions or lattice parameters.

transition can occur for crystalline solids as illustrated in
Fig. 2, and this rearrangement of atomic species into specific
periodic, lattice positions can have a profound effect on both
the physical and mechanical properties of solid-state material.
Figure 3 illustrates the perfect, periodic atomic structure
characterizing the packing of iridium atoms in an oriented (001)
face-centered cubic crystal lattice array accommodating an equi-
librium-related, pseudo-hemispherical geometry of a needle-pointed
end-form which is observed in a field-ion microscope (7,11,12).
Figure 4 illustrates the crystalline-non-crystalline regimes and
shows the disorder (non-crystalline)-to-order (crystalline)
transition which occurs in a so-called glassy metal or alloy.
The appearance of very short-range crystalline regions in the
glassy structure is especially noticeable by the appearance of
plane-edge "rings" of clusters of iron atoms shown in Fig. 4(d);
especially apparent when these image features are phenomenologi-
cally compared with those of the much larger, perfect crystal
surface shown in the field-ion image of Fig. 3.

It should be apparent from an inspection of Figs. 2 and 4
that temperature is an important factor in the determination of
specific phase equilibria as well as order-disorder phenomena,
and can have a profound effect on the structure and properties of
solar materials, especially at higher temperatures (300 - 600°C)
These effects can be predicted to some extent from phase diagrams,
as illustrated in Fig. 5 showing compound formation in a two-
component (binary) alloy system (13,14).

To a large extent, Fig. 4(a) and (b) are indicative of the
extremes in solid lattice energies which can characterize a
perfectly ordered crystal lattice and a non-crystalline, disordered array respectively composed of the same number of specific
atoms. A perfect crystal is a minimum energy configuration (8),
while any deviation from this configuration constitutes a degree
of disorder which requires additional energy, or is characteristic of a higher-energy state. Consequently, a higher-energy state

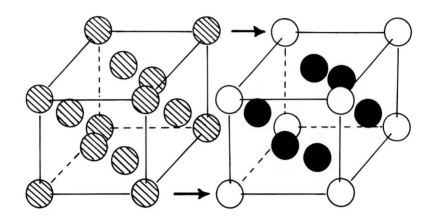

FIGURE 2. Disorder-to-order transition in a face-centered
cubic A_3B crystal unit cell such as Cu_3Au forming superlattices.

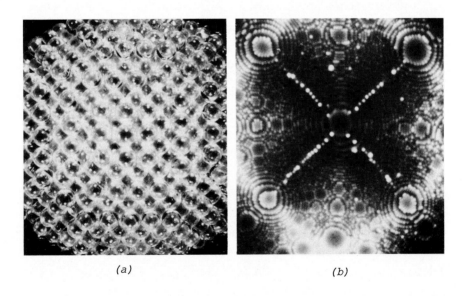

(a)   (b)

FIGURE 3. Perfect crystal packing of atoms to accommodate a
pseudo-hemispherical end form in the $[001]$ direction. (a) Ball
model. (b) Field-ion image of iridium $[001]$ end form showing
atomic structure. Note that the $[001]$ direction is normal to the
page. The symmetries observed arise from plane-packing and
crystal plane symmetries which are apparent from a $[001]$ stereo-
graphic projection. Individual image dots are single iridium
atoms.

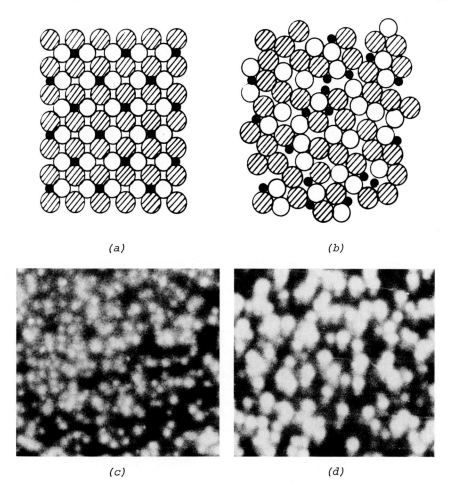

FIGURE 4.   Lattice order-disorder phenomena. (a) Perfect cubic lattice with long-range order. (b) Completely disordered structure in (a). There is no recognizable unit cell or unit cell periodicity. (c) Disordered (amorphous) $Fe_{80}B_{20}$ alloy surface observed in the field-ion microscope. (d) Same $Fe_{80}B_{20}$ as in (a) after partial ordering and segregation at $780°C$. The image is dominated by clusters of iron. Fe_2B and Fe_3B phases are only weakly imaged and suppressed in the background.

will, given the opportunity, revert to a minimum energy configuration, and such intrinsic instabilities suggest that essentially all materials are, to some degree, unstable. Consequently,
their structures and properties can be expected to change with

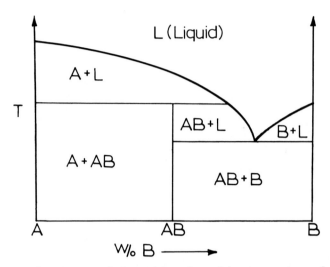

FIGURE 5. Compound formation in a binary system of elements A and B as a function of temperature and composition.

time and exposure to environments which provide an energy-related
mechanism for driving the system toward a lower energy state.
This can be achieved by alterations in the range of order or
variations in defect density and arrangement, or the degree of
disorder, and can include the imposition of temperature, pressure
(stress), and compositional changes as a result of environmental
reactions.

III.   CRYSTAL LATTICE DEFECTS AND SOLAR-RELATED MATERIALS
       STRUCTURES

With reference to the perfect, ordered crystal regime shown
in Figs. 2-4, the amorphous as well as the disordered regimes are
defect states. However, these defects are not readily defined as
systematic variations in the three-dimensional space lattices
shown in Fig. 1. Such variations can be geometrically characterized by variations in single atom periodicity or other interrup-
tions within the lattice as a result of the insertion of single
atoms, variations in linear arrays or chains of atoms, variations

in the periodicities over an area or plane of atoms or of the coincidence of separate crystals, and three-dimensional inter- ruptions in atomic periodicity.

A. *Point (Zero-Dimensional) Defects*

Point defects occur in a space lattice when atom positions are vacant (vacancies) (or anion-cation vacancy pairs in ionic solids in order to maintain charge neutrality), when atoms occur within the lattice interstices (interstitials), when (especially in ionic solids where charge neutrality must be preserved) a vacancy-interstitial pair is formed, or when a foreign or impurity atom substitutes for a normal lattice atom (substitutional impur- ity). The substitutional impurity can also be the same atom having a different valence, and an interstitial impurity can also be an atom of a different species or of different valence. An additional, basic modification of the vacancy in an ionic crystal can also occur when instead of forming a vacancy-interstitial pair, an electron is trapped at the site of an anion vacancy, thereby maintaining charge neutrality. Such a defect is referred to as a color center (or F-center) because of its selective effect upon optical absorption causing coloring in transparent crystals. Figure 6 illustrates the appearance of these different point defects in a simple cubic-crystal lattice section.

Point defects of the type shown in Fig. 6 can arise during growth and solidification, in response to energetic radiation of various types, and through diffusion. Their effects on a variety of electrical, optical, mechanical and other properties can be visualized to some extent by considering the effect they would have on an electron or related electromagnetic wave propagating through the lattice, and the variations such defects would impose upon lattice atom cohesion and coordination.

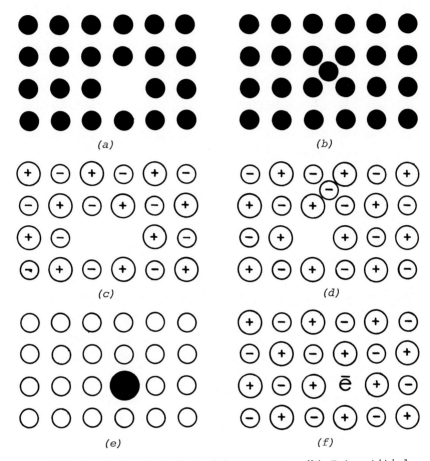

FIGURE 6. Point defects. (a) Vacancy. (b) Interstitial.
(c) Cation-anion vacancy pair maintaining charge neutrality.
(d) Vacancy-interstitial pair maintaining charge neutrality.
(e) Substitutional impurity. (f) Color center (anion vacancy
site occupied by an electron, $\bar{e}$).

B. Line (One-Dimensional) Defects

Figure 7 illustrates the simplest type of Volterra (15) or
line defects, consisting of an extra half-plane in the same
simple cubic crystal lattice section shown in Fig. 6(a). Crystal
lattice line defects are referred to as dislocations, and Fig. 7
shows an edge-type dislocation which has the property of glide on
a close-packed plane or slip plane, thereby accommodating lattice

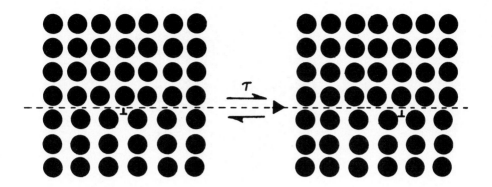

FIGURE 7.   Shear-induced motion of a simple edge dislocation.
The shear stress is denoted by τ.

shear.   Dislocations of the type shown in Fig. 7 allow for slip
as a systematic lattice translation which can be visualized by
the motion of the dislocation line, which is perpendicular to the
page in Fig. 7(a) and (b).   It can be observed in Fig. 7 that if
the dislocation line were to continue moving to the right (by
successive lattice atom translations) a step (slip step) would be
created.   Similarly, if a perfect crystal section is cut and
systematically sheared as shown in Fig. 8, an edge dislocation is
created within the lattice section and is connected with a screw
dislocation by a continuous dislocation line shown dotted.   The
lattice distortion creating the dislocation line is characterized
by a Burgers vector which is perpendicular to the edge dislocation line and parallel to the corresponding screw dislocation
line in Fig. 8.   Moving away from the two surfaces in Fig. 8 upon
which the edge or screw dislocation lines emerge (dislocation
lines are normal to the corresponding surfaces) results in a
systematic variation in dislocation type from pure edge (where
$\underset{\sim}{b}$, the Burgers vector is perpendicular to the dislocation line)
to pure screw (where $\underset{\sim}{b}$ is parallel to the dislocation line) (3).

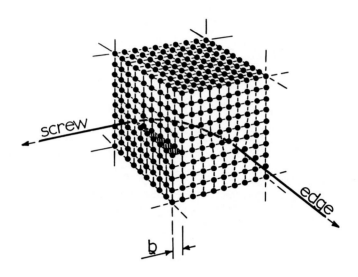

FIGURE 8. Total dislocation (having edge and screw-type character) in a section of a simple cubic crystal lattice.

Since dislocations, as shown in Figs. 7 and 8, are associated with stress (deformation), it is perhaps readily apparent that they arise in crystals in response to deformation, or accommodate externally applied stresses. Furthermore, it can be visualized from both Figs. 7 and 8 that because of the plane bending to accommodate the dislocation within the crystal lattice, a region of localized stress (or strain) occurs along the dislocation line. This disturbance would also be expected to have an effect upon wave propagation or the behavior of electrons in a crystal lattice section, and their movement could allow them to interact with the point defects shown in Fig. 6. Such interactions could impede the motion of the dislocation line, or allow for absorption of point defects or arrays of point defects along the dislocation line, within the so-called "core" region. For example, impurity atoms absorbed along a dislocation line would transform this linear array into a special regime which could profoundly effect physical and mechanical properties. In addition to the ability of dislocations to glide along a slip plane as illustrated

in Fig. 7, the selective absorption of vacancies at a dislocation
core can allow for the motion of a dislocation line in a direction normal to the slip plane shown, giving rise to a process
called climb. Since vacancy motion can occur by diffusion
resulting by thermal activation, climb is normally considered to
be associated with hot environments. Consequently, the application of both heat and stress can allow for alterations in dislocation arrangement in crystals as a result of both glide and
climb processes. The motion of dislocations can be impeded by
interactions with other dislocations and therefore as the density
of dislocations (line length per unit volume of crystal expressed
as cm^{-2}) increases, there is a marked effect upon deformation.
This gives rise to an increase in yield stress and ultimate tensile stress, as well as hardness; and physical and mechanical
properties of all crystalline materials are related to dislocation density. Consequently, the mechanical properties of solar-
related support materials are, as is well known, markedly
influenced by dislocations; especially fatigue, creep, and other
modes of deformation leading to eventual failure (3,5,10,15).

Dislocations in face-centered cubic materials are characterized by Burgers vectors of the type $\frac{a}{2}$ <110>, where a is the
lattice parameter. The dislocation line is characterized by
glide upon {111} planes in specific <110> directions. The energy
associated with the formation of a unit length of dislocation
line is proportional to the square of the Burgers vector and as a
consequence can be lowered by selective adjustments in lattice
translations (3). This process creates partial dislocations
which result by the systematic splitting up of total dislocations
into two partials having Burgers vectors different from the total
dislocation; resulting in a lower total energy. The partials so
created are connected by a region of systematic lattice distortion which is coincident with the slip plane, and which, as a
consequence, forms a planar (or two-dimensional) defect having an
area equal to the product of the dislocation length and the

FIGURE 9. *Dislocations in a 0.1 μm thick nickel electrodeposited layer. The dislocations appear as black lines in the
bright-field electron transmission image because plane bending
around the dislocation core as shown in Figs. 7 and 8 causes
electrons diffracted in this regime to be systematically excluded
in the image. Stacking fault and microtwin fringes are also
observed in the image area.*

distance separating the partials, or the area enclosed by a dis-

location loop. Dislocations in crystals either terminate upon

themselves (forming a dislocation loop), another dislocation

(which can lead to the formation of networks), or at a free sur-

face or interface. Dislocation lines do not end within a volume

of perfect crystal. These features are now well known and well

documented (2,3,15). Figure 9 shows dislocations in a thin

electrodeposit of nickel observed by transmission electron

microscopy (12).

In addition to applied stresses, dislocations (and stacking

faults) can arise in response to internal stresses or they can

accommodate stresses which arise in solids during growth and

related processes of lattice accommodation. Dislocations can

also be regarded as regions of different stoichiometry in many

solid crystalline materials, and can be altered in form to maintain charge neutrality in ionic solids. Because the stoichiometry (or charge regime) in the vicinity of a dislocation line
emerging on a surface can be different from the surrounding
matrix, and as a consequence of differences in local lattice
energy in the dislocation core (at the point where a dislocation
line emerges on a free surface), dislocations can act as preferential sites for incipient process initiation such as corrosion
(dissolution or solubilization) and related surface reactions; or
nucleation and growth. This is especially true for screw dislocations which, as shown in Fig. 8, give rise to a spiral ramp
allowing for crystal growth in the direction of the screw dislocation line by a spiral growth around the dislocation line. In
most cases, however, dislocations have little effect upon growth
processes such as the deposition and growth of one material upon
another in processes like vacuum vapor deposition, vacuum sputtering, or electrodeposition. Such processes of film growth upon
a substrate are mainly characterized by two-dimensional (layer)
growth and are controlled primarily by crystallographic orientation - initially that of the substrate, and then as the influence
becomes rapidly weaker as a result of continuing growth away from
the substrate, that of the overgrowth itself; including recrystallization (16,17). These features are illustrated in Fig. 10
which shows the growth of a nickel electrodeposit upon a polycrystalline copper substrate at various overgrowth thicknesses.

C. *Planar (Two-Dimensional) Defects*

We have briefly described one of the simplest types of planar
defects in crystalline materials in the discussion of stacking
faults above (Sect. III *B*). Stacking faults are by their nature
prominent in materials where the stacking-fault free energy is
low because dislocations are widely separated in such materials
and cannot cross-slip easily (move from one slip plane to another,

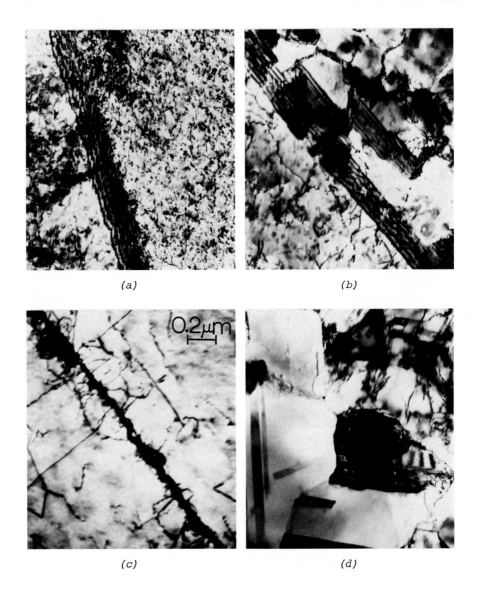

(a)   *(b)*

(c)   *(d)*

FIGURE 10. Sequence of bright-field electron micrographs showing the development of microstructure in electrodeposited nickel overgrowths upon annealed, polycrystalline copper. (a) 0.1 μm Ni overgrowth. Note different dislocation arrangements and density on either side of the grain boundary. (b) 0.5 μm Ni overgrowth showing stacking faults and twins developing. (c) 5 μm Ni overgrowth showing a reduction in dislocation density and a variation in grain and grain boundary structure. (d) 25 μm Ni overgrowth showing recrystallization phenomena.

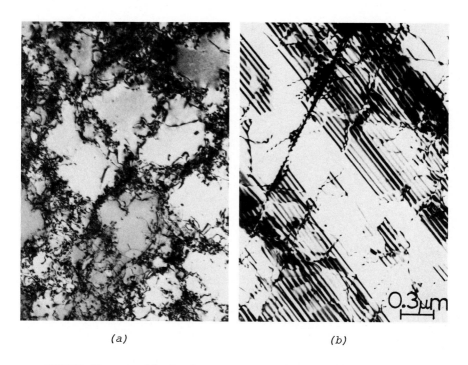

(a)   *(b)*

FIGURE 11. *Residual microstructures resulting from differ-*
ences in stacking-fault free energy, which governs the ability of
dislocations to extend forming stacking faults, or cross-slip.
(a) Dislocation cells in nickel resulting by extensive cross-slip
of dislocations. (b) Planar dislocation arrays and stacking
faults in stainless steel where cross-slip is effectively pro-
hibited. Both materials in sheet form were subjected to a shock
stress of 15 GPa magnitude and 2 μs duration.

parallel slip plane). Consequently, materials with a high stacking-fault free energy respond very differently to deformation:
those with high stacking-fault free energy forming complex dislocation arrays as a consequence of prominent cross-slip, as compared to those with low stacking-fault free energy forming
primarily planar dislocation arrays or stacking faults. These
features are illustrated in Fig. 11 for nickel (having a stacking-
fault free energy of 128 mJ/m^2) and type 304 stainless steel
(having a stacking-fault free energy of 21 mJ/m^2), subjected to
exactly the same deformation (3,8,15).

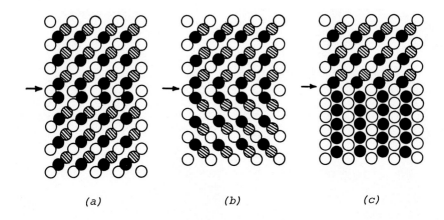

(a)   (b)   (c)

FIGURE 12. Schematic view of planar shear interfaces formed
in {111} planes of a face-centered cubic lattice having perfect
atomic packing periodicities A, B, C, denoted ○ , ● , ◍ . (a)
Single intrinsic stacking fault (arrow). (b) Overlapping stacking
faults (on every {111} plane) producing a twinned region separated by planar interfaces (arrow). (c) Systematic stacking-
fault formation and shear adjustments producing a new phase
separated by planar interfaces (phase boundaries) (arrow).

Figure 12 shows some simple schematic diagrams for the formation of stacking faults and related planar defects in face-centered cubic materials by the systematic shearing of {111} planes
through the propagation of partial dislocations. Deformation
twins are observed to result by the formation of intrinsic stacking faults on every (111) plane while more selective shear
operations result in a hexagonal close-packed regime with planar
interfaces separating the hcp phase from the unaltered fcc phase.
Such solid interfaces, separating twinned regions or different
phases (having differences in crystal structure, composition, or
both), are also planar defects because they interrupt the periodicity of any specific phase, and act as a region of transition
from one structure to another.

   1.   *Grain Boundaries*. When crystals of the same material are
joined together, the interface characterizing the transition from
one orientation to the other is called a grain boundary. Essentially all useful engineering materials are polycrystalline and
therefore contain grain boundaries separating the individual
grains or crystallites. In general, a grain boundary can be
characterized by five degrees of freedom (8,18). In the case of
very elementary boundaries characterized by only one symmetrical
rotational parameter, or the angle of misorientation between two
identical lattices, the boundary can be regarded simply as a
regular array of edge dislocations accommodating the lattice mismatch. This dislocation concept of a grain boundary usually
applies strictly to low angles of misorientation ($\Theta \leq 5^{\circ}$), however regularly spaced dislocations can be shown to characterize
the intrinsic or equilibrium structure of even larger-angle
boundaries. For symmetrical, low-angle boundaries, the dislocation spacing is given approximately by the ratio of the Burgers
vector to the misorientation angle (in radians) ($d \cong |b|/\Theta$).
Consequently, for very small misorientations the spacing can be
large (hundreds or thousands of nanometers) while for large misorientations, the spacings become very small. In addition, the
energy per unit area of a grain boundary is proportional to the
angle of misorientation for a simple, symmetrical boundary, and
increases rapidly with increasing misorientation between zero and
roughly 15° (8).

   Figure 13 demonstrates some of these features and shows
several, more realistic examples of grain boundaries. Figures
13(a) and (b) show the idealized grain boundary structures described above while Fig. 13(c) shows a surface view of the atomic
features of a non-symmetric, large-angle grain boundary in iridium as observed in the field-ion microscope (8,11,12). By systematically removing successive atom layers of the iridium end
form in Fig. 13(c) by field evaporation (11,12), the detailed
structure of the grain boundary plane can be reconstructed. It

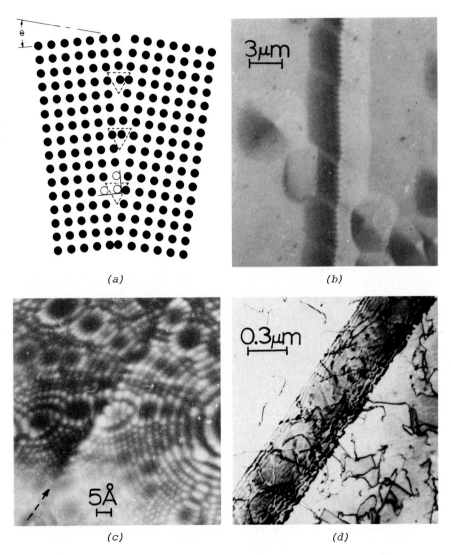

(a)   (b)

(c)   (d)

FIGURE 13. Examples of grain boundary structure. (a) Simple
(symmetrical) grain boundary schematic showing intrinsic dislocation structure (dotted triangles) which can also be regarded as
primitive ledges (open circles). The grain misorientation is
denoted by Θ. (b) Low-angle grain boundary in LiF characterized
by regularly spaced dislocation etch pits observed by optical
microscopy. (c) Asymmetrical grain boundary in an iridium field-
emission end form. The arrow denotes the interface. (d) Grain
boundary projection from a thin section in an electrodeposited Ni
film showing dislocations and more complex structure characterizing the interface. Bright-field transmission electron micrograph.

must be recognized that the area might only be 0.01 µm on a side.
By comparison, grain boundaries can be observed in projection by
electron transmission through thin polycrystalline films, and
this technique can allow for a complete characterization of many
representative grain boundaries in a specific solid-state material as illustrated in Fig. 13(d). It is especially significant
to observe that while the intrinsic dislocation structure is not
visible in Fig. 13(d), other so-called extrinsic dislocations are
visible within the boundary, and dislocations are observed to
terminate on the boundary plane as described previously. Extrinsic dislocation structures can accommodate deviations from idealized coincidence arrays (low energy configurations) or allow for
other structural phenomena. Extrinsic grain boundary dislocations
can also be regarded as ledges upon the boundary plane or boundary "surface", and can actually take the form of large steps
having a height of many matrix dislocation Burgers vectors. To
some extent, the concept of interfacial structural steps or ledges
is even apparent in Fig. 13(a), but extrinsic dislocations or
ledges are usually not associated with the equilibrium boundary
structure. Figures 14 and 15 illustrate some of the features of
intrinsic and extrinsic grain boundary dislocation structure
alluded to above, and Fig. 15 shows the interaction of the extrinsic structure (ledges) with the equilibrium dislocation structure
of a grain boundary in molybdenum. Figure 15(b) convincingly
demonstrates the very specific step character of large grain
boundary ledges as evidenced by the contrast fringe displacements
along particular ledge portions.

     Grain boundaries, because of their regular dislocation arrays
(intrinsic grain boundary dislocations) and other more complex
extrinsic or ledge structures become very prominent regimes adsorbing or absorbing impurities, creating a transitional or interfacial phase region which can have a profound effect upon the
physical and mechanical properties of solar materials. Figure 16
shows a more complete view of grain boundary structure which is

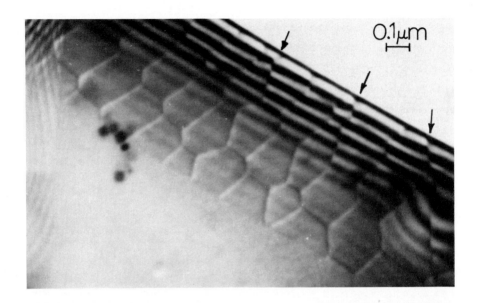

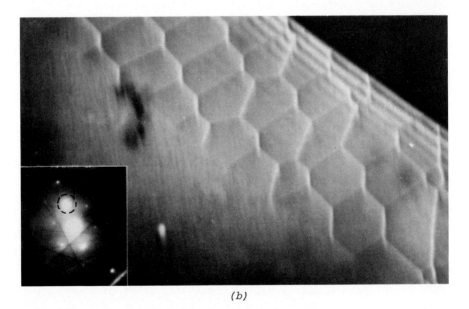

(b)

FIGURE 14. Grain boundary dislocation networks forming ledges
on the interfacial planes as evidenced by the boundary contrast
fringe displacements indicated by the arrows. (a) Bright-field
electron transmission image in 304 stainless steel. (b) Dark-
field image utilizing strong reflection shown circled in the
selected-area electron diffraction pattern insert.

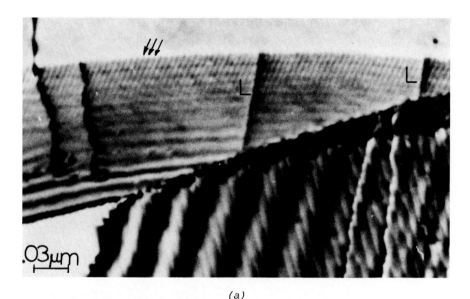

(a)

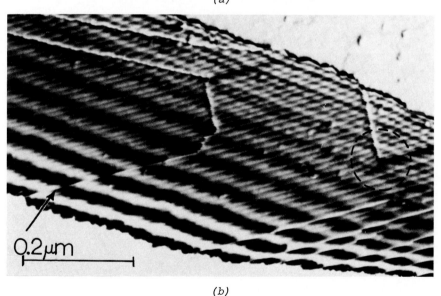

(b)

FIGURE 15.   Intrinsic (equilibrium) and extrinsic (ledge)
grain boundary dislocation (GBD) structure in molybdenum thin
films.   (a) Regularly-spaced intrinsic GBD's (arrows) and ledges
(L).   (b) Interaction of intrinsic and extrinsic GBD's (within
dotted area) and large ledge portions causing boundary contrast
fringe displacements (arrow).   (a) and (b) are bright-field
electron transmission micrographs.

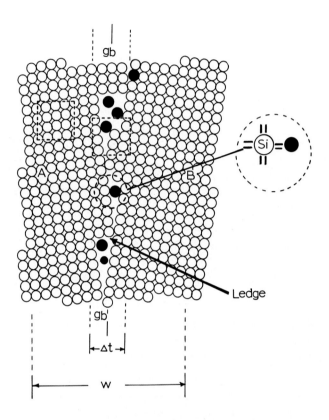

FIGURE 16. *General grain boundary schematic showing struc-*
tural features (especially larger ledges), impurity segregation,
and passivation effects in covalent semiconductors such as Si.
The grain boundary plane or Gibbs dividing surface is denoted
gb while the interfacial phase is denoted by Δt. *W denotes the*
range of space-charge region which can be associated with semi-
conductor interfaces. Dotted boxes in the upper portion illus-
trate variations in density, structure, and composition which can
be characteristic of a specific volume of the interfacial phase
as compared to the grain structure in perfect crystal portions A
or B.

nonetheless a very simplified schematic depicting only a somewhat

idealized, two-dimensional section. It is especially significant

to note, as discussed extensively by Murr (8,19,20), that grain

boundary ledges, because of their obvious, localized stress con-

centration are sources of dislocations when an external stress is

applied, allowing for translations and rotations of individual
grains in a polycrystalline regime, which Ashby (21) has described
as giving rise to non-homogeneous plastic deformation gradients.
These deformation gradients assume a periodicity or wavelength
equal to the grain size, D, so that residual yield stress, σ, for
polycrystalline metals and alloys can be expressed as

$$\sigma = \sigma_o + kD^{-1/2} \qquad (1)$$

where σ_o is a strain-dependent term which includes the so-called
friction stress, and k is a materials constant which can depend
upon the stacking-fault free energy or the grain boundary structure (8,22). Equation (1) is usually somewhat historically
referred to as a Hall-Petch law, but in a more general form we
can write

$$\sigma = \sigma' + k\lambda^{-n} \qquad (2)$$

where σ' is a strain-dependent term, λ is the deformation gradient wavelength (21), and n can vary between approximately 0.25
and 1, depending upon the nature of the microstructural units.
For example, if a polycrystalline material is further sub-refined
by the formation of additional interfaces having structural properties similar to grain boundaries, then it might be expected
that the yield stress could be expressed by a relationship similar to Eq. (1). Indeed, this is exactly what has been observed
for the formation of deformation twins in shock-deformed polycrystalline metals and alloys (23,24). Figure 17 illustrates
this phenomena for shock-loaded type 304 stainless steel sheet.
If the deformation twin spacing is Δ, the yield stress can be
expressed by

$$\sigma = \sigma' + k'\Delta^{-1/2} + kD^{-1/2} \qquad (3)$$

or if we let $\sigma_o = (\sigma' + k'\Delta^{-1/2})$, then Eq. (3) is identical to
Eq. (1). The friction stress also includes of course the impedance to dislocation motion caused by the presence of the

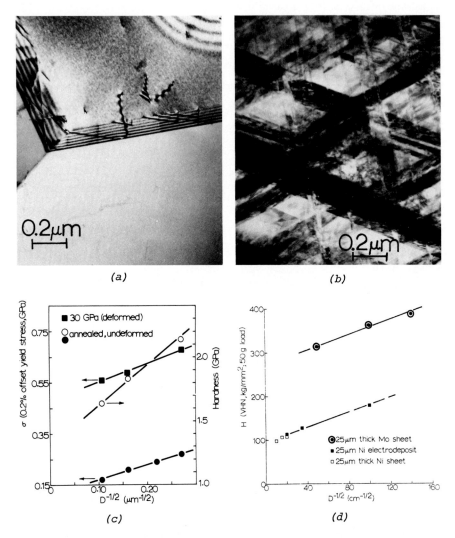

FIGURE 17. *Effect of grain size and grain-refinement-producing*
substructure on the mechanical properties (yield strength and hard-
ness) of thin metallic sheets and coatings. (a) Annealed 304
stainless steel grain structure (28 μm grain size, D). (b) Sys-
tematic deformation twin formation in (a) in response to shock
deformation at 30 GPa peak pressure (2 μs pulse duration). The
effective twin boundary spacing, Δ is much less than the original
grain size, D. (c) Residual yield stress and microhardness for
various grain sizes of annealed and shock-deformed 304 stainless
steel sheets as shown in (a) and (b). (d) Hardness versus recip-
rocal square root of the residual grain size in thin metal sheets
and electrodeposits.

shock-induced deformation twins.   These new interfaces also provide for additional ledge sources for dislocations when a stress
is applied.   Since hardness is usually simply related to yield
stress (8), there is a similar dependence of hardness on the deformation gradient wavelength (or grain size) as shown in Fig.
17(c) and (d).   Figure 17(d) suggests, on comparison with Fig. 10,
that not only would the mechanical properties of solar-related
overgrowths vary with thickness as a consequence of variations in
grain size with thickness, but optical properties might also be
expected to vary because electromagnetic wave interactions will
change with changes in the volume fraction of interfacial surface,
variations in interfacial thickness, and alterations in interfacial structure.   This is also true of free electron interactions in metals, and these features are illustrated in Fig. 18.

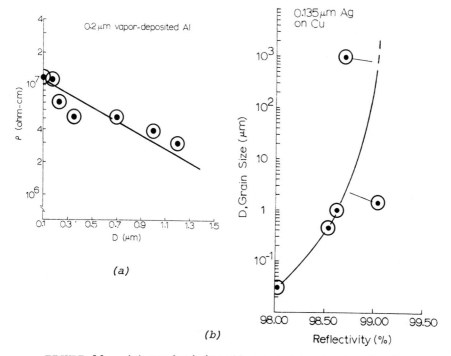

FIGURE 18.   (a) Resistivity (d.c.) versus grain size in vapor-
deposited aluminum films.   (b) Reflectivity (at 10.6 μm) versus
grain size for silver films vapor-deposited upon polished, polycrystalline copper substrates.

2. *Phase Boundaries and Related Interfaces.* In Fig. 12,
only the twin boundary shown schematically in Fig. 12(b) is a
grain boundary while the stacking fault is a special intrinsic
interface and the boundary shown in Fig. 12(c) is a phase boundary. Interfaces separating precipitates or dispersed phases in
solid matrices are also phase boundaries, and interfaces can also
be described as domain boundaries separating polarized or magne-
tized regimes (as in ferroelectric and ferromagnetic materials,
respectively), or as domain boundaries separating an ordered
regime from a disordered regime, or different arrangements of
superlattices (with reference to Fig. 2) (4). Interfaces can
also separate phase regions such as solid from liquid [although
this is sometimes a comparatively thick interfacial phase region
(8)], and a free surface can also be considered as an interface
separating the solid and vapor phases. In many respects, a free
surface (solid-vapor interface) is another special case of a
planar defect, however, surface structure can complicate this
rather idealized concept. Surfaces and surface structure are,
nonetheless, of major importance in the fabrication and utilization of a great variety of solar materials because most solar
coatings - absorbers, reflectors, photovoltaic conversion films -
are grown upon substrates, either structural members or other
types of deposits. Having formed a deposit, the substrate/deposit
interface then can be characterized as a special type of phase
boundary. Interfaces or grain boundaries in semiconductors or
photovoltaic heterojunctions are not only structurally very
important but also electrically important. While the interface
phase or geometrical and crystallographic structure can be defined within an interface phase region having a thickness Δt, the
electronic phase region associated with the interface is characterized by the width of the space-charge region having a dimen-
sion W as shown in Fig. 16. Figure 19 illustrates the idealized
structure of a free surface. It should be noted that if either
grain A or B in Fig. 16 is viewed separately, the "surface"

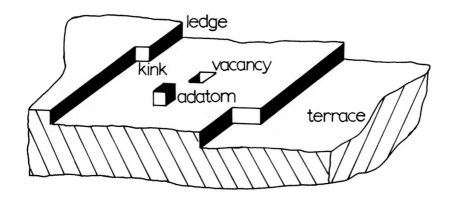

FIGURE 19.   Schematic view of idealized surface structure
(solid-vapor interface).   The terrace planes are parallel to
(001).

structure at the interface will be identical to Fig. 19, and the

only significant differences are the coordination, bonding, and

associated relaxations.   The absorption of certain atoms such as

hydrogen at grain boundaries in semiconductors can have a passi-

fying effect with regard to recombination and trapping of elec-

trons at "dangling bonds" as shown in Fig. 16, and this phenomenon

can have important implications in the development of more

efficient solar cells, to be discussed in more detail in subse-

quent chapters (c.f. chapters by Ginley and others).

Murr and Annamalai (25) and Murr and Inal (26) have recently

demonstrated that not only are dislocations in substrates of

little consequence in influencing the structure of an overgrowth

(as shown in Fig. 10), but also grain boundaries are also not

prominent in promoting specific features of nucleation and growth

of a deposited layer.   This particular feature is illustrated in

Fig. 20.   In Fig. 20(a) it can be observed that while very thin

films are not strongly influenced by the grain boundary intersec-

ting the surface, the grain orientations, as mentioned earlier,

have a very significant influence on the overgrowth structure.

Figure 20(b) shows that even in the initial stages of nucleation,

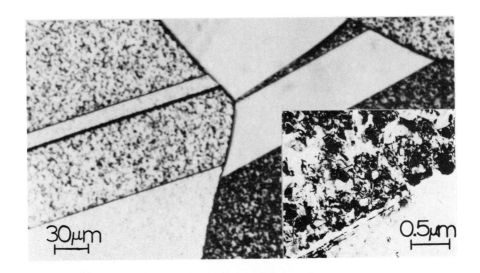

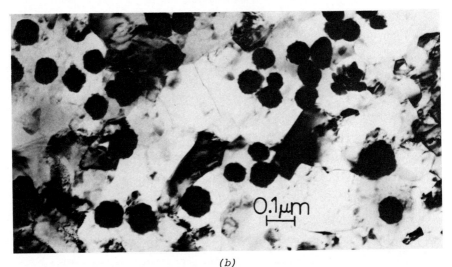

(b)

FIGURE 20. Microstructural features of deposit growth upon
a substrate. (a) Optical microscopic view of 1 μm thick nickel
electrodeposit upon a polycrystalline copper substrate. The insert shows a magnified view of a 0.1 μm thick Ni overgrowth in a
twinned orientation as observed in the transmission electron microscope. (b) Auto-electrodeposited copper nuclei (primarily
single-crystal clusters) on a polycrystalline thin-film aluminum
substrate observed in the transmission electron microscope.
There is no systematic or preferred coincidence of the copper
nuclei (black) with the substrate grain boundary planes which
intersect the substrate surface.

grain boundaries are not preferred sites, and nucleation occurs
somewhat heterogeneously over the surface. Nucleation and growth
of a deposit initially occurs epitaxially, and in the first few
atom layers even pseudomorphically (where the deposit atoms
are in exact registry with the substrate atoms) (26,27). Epitaxial growth is illustrated in the observations of copper electrodeposited upon a tungsten single-crystal field-emission end form
in Fig. 21(a) and (b). While, as shown in Fig. 20(b), many
initial nuclei are single crystals which assume some minimum
energy, faceted shape, many are initially short-range ordered and
actually composed of clusters of very small grains or crystalline
regions which pack as spherical or hemispherical clusters. This
is especially true in deposition processes involving complex
chemical reactions typical of most electrodeposition processes.
Figure 21(c) and (d) illustrate these features for very small
electrodeposited clusters upon field-emission end forms. Such
structures, like those described recently for very fine, polycrystalline iron whisker deposits (28) can incorporate oxides and
other reaction products, producing a kind of microstructural composite with metallic clusters surrounded by or imbedded in oxides
or other compounds which may be crystalline (with very short-range
order) or non-crystalline (amorphous). To a large extent, these
structures and microstructural composites can be altered by a
host of deposition parameters including chemical composition of
the bath in electrodeposition, the temperature (also of the substrate), rate of deposition (and current density in electrodeposition), substrate orientation and grain size, and overgrowth
thickness. Indeed, as we have discussed earlier, grain size and
related microstructural dimensions are altered with deposit thickness. Figure 22 illustrates some of these effects in relation to
residual absorptivity (α) for electrodeposited black-chrome
selective absorber deposits on polycrystalline nickel.

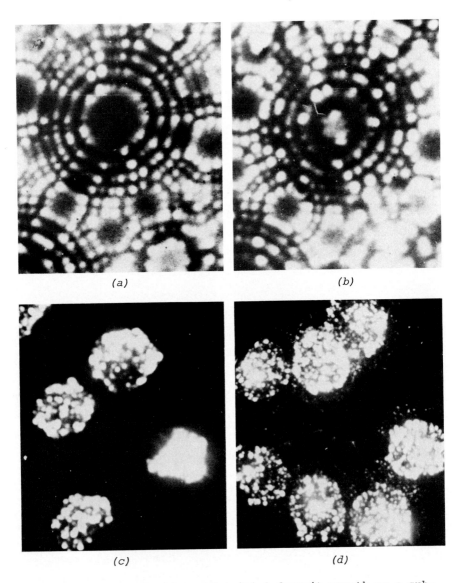

FIGURE 21. Nucleation and initial deposit growth on a substrate surface. (a) Perfect [011] oriented tungsten end form observed in the field-ion microscope. (b) Same surface as (a) following epitaxial growth of a few layers of copper upon the surface by auto-electrodeposition (electrochemical displacement reaction). (c) Partially ordered clusters of Ni electrodeposited upon a 304 stainless steel substrate observed in the field-ion microscope. (d) Partially ordered clusters of black chrome electrodeposited upon a nickel substrate (field-ion image).

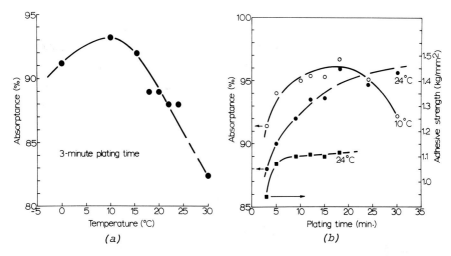

FIGURE 22.   (a) Absorptance of black chrome electrodeposits
on plane nickel substrates as a function of bath temperature during electrodeposition.   Constant plating time assures essentially
identical deposit thickness of ~ 2 μm.   (b) Absorptance and adhesive strength of black chrome electrodeposits on plane nickel
substrates at different (constant) bath temperatures as a function of plating time, which corresponds to varying deposit thicknesses.

D.   Volume (Three-Dimensional) Defects

When individual point defects such as those illustrated in
Fig. 2 agglomerate in a solid material, a three-dimensional defect arises.   This can take the form of a vacancy cluster or
void, a region of altered chemistry in the case of an element-rich
cluster or impurity cluster, precipitates or agglomerated dispersed phases, vacancy discs which allow for lattice relaxation to
produce a dislocation loop as referred to briefly above; and
related defects, including metal colloids, F-aggregate centers,
and the like.   Such defects can have a marked effect on physical
and mechanical properties and optical properties as well by altering electromagnetic wave propagation.   For example, the distribution and size of metal colloids can have a significant effect
on optical absorption and this is also true of color-center
aggregates as recently discussed by Murr (29,30), and Birsoy and

Murr (31). Furthermore, the production of complex color centers
by elemental substitutions and valence changes can have a significant effect on selective absorption of light in transparent
crystals (31). Figure 23 illustrates some of these defects.

Szilva and Murr (32) have demonstrated that void density in
thin, vapor-deposited, polycrystalline silver films strongly influences reflectivity (at $\lambda = 10.6\,\mu m$); reflectivity decreases
with an increase in void density. Furthermore, void density has
been shown to be temperature dependent in thin, vapor-deposited
silver films (32).

Void structures, such as those shown in Fig. 23(c) can form
during deposit growth by exclusion of certain crystal orientations
which are rapidly enveloped in the growing deposit. In addition,
porous structures can arise by variations in deposit growth morphologies and the development of crystallographic orientations.
This can lead to dendritic structures or columnar growth, botryoidal morphologies, and the like in deposits for specific solar
applications. These deposit structures, illustrated in Fig. 24,
are highly selective with regard to the absorption or reflection
of optical waves, and can, as noted earlier, be altered by altering reaction chemistries or other deposition or deposition-rela-
ted parameters.

IV. SUMMARY

After having defined crystal lattice arrays and described the
concepts of order, disorder, and the so-called amorphous state of
atomic structure, we have endeavored to systematically and chron-
ologically define and illustrate the principal crystal defects in
the context of their recognized and anticipated influence on a
variety of properties associated with solar materials performance,
properties, and utilization. The implications are that solar
materials science is simply an area of specialization within the
broadest framework of materials science, and that there exists a

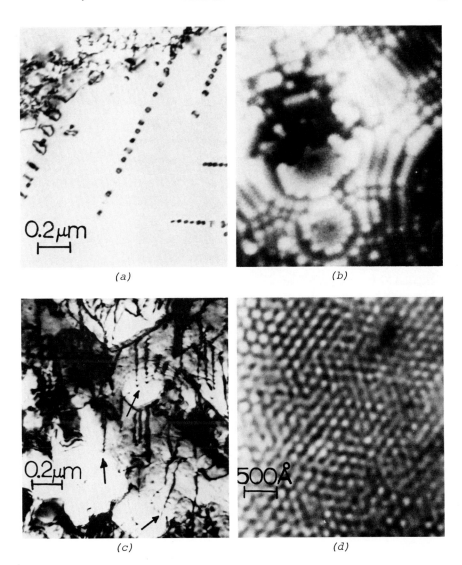

FIGURE 23.   Examples of three-dimensional crystal lattice de-
fects (voids).   (a) Dislocation loops (vacancy cluster discs) in
rapidly quenched nickel observed in the transmission electron mi-
croscope.   (b) Large vacancy disc or void layer in (102) plane of
vapor-deposited platinum observed in the field-ion microscope.
(c) Voids in electrodeposited nickel film observed in the trans-
mission electron microscope.   Note dislocations terminating on
the void surfaces (arrows).   (d) F-aggregate centers (forming a
special void lattice structure) in electron irradiated CaF$_2$.   The
(111) plane is parallel to the image surface (bright-field elec-
tron transmission image).

(a)

(b)

FIGURE 24. Thick, auto-electrodeposited (electrochemical displacement reaction) overgrowths of copper on iron (a) and aluminum (b) substrates observed in the scanning electron microscope.
(a) Shows botryoidal growth morphology resulting from crystal
agglomeration while (b) shows dendritic crystal morphologies.
(a) corresponds to deposition from a $CuSO_4$ solution at $30°C$, pH
2.2.   (b) Corresponds to deposition from a $CuSO_4$ solution at
$50°C$, pH 3.2.

very obvious relationship between crystal structure, particularly
defect microstructure, and the performance of materials in solar
applications. We have also briefly outlined the basic features
of materials growth and microstructural alterations during growth,
especially in the context of deposits grown upon a suitably defined substrate or polycrystalline support material. It has been
shown that microstructure has a very important if not controlling
effect on optical properties as well as associated physical and
mechanical properties, and that in many instances a compromise in
the microstructural manipulations must be considered in order to
optimize all of the desirable features of a specific solar materials application. For example, in the case of highly reflecting
films grain size should be large, but the associated mechanical
properties such as tensile strength and hardness decline with increasing grain size. Furthermore, grain size in most overgrowths
will vary with deposit thickness, and this variation can occur in
a complex way, being dependent upon a host of interdependent variables which can include deposit thickness, deposition rate,
temperature, deposition environment parameters such as vacuum or
bath composition, etc. All of these parameters can have a profound effect on the growth and regularity of deposit crystal
structure, perfection, and range of order. However, we have also
demonstrated that in many instances, especially in the growth of
deposits upon a substrate, the defect structure of the substrate
may have little if any effect on the development of deposit microstructure. Moreover, variations in crystal orientation, especially surface orientation, is a controlling feature in the
development of an overgrowth upon a substrate.

In describing crystal defects and illustrating their influence
in the structure and properties of solar materials, some instrumentation techniques in solar materials research were also illustrated with regard to the direct observation of crystal defects,
the range of lattice order or disorder, and related microstructural phenomena. In some respects it was intended to illustrate

not only specific defect microstructures but to demonstrate the
utility in a varied approach involving several techniques to
directly observe a material's microstructure at different dimensional levels. This interdisciplinary instrumental analysis
approach, involving direct observations of microstructure, can
allow for very precise structure - property relationships to be
established, and for the development of specialized processes for
the efficient and economical manufacture and further development
of materials for specific solar energy applications. A more
general, concise review of instrumentation might be found in the
November 1976 issue of Physics Today. Finally, while we have not
stressed the very complex role crystal defects can have in recombination contributing to loss of efficiency in the operation
of solar cells, especially dislocations and grain boundaries,
these features will be described in detail in later chapters. A
recent article by Boer (33) is also recommended.

REFERENCES

1. Damask, A.C., and Dienes, G,J., "Point Defects in Metals",
    Gordon and Breach, Science Publishers, Inc., New York,
    (1971).

2. Amelinckx, S., "Direct Observation of Dislocations", Academ-
    ic Press, Inc., New York, (1964).

3. Hirth, J.P., and Lothe, J., "Theory of Dislocations",
    McGraw-Hill Book Co., New York, (1968).

4. Barrett, C.S., and Massalski, T.B., "Structure of Metals",
    3rd ed., McGraw-Hill Book Co., New York, (1966).

5. Thomas, G., and Washburn, J. (eds.), "Electron Microscopy and
    Strength of Crystals", Interscience, New York, (1963).

6. Seitz, F., *Rev. Modern Phys.*, 26, 7 (1954).

7. Amelinckx, S., Gevers, R., Remaut, G., and VanLanduyt, J.
    (eds.), "Modern Diffraction and Imaging Techniques in Mater-
    ial Science", North Holland/American Elseview, Amsterdam/

New York, (1970).

8. Murr, L.E., "Interfacial Phenomena in Metals and Alloys",
    Addison-Wesley Publishing Co., Reading, Mass., (1975).

9. Arsenault, R.J. (ed.), "Defects and Defect Clusters in B.C.C.
    Metals and Their Alloys", Proceedings of an International
    Conference, Nuclear Metallurgy Vol. 18, National Bureau of
    Standards, Gaithersburg, Marylang, (1973).

10. Kuhlmann-Wilsdorf, D., Interactions Between Vacancies and
    Dislocations, Chap. 7 in "Frontiers in Materials Science",
    Murr, L.E. and Stein, C. (eds.), Marcel Dekker, Inc., New
    York, p. 253, (1976).

11. Müller, E.W., and Tsong, T.T., "Field-Ion Microscopy: Prin-
    ciples and Applications", Elsevier, New York, (1969).

12. Murr, L.E., "Electron Optical Applications in Materials
    Science", McGraw-Hill Book Co., Inc., New York, (1970).

13. Gordon, P., "Principles of Phase Diagrams in Materials Sci-
    ence", McGraw-Hill Book Co., New York, (1968).

14. Staudhammer, K.P., and Murr, L.E., "Atlas of Binary Alloys:
    A Periodic Index", Marcel Dekker, Inc., New York, (1973).

15. Cottrell, A.H., "Mechanical Properties of Matter", J. Wiley
    and Sons, Inc., New York, (1964).

16. VanDerMerwe, J.H., "Single-Crystal Films", Pergamon Press,
    Oxford, (1964).

17. Gündiler, I., and Murr, L.E., *Thin Solid Films,* 37, 387
    (1976).

18. Lange, F.F., *Acta Met.,* 15, 311 (1967).

19. Murr, L.E., *Met. Trans.,* *6A,* 505 (1975).

20. Murr, L.E., and Venkatesh, E.S., *Metallography,* 11, 61 (1978).

21. Ashby, M.F., *Philos. Mag.,* 21, 399 (1970).

22. McElroy, R.J., and Szkopiak, Z.C., *Int. Metall. Rev.,* 7, 175
    (1975).

23. Moin, E., and Murr, L.E., *Mater. Sci. Engr.,* 37, 249 (1979).

24. Murr, L.E., Moin, E., Wongwiwat, K., and Greulich, F., Effect
    of Grain Size and Deformation-Induced Grain Refinement on

the Residual Strength of Shock-Loaded Metals and Alloys, in
"Strength of Metals and Alloys", Vol. 2, Haasen, P., Gerold,
V., and Kostorz, G., (eds.), Pergamon Press, New York,
p. 801, (1979).

25. Murr, L.E., and Annamalai, V., *Met. Trans.*, *9B*, 515 (1978).

26. Murr, L.E., and Inal, O.T., *Phys. Stat. Sol. (a)*, 51, 345
    (1979).

27. Murr, L.E., Inal, O.T., and Singh, H.P., *Thin Solid Films*,
    9, 241 (1972).

28. Wilsdorf, H.G.F., Inal, O.T., and Murr, L.E., *Z. Metall-
    kunde*, 69, 701 (1978).

29. Murr, L.E., *Phys. Stat. Sol. (a)*, 22, 239 (1974).

30. Murr, L.E., Void-Lattice Formation in Natural Fluorite by
    Electron Irradiation, "Proc. Electron Microscopy Society of
    America", Bailey, G.W. (ed.), Claitor's Publishing Div.,
    Baton Rouge, La., p. 614, (1976).

31. Birsoy, R., and Murr, L.E., *Phys. Stat. Sol. (a)*, 52, 77
    (1979).

32. Szilva, W.A., and Murr, L.E., *Phys. Stat. Sol. (a)*, 40, 211
    (1977).

33. Boer, K.W., "The Physics of Solar Cells", *J. Appl. Phys.*,
    50(8), 5356 (1979).

CHAPTER 3

SURFACE AND INTERFACE STUDIES AND THE STABILITY
OF SOLID SOLAR ENERGY MATERIALS[1]

A.W. Czanderna

Solar Energy Research Institute[2]
Golden, Colorado

I. INTRODUCTION

The purpose of this chapter is to provide an overview of the
extremely important role that surface and interface studies of
solid materials must play if the United States is to achieve wide
scale commercialization and application of the various solar technologies. After introductory comments are made to provide a general perspective of the problems, an overview of surface and interface studies that are immediately applicable to solar energy technologies will be presented. For conciseness in this chapter, I
shall include both surfaces and interfaces in the term "surface".
Following the overview on surface studies, a number of problems in
solar technologies will be discussed briefly and in a general manner. The examples chosen will be representative, important problems in solar materials research, but they are not necessarily

[1]*This work is supported by the Division of Solar Technology, U.S. Department of Energy (DOE), under contract EG-77-
C-01-4042.*
[2]*U.S. DOE facility.*

those of highest priority and certainly not the only problems that
need to be studied. I will conclude the chapter with some recent
results obtained from studying the enhancement of the thermal
oxidative degradation of isotactic polypropylene on copper oxide
surfaces resulting from the catalytic effect of copper ions.
These results will serve as a detailed example of a problem caused by a basic incompatibility of two materials. This example will
also demonstrate the extensive prior work that is necessary to
secure an understanding of the problem. Improved materials and
processes can be most readily conceived from an understanding of
degradative processes. Thus, an underlying theme of this entire
chapter will focus attention on the mechanisms of degradation, or
surface reactions that cause a change in the desired properties of
a component of an entire solar energy system.

As indicated in the chapter by Butler and Claassen, the major
problem is not discovering how to collect radiant energy, but how
to collect it cheaply. The solar flux reaching the earth is low
in energy density, which means very large collection areas will be
required to capture significant amounts of solar energy.

It will be necessary to deploy large areas of heliostats,
photovoltaic arrays, or flat plate collectors to concentrate the
energy density or convert it into another energy form, as is accomplished by photovoltaic materials. The cost of materials utilized, production processes, and the operation and maintenance of
systems must be held to a minimum. This requires, for example,
using thin films for mirrors, for photovoltaic systems, for antireflection coating, etc., and that these films be made from inexpensive, durable, and easily processed materials. Inexpensive
long-life materials in flat plate collectors and durable, stable
absorber coatings are also necessary.

In general, the actual or conceptual design of solar collection systems requires large areas of contact between different

materials, i.e., metals, oxides, polymers, semiconductors, ceramics, composites, and transfer media. Reflectors, transmitters, absorbers (both thermal and quantum), and heat transfer materials must have low life cycle cost, reliable and durable performance, and remarkable long term stability. The latter is apparently counter to basic physics and chemistry, because atoms at a surface are necessarily in a higher energy state than if they were in the bulk. By their very presence, surface materials must be more reactive and thermodynamically less stable than bulk materials. Yet, the dilute energy density of solar radiation requires us to deploy systems with large surface areas! Fortunately, kinetic limitations frequently retard the effects of a thermodynamic driving force. Therefore, systems with large areas of different kinds of materials, can remain stable if they are chosen correctly.

Solar systems are subjected to a unique set of stresses that may alter their stability and, hence, their performance and life cycle costs. These stresses include UV, temperature, atmospheric gases and pollutants, the diurnal and annual thermal cycles, and in concentrating systems a high intensity solar flux. In addition, condensation and evaporation of water, rain, hail, dust, wind, thermal expansion mismatches, etc., may impose additional problems on the performance of a solar system. These stresses and factors must be considered not only individually, but also for synergistic degradative effects that may result from their collective action on any part of the system. The initiation of many degradative effects can reasonably be expected to occur at surfaces.

Heat transfer fluids and storage containers offer another set of problems after the solar energy is collected. While special factors such as biofouling in OTEC or the diurnal cycle play a role in some of these cases, there exists a new set of problems at the fluid/container surface. Again, an interaction at a surface may limit the performance of a component of a solar energy system.

It is fair to conclude that surface studies are of much
greater relative importance in solar materials science than in
materials science in general. After studying this chapter, please
consider the devices described and the collection processes presented in the remaining chapters, and try to recognize the surface problems that will arise in each case. It is likely that
every one of the solar devices or systems described in the following chapters has a surface-related problem, either directly or
indirectly.

The editor has demonstrated excellent insight into solar
materials science by requesting the preparation of a chapter dealing with surface phenomena. For example, how have you been exposed in your curriculum to the fundamental aspects of surface
phenomena? It is likely the bits and pieces of surface phenomena
that you have encountered have not been presented in an organized
overview course. In different courses, you may have touched on
the thermodynamics of interfaces, dislocations at grain boundaries, nucleation and growth phenomena, friction and wear, interfacial diffusion, faceting, grain boundary sliding, thermally
activated growth, and adsorption phenomena without really recog-
nizing that these are part of a surface science discipline. You
are fortunate if you have had a course on surface physics, surface chemistry, or surface phenomena. There is little question
that one of the most active research areas in materials science
is the study of surfaces. In the past, proper emphasis has been
placed in most materials science curricula on relating the bulk
properties to the structure and composition of the solid. In the
future, one should expect a similar emphasis to be placed with
respect to surfaces, because of the wide scale research efforts
that are in progress all over the world in which surface reactivity and stability are being related to the crystallographic

orientation and composition at the surface of the materials. What part of this type of research is of near-term importance to solving the surface problems in solar systems? After the discussions presented in the next two sections, you should be able to answer this question.

II.   SURFACE STUDIES APPLICABLE TO SOLAR MATERIALS

An overview of the most important phenomena in surface science related to studying solar materials is presented first in this section. The methods of characterizing surfaces and those deemed likely to have the largest near-term impact on solving the problems of surface degradation are then discussed.

If the ionic species in plasmas are neglected, the universe of interest consists of solid, liquid, and gas phases, none of which is infinite. The boundary region between these phases, i.e., the surface phase, has fundamentally different properties from the bulk. The possible boundaries are the solid/solid, solid/liquid, solid/gas, liquid/liquid, and liquid/gas surfaces (interfaces). These boundaries are studied in surface science to develop an understanding of phenomena and to develop theories that will permit prediction of future events. Some of the broad topical areas of study at the five possible surfaces are listed in Table 1. An understanding of these topics is enhanced by applying the methodologies of surface science. A list of acronyms used in this chapter is appended.

The S/S, S/G, and S/L surfaces, in that order, are of greatest interest to the solar materials scientist. As detailed in recent books (1-16), the experimental effort for studying these surfaces is now very extensive; the theoretical treatments are difficult. In science, it is customary to adopt a model based on an ideal situation and to compare the behavior of real systems

Table 1. Topical Study Areas at Different Surfaces Between the Solid (S), Liquid (L), and Gas (G) Phases

Surface		Topical Area of Study
Solid A Solid B	S/S	Corrosion, Adhesion, Epitaxial Growth, Friction, Diffusion, Nucleation and Growth, Thin Films, Solid State Devices
Liquid Solid	S/L	Wetting, Lubrication, Friction, Surface Tension, Capillarity, Electrochemical Properties
Gas Solid	S/G	Catalysis, Corrosion, Oxidation, Diffusion Surface states, Electron emission Thin films, Condensation and Nucleation
Gas Liquid	L/G	Diffusion, Surface tension Vapor pressure
Liquid A Liquid B	L/L	Immiscible Phases Spreading of films

with the ideal model. What is a realistic view of the boundary at a solid surface? It is not the ideal plane of infinite dimensions, but on an atomic scale consists of different crystal planes with composition, structure, orientation, and extent that are fixed by the pretreatment of the solid. As shown by Fig. 1, imperfections such as an isolated atom, a hole, an edge, a step, a crevice, a corner, and a screw dislocation may also coexist on the surface. Wide variations in the microscopic topography, as shown in Fig. 2, may also adversely influence the stability of the surface (Figs. 2(a) and (b) or the operation of a device because of voids (Fig. 2(c)).

During pretreatment of a solid, impurities may accumulate at or in the boundary in trace and larger quantities, and drastically alter the behavior of the boundary. The pretreatment of solid surfaces, which may involve outgassing, chemical reduction,

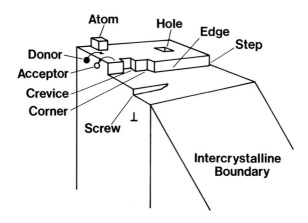

FIGURE 1. *The solid surface is not an ideal infinite single crystal.*

flashing a filament, ion bombardment, cleavage, field desorption, or depositing a thin film (17), are all made in an effort to minimize the uncertainty about the initial composition of the surface. Following controlled use of the solid, reexamination of the surface permits evaluation of the influence of the use on the measured properties of the surface. Characterization of the surface before, after, and if possible, during use of the material is clearly required.

For an overview of characterizing a solid surface, consider the questions: How much surface is there and where is it located? Is the surface real or clean? What solid form does the surface have? What is its topography and structure? What thermodynamic processes occur? How do surface species migrate? What is the equilibrium shape of surfaces? What is the depth of the surface phase? How much gas or liquid is adsorbed and where? What is the nature of the adsorbate-solid interaction? How should these phenomena be studied? The history of studying these effects in

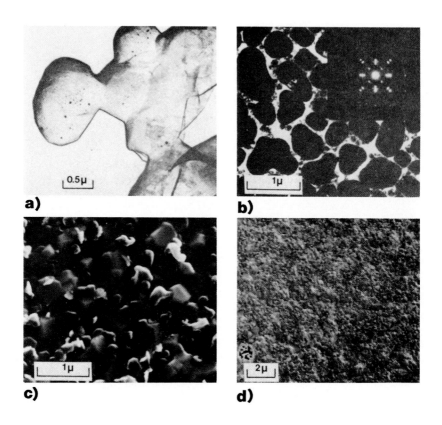

FIGURE 2. Real surfaces: a. Silver powder; b. a partially oxidized copper film on NaCl; c. sulfided copper film in H_2S/H_2 at $200°C$; d. as in c but at $73°C$.

gas adsorption on solid surfaces alone leads us to recognize that careful experimentation is the primary necessity in surface science. The experimental methods deemed most appropriate for solar materials will be indicated as each of the characterization questions are addressed in the following paragraphs.

A. *Surface Area*

Following Langmuir's pioneering work about 70 years ago, in
which the importance of structure, composition and bonding to
chemisorption were demonstrated, Brunauer, Emmett, and Teller
(BET) provided a means for deducing the surface area from multilayer physical adsorption isotherms (18). The BET method for
analyzing type II adsorption isotherms is used routinely by hundreds of laboratories with commercially available facilities (19).
When the surface is located internally, hysteresis is observed
between the adsorption and desorption branches of the isotherm,
forming type IV isotherms. The hysteresis results from capillary
filling of internal pores. A large number of methods have been
developed to determine the internal surface area from type IV
isotherms (19). An excellent treatise on surface area and porosity of solids is available (20), extensive examples from recent
studies of this subject have been cataloged (21), and the parameters used to differentiate between physical and chemical adsorption have been tabulated (21).

Visual observation of the topographical features of solids
is possible using electron microscopic (EM) examination of re-
plicas of surfaces, or by using a scanning electron microscope
(SEM). An assessment of the external surface area can be made
from SEM or EM photographs, but obviously the internal area is
not directly observable. Thus, qualitative and quantitative ans-
wers are provided to the questions of how much surface is there
and where is it located, except for the case of S/S.

B. *Real and Clean Surfaces; Solid Forms*

Broadly speaking, real surfaces are those obtained by ordinary laboratory procedures, e.g., mechanical polishing, chemical

etching, etc. Such a surface is covered by chemisorbed material, generally an oxide, and by physically adsorbed molecules from the surrounding environment. Real surfaces have been studied extensively because they are easily prepared, readily handled, and amenable to many types of measurements. It is the real surface that is encountered in most practical applications, and this will be particularly the case for solar materials. Figures 1 and 2 are illustrative of real surfaces.

Clean surfaces, which may be obtained by outgassing, chemical reduction, cleaving, field desorption, ion bombardment and annealing, preparing a thin film in ultra high vacuum (UHV) or flashing a filament (17) are more difficult to prepare and keep. Clean surfaces constitute the closest approximation to the true crystalline surface. They can be prepared on ordered single crystals, and can then be perturbed in carefully controlled experiments. Fundamental studies on clean well-ordered surfaces are increasing the basic understanding of S/G interactions in the modeling tradition of the kinetic theory, the dilute solution, the point particle, and the harmonic oscillator. Except for materials that can withstand solar concentration factors of about 1000, the large areas required for solar collectors mitigate against using clean single crystal surfaces in devices.

In surface studies, various solid forms are used, such as: (1) powders, (2) foils, (3) vacuum-prepared thin films, (4) coatings, (5) filaments, (6) cleaved solids, (7) field emitter tips, (8) single crystals, and (9) polycrystalline solids. These may be formed from metals, semiconductors, and compounds. The specialization of solid forms (5) through (8) preclude their widespread surface study for solar applications. Solid forms (3), (4), and (9) will be used widely in solar systems for reasons given in Section I.

C.   *Structure and Topography*

A clear distinction must be made between structure and topography, which are frequently used interchangeably (and incorrect-
ly) in the literature. Topography can be illustrated with a few
SEM photographs showing contouring, hills and valleys, and
superatomistic surface features. Structure refers to the repet-
itive spacing of atoms in a surface grating. The results of surface structural determinations show that the lattice spacing in
the bulk and at surfaces is generally the same. The most common
methods of deducing surface structure are by the diffraction of
low energy electrons, either elastically or inelastically (ELEED
and ILEED), by reflected high energy electrons (RHEED), and by
using the imaging techniques of the field emission (FEM) and
field ionization (FIM) microscopes (1,8-10,12,13). The value of
the techniques of LEED, RHEED, FEM, and FIM to fundamental surface science is evidenced by the extensive results secured using
them and published in the journal, *Surface Science*. The cost
effective deployment of large areas of solar collectors will most
probably be in the form of polycrystalline materials, with all
index planes emerging at the surface. Therefore, it is not expected that the structural determination of solar materials surfaces will be applied except for a few special cases. Rather,
the challenge will be to understand the changes in surface activity of heterogeneous real surfaces, such as those shown in Figs.
1 and 2.

D.   *Surface Composition or Purity*

The useful properties of real surfaces often depend on the
presence of chemical groups, e.g., impurities, that are extraneous
to the bulk composition. These tend to prevent the self-minimi-

zation of the surface energy of the solid. They also influence
the growth kinetics, topography, surface diffusion coefficients,
and residence time of adsorbed species. Depending on the surface
energy, impurities may concentrate at the surface or incorporate
themselves into the bulk. These processes involving mass transport to, from, and along the surface may produce significant time-
dependent changes in the properties of the material. Therefore,
it is extremely important to be able to identify the elemental
composition of solid surfaces at various stages of using solar
materials.

Until eleven years ago, experimental techniques did not
exist for identifying the elements on a surface. Since then,
commercial instruments have become readily available and have
been developed for the single purpose of measuring the elemental
composition of surfaces. The basic principles of the four most
widely used methods, Auger electron spectroscopy (AES), X-ray
photoelectron spectroscopy (XPS or ESCA), ion scattering spectroscopy (ISS), and secondary ion spectroscopy (SIMS), are compared in Table 2. As is seen, electrons in AES and XPS are
energy analyzed, and the scattered or sputtered ions in ISS and
SIMS are mass analyzed. As a result, advantages and limitations
are imposed on the information that can be secured from each of
these techniques. A listing of the most important characteristics of AES, XPS, ISS, and SIMS is given in Table 3.

Although the samples must be analyzed in vacuum in all four
cases, the analysis for the accumulation of impurities from the
gas and liquid phases can be accomplished using load lock cham-
bers. For the S/S surface, composition profiling can be accomplished by in situ ion etching. For this a 0.5 to 10 keV ion
beam is used to bombard a surface. The energy transfer causes
particles to be ejected (sputtered) from the surface, thus exposing increasing depths of the solid to probing with surface-

Table 2. Comparisons of Some Categories for
Four Methods of Surface Analysis

Method/ Category	AES	XPS(ESCA)	ISS	SIMS
Input "Particle"	Electrons	Photons (x-rays)	0.5-3.0 KeV Ions	0.5-20 KeV Ions
Damage to Sample by input particle	Moderate (focussed e-beam)	Minimal (x-radiation)	Moderate (defocussed ion beam)	Moderate and variable depending on the sample
Output Particle	Electrons	Electrons	Inert Ions	Secondary Ions
Measured Quantity of Output Particle	Auger Electron Energy	Auger and Core Level Electron Energy	Scattered Ion Energy after Binary Collision	Mess of sputtered surface Ions
Sampled Depth	2-10 Monolayers	2-10 Monolayers	Surface Monolayer	1-2 Monolayers

sensitive analysis techniques. Details of the ion-etching processes for probing the composition at increasing depths of the solid can be secured from a number of sources (22-24). The principal factors that influence ion yields during sputtering, i.e., energy, number, angle of incidence, and ion used in bombardment, as well as the binding energy of the solid, have been collected into a useful set of illustrations (25). The principal limitations result from cratering problems, different crystal faces having different sputtering yields, sputtering rates depending on bombardment angles, bombardment damage, and bombardment enhanced diffusion.

Further information about ISS   (25-28),   SIMS   (22,23,29), AES (30-32), and XPS (32-34) is readily available. For comparing these four surface analysis techniques, a continuation of Table 2 is useful. Factors that need to be compared include the influence of the sample on the results under such categories as vacuum worthiness, geometry, solid type (metal, semiconductor,

Table 3. Advantages and Limitations of Using AES, SIMS, ESCA, and ISS Surface Analysis Techniques

Auger Electron Spectroscopy

- is useful for chemical shifts for some elements when in different compounds
- is especially good for CIDP (composition-in-depth analysis)
- is especially sensitive to light elements
- will form carbon from polymers (e-beam cracking)
- has superb lateral resolution (0.2-5.0µ beam)
- is fastest of the methods
- is sensitive to 2-10 monolayers
- may have severe charging problems

Electron Spectoscopy for Chemical Analysis

- is most quantitative of major techniques
- is especially useful for chemical shifts from the same element in different compounds
- is useful for CIDP
- is especially sensitive to light elements
- is least destructive of all techniques (X-ray excitation)
- has poor lateral resolution
- is sensitive to 2-30 monolayers; metals, oxides, and polymers, in that order

Secondary Ion Mass Spectroscopy

- is sensitive to 1-2 monolayers
- can detect isotopes with unit resolution
- can detect H and D present (the only one)
- can be sensitive to 1 ppm
- is especially good for CIDP
- is responsive to chemical matrix effects
- has complex spectra
- has varying elemental sensitivity
- is quantitative with difficulty, at best
- requires destruciton of sample for the analysis

Ion Scattering Spectroscopy

- is sensitive to first monolayer
- can detect isotopes present ($O^{16} - O^{18}$, $N^{14} - N^{15}$, $S^{32} - S^{34}$)
- for some elements, may provide chemical information about surface compounds
- is useful for composition in-depth profiles
- can possibly be quantitative
- is useful for studying ordered surfaces
- has simple spectra
- has poor resolution
- is necessarily destructive

insulator, organic, compound), atomic number, isotopic and matrix effects, and surface structures. Then, factors should be compared that influence the results or data acquisition such as sputter rate, lateral resolution, vacuum requirements, time to obtain a spectrum, signal to noise, ability to depth profile, elemental identification, resolution, sensitivity, depth resolution, chemical effects, information about the matrix, and quantification potential.

There are many other methods applicable to surface studies (35). Some of these include Rutherford back scattering (RBS) (36), the electron microprobe (37), the atom probe (38), glow discharge mass spectrometry (GDMS) (39), glow discharge optical spectroscopy (GDOS) (40), and surface analysis of neutral and ion impact radiation (SCANIIR) (41). Some of these methods, e.g.,

GDMS, GDOS, and SCANIIR, are not commercially available or are very expensive, e.g., RBS and atom probe, so are currently of marginal value for widespread study of solar materials. The methods of surface analysis mentioned in this section are of obvious importance to the characterization of solar materials. A listing of techniques required for characterization studies of the surfaces of solar materials will be presented at the end of Section II.

E. *Surface Thermodynamics, Equilibrium Shape, and Diffusion*

Surface atoms are in a markedly different environment than bulk atoms. They are surrounded by fewer neighboring atoms, which are in a surface-unique anisotropic distribution, compared to those in the bulk. The surface phase has higher entropy, internal energy, work content (Helmholtz energy), and free energy (Gibbs energy) per atom than when in the bulk phase. For isotropic solids, surface free energy and surface tension are equivalent

8-10 . At equilibrium, the solid surface will develop the shape
that corresponds to the minimum value of total surface free energy. The surface free energy of solids can be calculated for surfaces of different structures. The relative magnitude of surface
free energy can be deduced from experiments involving thermal
facetting, grain boundary grooving, etc. The low index planes
are the most stable because they have the lowest free energy.
All small particles and all large flat polycrystalline surfaces,
which are characteristic of the the type planned for solar applications, possess a relatively large surface excess energy and
are thermodynamically unstable.

Based on calculations made for inert gases (42), ionic halides (43), and semiconductors (44), the depth of the surface phase
may range up to 5 nm. In the latter two cases, substantial deviations of the actual surface structures from those characteristic
of a truncated but otherwise bulk solid have been reported from
theoretical analyses of LEED intensities from single crystals.
The reconstruction or rearrangement of the surface involves move-
ments of up to 0.05 nm in the surface atomic plane and possibly in
the first bulk layer as well (44).

Surface atoms are restrained from interatomic motion by nearest neighbor bonding, but these bonds are weaker than for a corresponding position in the bulk. The potential barriers for surface diffusion are lower than for bulk diffusion, so less activation energy is required to produce surface diffusion processes.
The mechanism of surface diffusion may change with temperature
because the surface population of adions, vacancies, ledges, etc.
or other conditions such as structure and ambient atmosphere may
change.

F.   Amount Adsorbed and Nature of Adsorbate/Solid Interactions

This section applies primarily to the G/S surface, although the general concept is valid for the L/S surface. Adsorption is the accumulation of a surface excess of two immiscible phases. Except for multilayer physical adsorption, the forces of interaction limit the amount adsorbed to one atomic layer, i.e., up to a monolayer of chemisorption. For measuring the amount of gas adsorbed directly, mass gain, volumetric, and radiotracer techniques are used. Indirect techniques require relating the amount adsorbed to the increase or decrease in intensity of some "output particle" such as desorbed gas, scattered ions, ejected Auger electrons, or quanta of radiation. The direct techniques are limited in their applicability but there is a danger for the indirect techniques in assuming that the change in signal intensity varies linearly with adsorbate coverage (21).

Adsorbed gases may bind to the surface nondissociatively or dissociatively, remain localized or have mobility, form two dimensional surface compounds, or incorporate themselves into the solid, either as an absorbed entity or as part of a new compound. The interaction between a chemisorbed species and the solid depends on the geometric configuration, fractional coverage, and the electronic interaction. Understanding of the electronic bonding interaction is sought using a wide diversity of measurement techniques, such as infrared spectroscopy, magnetic susceptibility, electron spin and nuclear magnetic resonance, work function, conductance, LEED, impact desorption, FEM, FIM, electron energy loss, ultraviolet photoelectron spectroscopy, etc..

Most of the measuring techniques used to study the chemisorptive interaction require high vacuum or better. This is unfortunate for the solar surface materials scientist. Solar systems are required to operate in real environments at atmospheric pressure. Therefore, the chemisorption phenomena of interest are those that lead to degradation of solar materials surfaces. Thus, an understanding of the weak co-chemisorption processes (40-150 kJ/mol) on a surface with a prior adsorbed layer, carried out at from sub- to super-atmospheric pressures, will require pushing to the limit the currently available techniques, such as infrared, ellipsometric, microgravimetric, and magnetic measurements. At the same time, innovative methods of detection need to be developed. The problems of studying real surfaces of solar materials are similar to those encountered by corrosion and catalytic scientists.

G. *Methods of Studying Surfaces of Solar Materials*

The object of scientific study is to gain understanding by securing control of the variables. As indicated in the chapter by Claassen and Butler, the detailed study of particular solar materials is best undertaken when the selections can be made as a result of other research. The selection of the solid form (e.g., polycrystalline thin films), the specific S/S, S/L, or S/G surface (e.g., silver/glass in mirrors), and the preparation technique will be dictated by the cost and the performance properties of the components. Once the options have been narrowed, it is clear that in-depth, fundamental investigations of the surfaces between the materials need to be carried out. The consumer should be understandably impatient if a degradative failure in a solar system cannot be corrected because the proper long-term applied research has not been carried out to gain the required understanding of the cause. At the same time, only finite resources of research funds and personnel can be made available. What types of surface characterization measurements then are most important and what surfaces should be studied? The latter question will be addressed in Section III.

According to the compilation by Lichtman (35), there were
over 70 methods available in 1974 for relating some measured parameter with a S/G surface. For surfaces as defined in this paper,
the number of techniques one could apply easily exceeds 100 (2,3,
4,11,12). Most of the available methods require some to total
specialization for its use alone. Broad philosophical guidelines
need to be developed in order to narrow the choices of surface
analytical facilities required for general study of solar materials surfaces. These guidelines should be based on (1) studying
the surfaces of long-term interest to solar energy conversion
systems; (2) studying surfaces under (actual and simulated) conditions as close as possible to those encountered when operating
the system; (3) studying the fundamental processes at the S/S,
S/G, and S/L surfaces that impact (1) and (2); (4) employing,
initially, those methods where equipment is commercially available
or where custom equipment can be used or reproduced quickly because of available expertise; and (5) selecting facilities that
will have the optimum impact on problems encountered in thermal,
photovoltaic, biochemical, and ocean thermal energy conversion
systems. Since solar materials must remain stable for long periods, the surface properties that provide input data on the reactivity (and/or degradation) must be specifically studied. These
properties are the surface composition or purity, topography
(including defects and inclusions), structure, area, interaction
with ambient gases or vapors at various pressures and radiation
exposures, and electron energy distribution. The leading experimental methods for characterizing these properties are listed in
Table 4. A number of these, notably LEED, FEM, and FIM are not
practically applicable to the likely use of polycrystalline materials in mass produced solar systems. Thus, the initial characterization should not attempt to relate reactivity to structural
sensitivity per se, but only to the average structural distribu-

Table 4. Tabulation of Methods for Measuring Important Properties of "Surfaces"

Area	Topography	Purity or Composition	Ambient Atmosphere
Gravimetric	SEM		
Volumetric	EM	AES	UHV
EM	Optical Microscopy	ISS	VHV
	Optical Interferometry	SIMS	Modified Atmosphere
Electron Density of States	**Microdefects**	ESCA	
		RBS	
	EM	Atom Probe	**Structure**
UPS	TED	Electron Microprobe	LEED
	Optical Microscopy		FIM
			FEM
Amount	**Inclusions**	**Chemical Bonds**	TED
Adsorbed, Film Growth	AES/SAM	IR, FTIR	X-Ray
	ISS/SIMS (Imaging)		
Gravimetric Ellipsometry		**Valence State and Coordination**	
		ESCA	

tion on a polycrystalline surface. As the number of problems at the S/L surface increases, some of the methodologies discussed at the recent conference on non-traditional approaches for studying the solid/electrolyte interface will have to be considered (45).

III. SURFACE SCIENCE OF SOLAR MATERIALS SURFACES

In this section, an overview will be provided of the type of surface science research that is important to all energy technologies. These comments are based in part on a report prepared from 51 authorities in the field (46). Broad areas of surface science that will impact solar materials research will be indicated. Some specific problems dealing with layers of supported thin films or coatings in reflectors, absorbers (heat), photovoltaics (quantum absorbers), and transmitters will then be outlined.

A. *Surface Science in Energy Technologies*

It is clear that surface phenomena play a widespread and
dominant role in various energy technologies. Our ability to
prepare surfaces with the desired performance properties will ex-
ert a major control over the wide range of energy conversion technologies. Most of the problems are long range in nature. An
understanding is needed to correlate the molecular properties of
surfaces with their behavior in corrosion, grain boundary fracture, adhesion, radiation damage, catalysis, lubrication, and embrittlement. Studies which have been concentrated on the S/G interface must be broadened to the S/L and S/S interfaces. New
measuring techniques need to be developed and, when necessary,
national centers of surface science research should be established.

To improve communication between scientists, efforts should
be made to bring together scientists from different disciplines.
Understanding of surface phenomena and its applications require
knowledge of physics, chemistry, materials, mathematics, and the
various applied engineering sciences. Team research in applied
surface science is absolutely necessary because of the enormous
background required from the various disciplines. I strongly
encourage those who major in materials to secure at least inter-
mediate-level courses in physics, chemistry, and another applied
science area. The time spent in broadening your scope with more
course work will make you a highly sought member of an interdisciplinary surface science team.

The following are the types of surface science studies that
are expected to have the broadest impact on all energy technologies (40).

o Studies of film growth;

o Characterization of the compositional, structural, and
   electronic properties that influence behavior,

o Studies of reaction dynamics and elementary processes (adsorption, diffusion, bonding, and desorption) at surfaces,
   and related theoretical developments;

o Development of instrumental methods for studies of external and internal surfaces under operating conditions;

o Studies of the structure, composition, and electronic
   properties of surfaces between condensed media (S/S, S/L),
   of grain boundaries, and of related theoretical developments;

o Studies of the strength of surfaces, as in adhesion, grain
   boundary cohesion, lubrication, and environmental influences;

o Quantification of surface measurement techniques and measurement of sputtering yields;

o Studies of real surfaces, oxides, sulfides, and carbides;

o Studies of surface modifiers such as implanted species,
   poisons, promotors, and inhibitors for the structural and
   chemical stabilization of surfaces;

o Studies of the surface properties of new classes of materials such as bimetallic clusters, composites, ceramics,
   polymers, ion-implanted materials, intercalation compounds,
   etc.;

o Search for substitutes of precious metals and other surface active materials not available in the U.S.; and

o Characterization of surface defects.

B. *Areas of Surface Science and Solar Energy Technologies*

It is useful to divide surface science into specific areas

of study to consider how these will apply to solar materials.
These study areas are chemisorption and catalysis; semiconductor,
thin film and solar device surfaces; electrode, electrocatalysis,
and photon-assisted surface regions; corrosion; ion interactions
with near-surface regions; and grain, phase, and interfacial
boundaries. The impact each of these areas of surface science
can make on solar materials is given in the next six subsections.

   1. Catalysis and Chemisorption of Solar Materials. Catalysis and chemisorption problems are more important in other energy technologies than to solar. However, chemisorption, corrosion,
and catalyzed activation in solar device surfaces are of significant short range importance. This is particularly true when the
reaction degrades the performance of the solar device. Examples
of current problems include the surface reactions of mirrors,
thermal absorbers, and photovoltaic materials, such as cuprous
sulfide. Mechanisms and dynamics of the reaction must be sought.
A specific example is the copper catalyzed thermal oxidative
degradation of certain polymers. If the properties of polymers
used to protect surfaces are degraded by this type of catalyzed
reaction, the surface may not only be left unprotected, but the
characteristics of the degraded polymer overlayer may lead to an
accelerated reaction at the polymer/solid interface.

   There are several longer term problems. The first is concerned with the dynamics of growth and preparation of photoactive
surfaces via G/S deposition processes under well defined conditions. Chemical vapor deposition and plasma sputtering processes
are included here. Processes of this type can be cost competitive with electrochemical film preparation procedures and avoid
the multiprocess steps where introducing undesired contaminants
is difficult or costly to control. The second problem involves
the photocatalysis of water to produce hydrogen using such compounds as titanium dioxide and strontium titanate.

2. *Thin Film and Semiconductor Surfaces*. Thin films and
semiconductors are especially important in solar energy. Metallization films, protective films, antireflection coatings, chemically passivating films, abrasion resistant coatings, transparent
conducting coatings, and semiconductors for photovoltaics are of
near-term importance. In general, the growth processes of films
and coatings need to be understood. In the solar thermal area,
studies of reflector, transmitter, and absorber surfaces that are
stable under solar stresses are needed. Studies of the effect
of long-term irradiation of light from 300-2600 nm on metal/semi-
conductor, metal/polymer, and semiconductor/polymer surfaces are
needed. There is need to elucidate the chemical and structural
properties at S/S surfaces, and to determine as well the electronic properties for photovoltaic materials. Considerable detail
of long term research needs for reflector surfaces, glass surfaces, and absorber coatings is available in recent planning doc-
uments (47-49). The photovoltaic needs are covered adequately
in other chapters of this book. For longer-term photoelectrochemical applications, investigations of stable surfaces for the photoelectrolysis of water for the production of hydrogen are needed.
The selection of specific candidate surfaces for further study
requires more research.

3. *Electrodes, Electrocatalysis, and Photon Assisted Surface Reactions*. The surface research on electrodes and electrocatalysis will affect solar technologies primarily in studying
fuel cells, storage batteries, industrial electrolytic cells,
and industrial electrocatalysts. All these study areas are also
general to energy since they touch upon conservation, storage,
generation, and use, such as in vehicle propulsion. Many of the
problems in this area are at the electrode surface, e.g., insufficient catalytic activity, susceptibility to poisoning, and loss

of electrocatalyst area.   These are all part of the broad field
of heterogeneous catalysis.   As such, impurities on the solid
surface require careful characterization.

The overview of photon assisted surface reactions will be
covered in the chapter on photoelectrocatalysis.   This area has
long range promise as a solar energy technology, and it contains
solid surface/electrolyte interface problems.   Research is needed
to establish the factors controlling photostable surfaces, surface states, electrode sensitization, and the consequences of
light intensity and temperature.   As indicated in the chapter by
Butler and Claassen, this is a long range research area, detailed
surface investigations of materials probably should await a nar-
rower selection of candidate materials.   The characterization of
the various interfaces needs to unravel the key problem of electrode stability as a function of potential and temperature.   Since
many electrolytes are transmitting, optical spectroscopic techniques (IR, ATR and Raman) are being used to probe in situ the
species attached to the electrode surface.   The detection of elements using ISS, ESCA, AES and SIMS are of obvious importance to
characterizing materials at various stages of use, as are BET
surface area measurements to characterize the area and pore size
distribution.

4.   *Corrosion*.   The corrosive degradation of solar material
surfaces at the S/G and S/L surfaces cover broad categories of
applications of surface science.   There is obvious overlap between corrosion and the problems mentioned in the previous three
sections.   Fluids used in heat transfer systems, in flat plate
collectors, in storage applications, and in OTEC systems provide
a different class of L/S problems from those encountered in L/S
photoelectrochemical systems.   Studies of the degradation of reflector and absorber surfaces at atmospheric pressure constitute

a different class of corrosion reactions from those normally encountered because of the synergism of UV, diurnal temperature
cycles, and changing atmospheric contaminants. Except for the
reflector surfaces, the change in the heat transfer across the
new interface is of dominant importance in the early stages while
the lifetime of the system can be compromised at later stages.

The effort in corrosion studies should be concerned with
studying nucleation and growth of the new phases formed on real
surfaces and then studying the structural, reactive, and adhesive properties of the fully developed phase. As the layer grows,
the heat transfer properties will be affected and system performance may be oscillatory, particularly, if scaling and spalling
occur during diurnal temperature fluctuations. Characterization
of the phases in the early stages of growth for composition,
structure, and defect distribution will assist in deducing the
mechanism of growth; the characterization methods are required
for both the S/L and S/G surfaces. The emphasis of chemisorptive
corrosion studies for metals and polymers should be on the early
monolayer growth stages; whereas, for ceramics, composites, and
glasses, the emphasis should be on the role of the defect structure after the product layer has been formed. Surface modifica-
tion/protection research should emphasize the compatibility of
coatings and inhibitors on the base material. The coatings research here is similar to that discussed in Section III B. The
gas (or liquid) transport to the surface, diffusion along and
into the surface, and stresses set up in the coating and at the
S/S surface all need to be understood.

The current techniques of surface studies need to be applied
to a wider range of conditions. For example, the relationship
between structure and reactivity at atmospheric conditions need
to be addressed by combining such techniques as ellipsometry,

LEED, and the surface analysis spectroscopies (ISS, SIMS, AES, and ESCA). Surface characterization of nonmetallic materials needs to be done for covalent and ionic solids, oxide semiconductors, as well as complex systems, including polymeric materials. Finally, studies and new techniques for characterizing the S/S surface are required if adherent protective coatings are to be understood.

5. *Interaction of Ions with Surface and Near-Surface Regions.* The two major areas of application of ion beams to surfaces are ion-beam surface modification and ion-beam surface analysis. Ion implantation can be used to modify the near-surface region to provide better corrosion protection, erosion and wear resistance, and solar cell junction performance. Basic methods or phenomena here include sputtering, diffusion, corrosion, simulation of damage by ion beams, effect of ion-damaged surfaces on chemisorption, and the effect of impurities on sputtering. The systematic application of ion beams to surface analysis is a highly neglected area of research. Both ISS and SIMS are sensitive to the surface monolayer and to isotopic detection, but support for their development has been modest compared with the much better developed electron spectroscopies.

Modification of solar conversion systems or preparation of system components will involve the sputter deposition films or coatings for mirrors, absorbers, photovoltaic junction contracts, and the formation of passivating layers. The quality of compositional analysis of these components, other thin film stacks, and of the type of surfaces described in the preceeding sections will be improved by more research on SIMS and ISS.

6. *Grain, Phase, and Interfacial Boundaries.* The most difficult surface to study is probably the S/S interface. Nondestructive techniques for examining this interface are difficult to

conceive, let alone implement. The underlying scientific problems are to relate the structure, composition, defects, phases,
and dynamic behavior of interfaces to their structure, strength,
and interaction with various environments; to study the kinetic,
thermodynamic, and structural factors that control segregation
at interfaces; and to determine the effects on surface dissolution, and boundary decohesion, both at the surface and internally,
on the interaction with various environments. With the expectation of the deployment of large areas of polycrystalline thin
film systems in solar systems, this area of surface science is
of both short- and long-term interest.

C. *Surface Science Applied to Specific Solar Energy Materials*

As a broad summary of the two preceding sections, major research efforts should be carried out on surfaces to elucidate
mechanisms of degradation in solar stressed environments, in
order to guide modifications of materials to minimize the effect
of the of the detrimental processes on the performance of the
system; to understand the influences of structure, bonding, and
composition that provide interfacial stability; to devise materials preparation and fabrication processes that yield the desired
chemical, electronic, physical, or mechanical properties; and
to devise new experimental techniques to study and/or measure the
important properties of surfaces. In the initial work, the most
promising polycrystalline materials should be selected, and enhanced stability/reactivity should be studied, when exposed to a
solar-stressed environment and interfacial contact. The work
should strive to elucidate mechanisms of degradation and reactions at S/S, S/L, and S/G interfaces, of photoenhanced degradation of polymers, and of photoenhanced ion transport in transmitting materials. Gas/solid reactions and cochemisorption

processes need to be studied at atmospheric pressure. Diffusion in coatings and films, including atom transport and accumulation at grain boundaries, needs to be understood. Studying the structural, bonding, and compositional influences on adhesion, including the interactions between polymer/metal (oxide) surfaces and between dust and transmitting materials is important. Influences of interfacial effects on the optical properties and on the nucleation and growth of new phases at interfaces, as both relate to system performance, need to be assessed.

In the following sections, the component of a solar energy conversion system, a current surface problem with it, the approach for studying it, and some recent progress will be discussed. An overview of the specific materials and problem is provided in Table 5.

1. Silver Interfaces. As indicated in the first chapter, silver has the most desirable reflectance property of any element (~97%) and will require the least concentrator area to collect and concentrate a given amount of solar radiation. Although silver itself is relatively unreactive, a fractional monolayer of adsorbed oxygen enhances its reactivity to atmospheric gases,

Table 5. Some Current Problems with Solar Materials

System or Component	Material(s)	Problem
Reflectors	Silver/Glass	Degradation
Absorbers	Black Chrome	Degrades at T>300°C
Polymers (Encapsulant)	Acrylics	Photodegradation
Polymers (Protective)	Copper/Polymers	Catalyzed Reaction
Solar Cell Material	Cu_2S	Degradation
Solar Cooling	Silica Gel, Zeolites	Cycle Times
Thin Film Solar Cell	CdS/Cu_2S	Interdiffusion
OTEC	Titanium	Biofouling

such as water, carbon dioxide, sulfur dioxide, nitric oxide, etc.
(50). At room temperature and atmospheric pressure, nearly one
monolayer of oxygen can always be expected on silver. Therefore,
chemisorption of atmospheric gases initiates corrosive reactions
and a degradation of the reflectance.   The results of
these reactions have yielded visually transparent areas in mirrors
used in demonstration heliostat fields in time spans ranging from
several months to a few years. Current solar economics require
mirror designs to last 30 years.

The present mirror system in use is a glass second-surface
silver mirror backed with copper and paint, as shown in Fig. 3.
Interfacial degradation reaction may begin at the silver/glass
interface because of impurities at the interface. These may be
residual impurities resulting from the method of preparation, or
the impurities resulting from the method of preparation, or the
impurities may accumulate there because of radiation-induced

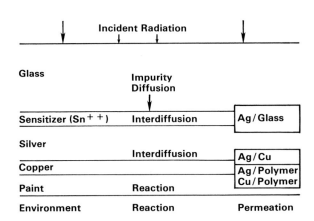

FIGURE 3.   Potential degradation surfaces in a typical
mirror system.

transport processes of various ions in the glass. Deterioration
of the mirror material may also result from interdiffusion of
copper and silver, and reaction of the copper and then the silver
with atmospheric gases. The rate of permeation of the paint backing by atmospheric gases may increase as the paint weathers in
the sun and elements. For this system, the characterization and
study of the glass/silver, silver/copper, and copper/paint interfaces before, and after various stages of use is clearly required.
The methods of characterization outlined in Section II, especially
those of ISS, ESCA, AES, and SIMS, are clearly applicable to this
problem. For solving the mirror degradation problem, it is important to prepare model mirror systems and secure commercially
made mirrors now in use, and to elucidate the mechanisms of the
reactions that result in deterioration of the reflectance of the
mirror. Characterization of the systems will include using optical, diffraction, adhesion, and corrosion resistance measurements
in addition to various methods of surface analysis (ISS, ESCA,
AES, SIMS, SEM, IR, etc.). The mechanisms of reactions will be
deduced by measuring the kinetics of pertinent surface reactions
(e.g. $Ag \cdot O_{ads} + SO_2$ (g) $+ H_2O$ (v) $\rightarrow AgS_xO_y$), identifying the
sources of contaminants that enhance the degradative reactions,
and carrying out accelerated tests of the interfacial reactions.
Both G/S and L/S interfacial reactions require study as well as
any surface degradation reactions at the polymer/copper (or other
S/S) interface.

Preliminary surface analysis on both polymer coated and uncoated degraded first-surface silver mirrors shows extensive accumulation of 0, S, and Cl on transparent silver regions and of
0 and S on visually reflecting regions (51). The transparent regions also contained silver, showing the corrosion product was
probably a silver oxy-sulfur compound; the shift in the sulfur

ESCA peak indicated the sulfur valence was of the order of +2
rather than the +6 required if silver sulfate were formed.

Extensive detailed study will be required to translate these
initial results into an understanding of the degradation of silver
in mirror systems. However, it is reasonable to anticipate that
a significant improvement in mirror stability will result from
using different glasses and/or backing materials and different
fabrication processes. The modifications of the materials and/or
processes will follow the understanding of the cause of the degradation.

2. *Black Chrome Degradation*. Black chrome has desirable
properties as an absorber coating, as indicated in several chapters in this book and in a recent review article (52). In demonstration solar systems, black chrome is overlaid on a nickel/cop-
per/heat exchanger material stack. At temperatures of 250-350°C,
black chrome loses its photothermal absorptivity to become a non-
absorber, but the mechanism of degradation is not known. A number
of studies for improving the stability of black chrome are underway (47). A key goal is to determine if the mechanism of degradation, which is presumably an interface oxidation of chromium particles in a chromium oxide matrix, is fundamentally different because of different substrates when overlaid on the heat exchanger
materials or if black chrome itself is kinetically unstable at
about 300°C.

The approach to isolating one possible cause of the degradation has been to prepare black chrome by reactive evaporation in
oxygen-18 on quartz substrates (53). Microgravimetric techniques
can then be used to measure the time-temperature reactivity of
the material in oxygen-16. Depth profiles using ISS and SIMS,
both of which can detect the presence of the two oxygen isotopes,
can be taken at various stages of the reaction as an additional

aid for elucidating the mechanism of reaction in oxygen.

Changes in the topography (by SEM), in the distribution and amount of Cr^o and Cr^{+3} (by ESCA), in the distribution of all elements (by AES), in the distribution of oxygen isotopes (by ISS and SIMS), and in the optical properties (% transmittance and reflectance), can all be correlated with the temperature dependence of the reaction measured microgravimetrically in oxygen and with other gases present. The power of the apparatus described recently (54) for measuring the mass change, optical transmittance and reflectance of thin films is enhanced by auxiliary surface analytical measurements.

In recent preliminary results, it has been shown by ESCA that reactively evaporated black chrome consists of only Cr^o and Cr^{+3} and by ISS that the preparation in oxygen-18 yields the desired isotopic labelling (53). Oxidation of black chrome on quartz in oxygen-16 becomes rapid at about 475°C and is completed below 700°C. It is not as yet known if the apparent increased stability (~475 vs 250-350°C) results from the absence of the substrate. stack or the preparative process.

3. *Polymer/Metal (Oxide) Interfaces.*   Polymeric materials are important to solar technologies for use as protective coatings, encapsulants, and backings for mirrors. In each of these cases, the "protected material" is known to degrade when exposed to a solar-stressed environment.

The problem is that the polymer/protected-material interface may experience a degradative reaction. The protective value of the polymer may deteriorate because the UV and/or environmental stresses change the properties of the polymer. The copper ion catalyzed thermal oxidative degradation of polypropylene, which will be discussed in technical detail in Section IV, is an example where the interface and/or material in the interface, e.g., copper,

results in a significant degradative reaction at lower temperatures than normal. Therefore, it is necessary to carry out studies
where the synergism of UV, temperature, atmospheric gases, and
the interface composition can be assessed.

One approach to studying polymers is to monitor the development of functional groups using an FTIR fitted with an environmental test chamber (ETC) and a UV simulator, as shown in Fig. 4.
The sample (1) and blank (2) in an ETC can be subjected to a flow
of gases (4) at the same temperature, while only the sample is
subjected to a UV flux (3). The IR beam is directed (5) at the
sample or reference blank and the reflected beam (6) is focused
onto the detector in the FTIR system. The polymer of interest
is coated onto both (1) and (2); by proper selection and sequenc-
ing the study of the variables, functional groups formed in the
polymer can be ascribed to UV, the polymer/S interface, temperature, or a particular gaseous constituent. (A description of
the overall apparatus is available [51, p. 7].) The significance
of this approach is that the simultaneous or sequential combinations of stresses can cause changes in the important mechanical,
chemical, physical, and interfacial properties of the polymer

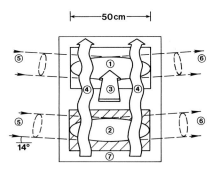

FIGURE 4. Scheme for in-situ photochemical studies using
a FTIR.

that adversely affect the functional performance of the component
material.

There are other surface and interface problems related to
candidate polymers for solar applications. For example, the
desirable transmittance of a polymer may be reduced by abrasion
from the environment or the adherence of dust. The latter is
primarily a problem in surface chemistry. The potential reactivity between a polymer and any other contact surface needs to be
studied to determine if an interfacial imcompatibility exists.
Reactions at the polymer/S interface may lead to delamination.
At the same time, UV enhanced degradation of the bulk polymer
may result in adsorption of a corrosive species at the polymer/G
interface and permeation of this species through the polymer to
the polymer/S interface, where an undesired reaction may actually
be accelerated rather than eliminated. The study of thin polymer
coatings on metal and metal oxide thin films of candidate solar
materials is certainly an area of broad opportunity in surface
science.

   4. Water Vapor Sorption by Desiccants. Desiccant materials
have potential applications for solar cooling of buildings by
using solar energy to desorb water vapor in regenerating a material used to sorb water vapor. The heat required to vaporize
water, which is sorbed by the regenerated desiccant, is extracted
from the room to be cooled. The problem is that the rates of
sorption and desorption are not known for candidate materials,
such as silica gels and zeolites. Furthermore, the effects of
several sorption-desorption cycles each day on the rate and
amount of water vapor adsorption are also not known. Microgravimetric and chromatographic methods are being used to secure the
desired data, which depend critically on the heat of adsorption
and on the stability of the internal surface area of porous materials. These methods will permit measuring changes in the surface

area and its location (Section IIA) at various stages of cycling
as a further technique to evaluate the potential of each candidate
desiccant.

The application of microgravimetric materials for studying
the cyclic effects of the reactions of a gas or vapors with a
solid for candidate G/S systems for storage applications is mentioned as an aside here. The diurnal solar cycle will subject
storage materials to total cycle times never before studied on
G/S systems. The long-term effects of cyclic de- and rehydration
or de- and re-ammoniation on the rate of reaction needs to be
assessed. Again, potential nucleation and growth problems may
be encountered, but the changes are induced by G/S interactions.

5. *Cuprous Oxide Degradation.* The Cu_2S/CdS system is one of
the leading candidates for the direct photovoltaic conversion of
solar radiation into electrical energy. A cross section of an
assembled solar cell is shown in Fig. 5. First, note the interfaces present are CdS/metal, CdS/Cu_2S heterojunction, Cu_2S/grid
material, epoxy/grid material, and epoxy/Mylar. Whether the materials shown in Fig. 5 actually are the best or not is unimport-
ant for the present discussion. First, the potential number of
interface degradative reactions is important and secondly, one of
the degradative reactions is that between cuprous oxide and
oxygen and/or water vapor. The epoxy and mylar are used to prevent this reaction, but potential interface and bulk polymeric

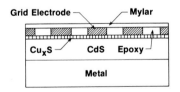

FIGURE 5. Cross section of a thin film Cu_xS/CdS solar cell.

reactions may permit both these gases to concentrate at the
Cu_2S/epoxy interface. Therefore, one interfacial study of interest is that of Cu_2S in an oxygen and water vapor environment. It
is known that cell efficiency drops sharply at small departures
from a pure Cu_2S stoichiometry. In his chapter, Dr. Kazmerski
will address another type of interfacial problem of interest,
i.e., changes at the interface because of electron transport
processes.

     A viable approach for determining the stability of Cu_2S in
potential reactive gases, and how much of the Cu_2S has reacted,
is to prepare the material and subject it to the degradative
components in situ. A microgravimetric system (54) has been used
for preparing Cu_2S by sulfiding copper films in an H_2S/H_2 atmosphere. Copper films were vacuum deposited onto quartz substrates
and then sulfided to Cu_2S while monitoring the mass gain, transmittance (T), and front-surface reflectance (R). After forming
Cu_2S, the sulfiding atmosphere can be replaced with the desired
reactive atmosphere, and again the mass change, T, and R of the
film can be measured as the reaction proceeds. Auxiliary surface
characterization measurements of the pure films included ESCA,
ISS, AES and SEM. The gravimetric data have shown that stoichiometric Cu_2S films 100 to 400 nm thick can be prepared in appropriate H_2S/H_2 mixtures and have identified the proper temperatures
and sulfiding times. From SEM data, the films are seen to be
topographically smooth enough (Fig. 2d) to permit using the T and
R data for deducing the optical constants of Cu_2S. These have
been found comparable to those reported by other investigators.
Furthermore, the control of the variables now permits using combinations of isotopically labeled oxygen, water, and other gases
for securing a small amount of reaction and confirming it with
ISS and/or SIMS depth profiles.

The critical amount of degradation can be inferred by correlating the extent of the reaction measured with a similar degradation of a functioning heterojunction. Future studies of the kinetics of degradation in reactive environments should lead to elucidating the mechanism of degradation and an understanding of the role of the important Cu_2S/G reactions.

6. *Biofouling of Heat Exchanger Surfaces in OTEC Systems.* The heat exchanger walls in OTEC systems are subjected to a continuous exposure to a marine environment. One of the key problems in OTEC systems is to find surfaces that will not permit the initiation of biological growths, but will withstand the attack of corrosive sea water. Further elaboration of this problem is given in the chapter by Pohlman. The approach for studying the L/S interactions and the nucleation and growth of the species that produces the undesirable biofouling products may well require employing some of the nontraditional approaches for studying the L/S interface (45).

IV. ROLE OF THE POLYPROPYLENE/COPPER OXIDE INTERFACE IN THE CATALYZED OXIDATIVE DEGRADATION OF POLYPROPYLENE.

A number of polymers, e.g., polyethylene and polypropylene, are known to degrade oxidatively in the presence of copper and its salts (55). The basic goal in this work was to determine the role of the polypropylene/copper oxide interface in the catalyzed thermal oxidative degradation of polypropylene. While a number of different kinds of polymers are proposed for contact with copper (oxide) in solar conversion systems, this problem has consequences of a generic nature in energy production and transmission and in other industrial applications. For example, it is known that polymer coatings on copper wire, such as polyethylene and polypropylene, degrade and lose their insulating properties. Failures in electrical and communication network systems cause problems in civilian and military ground, sea, and space equipment and vehicles. Therefore, research was carried out to determine the mechanism of the degradation. From this research, the needs could be determined for improving the lifetime and stability of systems where polymer/copper oxide interfaces are encountered. These currently include interconnectors in PV arrays, paint backing in mirror systems (Fig. 3), and interfacial contacts in flat plate collectors.

There were two parallel approaches taken during the study of the catalyzed oxidative degradation of polypropylene. These were a classical physicochemical approach and a surface analytical approach for probing the polymer/copper oxide interface.

In the physicochemical approach, 5 μm thick polypropylene films on different copper oxide (Cu, Cu_2O, $CuO_{0.67}$ and CuO) surfaces were heated in oxygen in a quartz spoon gauge reaction vessel. The oxygen consumption was measured as a change in pressure; the product gases, H_2O and CO_2, were gettered with P_2O_5 and/or KOH in the reaction vessel. The most important conclusions from these studies are that the oxygen consumption is fastest for the polymer on $CuO_{0.67}$ films and that the mechanism of polymer oxidation is not the same in the absence or in the presence of the product gases, H_2O and CO_2. The details are available for the results of oxygen consumption measurements in the presence and absence of product gas getters for combinations of polypropylene/glass, polypropylene/copper films, polypropylene/$CuO_{0.67}$ films, and polypropylene/CuO films as well as a discussion of the mechanisms of degradation (55).

In the surface analytical approach for probing the polypropylene/copper oxide interface, the key question was to determine if the copper oxide surface actually participated in the reaction or if it merely served as a source of copper ions that diffuse into the bulk of the polymer for catalytic activity there. Most of the work related to this question has been published (56-59), but an extended summary of it here will emphasize the importance of surface analytical tools for systematically isolating a part of the problem and will show the amount of prior work necessary if an understanding of the problem is to be achieved. For the prior work, the preparation and characterization of $CuO_{0.67}$ films will be discussed first without and then with an oxygen-18 isotopic label. A brief discussion of the preparation of poly- propylene coatings on the isotopically labeled $Cu^{18}O_{0.67}$ films will be followed by a summary of the ISS depth profile analysis of the polypropylene/$Cu^{18}O_{0.67}$ samples.

Copper films, vacuum deposited onto glass substrates up to thicknesses of 108 nm will oxidize in 13.33 kPa of oxygen to form the composition $CuO_{0.67}$ between 108 and 200°C (60). The films have optical and structural properties comparable to Cu_2O; these results combined with the stoichiometric, thickness, and magnetic properties indicate $CuO_{0.67}$ is a gross defect structure of cuprous oxide in which one copper ion, on the average, is missing in each unit cell of the oxide. The oxide, $CuO_{0.67}$, is paramagnetic, indicating that cupric ions or the equivalent are present in it (61). The material is stable in oxygen and air up to to 200°C, but oxidizes to cupric oxide at higher temperat- ures (60). The extensive optical transmittance data obtained during oxidation of copper films 20-165 nm thick can be used as an indicator of complete oxidation of the copper. In the present work, 20-nm copper films, deposited onto glass substrates, were oxidized to 44-nm thick $CuO_{0.67}$ films by heating at 140°C for two hours in air or for eight hours in 1.33 kPa of oxygen-18. Furth- er details of the preparation are available (56,57).

The 44-nm thick $Cu^{18}O_{0.67}$ films were overlaid with polypropylene coatings using a dip coating process. The glass slides
with copper oxide films on both sides were immersed in a solution
of the polymer in p-xylene at 110-120°C under nitrogen and with-
drawn after polymer thicknesses between 40 and 110 nm had been
reached. The thin polymer coatings were desired to reduce the
ion bombardment time for depth profiling to the polymer/copper
oxide interface. Each slide was cut into four pieces for the
experimental work. One was depth profiled as-coated, one was degraded in oxygen in the presence of getters, and one was degraded in oxygen in the absence of getters. Both degradation reactions were carried out at the same temperature. A series of
experiments were carried out at degradation temperatures of 90°,
100°, 110°, and 120°C at an oxygen pressure of 89.3 kPa.

Composition in depth profiles of the copper, partially
oxidized copper, and completely oxidized copper oxide films on
glass and of the undegraded and degraded polypropylene coatings
on the copper oxide were obtained using a 3-M ISS with a 90°
scattering angle. The details of the data acquisition and
interpretation of the results have been reported (56,58). In the
next three sections, the essence of the ISS results for unlabeled
and labeled copper oxide films is discussed for polypropylene on
copper oxide without degradation, and for degraded polypropylene
both in the presence and absence of getters.

A. ISS Results on $CuO_{0.67}$ and $Cu^{18}O_{0.67}$ Films

A representative ISS spectrum is shown in Fig. 6 for a 44-nm
$Cu^{18}O_{0.67}$ film on glass after profiling to the oxide/glass interface. The peaks correspond to Cu, ^{16}O, ^{18}O, Na and K. The appearance of ^{16}O in the isotopically labeled films is a result of the
preparation procedure, as has been discussed (55). Films oxidi-

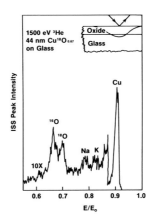

FIGURE 6.   ISS spectrum for a $Cu^{18}O_{0.67}$ *film at the oxide/ glass interface.*

zed in oxygen-16 yield spectra similar to that in Fig. 6, except no oxygen-18 peak is detectable.

The spectrum in Fig. 6 was selected to illustrate the elements detectable at the interface and the ^{16}O-^{18}O Peaks. Although the latter overlap, the individual peaks are readily discernible and are separated by 0.030 E/E_O, in agreement with that predicted from the binary collision theory for 90° scattering of ^3He ions. The quantitative potential for ISS has been discussed on the basis of careful analysis of the data (56,57).

Depth profiles of $CuO_{0.67}$ and $Cu^{18}O_{0.67}$ films on glass are shown in Fig. 7. These were obtained by plotting the peak height intensities on spectra [see Fig. 6] taken at various depths of ion milling through the copper oxide. An average Cu/O peak-intensity ratio of 16.7 ± 1.0 was obtained from nine different profiles of $CuO_{0.67}$ films, similar to the one shown in Fig. 7(a), and from partially oxidized copper films on glass substrates. For seven profiles taken with both ^{16}O and ^{18}O present, as illustrated by Fig. 7(b), a Cu/O ratio of 16.3 ± 1.3 was obtained when the contributions from both isotopes were summed to determine an

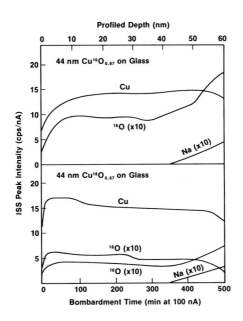

FIGURE 7. Depth profiles of copper films on glass after
complete oxidation in (a) air (^{16}O) and (b) ^{18}O in a stainless
steel chamber at approximately 190°C.

oxygen peak height. The Cu/O ratio for the isotopes was averaged from seven profiles with widely differing concentrations of
^{16}O and ^{18}O. The larger uncertainty for the isotopic mixture results from doubling the possible reading errors when summing the
contributions from the two peaks above the base noise level. The
possibility of carrying out quantitative ISS studies on a system
consisting of oxygen isotopes is supported by these preliminary
studies.

Other conclusions reached during the study of copper oxides
are: the Cu/O ratio is independent of the substrates used, viz.,
Au, Pt, or glass (56,58); the Cu/O ratio is slightly different
for CuO and $CuO_{0.67}$; and the sputtering rates for $CuO_{0.67}$, CuO,

and Cu films are 0.124 ± 0.029, 0.076 ± 0.016, and 0.124 ± 0.014 nm/min, respectively, when using a 100-nA, 1500-eV, 1-mm (FWHM) ^{3}He primary ion beam at $45°$ incidence (56,59).

B. *ISS results for Non-degraded Polypropylene on $Cu^{18}O_{0.67}$*

Originally it was hoped that the effects of the degradations could be studied in a straight forward manner by comparing the depth profiles of polymer coated samples with those of the uncoated copper oxide. However, the depth profiles of the nondegraded polymer coated copper oxides yielded quite unexpected results, as shown in Fig. 8. (For comparison, see the depth profile of the uncoated copper oxide films shown in Fig. 7). It can be noted here that the depth profiles of nearly all the polymer samples, both degraded and nondegraded were qualitatively similar to that illustrated in Fig. 8.

One unexpected characteristic of the polymer depth profiles was the virtual lack of an oxygen signal from the sample until the copper oxide/glass interface was reached by the ion beam. The absolute intensity of the copper signals were 25-35% as large as those obtained from the uncoated oxides, which resulted in disproportionately large Cu/O ratios of 100 to 1 or greater. These Cu/O ratios were five to six times larger than those obtained from the uncoated copper oxide films. This aspect of the depth profiles indicated that meaningful results could not be obtained directly from comparisons between polymer coated and uncoated copper oxide films. Furthermore, minor variations were found in profiles of the nondegraded samples that had been dip-coated for different times (producing different polymer thicknesses). Thus, even comparisons between samples with a similar history but different polymer overlayer thicknesses were

ambiguous. To circumvent these problems, a single glass slide
with a polymer coated copper oxide film was cut into several
pieces. One of the pieces was profiled to act as a control; the
remaining pieces were degraded under various conditions and then
depth profiled. Differences in the depth profiles could now be
attributed to the degradation and not to minor differences in
preparation or storage.

Though the oxygen signals were small, it was possible to
determine that little if any change occurred in the isotopic
composition of the copper oxide as a result of the dip-coating.
The extent to which the copper oxide might be altered by the
dip-coating process was examined by depth profiling a copper oxide film before dip-coating and after a polymer coating had been
dissolved from the copper oxide film by immersion in boiling
toluene. Only minor differences in the Cu/O ratio were noted
between the two depth profiles of the copper oxide, and it was
concluded that the dip-coating process alone did not significantly affect the copper oxide film. The minor changes in Cu/O
ratios that were observed could be explained by assuming that
the grain boundaries between $CuO_{0.67}$ crystallites of only the

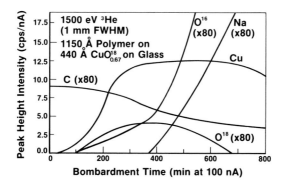

FIGURE 8.   Depth profile of 115 nm nondegraded polypropylene
coating on a 44 nm $Cu^{18}O_{0.67}$ film on glass.

atomic layer next to the polymer was reduced to free copper.

C. *ISS results on Degraded Polypropylene on* $Cu^{18}O_{0.67}$ *Films*

The results for degraded polymer films can be divided into two categories: those degraded with P_2O_5 and KOH present to getter the reaction product gases and those degraded without the getters present. The only differences between the preparation of the two sets of degraded samples was the presence or absence of the getters. In all but one case, the depth profiles of the degraded films were similar to the one illustrated in Fig. 8.

The effect of oxidative degradation was determined by comparing depth profiles of a nondegraded film with those of degraded polymers from the same set of samples. However, direct comparisons of peak height intensities were found to be less suitable than comparisons obtained from plotting the Cu/O ratios vs. sputtering time. The curves obtained from depth profiles of polymer samples degraded in the presence of the chemical absorbents P_2O_5 and KOH are illustrated in Fig. 9. The unmarked curve was obtained from the depth profile of a nondegraded polymer. The remaining curves in Fig. 9 were obtained from depth profiles of degraded polymers. The times and temperatures used yielded a 10% degration for curves a and b and 35% for curve c. It can be seen that the Cu/O ratios for the nondegraded polymer are substantially lower than those for the degraded polymer samples and that qualitatively there are few differences between curves a, b, and c. The larger Cu/O ratios for the degraded films result from larger copper concentrations in the "oxide-layer" after the degradation reaction, i.e., the polymer has reduced the oxide during degradation. From the actual depth profiles, no change could be seen in the isotopic composition of the oxide films resulting from the degradation process, showing that no exchange occurred between

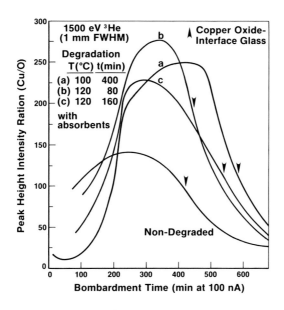

FIGURE 9. Plots of the Cu/O ratios obtained from nonde-graded and degraded polypropylene films after degradation with chemical absorbents present in the reaction vessel.

The oxygen of the degrading atmosphere and that in the copper oxide layer.

Other sets of films were degraded without the presence of P_2O_5 and KOH in the reaction vessel, thereby allowing substantial concentrations of product gases to accumulate in the polymer and in the reaction region during the degradation. The Cu/O curves for these samples also showed that reduction of the polymer occurred (56,59). However, for degradation at 100°C for 400 min., a smaller Cu/O ratio resulted both from a decrease in the copper signals and slightly larger oxygen signals. A depth profile from this degradation sample is shown in Fig. 10. This curve exhibits large differences from all other previously depth profiled polymer coated copper oxide samples. The Cu/O ratio throughout the profile was about 10 to 1, which was not only smaller than those

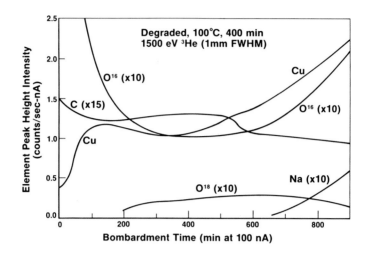

FIGURE 10.   Depth profile of a polypropylene coated copper oxide film that was degraded at 100°C for 400 min without chemical absorbents present in the reaction vessel.

observed from coated oxides analyzed previously, but it was also smaller than the Cu/O ratios that had been determined for the uncoated oxide films. The small Cu/O ratios resulted from an increase in the absolute oxygen-16 peak intensities and a decrease in the copper peak intensities. Most importantly, there was a marked change in the isotopic composition of the overlayers. The increase in the oxygen-16 in the film compared with the original composition (Fig. 7) is consistent with a process of reduction of the oxide by the polymer followed by reoxidation of the copper by an oxygen-16 environment. Other features in Fig. 10 have been discussed in more detail (56,59).

In conclusion, the ISS studies of polypropylene copper oxide samples show that the polymer/oxide interface participates in the degradative reaction. Reduction of the oxide by the polymer occurs, and, under conditions simulating actual use, copper can be

all or partially reoxidized. The relative importance of the
interface reaction versus the effect of copper ions in the bulk
polymer as catalytic sites is still being pursued.

REFERENCES

1. Hayward, D.O., and Trapnell, B.M.W., "Chemisorption",
    Butterworths, London, 1964; Tompking, F.C., "Chemisorption
    of Gases on Metals", Academic, N.Y., 1978; Rhodin T.N., and
    Ertl, G., "Nature of the Surface Chemical Bond", North-
    Holand, Amsterdam, (1979).

2. Flood, E.A. (ed.), "The Solid Gas Interface", Vols. 1 and 2,
    Dekker, N.Y., (1967).

3. Anderson, R.B. (ed.), "Experimental Methods in Catalytic
    Research", Academic, N.Y., (1968).

4. Green, M. (ed.), "Solid State Surface Science", Dekker, N.Y.,
    (1969).

5. Gould, R.F. (ed.), "Interaction of Liquids at Solid Sur-
    faces", ACS Publications, Washington, D.C., (1968).

6. Anderson, J.R. (ed.), "Chemisorption and Reactions on
    Metallic Films", Vols. I and II, Academic, N.Y., (1969 and
    1971).

7. Chopra, K.L., "Thin Film Phenomena", Krieger, N.Y., (1970).

8. Somorjai, G.A., "Principles of Surface Chemistry", Prentice-
    Hall, Englewood Cliffs, N.J., (1972).

9. Blakely, J.M., "Introduction to the Properties of Crystal
    Surfaces", Pergamon Press, Oxford, (1973).

10. Adamson, A.W., "Physical Chemistry of Surfaces", Wiley-
    Interscience, N.Y., (1976).

11. Kane, P.F., and Larrabee, G.B. (eds.), "Characterization of
    Solid Surfaces", Plenum, N.Y., (1974).

12. Czanderna, A.W. (ed.), "Methods of Surface Analysis",
    Elsevier, Amsterdam, (1975).

13. Hannay, N.B. (ed.), "Treatise on Solid State Chemistry",
     Vols. 6A and 6B, "Surfaces", Plenum, N.Y., (1976).

14. Vanselow, R. (ed.), "Chemistry and Physics of Solid Sur-
     faces", Vols. 1 and 2, CRC Press, Cleveland, (1977 and 1979).

15. Morrison, S.R., "The Chemical Physics of Surfaces", Plenum,
     N.Y., (1977).

16. Baglin, J.E.E., and Poate, R.E. (eds.), "Thin Film Phenom-
     ena-Interfaces and Interactions", Proc. Electrochem. Soc.,
     78-2, (1978).

17. Roberts, R.W., and Vanderslice, T.A., "Ultrahigh Vacuum and
     Its Applications", Prentice-Hall, Englewood Cliffs, N.J.,
     (1963).

18. Brunauer, S., Emmett, P., and Teller, E., *J. Am. Chem. Soc.*
     60, 309 (1938).

19. Robens, E., in Czanderna, A.W. and Wolsky, S.P. (eds.),
     "Microweighing in Vacuum and Controlled Environments",
     Elesevier, Amsterdam, p. 125, (1980).

20. Gregg, S.J., and Sing, K.S.W., "Adsorption, Surface Area and
     Porosity", Academic, N.Y., (1967).

21. Czanderna, A.W., and Vasofsky, R., *Prog. Surface Sci. 9*, 43
     (1979).

22. Honig, R.E., in West, A.R. (ed.), "Advances in Mass Spectro-
     metry", Vol. 6, Elsevier, Barking, England, p. 337, (1974).

23. Winters, H.F., in Kaminsky, M. (ed.), "Radiation Effect on
     Solid Surfaces", Amer. Chem. Soc., Washington, D.C, p. 1,
     (1976).

24. Wehner, G., in ref. 12, p. 1.

25. Czanderna, A.W., Miller, A.C., and Helbig, H.F., in Johari,
     O. (ed.), SEM/1978, Vol. 1, SEM Inc., AMF O'Hare, Il.,
     (1978).

26. Helbig, H.F., and Czanderna, A.W., *JEMMSE, 1*, 379 (1979).

27. Buck, T.M., in ref. 12, p. 75.

28. Rusch, T., and Erickson, R.L., in Tolk, N.H., Tully, J.C.,
     Heiland, W., and White, C.W. (eds.), "Inelastic Ion Surface
     Collisions", Academic, N.Y., p. 73, (1977).

29. McHugh, J.A., in ref. 12, p. 223.

30. Joshi, A., Davis, L.E., and Palmberg, P.W., in ref. 12, p.
     159.

31. Sickafus, E.N., *JEMMSE, 1,* 1 (1979).

32. Carlson, T.A., "Photoelectron and Auger Spectroscopy",
     Plenum, N.Y., (1975).

33. Riggs, W.M., and Parker, M.J., in ref. 15, p. 103.

34. Brundle, C.R., Baker, A.D. (eds.), "Electron Spectroscopy:
     Theory, Techniques, and Applications", Vols. 1-3, Academic,
     N.Y., (1977ff).

35. Lichtman, D., in ref. 12, p. 39.

36. Mayer, J.W., and Ziegler, J.F., "Ion Beam Surface Layer
     Analysis", Elsevier Sequoia, Luasanne, (1974).

37. Hutchins, G.A., in ref. 11, p. 441.

38. Müller, E.W., in ref. 12, p. 329.

39. Coburn, J.W., Taglauer, E., and Kay, E., *J. Appl. Phys. 45,*
     1779 (1974).

40. Greene, J.E., and Whelan, J.M., *J. Appl. Phys. 44,* 2509
     (1973).

41. Tolk, N.H., Simms, D.L., and Foley, E.B., *Radiation Effects
     18,* 221 (1973).

42. Shuttleworth, R., *Proc. Royal Soc. 62A,* 167 (1949).

43. Benson, G.C., and Yun, K.S., in Flood, E.A. (ed.), "The
     Solid-Gas Interface", Vol. 1, Dekker, N.Y., p. 227, (1967).

44. Duke, C., *Crit. Rev. Solid State Matls. Sci. 8,* 69 (1978).

45. Non-Traditional Approaches to the Study of the Solid-
     Electrolyte Interface, Snowmass, Co., Sept. 24-27, (1979).
     Proceedings to appear in a special issue of *Surface Science.*

46. Somorjai, G.A., Yates, J.T., Jr., and Clinton, W., Report
     of the Surface Science Workshop, LBL-6658, Lawrence Berkeley
     Laboratory, U. California, Berkeley, March (1977).

47. Call, Patrick J., National Program Plan for Absorber Surfaces
     R&D, SERI/TR-31-103, January (1979).

48. Solar Optical Materials Program Activity Committee, SERI/
     PR-31-137, July (1979).

49. Solar Glass Mirror Program, SERI/RR-31-145, January (1979).

50. Czanderna, A.W., in Czanderna, A.W. and Wolsky, S.P. (eds.),
     "Microweighing in Vacuum and Controlled Environments",
     Elsevier, Amsterdam, p. 175, (1980).

51. Czanderna, A.W., in Butler, B.L. (ed.), Basic and Applied
     Research Program, SemiAnnual Report, SERI/TR-334-244,
     p. 12, December (1979).

52. Ignatiev, A., O'Neill, P., and Zajac, G., *Solar Energy
     Materials 1*, 69 (1979).

53. Czanderna, A.W., Prince, E.T., and Helbig, H.F., SERI/TP-
     334-352, October (1979).

54. Prince, E.T., Helbig, H.F., and Czanderna, A.W., *J. Vac.
     Sci. Technol. 16*, 244 (1979).

55. Jellinek, H.H.G., Kachi, H., Czanderna, A.W., and Miller,
     A.C., *J. Polymer Sci. 17* 1493 (1979). A detailed list
     of references on the oxidation of polypropylene is given
     in ref. 1 of this paper.

56. Miller, A.C., Ph.D. Thesis, Clarkson College, Potsdam, N.Y.,
     (1977).

57. Czanderna, A.W., Miller, A.C., Jellinek, H.H.G., and Kachi,
     H., *J. Vac. Sci. Technol. 14*, 227 (1977); *Industrial Res.
     20*, 62 (1978).

58. Miller, A.C., and Czanderna, A.W., *Appl. Surface Sci.* (1980).
     In Press. SERI/TR-31-393, July (1979).

59. Czanderna, A.W., Jellinek, H.H.G., Kachi, H., and Miller, A.C., Annual Reports, Incra Project No. 224, July (1974); July (1975); July (1977); International Copper Research Association, N.Y.

60. Wieder, H., and Czanderna, A.W., *J. Phys. Chem. 66,* 816 (1962).

61. Czanderna, A.W., and Wieder, H., *J. Chem. Phys. 39,* 489 (1963); *40,* 2044 (1964).

APPENDIX I

AES	Auger Electron Spectroscopy
ATR	Attenuated Total Reflectance
BET	Brunauer, Emmett, and Teller
ELEED	Elastic Low Energy Electron Diffraction
EM	Electron Microscopy
ESCA	Electron Spectroscopy for Chemical Analysis (XPS)
FEM	Field Emission Microscopy
FIM	Field Ion Microscopy
FTIR	Fourier Transform Infrared
GDMS	Glow Discharge Mass Spectroscopy
GDOS	Glow Discharge Optical Spectroscopy
ILEED	Inelastic Low Energy Electron Diffraction
IR	Infrared (Spectroscopy)
ISS	Ion Scattering Spectrometry
LEED	Low Energy Electron Diffraction
RBS	Rutherford Back Scattering (Spectroscopy)
SAM	Scanning Auger Microscopy
SCANIIR	Surface Composition by Analysis of Neutral and Ion Impact Radiation
SEM	Scanning Electron Microscopy
S/G	Solid Gas Interface (G/S)
SIMS	Secondary Ion Mass Spectrometry

S/L   Solid Liquid Interface (L/S)

S/S   Solid Solid Interface (S/S)

TED   Transmission Electron Diffraction

TEM   Transmission Electron Microscopy

TGA   Thermogravimetric Analysis

UHV   Ultra High Vacuum

UPS   Ultraviolet Photoelectron Spectroscopy

VHV   Very High Vacuum

XPS   X-ray Photoelectron Spectroscopy (ESCA)

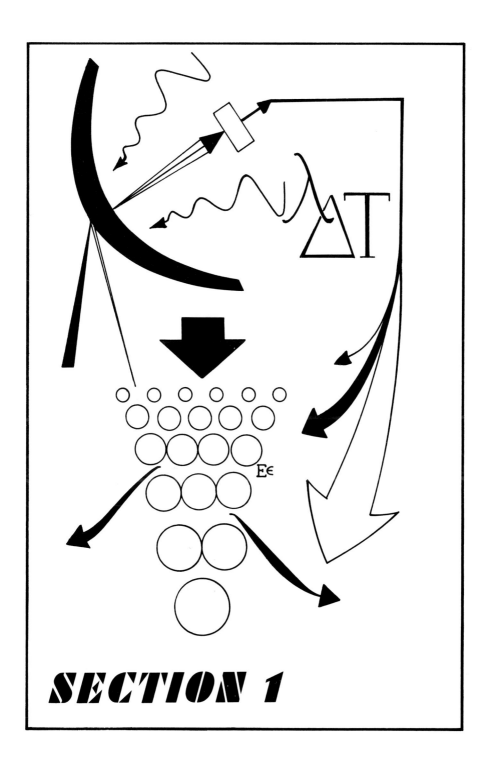

SECTION 1

1. SOLAR COLLECTOR (PHOTOTHERMAL) MATERIALS

The utilization of solar energy ideally begins, in the simplest cases, with a collection scheme in which sunlight is effectively converted to heat. Since this process is the most common and most important feature of home solar passive and active systems, the component materials requirements must ideally involve high efficiency, low cost, and long life. These requirements are certainly not unique to solar energy applications but are requisites in any successful engineering design application. To a large extent, requisite materials properties and performance in solar collection applications can be, or can hope to be, altered by the manipulation of specific microstructural features, even including the development of new materials. In the development of successful solar collectors, the structure, composition, or microstructure of selective absorbers, reflectors, concentrators, and associated support structures must be considered. Specific functional efficiencies, low cost, and long life will depend to a large extent upon structural stability and the interaction of a material with environmental phenomena, including other system components and materials. An understanding of the science of these materials is therefore a prerequisite to successful design implementation. This understanding must also include a knowledge of the relationship of the materials properties to the actual system performance where operationally and environmentally synergistic phenomena are prevalent over long periods of operation.

This section begins with a proposal for tailor-making selective solar absorber coatings based upon the effect of microstructure and microstructural distributions upon optical response, especially absorptivity. Solar mirror materials are discussed, especially their properties and structure, and the performance and degradation of solar mirror materials in the environment (outdoors) are described. Two chapters describe emissivity of metals and the role of emissivity in solar collection. In addition, the effect of microstructure on emissivity is described utilizing the concept of composite materials systems, and one additional chapter describes metal-insulator or metal-oxide composite systems or cermets and the selective manipulation of these microstructures in the development of high-temperature selective absorbers. Finally, the corrosion, erosion, and related degradation of solar collector and related solar materials systems and components are described, beginning with a review of principal corrosion mechanisms in metals and alloys.

The chapters contained in this section are aimed at providing the reader with an overview of solar collector materials, materials problems and design challenges, as well as an understanding of the specific optical features and physical phenomena which control the efficient conversion of sunlight to heat energy. This section is also intended to demonstrate the properties of materials, particularly microstructural properties, which are responsible for their efficient utilization in photothermal processes.

CHAPTER 4

THE OPTICAL PROPERTIES-MICROSTRUCTURE RELATIONSHIP
IN PARTICULATE MEDIA: OPTICAL TAILORING OF
SOLAR ABSORBERS[1]

Alex Ignatiev

Department of Physics
University of Houston
Houston, Texas

I. INTRODUCTION

Historically (1,2) and more recently (3-11) due to revived
interest in solar radiation absorption, studies of solar energy
absorbers have attributed the optical response of absorbers to
be due to the particulate nature of the material. This relationship between optical response and microscopic structure will be
examined in some detail in the following pages, however, it is
informative to first view the basic principles of electromagnetic
radiation absorption by a medium so as to obtain a better conceptual view of solar radiation absorption.

Absorption of electromagnetic radiation in solids is principally governed by the displacement of charges in the material.
Figure 1 illustrates the main electromagnetic radiation absorption mechanisms in materials. The principal contributions to
absorption are:

(a) Molecular dipole absorption (microwave region): absorption due to orientation changes of a polar molecule.

[1]*Sponsored in part by the Department of Energy through the
Solar Energy Research Institute and by the University of Houston
Energy Laboratory.*

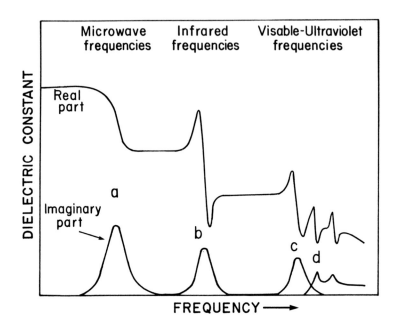

FIGURE 1. *Real and imaginary parts of the dielectric*
response in materials. Note the frequency dependence for the
various absorption mechanisms.

(b)   Ionic absorption (infrared region):   absorption due to
the displacement of ions in a material.

(c)   Electronic absorption (visible–ultraviolet region):
absorption due to the displacement of "free" or weakly bond
electrons with respect to nuclear charges.

(d)   Interband absorption (visible–ultraviolet region):
absorption due to the excitation of electrons in a material to
higher energy states.

The absorption of solar radiation (mainly in the visible
region) will therefore principally involve mechanisms (c) and
(d) and as a result of this and the opening remarks covering
particulates the broad conditions for a good solar energy
absorber can be defined as:

i.) The absorber should have a metallic or semimetallic component.

ii.) The absorber should be composed of small particles.

The understanding of the interaction of an array of small conducting particles with electromagnetic radiation is therefore of interest in the study of solar energy absorption.

II. EFFECTIVE MEDIUM THEORY

The analysis of the interaction of an array of small metal particles with an electromagnetic (EM) field requires a self consistant approach for completeness due to the possible multiple scattering of the incident EM wave by the medium. Within the restriction of small packing fraction of the particles in the host medium and small particle size, self consistant analyses of the multiple scattering have been formulated (12-15) which desscibe the response of the particulate medium to be EM field. A simplified approach to the problem was, however, presented over 70 years ago by Maxwell Garnett (MG) (16). MG equated the interaction of an array of small particles with the EM field to the interaction of a homogeneous "effective medium" with the EM field. In this approach, the particles are assumed to be small enough and separated enough that they are not sensitive to the dynamic characteristics of the incoming wave, i.e., that they exist in a *constant field* region in the material. The metallic particles are taken to be imbedded in a dielectric (Figure 2) and the constant effective field in the material $\overline{E}$ is taken as the volume average of the induced field in the particles E_i, and the field in the host dielectric E_m;

$$\overline{E} = fE_i + (1 + f)E_m \tag{1}$$

where f is the packing fraction of the particles in the dielectric. The effective polarization field in the medium can

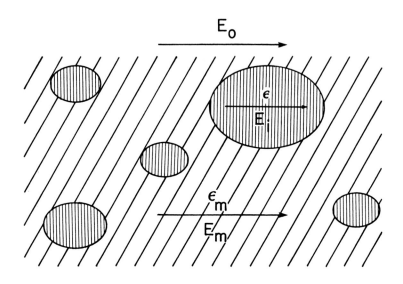

FIGURE 2. Structural model of a particulate analyzed by
Effective Medium Theory--Maxwell Garnett approach.

therefore be written as

$$4\overline{P} = (\overline{\varepsilon} - 1)\overline{E} = f(\varepsilon - 1)E_i + (1 - f)(\varepsilon_m - 1)E_m . \tag{2}$$

Under the assumption of spheroidal particles oriented with
respect to the field

$$E_i = \frac{\varepsilon_m}{L_\varepsilon + (1 - L)\varepsilon_m} \tag{3}$$

where L is a depolarization factor (17) corresponding to a
specific shape and orientation spheroid. The depolarization
factor defines the polarizability of the spheroid in question.

It is well to note that for a spherical particle L = 1/3 and
thus Equation (3) reduces to the well known Clausius-Mossotti

equation for the field in a spherical particle.

Systematic reduction of Equation (2) under the assumption of spherical particles gives the MG expression:

$$\bar{\varepsilon} = \varepsilon_m \frac{\varepsilon(1 + 2f) + 2\varepsilon_m(1 - f)}{\varepsilon(1 - f) + \varepsilon_m(2 + f)} . \tag{4}$$

Generalizing Equation (2) to spheriodal particles, however yields

$$\bar{\varepsilon} = \varepsilon_m \frac{(1 + 2/3\ f\alpha)}{(1 - 1/2\ f\alpha)} \tag{5}$$

where

$$\alpha = \frac{1}{3} \sum_{i=1}^{3} \frac{(\varepsilon - \varepsilon_m)}{\varepsilon_m + L_i(\varepsilon - \varepsilon_m)} \tag{6}$$

with L_i the depolarization factors for given shape spheroids aligned along three orthogonal directions to the electric field.

Although MG is the most widely used effective medium theory, others have been produced by Bruggeman (14) and by Hunduri (15). Both of these are self consistant in that in expressions similar to (6) ε_{me} is replaced by $\bar{\varepsilon}$. They also have their basis in topological similarities between the particles in the medium and the host dielectric. This results for the case of spherical particles, such particles imbedded in a dielectric composed also of spherical particles. Such topological restrictions lend to limited applicability of these approaches in defining optical properties of particulares (3,18).

The applicability of the effective medium approach in general has certain restrictions which historically were defined by a rule of thumb:

$$ka \ll 1.0 \quad \text{and} \quad f \ll 1 \tag{7}$$

where $k = 2\pi/\lambda$ and a = particle radius.

This would restrict the use of EMT in analyzing solar absorbing particulates to those with particle sizes less than $\sim$ 100 Å. EMT seems, however, to have applicability to systems with much larger particle sizes (19). Such results have led to a recent reevaluation of EMT applicability criteria to particulates by Smith (18). His analysis is based not so much on the requirement of a uniform field at the particle, but on the requirement of *no scattering* of the incident wave by the particle plus an envelope of medium surrounding and correlated with the particle. His results yield the expanded EMT applicability criteria of:

$$a \underset{\sim}{<} 0.2\,\lambda$$

and (8)

$$(ka)f^{1/3} \underset{\sim}{\sim} 1.0 \;.$$

This means that for solar absorbing particulates, EMT can safely be applied to absorbers with particle sizes of $\sim$ 1000 Å and packing fractions of $\sim$ 0.5, e.g., the majority of currently relevant solar absorbers.

The above noted EMT formalism has been given for the general case of spheroidal particles imbedded in a dielectric. That this assumption of spheroidal particles is realistic is, however, yet to be established. Recent scanning and transmission electron microscopy (SEM and TEM) studies (3-11,20-22) on several different particulate solar absorbers (both vacuum evaporated and electrodeposited) have indicated that the absorbers are in fact composed of small particles (Figure 3). The particles in most cases do well approximate oblate and/or prolate spheroids, and in some cases chains of particles have been observed which can be approximated by large eccentricity prolate spheroids, e.g., sausages (Figure 4). The spheroid approximation is therefore acceptable and although it cannot cover all possibilities of particle shape, it can cover the possibilities ranging from platelet to sphere to rod.

FIGURE 3. Scanning electron micrograph of electrodeposited black chrome after the removal by sputter etching of several hundred angstroms of material. Note the various particle shapes.

In real particulate absorbers, micrograph observations (Figure 3) indicate not only that the particles are well approximated by spheroids, but that under this assumption one would require a distribution of spheroid shapes to approximate the actual particulate structure of most absorbers. The power of the spheroid approximation for the analysis of the optical response of such real absorbers can now readily be seen since the possibility exists in the approximation for the use of a distribution of spheroid shapes to model the distribution of particle shapes observed in real absorbers. This approach has been adopted by several authors to analyze the optical response of particulate gold (20,23), chromium (4,24), nickel (24) and gold cermet (25) vacuum deposited films and black chrome (3,7)

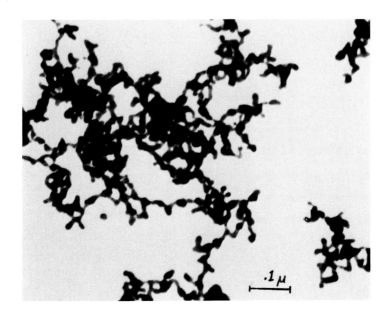

FIGURE 4. Transmission electron micrograph inert-gas evaporated gold black. Note the spherical particles linked together in chain-like fashion.

electrodeposited films. The approach of O'Neill and Ignatiev (26) will be expanded upon here to illustrate the applicability of defining distributions of spheroidal particles in the modeling of the optical response of particulate films.

III. SPHEROID MODEL

Within the EMT approach of O'Neill and Ignatiev, the "effective" dielectric permeability of a particulate material with a specific distribution $\rho(m)$ of randomly oriented prolate spheroids with semimajor to semiminor axis ratio m, (m = b/a), is given by:

$$\bar{\varepsilon} = \frac{\varepsilon_{me}(1 - f) + \varepsilon \frac{f}{3} \sum_{m} g(m) \rho(m) \, \Delta m}{(1 - f) + \frac{f}{3} \sum_{m} g(m) \rho(m) \, \Delta m} \tag{9}$$

where

$$g(m) = \sum_{i} \frac{1}{1 + [\varepsilon/\varepsilon_{me} - 1] L_{i}(m)} \tag{10}$$

with $L_{i}(m)$ = depolarization factor for spheroids of axis ratio m aligned in three orthogonal directions with respect to the incident electric field (Figure 5) to stimulate random orientation of spheroids.

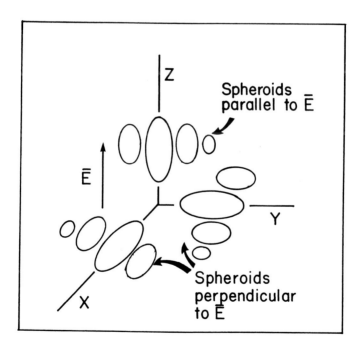

FIGURE 5. Structural model for spheroid orientation in the Spheroid Model analysis. Averaging over spheroids aligned in three orthogonal directions simulates random orientation of spheroids.

Here ε_{me} is the dielectric permeability of the host dielectric and ε is the permeability of the metallic particles.

It should be realized that $\bar{\varepsilon}$ will be complex since in general ε_{me} and ε are complex, and that $\bar{\varepsilon}$ will be a function of both λ and m.

Under the conditions of $f << 1$ which is satisfied for many solar blacks, Equation (9) reduces to:

$$\bar{\varepsilon}_1 \simeq 1$$

$$\bar{\varepsilon}_2 \simeq \varepsilon_2 \frac{f}{3} \sum_m [g||(m) + 2g\perp(m)]\rho(m)\,\Delta m \qquad (11)$$

where $\bar{\varepsilon} = \bar{\varepsilon}_1 + i\,\bar{\varepsilon}_2$ and $g||(m)$ and $g\perp(m)$ are the functions of Equation (10) for spheroids oriented parallel and perpendicular to the electric field, respectively.

The absorption coefficient for electromagnetic radiation is defined by:

$$\alpha(m,\lambda) = \frac{2\pi}{\lambda}\,\bar{\varepsilon}_2(m,\lambda) \qquad (12)$$

and under the conditions of Equation (11) the transmittance through the particulate film can be defined by:

$$T = e^{-\alpha(\lambda)t} \qquad (13)$$

where t is the film thickness and

$$\alpha(\lambda) = \sum_m \alpha(m,\lambda). \qquad (14)$$

It should now be pointed out that not only is the absorption coefficient dependent on m and λ, but it is also governed by material dependent quantities such as free electron density and interband absorption which affect ε and ε_{me}. These effects on $\alpha(m,\lambda)$ can be readily seen in Figs. 6 and 7 which compare $\alpha(m,\lambda)$ for low density nickel and chromium particulates.

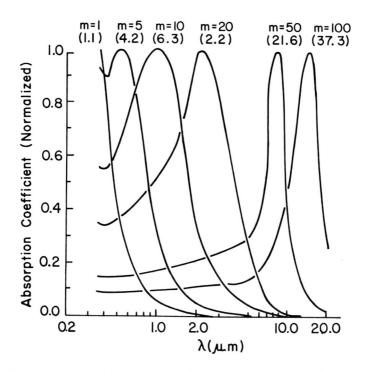

FIGURE 6. *Normalized absorption coefficient* $\alpha(m,\lambda)$ *as a function of m factor and wavelength for particulate nickel.*

The applicability of the Spheroid Model to various particulate media has been previously shown (7,26) and can be seen in Figs. 8 and 9. Figure 8 gives the correspondence between Spheroid Model calculations for the transmittance of an inert gas evaporated particulate chromium black and experiment. The particle shape distribution for the black has been defined as log-normal (26) with thickness and packing fraction measured. Figure 9 gives the correspondence between Spheroid Model calculations for the reflectance of electrodeposited black chrome and experiment (7). The particulate structure used in the calculations was determined by scanning electron microscopy (SEM) and X-ray photoemission spectroscopy (XPS) and principally involves one to several layers of ~ 1000 Å chromium particles on the substrate. The good agreements in Figs. 8 and 9 between

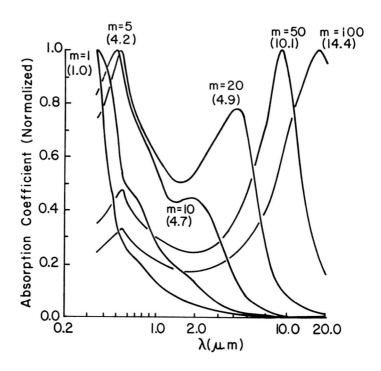

FIGURE 7. Normalized absorption coefficient α(m,λ) as a function of m factor and wavelength for particulate chromium.

calculations and experiments indicate direct applicability of the Spheroid Model to the analysis of the optical response of particulates.

IV. OPTICAL TAILORING

Reexamination of Figs. 6 and 7 reveals that the absorption coefficients α(m,λ) are peaked with peak positions and intensities very strongly dependent not only on wavelength, but on spheroid shape (m factor). The wide spectral range of the absorption peaks resulting from the shape dependence (m dependence) clearly indicates the possibility that almost any spectral response can be obtained for a particulate absorber through a

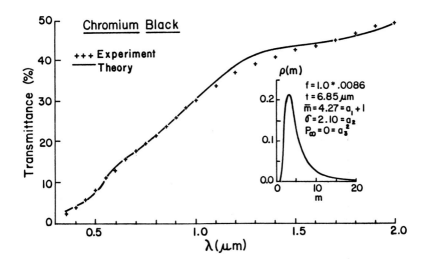

FIGURE 8. *Comparison of transmittance from Spheroid Model calculations and experiment for a particulate inert gas evaporated chromium black.*

judicious choice of particle shapes comprising the material. That is, the optical properties of a particulate medium can be manipulated through the choice of a specific particle shape distribution, i.e., modification of the microscopic structure of the particulate.

Modification of the optical response of a material has been long sought after and has at best been only achieved through interference effects due to the application of antireflection coatings to optical systems. Figure 1 indicates that additional methods for altering optical response could include electron density modification and band structure modification. Except for semiconductors, these possibilities have not proven very successful (27) and it is of interest to see if modification of the microscopic structure of a particulate could result in the manipulation of its optical properties.

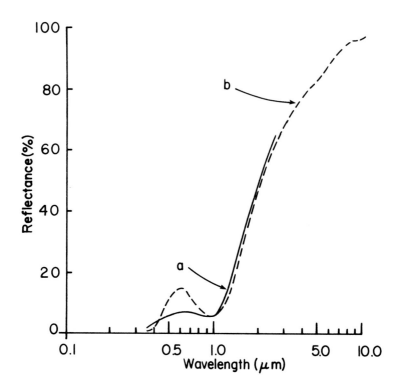

FIGURE 9. Comparison of hemispherical reflectance from Spheroid Model calculations (b) and experiment (a) for a particulate electrodeposited black chrome film.

To investigate this point, the Spheroid Model has been used in a procedure aimed to modify the absorptance of a particulate film to approach that of the optimal solar selective step function absorptance. The procedure minimizes the difference between the absorptance of the film and the solar selective step function absorptance by varying the particle shape distribution in the film. The step function is set at 3 μm for operation of the absorber at a nominal 120°C. The step function cannot be a sharp step as depicted in most previous literature, but must be sigmoid shaped due to the fact that thermal broadening of the absorptance profile will exist (28) (Figure 10).

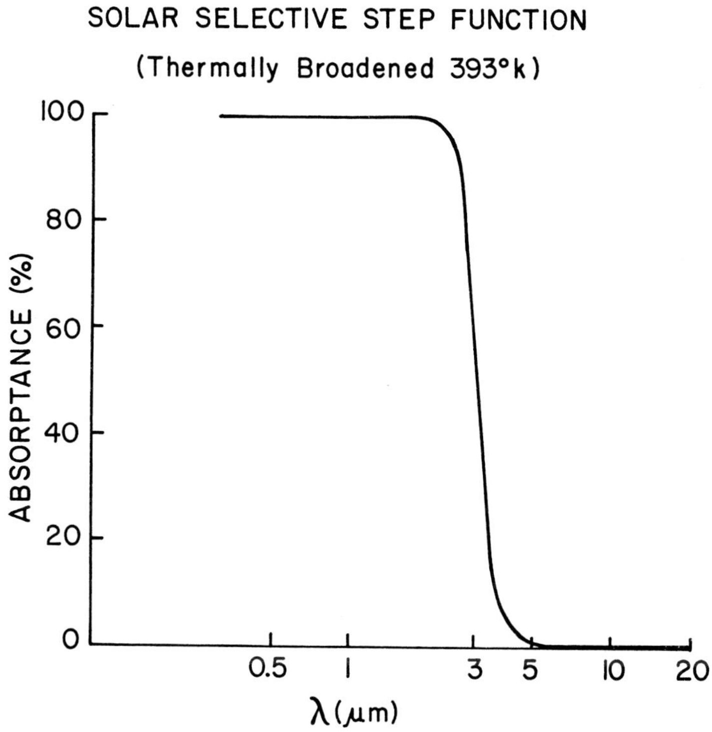

SOLAR SELECTIVE STEP FUNCTION
(Thermally Broadened 393°k)

FIGURE 10. Optimal solar selective step function response for the absorptance of a material. The step is thermally broadened due to the finite temperature of operation of the absorber (120°C in this example) of each Gaussian.

A least square approach was taken for the minimization procedure where J was minimized in the expression involving the absorptance $A(\lambda)$:

$$J = \sum_{\lambda} [A(\lambda)_{IDEAL} - A(\lambda)_{CALC}]^2. \tag{15}$$

Under the assumption of a perfectly reflecting substrate for the absorbing film

$$A = 1 - T^2. \tag{16}$$

Therefore, the following J was minimized by the optimization of a particle distribution $\rho(m)$ in the film.

$$J = \sum_{\lambda} \left[(1 - T^2_{IDEAL}) - (1 - T^2_{CALC}) \right]^2 \tag{17}$$

For simplification of the minimization procedure and since it is not clear whether the $\alpha(m,\lambda)$'s are linearly independent for each m, three Gaussians were used to depict the particle distribution instead of a histogram element for each m value present in the distribution. The first Gaussian was centered at m = 1, the third at m = cutoff value with no intensity above the cutoff. The cutoff value of m for each material studied (gold, nickel, and chromium) was arbitrarily defined as the largest m for which the absorption coefficient at 3 μm (the step position in the absorptance) was 10% of its peak value. The second Gaussian was centered between the first and third with the standard deviation σ of the Gaussians defined such that under the condition of equal intensities, the Gaussians would overlap at half their peak values.

The particulate films modeled were composed of small particles of either gold, nickel, or chromium imbedded in air with packing fractions of 0.06. The thicknesses of the films were defined in the absorptance optimization procedure by the conservation of particles criterion:

$$\sum_{m} \rho(m) \Delta m = 1 \tag{18}$$

and the absorptance was optimized to approach the solar selective thermally broadened step function by varying the intensities of the three Gaussians.

Figure 11 gives the results of the optimization. It is seen that absorptances approaching the step function response are obtained and that the particle shape distributions required for

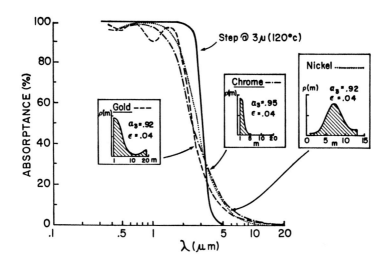

FIGURE 11. Optimized selective absorptance of particulate gold, nickel, and chromium films obtained by varying particle shape distributions.

the three metals are quite different. This latter point is principally due to the differences in interband and free electron absorption in the three materials. As can be noted, solar absorptances of 0.92 to 0.95 (AMII) are realized with infrared emittances of ~ 0.04 obtained. The solar absorptance values though not extremely high, are significantly higher than the 0.6 to 0.7 values for equivalent density *non-particulate metallic* films. It is not clear from this exercise that solar selectivity can be induced in a material by the modification of its microscopic structure.

V. CONCLUSIONS

In conclusion, the main points resulting from the above noted work are that a) through effective medium theory it has been clearly shown that the optical response of solar absorbers is strongly dependent on the microscopic structure of the absorber;

and b) that as a result of a), optical tailoring of a material could be accomplished through the manipulation of its microscopic structure.

As to how the manipulation of microstructure is to be accomplished on the required ~ 1000 Å size scale is currently being studied. Electrodeposition, surface damage by ions or other projectiles, and evaporation are being considered as possible techniques which may yield the required particle distributions. If such distribution modification can be generated, it will still be left to see how thermally and photothermally stable such particulate coatings are. This brings up the final point which is that the current understanding of a material's response to concentrated solar radiation is wanting. Since energy materials research is in its very infancy, it is clear that there is still much to be done in this research area before our base of understanding will be expanded to the point where systematic development of new energy materials can be undertaken.

ACKNOWLEDGMENTS

Close collaboration with G. Zajac, P. O'Neill, and G.B. Smith in all phases of the work is gratefully acknowledged.

REFERENCES

1. Harris, L., "The Optical Properties of Metal and Carbon Black", (Epply Foundation, 1967).

2. Pfund, A.H., *J. Opt. Soc. Am.*, *23*, 275 (1933).

3. Window, B., Ritchie, I., and Gathro, K., *Applied Optics*, *17*, 2637 (1978).

4. Grandqvist, C.G., and Hunderi, O., *J. Appl. Phys.*, *50*, (1978).

5. Inal, O.T., and Torma, A.E., *Thin Sol. Films*, *54*, 161 (1978).

6. Pettit, R.B., and Sowell, R.R., *J. Vac. Sci. Technol.*, *13*, 596 (1976).

7. Ignatiev, A., O'Neill, P., and Zajac, G., *Sol. Energy Mat.*, *1*, 69 (1979).

8. Doland, C., O'Neill, P., and Ignatiev, A., *J. Vac. Sci. Technol.*, *14*, 259 (1977).

9. Sievers, A.J., in "Solar Energy Conversion: Topics in Applied Physics", Seraphin, B.O. (ed), 31, 57, (1979).

10. Craighead, H.G., and Buhrman, R.A., *J. Vac. Sci. Technol.*, *15*, 269 (1978).

11. Lampert, C.M., and Washburn, J., *Sol. Energy Mat.*, *1*, 81 (1979).

12. Smith, G.B., *J. Phy. D.*, *10*, 139 (1977).

13. Wood, D.M., and Ashcroft, N.W., *Phil. Mag.*, *35*, 269 (1977).

14. Bruggman, D.A.G., *Ann. Phys. (Leipz.)*, *24*, 636 (1935).

15. Hunderi, O., *Phys. Rev. B*, *7*, 3419 (1973).

16. Maxwell Garnett J.C., *Philos. Trans. Roy. Soc. Long.*, *203*, 385 (1904), *205*, 237 (1906).

17. Stratton, J.A., "Electromagnetic Theory", McGraw-Hill, New York, Chap. 3, (1941).

18. Smith, G.B., *Appl. Phys. Letters*, (1980).

19. Ignatiev, A., O'Neill, P., Doland, C., and Zajac, G., *Appl. Phys. Letters*, *34*, 42 (1979).

20. Zajac, G., and Ignatiev, A., *J. Vac. Sci. Technol.*, *16*, 233 (1979).

21. Zajac, G., Doland, C., and Ignatiev, A., (to be published).

22. Pettit, R.B., and Sowell, R.R., Proc. 2nd Conf. on Absorber Surfaces for Solar Receivers, (SERI, Boulder, 1979).

23. Norrman, S., Anderson, T., and Grandqvist, C.G., *Phys. Rev. B*, *18*, 674 (1978).

24. Zajac, G., and Ignatiev, A., *Sol. Mat.*, (1980).

25. Grandqvist, C.G., *J. Appl. Phys.*, (1979).

26. O'Neill, P., and Ignatiev, A., *Phys. Rev. B*, *18*, 6540 (1978).

27. Enrenreigh, H., and Seraphin, B.O., Report on Symp. on
     Fundamental Optical Properties of Solids Relevant to Solar
     Energy Conversion, (Tucson, 1975).
28. Sievers, A.J., Proc. 25th Jubilee ISES, (Atlanta, 1979).

CHAPTER 5

SOLAR MIRROR MATERIALS: THEIR PROPERTIES AND
USES IN SOLAR CONCENTRATING COLLECTORS[1]

R.B. Pettit
E.P. Roth

Sandia Laboratories[2]
Albuquerque, New Mexico

I. INTRODUCTION

Solar mirror materials are used in a variety of solar collectors in order to redirect the incident sunlight onto a receiver
surface. Applications range from augmented flat plate collectors
to high concentration tracking parabolic dish concentrators (1).
The primary advantage in using solar mirrors to concentrate sunlight is either to increase the system efficiency (e.g. by reducing thermal losses) or to reduce the system cost where relatively expensive receiver materials are utilized (e.g. photovoltaic cells). In most applications, the total mirror surface
area deployed is large; thus the mirrors must be manufactured at
a relatively low cost. Because of the variety of solar applications, the optical requirements of solar mirrors vary greatly.
All applications are sensitive to the solar averaged reflectance
properties, R_s, which should be as close to unity as possible.
However, the requirements on the distribution of the reflected
sunlight (defined as specularity) can span a wide range from
very diffuse all the way to highly specular depending upon the

[1]This work is supported by the Division of Solar Technology,
U.S. Department of Energy (DOE), under contract DE-AC04-76-DP00789.
[2]A U.S. DOE facility.

application. Because of these variations in the required optical properties, it is important that optical characterization
of solar mirrors be consistent with the intended application.

Because all mirror materials must be supported or held in
a particular orientation with respect to the receiver, the
effect of the supporting structure on the reflected beam location (focal point) must also be carefully considered. The
shape of the mirror surface can be either flat as in a heliostat, curved with one dimensional curvature as in a parabolic
trough, or with two dimensional curvature as in a parabolic
dish. Any change in the reflectance properties of the material
due to manufacturing, processing, or forming to the desired
shape must also be determined.

In addition to the reflectance properties, there are many
other important properties of mirrors when used in solar concentrators. Of primary importance is the environmental
stability of the material. This includes resistance to ultraviolet radiation, moisture, temperature cycling, abrasion,
etc. Another important factor is the resistance of the mirror
and support structure to hail impact unless special storage
configurations have been designed (2). The accumulation of
dust particles and other particulates on the mirror surface
can drastically change the reflectance properties; details of
this subject are covered in the next paper in this text. Other
properties of importance include the mirror cost, mechanical
properties, manufacturing tolerances, etc.

The purpose of this paper is to review the current state of
mirror materials with application to solar concentrators. After
specular reflectance is defined, the optical measurement techniques developed specifically for these materials are discussed.
Next, the solar reflectance properties of mirror materials,
divided into categories of glass, metallized plastics, polished

aluminum, and protective coatings, are summarized. Finally,
current problem areas and future areas of research are discussed.

II. OPTICAL MEASUREMENTS

Loss of redirected sunlight from a solar mirror can result
from surface contour variations (sometimes called slope errors)
and from scattering due to the mirror material (3). Both effects
are due to surface irregularities at the mirror surface. For the
purposes of this paper, scattering from surface variations with a
characteristic spacing less than approximately 1 mm will be referred to as the surface "specularity" (sometimes referred to as
specular reflectance properties). On the other hand, the effect
of the surface contour variations on a scale greater than approximately 10 mm will be referred to as the surface "figure." Contour
variations typically result from local deviations in the surface
normal about the ideal surface shape (see Fig. 1). Thus the surface figure determines the direction of the reflected radiation
while the surface specularity determines the angular spread and
intensity of this radiation.

Because of the large difference between the scales of these
surface variations, two separate and distinct measurement techniques have been used for their determination. Currently a laser
ray trace technique is used for the surface figure measurements

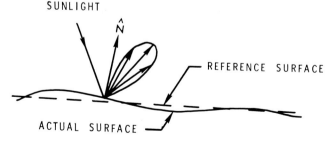

FIGURE 1. Schematic representation of a mirror surface
showing the difference between slope errors and the mirror
specularity.

while a bidirectional reflectometer is used to measure the specularity (3,4). The overall quality of the particular reflector surface is obtained by combining these two results.

A measurement of the surface specularity is determined by irradiating a small area of a flat mirror with a collimated beam of light and measuring the amount of energy reflected into a specified solid angle (angular aperture) centered around the specular beam direction (4). The area irradiated depends upon the characteristics of the instrumentation, but is typically 1-5 cm^2. Surface variations with a scale in the range 1-10 mm may affect either the surface figure or the surface specularity or both depending upon the details of the measurement technique utilized. Therefore, care must be exercised when surface irregularities within this range are present. For curved surfaces (e.g. parabolic troughs), any effect of the surface curvature must be removed from the specularity measurements. Because of these problems, current measurement techniques used for surface specularity and figure determinations may need additional refinement either in the type of instrumentation or data analysis procedures used.

Besides the specularity, another optical quantity of interest is the hemispherical reflectance (5), which measures all of the reflected radiation from a surface independently of its angular distribution. This measurement is typically performed as a function of wavelength from 350-2500 nm using an integrating sphere reflectometer (6) (see Fig. 2). These instruments are commercially available (7). A solar averaged reflectance value for a mirror is usually determined by averaging the spectral hemispherical reflectance data (8), since these data cover the complete solar spectrum. This value represents the maximum available reflected solar energy for a particular mirror; however, depending on the angular distribution of this radiation, only a portion of this reflected radiation may be utilized in a solar collector.

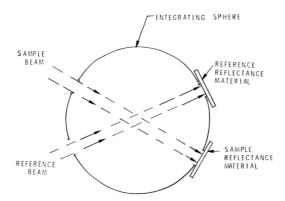

FIGURE 2. Typical double-beam integrating sphere reflec-tometer with wall mounted sample and reference reflectance materials.

The diffuse reflectance is another quantity that can be meas-ured using an integrating sphere reflectometer. For this measure-ment, the mirror is typically aligned normal to the incident radia-tion in the integrating sphere so that a portion of the normally reflected radiation passes back out the entrance aperture (see Fig. 2) (9). The angular size of this aperture as seen by the sample is usually very large (approximately 5-10°). By subtracting the diffuse reflectance from the hemispherical reflectance, an indi-cation of the specular reflectance properties can be obtained but with very poorly defined incident and collection solid angles.

The specularity properties of a solar mirror material are measured using a bidirectional reflectometer. Since these instru-ments are not available commercially, several different designs have been built (4,10-13). A schematic of a typical instrument is shown in Fig. 3. The instrument consists of source optics and detection optics. The source optics produce a collimated beam of radiation which is reflected off the mirror and directed into the detection optics. The source beam is typically collimated to a few milliradians (mrad) or to a value that is equal to the angular size of the sun (~ 10 mrad). The degree of collimation is determined by the size of the source aperture divided by the focal

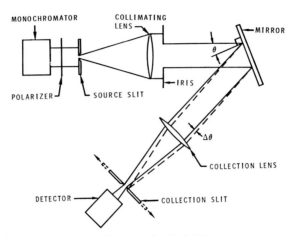

FIGURE 3. Schematic of a typical bidirectional reflectometer showing the collimation and collection optics.

length of the collimation lens. The reflected beam is imaged onto
either a circular or slit collection aperture. For a perfectly
specular mirror, all of the reflected radiation is imaged onto the
collection aperture. Reflected radiation that deviates from the
specular direction is imaged adjacent to the collection aperture.
The reflected beam intensity profile is then recorded in one of two
ways: First, the collection aperture size can be increased,
starting at the same size as the source aperture, so that an integration of the reflected beam profile is recorded. Alternatively,
a fixed collection aperture (whose size is the same as the source
aperture) can be swept through the focused beam so that the reflected beam profile is recorded directly. For either technique,
the instrument is usually calibrated by aligning the collection and
source optics to measure a straight-through beam, although standard
mirrors can also be used for calibration. The accuracy of present
instruments is typically ± 0.01 reflectance units (100% reflectance
= 1.00 reflectance units). The wavelength range covered by the reflectometer is determined either by the light source (e.g. laser),
by a monochromator or filters to define a narrow wavelength region,
or by the source optics and detector spectral response characteristics (usually broad band).

Results obtained in this laboratory are shown in Fig. 4 for
silvered float glass (2.7 mm thick), a roll polished aluminum sheet
(Alzak, 0.64 mm thick), and aluminized teflon and acrylic (3M
Scotchcal 5400) films laminated to an aluminum backing sheet.
For these measurements the incident beam was collimated to 1 mrad
and the collection aperture was varied from 1 mrad to 18 mrad.
Note that while the glass and plastic materials reach an approximately constant reflectance value at an angular aperture of 10 mrad,
the specular reflectance curves for the polished aluminum continue
to increase at the maximum angular aperture of 18 mrad.   The hemispherical reflectance values for these materials measured at the
same wavelength of 500 nm are also shown in Fig. 4.   Note that at
18 mrad, the specular reflectance values for the silvered glass and
aluminized acrylic mirrors are equal to the hemispherical reflectance values; however the specular reflectance values for the
aluminized teflon and polished aluminum mirrors at 18 mrad are from
0.05 to 0.17 reflectance units below their respective hemispherical
reflectance values even though the reflectance profile for the
aluminized teflon mirror appears to have reached a plateau or constant value.   The difference between the specular and hemispherical

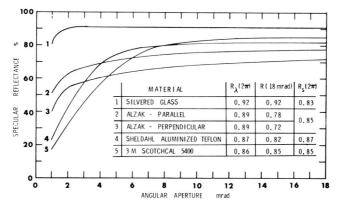

	MATERIAL	$R_\lambda(2\pi)$	R(18 mrad)	$R_s(2\pi)$
1	SILVERED GLASS	0.92	0.92	0.83
2	ALZAK - PARALLEL	0.89	0.78	0.85
3	ALZAK - PERPENDICULAR	0.89	0.72	
4	SHELDAHL ALUMINIZED TEFLON	0.87	0.82	0.87
5	3 M SCOTCHCAL 5400	0.86	0.85	0.85

*FIGURE 4. Specular reflectance properties for several solar
mirror materials. The table lists the hemispherical reflectance
[$R_\lambda(2\pi)$] and the specular reflectance at 18 mrad [R(18 mrad)]
measured at 500 nm. Also listed is the solar averaged hemispherical reflectance [$R_s(2\pi)$].*

reflectance values is due to large angle scattering at the mirror
surface. The reflectance profiles for these materials illustrate
the importance of measuring both the specular reflectance and hemispherical reflectance properties at the same wavelengths for all
mirror materials.

While the curves shown in Fig. 4 contain all of the information
needed to properly compare the different materials, a data analysis
procedure has been developed which reduces these curves to a form
more easily utilized in collector performance calculations. For
this analysis it is assumed that the mirror will scatter the reflected radiation into angles centered aroung the specular direction according to a normal distribution of the form

$$R(\Delta\theta) = \frac{R_o}{2\pi\sigma^2} \exp\left[\frac{-\Delta\theta^2}{2\sigma^2}\right]$$

(1)

where $\Delta\theta$ is the deviation of the reflected beam from the specular direction, σ is the dispersion of the distribution, and R_o
is the integrated intensity. Since the specular reflectance
properties shown in Fig. 4 represent the integration of this
distribution as a function of angular aperture, the integration,
when carried to large angles, must approach the hemispherical
reflectance value. Therefore R_o in Eq. (1) is usually set equal
to the hemispherical reflectance value, R_λ (2π), measured at
the same wavelength. The specular beam profile measured using
a particular bidirectional reflectometer is given by the convolution of the source aperture intensity distribution with the
normal distribution of the mirror scattering function. By integrating this profile as a function of the angular collection aperture and comparing the results with the measured curve, a best-
fit value for σ (and R_o if it is also varied) can be obtained.
Typical results are shown in Fig. 5a for the aluminized acrylic
mirror. In this case R_o was set equal to the hemispherical reflectance value of 0.86 measured at 500 nm. Note that although

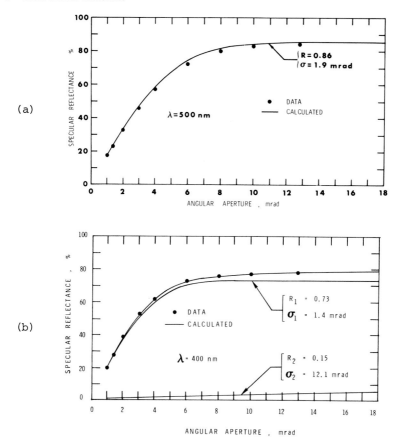

FIGURE 5a. *Comparison between the analytical fit and measured reflectance values for (a) an aluminized acrylic film (3M Scotchcal 5400) at 500 nm using a single normal distribution, and (b) for an aluminized teflon film (Sheldahl, Inc.) at 400 nm using two normal distributions.*

the calculated curve does not reproduce the measured data exactly, it is within the measurement error of ± 0.01 reflectance units at each angular aperture.

The reflected beam profile for some of the solar mirror materials investigated could not be described by a single normal distribution. For these materials the specular beam appeared to be composed of a sharp central peak together with a broad, diffuse background (see Alzak in Fig. 4). In this case, it was

assumed that the reflected beam is composed of the sum of two
normal distributions. Thus the specular reflectance is given by

$$R(\Delta\theta) = \frac{R_1}{2\pi\sigma_1^2} \exp\left[\frac{-\Delta\theta^2}{2\sigma_1^2}\right] + \frac{R_2}{2\pi\sigma_2^2} \exp\left[\frac{-\Delta\theta^2}{2\sigma_2^2}\right] \qquad (2)$$

The division of the reflected beam into distinct components
appears to be a reasonable assumption since areas of the mirror
may be relatively smooth and flat (highly specular component)
while other areas have scratches, bumps, etc. which can cause
large angle scattering (diffuse component). With the assumption
of two normal distributions, four parameters are necessary to
specify the specular reflectance: (R_1, σ_1) and (R_2, σ_2).
The constraint that the sum $(R_1 + R_2)$ should equal the hemispherical reflectance value at the same wavelength is usually
employed, although this constraint can be relaxed so that all
four parameters are varied independently. An example of the
curve fitting results is shown in Fig. 5b for an aluminized
teflon film. The parameters obtained at 400 nm were $R_1 = 0.73$,
$\sigma_1 = 1.4$ mrad and $R_2 = 0.15$, $\sigma_2 = 12.1$ mrad $[(R_1 + R_2) =$
$R_\lambda(2\pi) = 0.88]$.

It should be noted that the agreement found between the
measured curve using the assumption of two normal distributions
for the specular beam does not imply that there are only two
scattering mechanisms associated with these materials. In fact
there are probably more. Therefore without further verification,
the parameters (R_1, σ_1) and (R_2, σ_2) cannot be used to define
the surface structure of these materials, but only the specularly
reflected beam profile over the range covered by the measurements.
If the reflectance measurements are obtained for angular apertures
which range from the incident beam width up to 2π radians
(hemisphere), the assumption of two normal distributions may not
be adequate. However, for measurements obtained in this laboratory, the specular beam profile for a wide variety of solar mirror

materials could be characterized by either Eq. (1) or Eq. (2) over the range covered by the measurements (18 mrad in this case).

It should also be pointed out that for some materials the specular reflectance values are not very sensitive to small changes in the calculated parameters, R_1, σ_1 and R_2, σ_2. This is especially true if one distribution R value is small and its corresponding σ value is large. Therefore no real significance should be attached to the calculated parameters in terms of the surface characteristics of each material or to small changes in these parameters with wavelength.

There are many advantages to the data analysis procedure outlined above. These include: 1) The results are not dependent on the measurement characteristics of the bidirectional reflectometer, 2) scattering profiles down to approximately 0.2 mrad can be determined even though the incident beam is collimated to only 1.0 mrad, 3) the results can be easily related to other beam spreading errors such as tracking errors, slope errors, etc, and 4) the results are in the proper form for ray-trace computer programs (14).

The reflected beam profile that would be measured by sweeping a fixed collection aperture through the focused beam is plotted in Fig. 6 for an aluminized teflon mirror at 400 nm. Note that the ratio of the reflected beam intensity of the two normal distributions at $\Delta\theta = 0$ is 320 to 1 while the ratio of the beam intensity at $\Delta\theta = 0$ and $\Delta\theta = 14$ mrad is almost 1000 to 1. These large ratios illustrate an advantage in measuring the integral of the specular reflectance curve by increasing the collection aperture as opposed to sweeping the small aperture through the focal plane and measuring the reflected beam profile directly.

The bidirectional reflectometers used for solar applications incorporate either a slit aperture system or a circular aperture system. There are two advantages in a slit aperture system: 1) Using a slit aperture, asymmetrical distributions can be determined

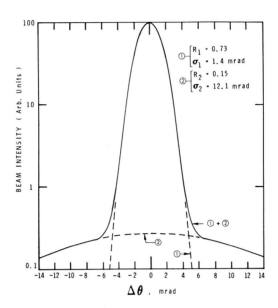

FIGURE 6. The reflected beam profile for an aluminized teflon film (Sheldahl, Inc.) at 400 nm as a function of angular aperture (compare with Fig. 5b).

(e.g. Alzak). 2) Because the area of a slit can be made much larger than the area of a circular aperture which has the same width, there is more intensity in the slit aperture instrument. The specification of a "standard" measurement technique for specularity determinations is currently being considered by the American Society of Testing Materials (ASTM).

Recently, portable bidirectional reflectometers have been developed for use in the field (15). These instruments are not as versatile as the laboratory instruments, but are designed primarily to determine changes in the specular reflectance due to dust accumulation after outdoor exposure. By comparison with standard samples, the specular reflectance changes for a fixed collection aperture can be determined before and after cleaning.

III.   EFFECTIVE SUNSHAPES

A calculation of the amount of solar radiation that is incident on a receiver surface as a function of the mirror specularity
can be illustrated with a simple example. Consider two mirror materials which have the following optical properties: Material A
-- R_s = 0.80, σ_A = 1.0 mrad and Material B -- R_s = 0.95, σ_B =
4.0 mrad. Thus Material A has a smaller dispersion (i.e., more
specular) than Material B but it also has a lower solar reflectance value (see Fig. 7a). For simplicity we will assume that the
incident sunshape (i.e., the angular distribution of rays from the
sun) can be approximated by a normal distribution with a dispersion
σ_{SUN} = 3.5 mrad, which is characteristic of a "clear" day [14].
The angular distribution of solar rays after reflection from each
mirror, or the effective sunshape, is given by the convolution of
the incident sunshape with the reflected beam profile of each mirror. In this case, the convolution of the incident sunshape and
the reflected beam profile gives another normal distribution
(effective sunshape) whose dispersion is given by the following
equation

$$\sigma_{A,B}(\text{effective}) = \left[\sigma_{SUN}^2 + \sigma_{A,B}^2\right]^{\frac{1}{2}} \tag{3}$$

which gives σ_A(effective) = 3.64 mrad and σ_B(effective) = 5.32
mrad. The integral of these effective sunshapes together with the
integral of the incident sunshape are shown in Fig. 7b as a function of the angular acceptance aperture. Note that even for a perfect mirror (R_s = 1.0, σ = 0), 95% of the incident solar radiation
is not obtained until the angular size of the receiver is approximately 8.6 mrad. For receiver angular sizes below approximately
10 mrad, Material A reflects more of the sun's energy onto the receiver; for larger receivers, mirror Material B delivers more reflected sunlight onto the receiver. For a complete calculation,

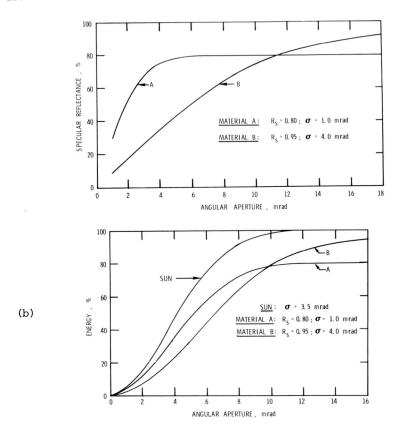

(b)

FIGURE 7. (a) Specular reflectance as a function of angular aperture for two hypothetical mirror materials A and B. (b) The integrated energy as a function of angular aperture for the assumed incident sunshape (σ = 3.5 mrad) and the effective sunshape after reflecting from mirrors A and B.

other beam spreading errors must be considered as well as the detailed shape of the incident sunlight.

IV. REFLECTANCE PROPERTIES OF VARIOUS MATERIALS

Of all of the metals that could be used for solar reflectors, only silver and aluminum have solar reflectance values above 0.90. All other metals, including gold, nickel, chromium, stainless steel, rhodium, and copper, have lower solar reflectance

values. The spectral reflectance properties of silver and aluminum, together with gold, measured for the metal/vacuum interface are shown in Fig. 8. The solar averaged reflectance values
are calculated to be: silver -- 0.98, aluminum -- 0.92 and gold
-- 0.85. These values represent the practical upper limit of
solar reflectance for these materials. The reflectance of aluminum is reduced over the solar region primarily due to an interband absorption centered at approximately 800 nm. It has been
suggested that this absorption band may be eliminated by using
amorphous aluminum; this would increase the solar reflectance
several percent (16). In almost all solar applications, the
metal reflecting layer is protected by a transparent coating
such as an oxide (e.g., Al_2O_3), glass, plastic film, etc. The
index of refraction of most coating materials is typically 1.5
through the solar spectral region. With a nonabsorbing dielectric layer applied over the metals, the solar average reflectance
is reduced by 0.01 reflectance units for silver, 0.03 for gold,
and 0.04 for aluminum (see Fig. 8). Thus, in most applications,
silver mirrors will have at most a solar reflectance of 0.97,
while the solar reflectance of aluminum will be closer to 0.88.
Very little effort is currently applied to using gold mirrors,

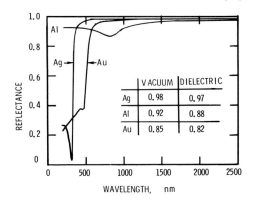

*FIGURE 8. Reflectance properties of silver (Ag), aluminum
(Al), and gold (Au) as a function of wavelength for the metal/
vacuum interface. The table lists the solar averaged reflectance for a metal/vacuum interface and a metal/dielectric
interface with an assumed index of refraction of 1.5.*

although it has been suggested that in certain photovoltaic applications, gold mirrors offer some advantages (17).

In order to understand the specularity and the reflected beam intensity of various mirror materials, it is instructive to consider the construction of a typical reflector. The typical construction of both a front-surface and second-surface reflector is shown in Fig. 9. In a front-surface mirror, the reflecting layer is applied to the substrate and then overcoated with a protective coating. In a second-surface mirror, the reflecting layer is applied to a transparent superstrate (e.g. glass, plastic film) which is then bonded to a support structure. Also shown in Fig. 9 is an adhesive layer and back protection layer as well as the supporting structure. The entire composite shown in Fig. 9 is termed a "reflector," while the outer layers which include the silver or aluminum film are termed the "mirror".

Both the specularity and the reflected beam intensity can be affected by all of the components shown in Fig. 9. Depending upon its optical properties, the outer protective layer can modify the reflected beam intensity at the metal/dielectric interface as previously shown. In addition the outer layer can absorb radiation and thereby reduce the solar reflectance. The specularity can be affected by scattering within this layer or at the inner or outer surfaces. If the outer surface is not parallel to the reflecting surface, multiple reflected images are formed.

The reflecting surface itself can scatter radiation if it has a rough surface. In addition, the reflected beam intensity may

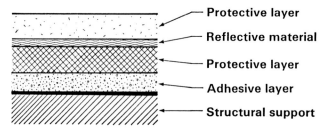

FIGURE 9. The construction of a typical reflector structure.

depend upon the purity of the metal layer as well as the deposition process (vacuum evaporation, chemical reaction, ion plating,
etc.). The surface texture of the backing layer, which is typically a thin metal sheet, can be important in affecting the specularity of the reflector, as well as the lamination or bonding
technique that is used to attach the reflecting surface to the
backing layer. The same is true for the bonding of the backing
layer to the support structure.

The different types of reflector materials can be conveniently
divided into four classes: 1) silvered glass, 2) metallized plastic films, 3) polished aluminum sheet, and 4) protective coatings.
The typical hemispherical and specular reflectance properties and
the environmental characteristics of materials within these groups
are discussed below:

A. Silvered Glass

Glass is usually silvered using a commercial chemical process
that has been developed for the mirror industry (18). The solar
reflectance of the silver/glass interface is generally in the
range 0.95 to 0.97, which agrees with the previous calculations.
However, it has been suggested that vacuum evaporated silver may
have a slightly higher (~2%) solar reflectance than chemically
deposited silver. This remains to be demonstrated.

The solar reflectance values obtained for typical silvered
glass mirrors are usually less than the theoretical value of 0.97.
This results primarily from absorption within the glass due to impurities, mainly iron. Iron in the Fe^{+2} oxidation state has a
large absorption band centered at approximately 1000 nm (19). For
a typical 3 mm thick float glass with 0.1 weight percent iron content, the solar reflectance when chemically silvered will be approximately 0.83. Because of variations in the total amount of
iron within different glasses as well as the amount of this iron
that is in the Fe^{+2} oxidation state, the solar reflectance

properties of silvered glass mirrors from different glass manufacturers and for different types of glass (float, sheet, etc.)
must be determined individually.

The solar reflectance properties of silvered glass mirrors
can be improved above the value of 0.83 for 3 mm float glass
in four ways: 1) by reducing the amount of iron in the glass
(low iron glass), 2) by reducing the glass thickness (e.g.,
using microsheet glass which is approximately 0.1 mm thick),
3) by controlling the oxidation state of the iron so that it
is all in the Fe^{+3} state instead of the Fe^{+2} state, or 4) by
addition of chemicals to the glass which form a compound(s)
with the iron and thereby modify the optical absorption. All
of these procedures have been or are currently under study (20).

The specularity of glass mirrors is described by σ values
which are typically less than 0.5 mrad. However, surface slope
errors and thickness variations, which lead to multiple reflected
images, are a function of the glass manufacturing process; typically the surface slope deviations are less than a 2-3 mrad (21).

Environmentally, glass samples exposed for over 10 years outdoors exhibit transmittance values that are typically within a 2-3
percent of the original values (21,22). Glass has good abrasion
resistance, which is important for both resistance to blowing dust
particles and mechanical abrasion resulting from cleaning, etc.
Since a meaningful abrasion test for solar mirrors has not been
developed, quantifying this property is not currently possible.

Recently, environmental protection of the chemically deposited silver layer has become an important problem (23). Degradation of the silver, which appears as black spots or streaks, has
been found after approximately 6 months of outdoor exposure in
some locations for some mirror configurations. Both edge protection and back surface protection of the silver layer using edge
sealants and paints are currently being investigated. In addition, the corrosion products and mechanisms are being studied.

Corrosion of the glass surface after extended outdoor exposure may be important, especially in the presence of moisture and
accumulated dust. Although there are some long term samples that
have exhibited little corrosion problems, new glass formulations
for solar applications and the use of glass in non-vertical
orientations may result in corrosion problems. Recent high humidity accelerated testing of different types of glass has shown
significantly different corrosion characteristics, although
interpretation and application of the results to real outdoor environments remains to be shown (22). A compilation of mechanical,
optical and environmental properties of glasses for solar applications has recently been published by F. L. Bouquet (23).

B. *Metallized Plastics*

There are several metallized plastic mirrors available for
solar applications (4). The reflecting aluminum or silver
layer is usually vacuum deposited. Hemispherical reflectance
values for silver materials are typically near 0.95 while
materials that use aluminum are typically near 0.87-0.85 in
agreement with the previously calculated values for a metal/
dielectric interface. Absorption within the plastic film,
which usually occurs in the near infrared region (greater than
1000 nm), can reduce the solar reflectance below these values.
However, since the films are usually thin (~0.1 mm), this
absorption reduces the reflectance by only a few percent.

The specularity of the reflected beam from the metallized
plastic films is primarily controlled by both the surface roughness of the metallized side of the plastic film and the lamination procedure used to bond the film to the substrate. The
surface smoothness of a variety of plastic films (including
polyesters, fluorocarbons and polypropylene) have been measured
by Boeing (25). Factors important for the lamination procedure
include the adhesive thickness, adhesive application technique,

substrate roughness, curing temperatures, etc. (26). It is interesting to note that a commercially available aluminized plastic film with a pressure sensitive adhesive has a dispersion less than 1.0 mrad when bonded to a smooth metal substrate. This dispersion is much less than the dispersion of the incident sunlight.

Environmentally, plastics have less abrasion resistance than glass, but again a definitive abrasion test does not exist for solar applications. Unless special UV stabilizers are added, plastics can suffer from ultraviolet degradation which causes increased absorption, crazing or embrittlement. Long term data exist for some plastic materials after outdoor exposure; the results show only a few percent decrease in specular reflectance after two years (27). A recent review of the outdoor aging characteristics of plastic materials indicates that for the most part applicable data are not available (28).

C. *Polished Aluminum*

Polished aluminum mirror materials are generally protected by the application of an anodized film which is several microns thick. Solar average hemispherical reflectance values of 0.85 to 0.87 for these materials are typical of aluminum reflectors. Most of these materials are mechanically and chemically polished in order to increase their specularity; however, because aluminum is a soft metal, it is very difficult to remove all scratches and pits from the surface during the mechanical polishing. In addition, the anodizing process may introduce small scattering centers which decrease the specularity. Because of these surface defects, the specular reflectance for angular apertures in the range of 15 mrad are usually from 5 to 15 percent below the hemispherical reflectance values. Environmentally, these materials have good abrasion resistance and ultraviolet stability and have maintained their reflectance properties after extended outdoor exposure (27).

D.   Protective Coatings

Protective coatings are being developed to improve the abrasion resistance of metallized plastic mirrors and offer the advantages of low cost, light weight, and improved optical performance over currently available materials. Protective coatings that are being considered for front surface mirrors are in an exploratory stage of development (29). Several existing coatings are listed in Table I; additional experimental data are needed in order to determine if any of these materials are suitable for solar mirrors.

V. INCIDENT ANGLE PROPERTIES

Measurements of the specular and hemispherical reflectance properties of solar mirror materials as a function of incident angle have recently been performed (30). In most applications

Table I.   Some Commercially Available Protective Coatings

Manufacturer	*Coating*	*Comments*
3M Company	*(Experimental)*	*Coating under development for application to FEK-244 acrylic film.*
Dow Corning	*Vestar*	*High silica coating; needs high temperature cure (>100°C).*
Dupont	*Lucite SAR*	*Abrasion resistant coating applied to acrylic sheet material.*
General Electric	*Alglas Reflector Finish*	*High silica coating for sheet aluminum; needs high temperature cure (~100°C).*
Optical Coating Lab, Inc.	*Type II-R Enhanced High Reflector*	*Multilayer protective coating for sheet aluminum.*
Owens-Illinois	*Glass Resin Type 650*	*Thermoset silicone coating for copper and aluminum.*

the incident angle of solar radiation on the mirror will be no greater than 45 to 50 degrees from normal. However, there are some applications where the incident angle reaches more than 60°, especially for non-tracking systems. As the incident angle is increased, it is reasonable to expect that both the reflected beam intensity and the reflected beam width may change. Measurements of these properties for silvered glass and metallized plastic films show that the solar averaged hemispherical reflectance decreases less than 0.04 reflectance units at 65° from normal (30). The reflectance values as a function of incident angle for these materials could be accurately calculated using a multiple beam reflectance model (30). The complex index of refraction values of the outer protective coating and the metal reflecting layer are needed for the calculations. For the metal layer (either silver or aluminum) the optical constants can be found in handbooks, while the optical constants of the outer protective layer are easily determined from standard optical measurements at normal incidence (31).

For a silvered glass sample, the specular beam profile does not change with incident angle. However, the beam dispersion for an aluminized acrylic film increased from 1.5 mrad to 1.8 mrad as the incident angle was varied to 65° from normal (30).

VI. CURRENT AND FUTURE RESEARCH AREAS

Optical measurement techniques have been developed so that meaningful and reliable data on solar materials are currently being obtained. The National Bureau of Standards is preparing specular reflectance standards so the calibration of laboratory instrumentation can be certified (32). There is a need to standardize the measurement instrumentation and data analysis; an ASTM committee is presently active in this area.

There are currently a variety of commercial mirror materials, including silvered glass, metallized plastics and polished aluminum, that can be used in solar collectors. Although some new material development in the area of thin glass, low iron float glass and protective coatings is currently being pursued, the major effort in current research on solar mirrors involves the effect of outdoor aging on the optical properties. Testing has generally followed the pattern of exposing materials at various locations throughout the country and periodically returning the samples to the laboratory for measurements. Besides the standard 45° south exposure geometry, samples have been exposed to multi-sun environments through the use of tracking concentrators, in order to accelerate degradation processes. Handling of the sample deserves careful attention since surface scratches and stressing of the samples can adversely affect the optical properties. Samples exposed outdoors will have accumulated dust and pollutants on the outer surface which can strongly influence the specular reflectance properties. Recent studies have been directed toward characterizing the effect of dust accumulation on the specular reflectance properties of various mirrors. Areas of particular interest include determining dust accumulation rates, investigating effective mirror storage strategies (vertical or inverted) and defining cost effective cleaning intervals and procedures. For long term outdoor exposure studies, it can become very difficult to separate the effects of surface abrasion, dust accumulation, ultraviolet aging, surface crazing, or metal degradation since they can have similar effects on the reflectance property. Although limited data exist, some data are subject to misinterpretation due to improper sample handling, measurement technique limitations, or sample cleaning variations.

In addition to outdoor exposure testing, some accelerated laboratory testing has been used to help characterize different materials. These tests have included temperature and humidity cycling through a freeze-thaw cycle, constant temperature at high relative humidity, weatherometer, ultraviolet lamp exposure at room temperature or elevated temperatures, salt spray, various abrasion tests, etc. Depending upon the material being studied, the type of environment being simulated and the suspected aging mechanism, these tests must be carefully chosen and correlated with real time outdoor exposure results. Specific areas of current investigation include degradation of the silver/copper reflecting layer of chemically silvered glass mirrors (24), improvement of the reflectance of aluminum by alloying with silver (16), long term effects of dust accumulation, cleaning frequency and cleaning techniques, use of electrostatic coatings to reduce dust accumulation and the development of accelerated environmental tests. A detailed discussion of these areas is contained in the solar reflector materials research and development program plan recently published by the Solar Energy Research Institute (20).

REFERENCES

1. Rabl, A., "Comparison of Solar Concentrators," *Solar Energy 18*, 93 (1976).

2. Miller, D. W., "A Simulated Hail Test Facility," Sandia Labs Report SAND-79 1114 (June, 1979).†

3. Butler, B. L. and Pettit, R. B., "Optical Evaluation Techniques for Reflecting Solar Concentrators," SPIE Proceedings Optics Applied to Solar Energy Conversion 114, 43 (1977).

4. Pettit, R. B., "Characterization of the Reflected Beam Profile of Solar Mirror Materials," *Solar Energy 19*, 733 (1977).

5. Judd, D. B., "Terms, Definitions, and Symbols in Reflectometry," *J. Opt. Soc. Am. 57*, 445 (1967).

6. "Thermal Radiative Properties - Metallic Elements and Alloys," Vol. 8, Thermophysical Properties of Matter, (Ed. by Y. S. Touloukian and D. P. DeWitt), pp. 32a-37a, Plenum, New York (1970).

7. Commercial instruments are currently available from: (1)
   Beckman Inst., Scientific Inst. Div., 2500 Harbor Blvd., Ful-
   lerton, CA 92634.   (2) Gier-Dunkle Inst., 1718 21st Street,
   Santa Monica, Monica, CA 90404.   (3) Perkin-Elmer Corp., Inst.
   Div., Box 730, Main Avenue, Norwalk, CT 06856.   (4) Varian
   Associates, Inst. Div., 9901 Paramount Blvd., Downey, CA 90240.
   (5) Carl Zeiss, Inc., 444 Fifth Ave., New York, NY 10018.

8. Lind, M. A., Pettit, R. B., and Masterson, K. D., "The Sensitivity of Solar Transmittance, Reflectance and Absorptance
   to Selected Averaging Procedures and Solar Irradiance Distributions," to be published *J. Solar Energy Engineering.*

9. Alternately, a small port in the sphere wall can be removed
   so that the normally reflected beam exits the sphere.

10. Lind, M. A., Hartman, J. S., and Hampton, H. L., "Specularity
    Measurements for Solar Materials," SPIE Proceedings Optics
    Applied to Solar Energy Conversion 161, 98 (1978).

11. Zentner, R. C., "Performance of Low Cost Solar Reflectors
    for Transmitting Sunlight to a Distant Collector," *Solar
    Energy 19,* 15 (1977).

12. Stickley, R. A., "Solar Power Array for the Concentration
    of Energy (SPACE), Semi-Annual Progress Report," NSF/RANN/
    SE/GI - 41019/PR/74/12, Sheldahl, Inc., Northfield, MN,
    Jan 1-June 30, 1974, pp. 156-159.

13. Sanchez, J., General Electric Co, Valley Forge Space Center,
    P. O. Box 8555, Philadelphia, PA 19101, private communication.

14. Biggs, F. and Vittitoe, C. N., "The Helios Model for the
    Optical Behavior of Reflecting Solar Concentrators," Sandia
    Labs Report SAND 76-0347, (March, 1979).

15. Freese, J. M., "The Development of a Portable Specular Reflec-
    tometer for Field Measurements of Solar Mirror Materials,"
    Sandia Labs Report SAND 78-1918 (Oct., 1978).†

16. Trotter, D. M., et al., "Enhanced Reflectivity Aluminum," in
    Proceedings of Solar Reflective Materials Technology Workshop,
    Battelle PNL Report PNL-2763, (Oct., 1978), p. 75.

17. Saylor, W. and Sanchez, J., "The Effect of Different Reflec-
    tor Surfaces on Photovoltaic Concentrator Performance," Pro-
    ceedings of 1979 Joint Meeting of American Section of
    ISES, May 28-June 1, 1979, Atlanta, GA.

18. Lind, M. A. et al, "Heliostat Survey and Analysis," Battelle
    Pacific Northwest Lab Report - to be published; and Taketani,
    H., and Arden, W. M., "Mirrors for Solar Energy Applications,"
    McDonnell Douglas Report MDC G7213 (Sept., 1977).

19. Vitko, Jr., J., "Optical Studies of Second Surface Mirrors
    Proposed for Use in Solar Heliostats," Sandia Labs Report
    SAND 78-8228-0U0.†

20. "Solar Optical Materials Program Activity Committee -
    Progress Report, July 1, 1978 to Jan. 31, 1979," Solar
    Energy Research Inst. Report SERI/PR-31-137 (July, 1979)
    and "Solar Glass Mirror Program - A Planning Report on
    Near-term Mirror Development Activities," Solar Energy
    Research Inst. Report SERI/PR31-145 (Jan., 1979).

21. Lind, M. A. and Rusin, J. M., "Heliostat Glass Survey
    Analysis," Battelle Pacific Northwest Labs Report PNL
    2868/UC62 (Sept., 1978).

22. Vitko, Jr., J. and Shelby, J. E., "Solarization of Helio-
    stat Glass," Sandia Laboratories Report, to be published.†

23. Bouquet, F. L., "Glass for Solar Concentrator Applications,"
    Jet Propulsion Lab. Report 5102-105 (April, 1979).

24. Burolla, V. P. and Roche, S. L.," Silver Deterioration in
    Second-Surface Solar Mirrors," Sandia Labs Report, to be
    published.†

25. "Solar Central Receiver Prototype Heliostat - Final Techni-
    cal Report," Vol. 1, Boeing Eng. & Construction Co., Report
    SAN/1604-1 (June, 1978), and "One Piece Domb Heliostat
    Studies," Final Report under Sandia Labs Contract 18-7830,
    Sandia Labs Report SAND 78-8184 (July, 1979).

26. Stickley, R. A., Sheldahl Inc., Northfield, MN 55057 –
     private communication.

27. Rausch, R. A. and Gupta, B. P., "Exposure Test Results
     for Reflective Materials," Proceedings of Inst. of Env.
     Science Seminar on Testing Solar Energy Materials, Gaithers-
     burg, MD, May 22-24, 1979, pp. 184-187.

28. Hampton, H. J. and Lind, M. A., "Weathering Characteristics
     of Potential Solar Reflector Materials: A Survey of the
     Literature," Battelle Pacific Northwest Labs Report PNL
     2824/UC62 (Sept., 1978).

29. Bieg, K. W. and Wischmann, K. B., "Plasma-Polmerized Organo-
     silanes as Protective Coatings for Front-Surface Solar Mir-
     rors," Sandia Laboratories Report, to be published.†

30. Glidden, D. N. and Pettit, R. B., "Specular Reflectance
     Properties of Solar Mirrors as a Function of Incident Angle,"
     Proceedings of 1979 Joint Meeting of American Section of
     ISES, May 28-June 1, 1979, Atlanta, GA.

31. Pettit, R. B., "Hemispherical Transmittance Properties of
     Solar Glazings as a Function of Averaging Procedure and
     Incident Angle," *Solar Energy Materials 1*, 125 (1979).

32. Richmond, J. C., "Evaluation of Solar Absorbers, Reflectors
     and Transmitters Physical Reflectance Standards," Proceedings
     of the 25th Annual Technical Meeting of the Inst. of Env.
     Sciences, Seattle, WA, (May, 1979), p. 2.

†Available from: NTIS, U.S. Dept. of Commerce, 5285 Port Royal
                  Rd., Springfield, VA 22161.

CHAPTER 6

THE EFFECT OF SOILING ON SOLAR MIRRORS AND TECHNIQUES USED TO MAINTAIN HIGH REFLECTIVITY[1]

E.P. Roth
R.B. Pettit

Sandia Laboratories[2]
Albuquerque, New Mexico

I. INTRODUCTION

Solar mirrors are used to concentrate low-level solar radiation to power levels which are practical and efficient for consumption. Any interference with the collection of that energy not only decreases the power level but also increases the cost of the energy available from a solar power system. Solar mirrors are designed to initially achieve the maximum possible reflectance. However, outdoor exposure subjects the mirror materials to environmental conditions which can quickly degrade their efficiency. One of the most immediate and drastic effects of outdoor exposure is the reflectance loss due to the accumulation of foreign particles on the mirror surface. Specular reflectance losses as great as 25% have been observed for mirrors exposed for only a few weeks. The effect of the deposited particles is to reduce the reflected energy by both absorbing and scattering light[1,2]. The degree to which the particles reduce the collection of reflected energy depends on their composition, number and size distribution[1,2]. An additional factor is the optics of the collection system. The angular acceptance aperture of the system, defined as the angle subtended by the receiver as the concentrator surface, determines the

[1]This work is supported by the Division of Solar Technology, U.S. Department of Energy (DOE), under contract DE-AC04-76-DP00789.
[2]A U.S. DOE facility.

relative importance of the scattering due to dust accumulation.
For flat plate thermal and photovoltaic collectors which have
essentially a 180° angular acceptance aperture, scattering of the
incident light is not critical but absorption can be an important
factor in the loss of energy. For concentrating collection
systems, such as line focus collectors and central receivers,
angular acceptance apertures of a few degrees make scattering at
the concentrator surface much more important and can result in
severe energy losses. Thus, from an economic point of view,
periodic cleaning or reduction of soil accumulation is a practical
necessity.

II. EFFECT OF NATURAL SOILING ON MIRROR REFLECTANCE

   Potential methods for controlling the reflectance loss due
to soiling must be based on both measurements of actual particulate accumulation in an outdoor environment and an understanding
of the basic physical mechanisms of adhesion and light scattering.
In order to establish a data base for the reflectance loss of
exposed mirrors, a field test study was initiated simulating some
of the operational configurations of solar mirrors.

A. *Long Term Soil Accumulation Study*

   Solar mirror materials have been exposed to natural weathering
in Albuquerque, NM for periods exceeding one year.[3] The mirror
materials used were second-surface silvered glass obtained from a
heliostat panel at the 5MW Central Receiver Test Facility (CRTF)
at Kirtland AFB, Albuquerque, NM.[4] These samples are typical
of the type of materials used in many solar thermal power systems.
The specular reflectance of the mirrors was measured with a bidirectional reflectometer over a wavelength range 400-900 nm and
over a 3-15 mrad range of angular acceptance apertures.[5] Figure
1 shows the specular reflectance data for a mirror exposed for

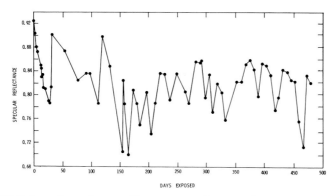

FIGURE 1. Specular reflectance of second-surface silvered glass mirrors exposed to natural weathering for 480 days. Reflectance measured at 500 nm and 15 mrad angular aperture.

480 days to natural weathering.[6]   The data shown are for a wavelength of 500 nm and a 15 mrad angular aperture. The data show an initial rapid drop in reflectance of approximately 0.006 reflectance units per day (100% reflectance = 1.00 reflectance units) followed by large fluctuations in reflectance which are induced by variations in weather conditions. The weather condition at a particular mirror location is one of the critical factors affecting the rate of dust accumulation and the eventual long term reflectance loss of an exposed mirror. Daily reflectance losses as great as 0.144 reflectance units have been measured under certain conditions (light rain followed by a wind and dust storm) while increases as large as 0.121 reflectance units have been measured at other times (snow-rain weather conditions).[6] Because of the variation in local weather, it is very difficult to predict long term reflectance losses for a given location and more difficult to apply those results to other locations. In general, uncleaned mirrors in the Albuquerque area show a long term decrease in specular reflectance of approximately 0.10-0.15 reflectance units with large fluctuations about the average.[3] Larger reflectance losses can occur in other geographic locations, especially in urban environments where optically absorbing particles from pollutants can lead to additional energy losses.[7] Additional outdoor

exposure studies at other geographic locations are required to
obtain a more general understanding of mirror soiling.

B. *Cleaning Cycle Experiment*

When the reflectance of a solar mirror drops sufficiently,
cleaning the mirror surface becomes economical. Increasing the
cleaning frequency should raise the average long-term reflectance
of the mirror, as depicted in Figure 2, although some long-term degradation may result from the cleaning procedures. Figure 3 shows
the results of actual cleaning cycle tests in which three sets of
mirrors were exposed and cleaned on 2-, 6- and 12-day cycles.[3]
The mirrors were measured every two days to show any fluctuations
due to weather conditions. The results show that laboratory
cleaning (three minutes in an ultrasonic bath of distilled water
and wiped dry with a soft tissue) essentially restores the reflectance of each mirror to its initial value. Subsequent exposure
results in a rapid nonlinear drop in reflectance for each set of
mirrors. The average daily reflectance loss for, respectively,
the 2-, 6- and 12-day cycle mirrors was 0.0085, 0.0061 and 0.0051
reflectance units. Thus, the rate of dust accumulation decreases
as the amount of accumulated dust increases. The long term average
reflectance loss for the 2-, 6- and 12-day cycled mirrors is, respectively, 0.0085, 0.018 and 0.031 reflectance units, indicating
that increased frequency of cleaning does raise the average reflectance of the mirrors.

The level of dust accumulation is also seen to affect the response of the mirrors to weather conditions.[3] For example, mirrors which had an appreciable accumulation of dust were cleaned by
a light rain while newly cleaned mirrors experienced a loss in reflectance under the same conditions. These results show that
weather and mirror conditions can significantly affect the reflectance of exposed mirrors and that these conditions must be fully
considered in any method to predict long-term reflectance loss.

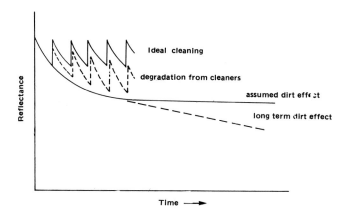

FIGURE 2. *Expected long-term reflectance of an exposed mirror with and without periodic cleaning.*

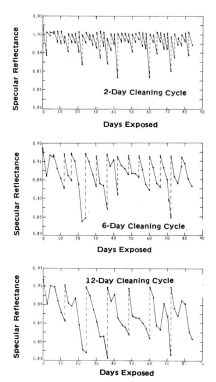

FIGURE 3. *Specular reflectance of mirrors undergoing (a) 2-day, (b) 6-day and (c) 12-day cleaning cycles. The dashed lines show the reflectance increase due to laboratory cleaning; the solid lines show the reflectance loss due to natural soiling.*

C. *Orientation Angle Experiment*

Several operational parameters can affect the rate of soiling of exposed mirrors. For example, the orientation angle of a mirror during periods of nonoperation can affect the rate of particulate settling on the surface and can maximize the effect of natural cleaning forces such as wind and rain. To investigate the effect of stowage angle on mirror degradation, a set of five mirrors was exposed on a test rack at different angles with respect to the horizontal: 0°, 30°, 45°, 60° and 180° (inverted).[3] These samples were exposed only during daylight hours of good weather, thus representing soiling strictly due to dry deposition. The results of this experiment are shown in Figure 4. Generally, the drop in specular reflectance decreased as the orientation angle increased; however, only the inverted (180°) mirror showed a significant reduction

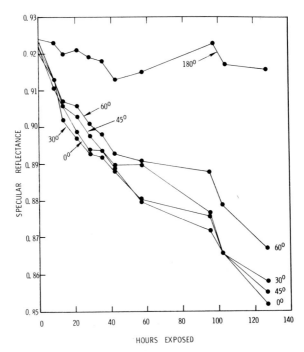

FIGURE 4. Specular reflectance of 5 mirrors as a function of mounting angle (0° = face up; 180° = inverted). The mirrors were exposed for a total of 127 hours (≈5 days) during daylight hours of good weather.

in soiling.   In other experiments, a 90° stowage angle has also resulted in a significant decrease in the rate of soiling.[8]   These
experiments show that any method formulated to maintain high reflectivity of solar mirrors must involve the optimization of both
the cleaning cycle and the stowage position.

III.   EFFECT OF ACCUMULATED DUST ON SPECULAR REFLECTANCE

The detailed interaction of accumulated dust with the incident
solar radiation is important in understanding and predicting the
loss in collected energy.   Measurements were performed on a set of
solar mirrors exposed to natural weathering for a period of five
weeks to determine the wavelength dependent scattering by the particulates and the effect of the dust on the beam shape of the
scattered light.[9]

A. Hemispherical Reflectance
Initially, the hemispherical reflectance of a clean mirror was
measured over the wavelength interval 320-2500 nm using an integrating sphere reflectometer.[10]   This device allows collection of
both the specular and diffuse component of the reflected beam over a
solid angle of 2π steradians.   Typical data are shown in Figure 5.

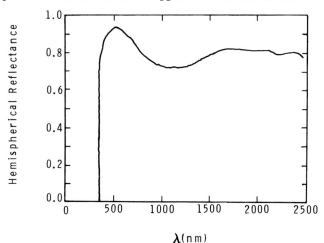

FIGURE 5.   *Hemispherical reflectance for a clean, second-*
surface, silvered float glass mirror.

The large dip in reflectance at 1000 nm is due to absorption in the glass by Fe^{+2} impurities.[11] The cutoff below 400 nm results from from losses in the glass and in the silver reflector layer.[12] Subsequent measurements of hemispherical reflectance after soiling showed no appreciable decrease in reflectance for specular reflectance losses up to ≈ 0.05 reflectance units, indicating that the energy lost from the specular component went into the diffuse scattering background with no measurable loss due to absorption. These results are consistent with the type of losses expected from dielectric (nonconducting) particles which are usually found in a desert environment. However, outdoor exposure to urban environments could lead to contamination by absorbing pollutants which could cause a decrease in the net hemispherical reflectance.[7]

B. Wavelength Dependence

The specular reflectance of mirrors with increasing levels of dirt accumulation is shown in Figure 6 over the wavelength range of 400–900 nm measured at a 15 mrad aperture.[9] The dominant effect of the accumulated dust is the decrease in specular reflectance over the entire wavelength range with increasing level of dust accumulation. At 500 nm the specular reflectance loss varied from 0.065 to 0.24 reflectance units. The wavelength dependence of the specular reflectance loss is shown in more detail in Figure 7. In this figure, the specular reflectance loss, R_D-R_C, is normalized by the reflectance value of the clean mirror, R_C, at each wavelength. The wavelength dependence of the reflectance loss is directly proportional to the scattering cross section of the dust particles. The data show that the net scattering by the accumulated particles increases with decreasing wavelength and with increasing level of soiling, with the scattering amplitude still increasing at 400 nm. Since the soil particles accumulated in this experiment do not result in any appreciable absorption, the light lost from the specular beam forms the diffusive reflection background. The wavelength

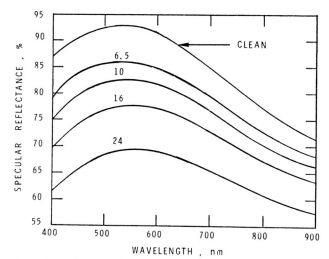

FIGURE 6. *Specular reflectance as a function of wavelength for an initially clean mirror (upper curve) and after increasing levels of dust accumulation (lower curves). Values listed are the specular reflectance loss measured at 500 nm for a 15 mrad angular aperture due to dust accumulation.*

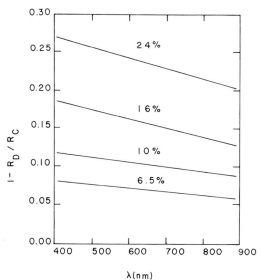

FIGURE 7. *Specular reflectance loss due to soiling normalized by the specular reflectance of the clean mirror as a function of wavelength.* (R_D = *Dirty mirror,* R_C = *Clean mirror*). *The functional dependence of the normalized reflectance loss is directly proportional to the scattering strength of the accumulated dust. The data are obtained from the curves in Fig. 6 for the various levels of dust accumulation. The values listed refer to the specular reflectance loss at 500 nm.*

dependence of the diffusive background is shown in Figure 8 for the
levels of soil accumulation shown originally in Figure 6. The
diffusive scattering was measured with the integrating sphere reflectometer over the wavelength range 320-2500 nm. Because of differences in the beam sizes, collection apertures and measurement
regions, the loss in specular reflectance and the increase in diffuse reflectance are not in exact agreement. However, by norma-
lizing the diffuse reflectance by the value at 500 nm, the wavelength dependent scattering can be approximately determined,
independent of the degree of soiling. The resulting normalized
curve is shown in Figure 9. The HIGH and LOW curves represent
the maximum and minimum normalized loss values respectively from
all regions measured. This figure shows that within the accuracy
of the data the wavelength dependence of the scattering is independent of the concentration of accumulated particles. This
result is useful since it allows a solar-averaged reflectance loss
for this silvered glass mirror to be calculated from a measurement
at a single wavelength. For the type of mirrors used in this experiment, the solar averaged reflectance loss is equal to 0.78
± 0.04 times the specular reflectance loss measured at 500 nm.

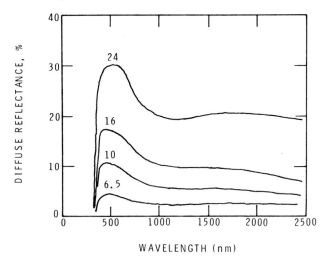

WAVELENGTH (nm)

*FIGURE 8. Diffuse reflectance as a function of wavelength for
the soiled mirrors shown in Fig. 6. The values listed refer to
the specular reflectance loss at 500 nm.*

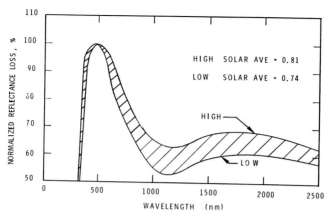

FIGURE 9.   Diffuse reflectance normalized by the diffuse reflectance at 500 nm as a function of wavelength for all levels of soil
accumulation shown in Fig. 8. The HIGH/LOW curves represent the limits of the normalized curve due to the uncertainty in each measurement.   The HIGH/LOW solar average values refer to the limits of the
solar averaged specular reflectance loss as discussed in the text.

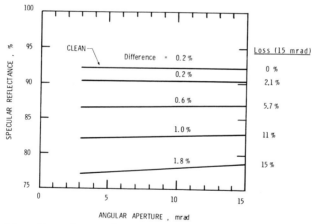

FIGURE 10.   Specular reflectance as a function of angular aperture for a clean mirror (upper curve) for increasing levels of dust
accumulation.   The difference values listed are the difference in
specular reflectance between 3 mrad and 15 mrad.   The losses listed
at the right are the specular reflectance losses measured at 15
mrad.   All data were measured at 500 nm.

C.   Effect of Dust Particles on Beam Shape

Accumulated dust particles can have a significant effect on

the performance of a solar collector with a small angular accep-

tance aperture by affecting the shape of the reflected beam.   The

effect of accumulated particles on beam shape was measured using

the laboratory bidirectional reflectometer over an angular aperture range of 3-15 mrad.[9] The data are shown in Figure 10 at the
standard wavelength of 500 nm for increasing levels of dust accumulation. The values listed in the figure are the differences in
specular reflectance between the 3 and 15 mrad measurement points.
The data show that the main effect of accumulated dust is to decrease the overall intensity of the reflected beam and not to significantly change the profile. Wide-angle scattering by the accumulated particles (scattering at angles much greater than the
acceptance aperture of the collection optics) can account for this
effect and result in comparable losses for both central receiver
and distributed power systems which both have apertures $\lesssim$ 2°.

IV. SCATTERING THEORY

The detailed scattering of light by particles is a complex
function of the optical properties of the particles, the size and
number distribution of the particles, and the wavelength of the
incident light.[1,2] For solar power systems, the incident light
comes from direct radiation by the sun. The wavelength distribution of the solar radiation may be modeled as a black body spectrum
corresponding to a temperature of $\approx$ 5800 K, modified by absorption in both the solar and terrestrial atmospheres.[1,3] The peak
in the atmospheric spectrum occurs at approximately 500 nm, with
a lower cutoff at 300 nm and an upper cut off at 3500 nm.

A. *Extinction Coefficient and Angular Scattering Function*

The scattering of light by a single particle is a function of
the particles' complex index of refraction, the particle shape and
the size of the particle compared to the wavelength of the incident light.[1,2] The efficiency of a particle in removing energy
from incident light is derived from Mie scattering theory and is
given by its extinction coefficient:

$$K_{EXT} = K_{SCAT} + K_{ABS} \tag{1}$$

$$= \sigma_{SCAT}/\sigma_A + \sigma_{ABS}/\sigma_A \qquad ,$$

where K_{SCAT} is the ratio of the effective scattering cross section (σ_{SCAT}) of the particle to its actual geometric cross section (σ_A) and K_{ABS} is the ratio of the effective absorption cross section (σ_{ABS}) to the geometric cross section. The extinction coefficient for a spherical dielectric particle (no absorption) is shown in Figure 11 as a function of the particle circumference/wavelength ratio.[1,2] This curve is valid for most particles of interest for which $1 \lesssim |m| \lesssim 2$ where m is the complex index of refraction. The figure shows that the extinction coefficient drops off rapidly for particles small compared to the wavelength of the incident light, peaks at a value where the particle size is comparable to the wavelength and then undergoes oscillations of decreasing amplitude about a value of ≈ 2 with increasing particle circumference/wavelength ratio. For increasing magnitude of the index of refraction, the peak in the extinction coefficient shifts to longer wavelengths.[1,2]

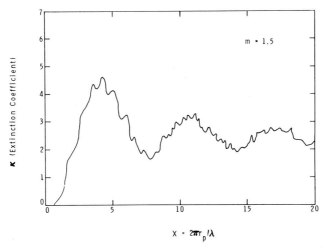

FIGURE 11. *Extinction coefficient calculated from Mie scattering theory for a spherical particle with index of refraction* $m = 1.5$ *as a function of the size parameter* $x = 2\pi r_p/\lambda$ *where* r_p = *particle radius and* λ = *wavelength of incident light.*

The angular distribution of the scattering energy is a complicated function of the relative particle circumference/wavelength
ratio, particle index of refraction and polarization of the incident light.[2] Figure 12 shows the scattering amplitude as a function of angle for a particle with 1.55 index of refraction and with
circumference/wavelength ratio $2\pi r_p/\lambda = 3.0$ (r_p = particle radius).
As the size of the particle becomes equal to or larger than the
wavelength of the incident light the scattering amplitude becomes
peaked in the forward direction with weaker lobes occurring at
larger angles. However, for most naturally occurring particles, the
majority of the scattered energy still occurs at angles greater than
the few degree angular acceptance apertures of most concentrated
power systems. These calculations agree with the large-angle scattering which previously accounted for the negligible effect of accumulated dust on the shape of the specularly reflected beam profile.

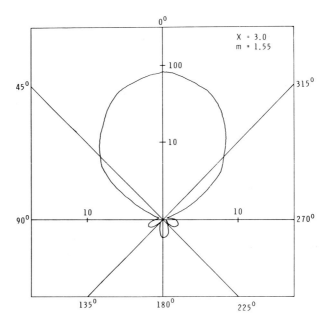

FIGURE 12. Amplitude of scattered light as a function of
angle for a spherical particle with index of refraction m = 1.55
and size parameter $x = 2\pi r_p/\lambda = 3.0$. The angle of the incident light is 0° and the polarization normal to the plane of
the paper.

B.   *Loss in the Specular Reflection Component*

The loss in the specular component of reflected solar energy due to scattering by dust particles can be calculated by convoluting the particle extinction coefficients, particle size distribution and solar spectral distribution over all particle sizes and solar wavelengths.[1]   The expression for this loss is given as

$$\Delta I_m/I_o = 2 \int_o^\infty dr \int_o^\infty d\lambda \left\{ \pi r^2 K_{EXT}\left(\frac{2\pi r}{\lambda}, m\right) n(r,m) f(\lambda) \right\}$$

$$= 2 \int_o^\infty dr \left\{ \pi r^2 n(r,m) \int_o^\infty K_{EXT}\left(\frac{2\pi r}{\lambda}, m\right) f(\lambda) d\lambda \right\},$$

(2)

where r = particle radius, $n(r,m)$ = number of particles/unit area-unit radius, m = complex index of refraction, I_o = solar spectral intensity and $f(\lambda)$ = wavelength function of the solar spectrum. In this expression, the incident solar radiation is assumed to have interacted with the surface layer of dust particles twice as shown schematically in Figure 13a. The result of convoluting the wavelength dependent extinction coefficient for a spherical particle with $m = 1.5$ and the solar distribution function is shown in Figure 13b. The net loss in intensity of the specular beam is then obtained by further convoluting the function shown in Figure 13b with the particle area and the particle size distribution function shown in Figure 13c. The particle size distribution function shown in this figure is representative of the distribution function actually measured on exposed mirrors. This function will be discussed in more detail in Section V, B. The resultant energy lost from the specular beam per unit particle diameter; i.e., the integrand in the second line of Eq. (2), is shown in Figure 14, assuming a particle distribution function of the form $n(r,m) \propto r^{-3}$ for ease of calculation. The peak in energy loss occurs near 500 nm, corresponding to the peak in the solar spectrum. This analysis emphasizes the importance of the small particle (0.05 μm $\lesssim r_p \lesssim$ 1 μm) in

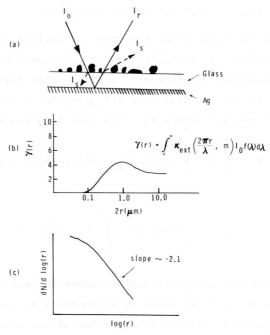

FIGURE 13. (a) Schematic of light scattering by dust particles on a second-surface mirror where I_o = incident light, I_s = scattered light and I_r = reflected light. (b) Convolution of wavelength dependent extinction coefficient and the solar spectral distribution as a function of particle diameter. (c) Particle size distribution function for a naturally soiled mirror.

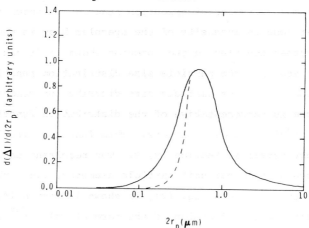

FIGURE 14. Energy loss as a function of particle size assuming a particle size distribution function $n(r) \propto r^{-3}$ (solid curve). Energy loss assuming a drop off in the particulate concentration for $r_p \lesssim 0.2$ μm (dashed curve). The incident radiation is assumed to have the standard solar distribution at sea level.

the scattering loss. A decrease in the number density of small particles ($r_p \lesssim 0.2$ μm), which has been measured by some investigators, would cause a sharper cutoff in the energy loss function for the small particle diameters, as shown by the dashed curve in Figure 14. However, the major loss of energy still results from particles in the submicron range.

V.   DEPOSITION AND ADHESION

The deposition of particles on a mirror surface is controlled by the complex fluid mechanical interaction of the dust-laden airstream with the entire mirror structure.[14] Processes such as convective diffusion, impaction and sedimentation play important roles in the deposition process depending on particle size and wind velocity. In general, particles whose Stokes velocity is less than the ambient wind velocity will be carried to the mirror surface and can be subsequently deposited. Particles with diameters $\lesssim$ 100 microns will be suspended by wind velocities of only a few miles per hour, resulting in a broad size spectrum of deposited particles.

A. *Forces of Adhesion*

A wide range of forces are responsible for the adhesion of the particles to the surface, as listed in Table I. The magnitude of

Table I. Mechanisms of Dust Adhesion

Mechanism	Affecting Material Property
Gravity	Mass
Electrostatic	Surface (coating) conductivity
Charge Double Layer	Contact potential (difference in work functions)
VanderWalls Force	Particle size; Surface roughness
Surface Energy	Solid surface relaxation
Capillary Force	Fluid surface relaxation
Chemical/Physical Bond	Chemical activity

these forces depends strongly on the nature of both the particles
and the mirror surface, varying from a fraction of the gravitational
force on the particle to several orders of magnitude greater than
the gravitational force.[15] The details of these different mechanisms are not sufficiently well understood to permit accurate esti-
mations of the type and magnitude of forces responsible for particle
adhesion. However, the initial forces of adhesion are probably dominated by electrostatic forces and surface energetics, while after
sufficiently long periods of time stronger chemical and physical
bonds can develop. The few experiments that have been performed show
that the forces of adhesion in general increase with decreasing particle size and with increasing time of surface contact.[15] The development of the stronger chemical bonds will depend strongly on the
amount of moisture present at the particle mirror interface and are
thus affected by such parameters as relative humidity and rainfall.

B. Particle Distributions

As discussed in Section IV, B, the small particles ($r_p \lesssim 1$ μm)
are the most important source of scattering for the solar spectrum
and, as stated in the previous section, experience the greatest surface adhesion. Measurements of the actual particle size distribution on weathered mirrors can yield information on the relative
significance of the various particle sizes and how different environmental conditions can affect their rates of accumulation.

Particle size distributions have been measured using a Quantimet particle sizer.[16] This instrument measures the number of
particles in selected size intervals from direct optical images of
the mirror surface and from high magnification micrographs taken on
a scanning electron microscope. Overlapping particle size measurements are made at different magnifications at several random locations on the surface to obtain a representative characterization
of the entire mirror surface. An average of 60-70 different locations are measured using five different magnifications covering

particles with diameters $\gtrsim$ 0.3 μm. A typical particle distribution for a mirror subjected to several months of outdoor exposure
is shown by the open circles in Figure 15. For convenience in
comparing to atmospheric aerosol distributions reported in the
literature, the size distribution is presented as a logarithmic
function,

$$dN/Ad(\log r) \quad (cm^{-2}) \qquad , \qquad (3)$$

where N is the number of particles with radii $\leq$ r, r is the particle radius and A is the unit area of mirror surface.[17-19] In
general, the size distribution is described as

$$dN/Ad(\log r) \propto r^{-k} \qquad , \qquad (4)$$

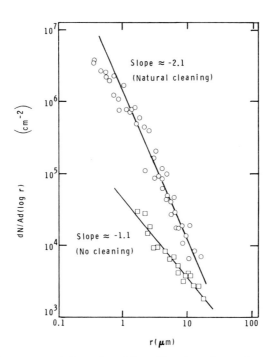

FIGURE 15. *Particle size distribution functions for naturally
soiled mirrors. Open circles: mirror exposed to natural cleaning
(wind and rain) for several months. Open squares: mirror exposed
only to "good" weather conditions for a period of 2 days.*

where $k \approx 2$ for this sample. The particle distribution measured
in atmospheric aerosols for particle radii $\gtrsim 1$ μm is usually
described by a power law distribution. The actual logarithmic
slope can vary significantly depending on location, weather and
time of year.[17-19] Below 1 μm the atmospheric particle distribution function usually levels off or actually decreases. The
entire distribution is often referred to as a lognormal distribution which is modeled with a logarithmic Gaussian function plus
a power law background distribution.[18] The distribution of
particles accumulated on the exposed mirror shows some deviation
from the power law distribution function below 1 μm but the
decrease is smaller than observed in aerosols, indicating some
preferential adhesion of the small particles out of the atmospheric
distribution. This result is consistent with experiments which
have shown increased adhesion for small particles.[15]

If indeed the small particles adhere more strongly to the mirror surface, then the small particles should likewise be more difficult to remove. The result of preferential adhesion of the small
particles can be seen by comparing the particle distribution of the
weathered mirror to the distribution of a mirror which has undergone only dry deposition. The open squares in Figure 15 show the
particle distribution for a mirror which has undergone only two
days of exposure during dry weather. The measured slope is ≈ -1.1
compared to ≈ -2.1 for the weathered mirror. The greater magnitude
of the slope of the weathered mirror occurs because of "natural"
cleaning conditions, such as wind and rain, which preferentially
remove the larger particles while the relative number of smaller
particles continue to increase. This result has been confirmed by
measurements on other mirrors which have undergone varying lengths
of outdoor exposure to both "dry" and "wet" environments.

VI.   ACCELERATED DEPOSITION STUDY

Variations in weather conditions cause such large fluctuations in the reflectance of exposed mirrors that long-term predictions of reflectance loss are difficult to make. Measurement of reflectance loss and particle accumulation under controlled conditions can yield a better understanding of the effect of such parameters as wind velocity, particle flux and humidity on the rate of reflectance loss. In addition, controlled particle deposition allows an accurate comparison of soiling rates for various mirror materials and cleaning techniques.

A. Wind Tunnel

Representative mirror materials have been exposed to accelerated dust deposition in a low-velocity wind tunnel equipped with a dust injector/disperser unit and laser optical systems capable of monitoring the flux rate of the incident particles and the real-time reflectance loss of the exposed mirrors. The mirrors are exposed to controlled amounts of a well-defined Arizona Desert Dust[20] over velocity ranges of ≈10-30 MPH. The samples are mounted normal to the incident airstream to achieve the maximum rate of dust accumulation. The dust injector/disperser unit is capable of injecting particles at densities ≈10^4 times greater than the particle densities present in normal atmospheric aerosols. The injector has also been designed to maintain a constant injected particle size distribution over the period of deposition. Specular reflectance losses observed after several months of outdoor exposure have been simulated in approximatly 30 minutes in the wind tunnel. Although this accelerated deposition system is not intended to exactly duplicate outdoor exposure, it does allow a comparative measurement of dust accumulation on various materials under a wide range of exposure conditions.

B. *Laser Optics*

The dust-laden airstream is monitored using a laser velocimeter
apparatus and a multichannel analyzer to record the flux of incident particles during the deposition period, as shown in Figure 16.
The green beam (λ = 0.5145 nm) from an argon laser is split into
two components and recombined in front of the exposed mirror to
form a region of interference fringes. As particles pass through
the sampling volume, they scatter light with an intensity pattern
characterized by the spacing of the interference fringes and the
particle velocity, thus allowing the particle transit to be distinguished from background noise in the phototube[21]. Typical flux
levels levels range from 5 x 10^4 - 5 x 10^5 particles/cm^2-sec.

During the deposition, the reflectance of the mirror is monitored using a He-Ne reflectometer. A He-Ne laser beam is expanded
to approximately 1 cm diameter and is split into a sample beam
and a reference beam. The two beams (45° to the mirror surface)
follow identical optical paths through the wind tunnel so that
any loss in intensity due to the dust-laden airstream is equal
for both beams. The ratio of the intensity of the sample beam
to the reference beam yields the normalized specular reflectance
of the mirror independent of fluctuations in the laser beam intensity. The reflectance losses measured with this system have been
compared with specular reflectance losses measured with a laboratory bidirectional reflectometer at 633 nm and agreement has been
found to be within 0.013 reflectance units over a loss range of
0.05-0.9 reflectance units. A typical reflectance loss curve as
a function of deposition time is shown in Figure 17 for a second-
surface silvered glass mirror (mirror A). This particular deposition was performed at a wind velocity of 20 MPH and a flux rate
of $\approx$1.2 x 10^5 particles/cm^2-sec. Note that the reflectance loss
is approximately a linear function of deposition time. An identical mirror (mirror B) was exposed to the same particle flux but
at a velocity of 25 MPH, also shown in Figure 17. At this higher

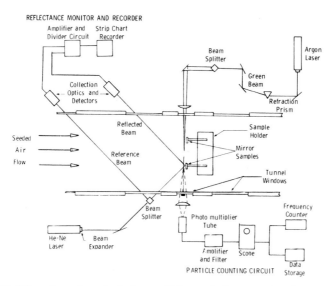

FIGURE 16.   Accelerated deposition wind tunnel with laser op-tics for monitoring particle flux and measuring real-time reflec-tance loss of exposed mirrors.

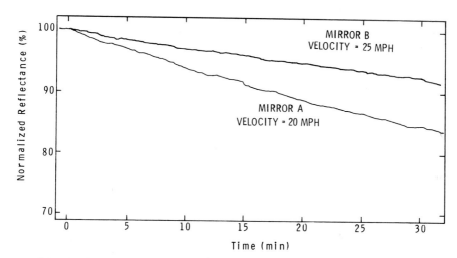

FIGURE 17.   Normalized real-time reflectance loss for mirrors exposed to a particle flux of $\approx 1.2 \times 10^5$ particles/cm^2-sec at wind velocities of 20 MPH (Mirror A) and 25 MPH (Mirror B).

velocity, the rate of reflectance loss decreased by a factor of ≈ 1.8. The variation of this single parameter shows that the reflectance loss rate is a rather strong function of wind velocity

under conditions of dry deposition. This effect results from the
increased kinetic energy of the particles at higher wind velocities
which causes the particles to rebound from the surface rather than
be held by the acting forces of adhesion.[15] A change in velocity
should have the greatest effect on the small particles which undergo the greatest deacceleration along the stagnation line of the
mirror. Increasing the kinetic energy of these particles raises
their energy above the effective "capture threshold" energy of the
mirror and results in a drop in the effective "sticking coefficient" of the small particles. The energy of a significant number
of the larger particles at low velocities already would exceed the
"capture threshold" of the mirror so that they would be less affected by a change in velocity. Initial measurements of the particle distribution on these two mirrors indeed show a significant
increase in the relative numbers of small particles ($r_p \lesssim 5$ µm)
for mirror A (V = 20 MPH) compared to mirror B (V = 25 MPH). The
1 µm particle density ratio of mirror A to mirror B was ≈ 3.1
while the ratio was ≈ 1.3 at a radius of 10 µm. These results
again point to the significant role of the small particles in
determining the reflectance loss of exposed mirrors. Experiments
are currently being conducted to extend the scope of these controlled depositions to include other mirror materials, coatings
and exposure conditions.

VII. CLEANING STRATEGIES

An understanding of the mechanisms of dust deposition and adhesion can lead to the development of techniques to maintain high
reflectivity under outdoor exposure conditions. Current cleaning
strategies can be generalized into the following categories:

(1) Keep dirt from settling and adhering to the surfaces.

(2) Wash off dirt with water or low surface energy detergent-
type solutions before strong chemical or mechanical
bonding can develop.

(3)   Wash off dirt with chemically or mechanically active
      cleaning techniques capable of breaking the chemical
      and mechanical bonds that have developed.

(4)   Modify the surface so that strong bonding cannot develop.

The above strategies can be divided generally into either
active or passive cleaning methods. Active cleaning methods
(strategies 2 and 3) are labor intensive techniques which can have
serious economic restrictions on the operation of a solar power
system, while passive techniques (strategies 1 and 4) are primarily capital intensive and can possibly result in lower, long-
range cleaning costs. Currently, strategies 1 and 4 are being
investigated as possible approaches to the soiling problems, en-
compasing such techniques as ultrasonic vibration, electrostatic
biasing and antistatic, antisoiling surface coatings.[22,23]

Investigation of strategies 2 and 3 has indicated that glass
mirrors can be cleaned to within 2% and acrylic mirrors to within
8% of their initial reflectance using a high-pressure (1000 psi)
tap water spray. In locations containing hard water, a final rinse
with deionized water or tap water containing a sheeting agent may
be required. Mechanically or chemically active cleaning is required to restore 100% of the initial mirror reflectance. However
it is not clear if there may be some long-term buildup of nonremov-
able soil or degradation of the mirror surface due to cleaning.[24]

Preliminary tests using conducting oxide coatings, [(SnO_2)
coupled with an electrostatic field] have resulted in the reduction
of dust accumulation during wind tunnel exposures.[14] More extensive experiments are planned using this technique to characterize
the independent effects of the coatings and the applied fields on
the rate of dust accumulation and the particle size distributions.
Eventually this technique will be applied to field test experiments.

VIII.   CONCLUSIONS

The accumulation of dust and the resulting loss in specular reflectance of exposed mirrors is a complex function of mirror material, weather conditions, geographical location and operational
methods. Some general conclusions based on natural and artificial
soiling of solar mirrors are:

(1)   Specular reflectance of a freshly exposed mirror under-
      goes an initial rapid drop (0.0085 reflectance units/day
      from 2-day cycle exposure) followed by a decreasing loss
      rate as the accumulated dust level increases.

(2)   The long-term reflectance loss of uncleaned silvered
      glass mirrors in Albuquerque is approximately 0.10-0.15
      reflectance units with large fluctuations about the
      average. Similar data at other locations and for other
      materials are needed.

(3)   Increased cleaning frequency raises the average reflec-
      tance of the mirror.

(4)   Inverted or vertical storage of the mirrors can signi-
      ficantly reduce the rate of dust accumulation.

(5)   The effect of weather on the specular reflectance of a
      mirror depends on the mirror's level of dust accumulation.

(6)   Dust accumulated upon exposure in the Albuquerque area
      results in wide-angle scattering of the incident light.
      The effect on the specular reflectance is primarily to
      reduce the intensity of the reflected beam while essen-
      tially maintaining the shape of its intensity profile.

(7)   Dust accumulated in the Albuquerque area results in little
      absorption. The specular reflectance loss to hemispheri-
      cal reflectance loss ratio is approximately 5 to 1.

(8)   Scattering caused by accumulated particles increases with
      decreasing wavelength and increasing level of soiling,
      with the scattering amplitude increasing below 400 nm,

and with the wavelength dependence of dust particle scattering independent of particulate concentration.

(9)   Small particles ($0.3 \ \mu m \lesssim r_p \lesssim 1 \ \mu m$) are the most significant source of scattering for the solar spectrum.

(10) The concentration of small particles ($r_p \lesssim 5 \ \mu m$) tends to increase more rapidly than the concentration of larger particles for mirrors exposed to natural weathering.

(11) Decreasing wind velocity increases the relative rate of accumulation of small particles ($r_p \lesssim 5 \ \mu m$).

(12) Surface coatings and electrostatic biasing can possibly reduce the rate of dust accumulation.

The development of any technique to reduce the rate of soiling of exposed solar mirrors must necessarily involve the optimization of both the operation and design of the mirrors. Long term field test studies will help determine the eventual technique or combination of techniques used.

REFERENCES

1.   Friedlander, S. K.,"Smoke, Dust and Haze: Fundamentals of Aerosol Behavior"(John Wiley and Sons, New York, 1977).

2.   Van de Hulst, H. C.,"Light Scattering by Small Particles" (John Wiley and Sons, New York, 1957).

3.   Freese, J. M., Effects of Outdoor Exposure on the Solar Reflectance Properties of Silvered Glass Mirrors, Sandia report SAND 78-1649, Sept. 1978.†

4.   Arvizu, D. E., "The Solar Central Receiver Test Facility Heliostat Development," presented at Solar Energy Research Institute International Symposium on Concentrator Solar Collection Technology (June 14, 1978), Denver, CO.

5.   Pettit, R. B., *Solar Energy 19*, 733 (1977).

6.   Freese, J. M., "Effects of Outdoor Exposure on the Solar Reflectance Properties of Silvered Glass Mirrors,"

presented at the 1979 International Solar Energy Society
Congress (May 28 - June 1, 1979), Atlanta, GA.

7.   Forman, S. E., "Endurance and Soil Accumulation Testing
     of Photovoltaic Modules at Various MIT/LL Test Sites,"
     Report #COO-4094-23 (September, 1978).†

8.   Dave King, Sandia Labs., personal communication.

9.   Pettit, R. B., Freese, J. M. and Arvizu, D. E., "Specular
     Reflectance Loss of Solar Mirrors Due to Dust Accumulation,"
     Proceedings of the Institute of Environmental Sciences
     Solar Seminar on Testing Solar Energy Materials and Systems
     (May 22-24, 1978), at the National Bureau of Standards,
     Gaithersburg, MD, pp. 164-168.

10. Edwards, D. K., et al., *J. Opt. Soc. Am.* *51*, 1279 (1961),
     and Beckman Instructions 1220-B (1962).

11. Taketani, H. and Arden, W. M., "Mirrors for Solar Energy
     Applications," Proceedings of the 1977 Annual Meeting
     American Section of ISES (June 6-19, 1977), Orlando, FL,
     Vol. 1, pp. 5-15 to 5-19, and Vitko, J., Optical Studies
     of Second Surface Mirrors Proposed for use in Solar Helio-
     stats, Sandia report SAND 78-8228 OUO, April 1978.

12. Huebner, R. H., Arakana, E. T., MacRae, R. A. and Hamm, R. N.,
     J. Opt. Soc. Am. *54*, 1434 (1964).

13. Meinel, A. B. and Meinel, M. P.,"Applied Solar Energy: An
     Introduction"(Addison-Wesley Publishing Co., Menlo Park, CA,
     1977).

14. Berg, R. S., Heliostat Dust Buildup and Cleaning Studies,
     Sandia report SAND 78-0510, March 1978.†

15. Zimon, A. D.,"Adhesion of Dust and Powder"(Plenum Press, New
     York, 1969).

16. Quantimet 720, Image Analyzing Computers Ltd., Subsidary of
     Metals Research Ltd., 40 Robert Pitt Drive, Monsey, NY 10952.

17. Gillette, D. A. and Blifford, Jr., I. H., *J. Geophys. Res. 79*,
     4068 (1974).

18. Quenzel, H., *J. Geophys. Res. 75,* 2915 (1970).

19. Blifford, Jr., I. H. and Ringer, L. D., *J. Atmos. Sci. 26,* 716 (1969).

20. Standardized Test Dust obtained from A. C. Spark Plug Div., GMC, P. O. Box 1001, Flint, MI   48501.

21. Lapp, M., Penny, C. M. and Asher, J. A., Application of Light-Scattering Techniques for Measurements of Density, Temperature, and Velocity in Gasdynamics, Aerospace Research Laboratories Report ARL 73-0045 (April, 1973).

22. Schumaker and Associates, Inc., "Preliminary Design Report on New Ideas for Heliostat Reflector Cleaning Systems," Sandia Report SAND 79-8181, Aug. 1979.

23. Electrostatic repulsion of dust using a corona discharge system has been investigated by Prof. Stuart A. Hoenig, Dept. of Electrical Engineering, Univ. of Arizona, Tucson, AZ   85721

24. McDonnell Douglass Astronautics Co., Final Report: Cleaning Agents and Techniques for Concentrating Solar Collectors, Sandia report SAND 79-1979, Sept. 1979.†

†Available from: National Technical Information Service (NTIS), U. S. Department of Commerce, 5285 Port Royal Road, Springfield, VA   22161.

CHAPTER 7

THE EMISSIVITY OF METALS[1]

A.J. Sievers

Laboratory of Atomic and Solid State Physics
and
Materials Science Center
Cornell University
Ithaca, New York

I. INTRODUCTION

Although radiant heat transfer processes have been studied
for many years, with entire textbooks and journals devoted to the
subject, the high temperature thermal radiative properties of
metals was not described correctly until 1978 (1). It is the
purpose of this chapter firstly to show what is wrong with the
early radiant heat transfer models of metals and secondly to
provide the physical insight behind the new model, which does
correctly describe the temperature dependent thermal radiative
properties of metals.

In 1915, Foote (2) calculated the thermal radiation emitted
normally from a heated metal surface using the Planck radiation
law. By comparing this amount to that expected for a black body,
he obtained an expression for the total normal emissivity of the
metal. Foote's equation, which has been only slightly modified
in the intervening years, is (3):

[1]*This work was supported by the Solar Energy Research Institute Contract No. XH-9-8158-1. Additional support was received
from the National Science Foundation under grant No. DMR-76-81083
through the Cornell Materials Science Center. MSC Report #4190.*

$$\varepsilon_{N} = 0.578(\rho T)^{1/2} - 0.178(\rho T) + 0.0584(\rho T)^{3/2} \, , \tag{1}$$

where ρ is the resistivity of the metal in ohm-cm and T is the
absolute temperature.

Somewhat later, Davisson and Weeks (4) compared the amount of
thermal radiation emitted at all angles from a heated platinum
surface with their calculated values using Planck's law. This
thermal radiation expression of Davisson and Weeks, in which the
hemispherical emissivity also is written as a power series in the
square root of the metal resistivity times the temperature is (3):

$$\varepsilon_{H} = 0.766(\rho T)^{1/2} - [0.309 - 0.0889 \ln(\rho T)]\rho T$$

$$- 0.0175(\rho T)^{3/2}. \tag{2}$$

The measured hemispherical emissivity of Pt versus temperature is fairly close to the values predicted by the Davisson-
Weeks expression (Eq. 2), as demonstrated in Fig. 1. The data
measured by Abbott (5) are represented by the dots and the calculated values from Eq. 2 are given by the dashed curve. The
measured temperature dependent d.c. resistivity (6) which is used
in Eq. 2 is represented by the lowest solid curve in Fig. 1. The
other curves are described later in the text.

Both Eqs. 1 and 2 have been used extensively in the literature since they relate the total emissivity to one characteristic
property of the metal, namely the d.c. electrical resistivity.
The series expansions are described in some detail in the classic
text by M. Jakob (7) and in the review by G. Rutgers (8). More
recently, they are reviewed in Touloukian's "Thermophysical Properties of Matter", Vol. 7 (3).

In 1965, Parker and Abbott (9) noted that the assumption
$\omega\tau << 1$, which had been used in all of the above derivations, was
not completely valid throughout the frequency region in which
thermal radiation was emitted. They calculated the total normal

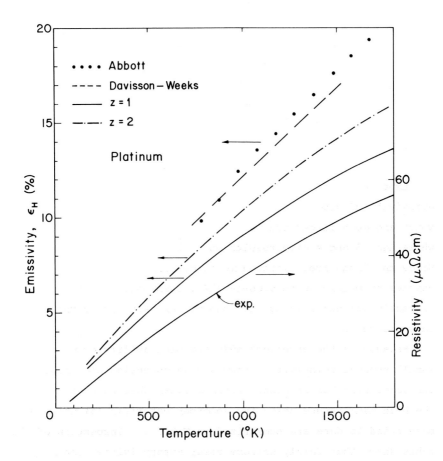

FIGURE 1. *Temperature dependence of the hemispherical emissivity of platinum. The data of Abbott (5) was represented by
the dotted curve and the Davisson-Weeks result by the dashed
curve. The dot, dashed and solid curves describe the emissivity
within the framework of the Drude model as detailed in the text.
z indicates the number of free electrons per atom, assumed for
each case. The temperature dependence of the d.c. resistivity,
in μΩcm, is also shown.*

and total hemispherical emissivity by simulating the infrared

properties of the metal with a series of exponential functions

and then expanded the result as a power series in $(\rho T)^{1/2}$. For

the total normal emissivity they found

$$\varepsilon_N = 0.578\, p_1\, (\rho T)^{1/2} - 0.178\, p_2\, (\rho T) + 0.0584\, p_3\, (\rho T)^{3/2} \qquad (3)$$

while for the total hemispherical emissivity they found

$$\varepsilon_H = q_1\, (\rho T)^{1/2} - (q_2 - q_3\, \ln\, \rho T)\rho T + q_4\, (\rho T)^{3/2} \qquad (4)$$

where p_1, p_2, p_3, q_1, q_2, q_3 and q_4 are nonlinear functions of the electron relaxation time.

Experimental studies of the temperature dependence of the emissivity of good conductors shows that Eqs. 1 and 2 give results which are much larger than experimentally measured values (10,11), while Eqs. 3 and 4 give results which are much smaller (1,10,11). Since the Drude free electron model describes the infrared properties of metals quite closely (12), then Eqs. 1-4 do not correctly reflect this model. One of our tasks is to understand this discrepancy.

Because of the excellent heat trapping ability of metals (small thermal emissivity), they are an essential component of the solar absorber in weakly focused solar thermal systems (13). The high temperature selective absorber materials (5) which have been tried to date are summarized in Table 1. Inspection of this table shows that fairly uniform solar absorptivities can be obtained with a variety of materials but that the hemispherical emissivities range from 0.07 to 0.50 at elevated temperatures. This large variation in the emissive properties of selective surfaces indicates that this physical parameter is not yet under control. Until 1978 one could not even determine a limit to the emissivity of a selective surface at a given temperature since the emissivity of a bare metal surface had not yet been properly characterized. One motivation, then, for obtaining accurate expressions for the emissivity of bare metals is to construct a solid foundation for the future determination of the minimum emissivities of the more intricate and complex surface configurations appropriate for selective radiators.

TABLE I.   High Temperature Selective Surfaces

Coating Material	Fabrication Technique	Solar Absorptivity $(T^{\circ}C)$	Hemispherical Emissivity $(T^{\circ}C)$	Stability	Ref.
GE	Gas evap.	0.91	0.5(350)	–	15
Si(on AG)	CVD+AR coat	0.80	0.07(500)	500	16
PbS	Vac. Deposit.	0.98	0.3(300)	300	15
Al_2O_3+Ni	Evap. Composite	0.94	0.35(500)	600	17
Al_2O_3+Pt	Evap. Composite	0.94	0.3(500)	600	17
CrO_x-Cr	Electrodeposit	0.94	0.2(350)	–	18
Cr_2O_3+Co	Plasma spray	0.90(800)	0.5(800)	800	15
NiS-ZnS	Electrodeposit	0.88	0.16(300)	<220	15
WC+Co	Plasma spray	0.95(600)	0.4(600)	>800	15
Al_2O_3/Mo/ Al_2O_3	Vac. Evap.	0.85-0.95	0.11(500)	>500	19

II.   SPECTRAL EMISSIVITY

The infrared properties of good conductors are determined by intraband transitions. These optical transitions, which are characteristic of a free electron metal, involve the excitation of an electron from one state to another in the same band. Experimentally, it has been found that the free electron model is a good first approximation for most metals in the infrared. Drude (20) was first to treat a metal as a free electron gas and, although the logic behind the model has changed over the years, Drude's name is still associated with it. Here we shall only review the details required to obtain the spectral emissivity. For a more complete discussion, the reader is referred to Reference (21).

The metal is modeled by a uniform distribution of zn electrons/cm^3 where n is the number of atoms per cm^3 and z is the number of free electrons per atom and an equal and opposite

uniform positive charge density of infinite mass. Since the
velocity of light is much larger than the electron velocity, only
the long wavelength limit need be considered in describing the
interaction of the electron polarization with electromagnetic
radiation. A shift of zn electrons by a position vector $\vec{r}$ with
respect to a positive background produces a polarization $\vec{p}$, where

$$\vec{p} = zne\vec{r} . \tag{5}$$

This polarization can be related to the applied electric field by
means of Newton's second law, which for no damping of the electron motion is

$$nm\ddot{\vec{r}} = (zn)e\vec{E} \tag{6}$$

The polarization and E field must also satisfy Maxwell's equations, which can be written in terms of the wave equation:

$$- \nabla^2 \vec{E} + \frac{1}{c^2}\frac{\partial^2 \vec{E}}{\partial t^2} - \frac{4\pi}{c^2}\frac{\partial^2 \vec{p}}{\partial t^2} = 0. \tag{7}$$

To solve this equation we try travelling waves

$$\vec{E} = \vec{E}_o e^{i(\vec{k}\cdot\vec{r} - \omega t)} , \qquad \vec{p} = \vec{p}_o e^{i(\vec{k}\cdot\vec{r} - \omega t)} \tag{8}$$

which give a solution when

$$k^2 = \frac{\omega^2}{c^2}\varepsilon_m \tag{9}$$

where

$$\varepsilon_m = 1 - \frac{\omega_p^2}{\omega^2} \tag{10}$$

with $\omega_p^2 = \frac{4\pi zne^2}{m}$. Although the dielectric function in Eq. 10 is
real, the real and imaginary parts of the dielectric function are

not independent but are related by a Kramers-Kronig transform (22). In general

$$\hat{\varepsilon}_m = \varepsilon_{m1} + i\varepsilon_{m2} \tag{11}$$

and if

$$\varepsilon_{m1} = 1 - \frac{\omega_p^2}{\omega^2} , \tag{12}$$

then for no damping

$$\omega\varepsilon_{m2} = \frac{\pi}{2} \omega_p^2 \, \delta(\omega) . \tag{13}$$

The frequency dependency of ε_{m1} and ε_{m2} are shown in Fig. 2(a). Both curves are characterized by one parameter of the metal, ω_p, which is directly related to the electron density.

An electromagnetic wave of frequence ω incident on a metal responds differently depending on whether $\omega < \omega_p$ or $\omega > \omega_p$. For $\omega < \omega_p$, k in the metal is imaginary and a propagating wave cannot exist. Since absorption has been excluded, a wave incident on the metal must be completely reflected. For $\omega > \omega_p$, k is real and a wave can propagate in the metal, part of the incident intensity is transmitted and part is reflected.

In general, scattering of the electrons in the metal cannot be ignored. This effect is usually included by introducing a scattering time, τ, between electron collisions. Newton's second law now becomes

$$n m \ddot{\vec{r}} + \frac{n m \dot{\vec{r}}}{\tau} = z n e \vec{E} . \tag{14}$$

The dielectric function which describes the wave in the metal is

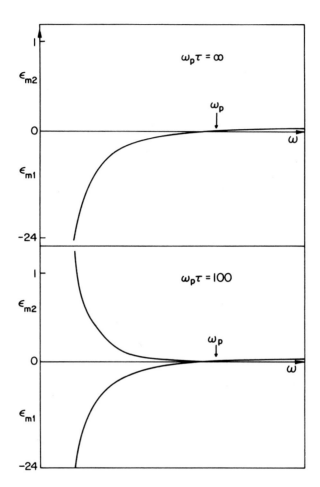

FIGURE 2. *Frequency dependence of the dielectric function of a Drude electron model. Both the real and imaginary parts are shown. (a) No damping ($\omega_p\tau = \infty$). (b) Damping ($\omega_p\tau = 100$). This value is appropriate for a good conductor such as copper.*

$$\hat{\varepsilon}_m = 1 - \frac{\omega_p^2}{\omega(\omega + \frac{i}{\tau})} \quad , \tag{15}$$

where

$$\varepsilon_{m1} = 1 - \frac{\omega_p^2 \tau^2}{1 + \omega^2 \tau^2} \tag{16}$$

and

$$\varepsilon_{m2} = \frac{\omega_p^2 \tau^2}{\omega\tau(1 + \omega^2\tau^2)} \quad . \tag{17}$$

The frequency dependences of ε_{m1} and ε_{m2} are shown for $\omega_p\tau = 100$ in Fig. 2(b). Two parameters, ω_p and τ, completely determine the infrared response of the metal.

Equation (15) is the dielectric function of the Drude model for one kind of an electron. For more than one kind of free carrier the model is readily generalized:

$$\hat{\varepsilon}_m = 1 - \sum_j \frac{\omega_{pj}^2}{\omega(\omega + \frac{i}{\tau_j})} \quad . \tag{18}$$

Once the dielectric function has been found, the reflection coefficient of the electromagnetic wave can be obtained by matching boundary conditions on the $\vec{E}$ and $\vec{H}$ fields at the air-metal interface. This is usually done by writing the dielectric function in terms of the optical constants, n and κ, or in terms of the surface impedance, r and x,

$$\hat{\varepsilon}_m = (n + i\kappa)^2 = \frac{1}{(r - ix)^2} \tag{19}$$

then the boundary conditions give

$$R(\omega) = \frac{(n - 1)^2 + \kappa^2}{(n + 1)^2 + \kappa^2} = \frac{(1 - r)^2 + x^2}{(1 + r)^2 + x^2} \tag{20}$$

where both r and x are normalized with respect to the impedance of free space.

In the infrared region for metals thicker than $\sim 400 \overset{\mathrm{o}}{\mathrm{A}}$, the transmission coefficient of the electromagnetic wave can be set equal to zero. The spectral absorptivity, $A(\omega)$, which is the fraction of incident energy adsorbed by the metal, then is

$$A(\omega) = \frac{4n}{(n+1)^2 + \kappa^2} = \frac{4r}{(1+r)^2 + x^2} \qquad (21)$$

also in the infrared n, $\kappa \gg 1$ and r, $x \ll 1$ so

$$A(\omega) \cong \frac{4n}{n^2 + \kappa^2} = 4r . \qquad (22)$$

By Kirchhoff's law, the spectral absorptivity is equal to the spectral emissivity, so Eq. (21) also describes the normal spectral emissivity of a Drude metal. Notice that in one representation both optical constants are required to obtain the spectral emissivity, while in the second, only the surface resistance need be calculated. As long as $\omega_p^2 \gg \omega^2$, Eq. (22) gives

$$A(\omega) = \varepsilon(\omega) = (\frac{2}{\omega_p \tau}) (2\omega\tau)^{1/2} \left[(1 + \omega^2\tau^2)^{1/2} - \omega\tau \right]^{1/2} . \qquad (23)$$

At small frequencies where $\omega\tau \ll 1$, the degree of electron polarization is controlled by τ (see Eq. 14) and the spectral emissivity is

$$\varepsilon(\omega) = (\frac{2}{\omega_p \tau}) (2\omega\tau)^{1/2} \qquad (24)$$

which is the Hagen-Rubens limit (1). At large frequencies, $\omega\tau \gg 1$, the accelerating electric field reverses the electron motion before a scattering event can occur so it is the inertia of the electrons which limits the electron polarization in this frequency range. The spectral emissivity approaches

$$\varepsilon(\omega) = \frac{2}{\omega_p \tau} , \qquad (25)$$

which is the frequency independent Mott-Zener limit. Equation

(23) and its limiting value, Eqs. 24 and 25, all depend on temperature because τ depends on temperature. For temperatures above room temperature, τ is determined by the electron-phonon interaction. Since the number of phonon scatterers in this temperature regime is proportional to $k_B T$,

$$\frac{1}{\tau} \sim T .\qquad (26)$$

A more exact expression can be obtained from the d.c. resistivity. Because the electron density, or plasma frequency, is almost temperature independent, the temperature dependence of τ is directly reflected in the temperature dependence of the d.c. resistivity, ρ, where

$$\frac{1}{\tau(T)} = r = \frac{\omega_p^2}{4\pi} \rho(T) \qquad (27)$$

where r is the relaxation frequency.

The normal spectral emissivity given by the Drude model with parameters appropriate to copper at 600°K is shown in Fig. 3. At the relaxation frequency, r, the emissivity is over 90% of its final, asymptotic value. Frequencies are given in cm^{-1} where $\omega = 2\pi c \tilde{\nu}$ so $\tilde{\nu} = \frac{1}{\lambda}$, where λ is the electromagnetic wavelength. As the metal temperature is increased, the frequency where $\omega\tau = 1$ moves to large frequencies in direct proportion to the temperature.

III. TOTAL EMISSIVITY

The radiant spectral power emitted by a surface element of a black body in a given direction per unit frequency interval per unit solid angle is given by Planck's law

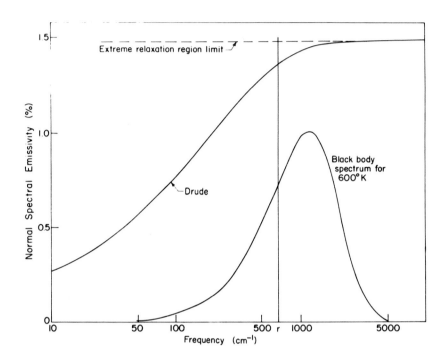

FIGURE 3. The normal spectral emissivity of the Drude free electron model as a function of frequency. The parameters are chosen to illustrate the infrared properties of copper at 600°K. Also shown is the spectral distribution appropriate to a 600°K black body radiator. The ordinate for this curve is not shown.

$$L_\omega = \frac{1}{\pi} \left(\frac{1}{2\pi c}\right)^2 \frac{\hbar \omega^3}{\left(\exp\left(\frac{\hbar \omega}{k_B T}\right) - 1\right)} \tag{28}$$

where ω is the angular frequency, c the velocity of light in vacuum, $\hbar$ is Planck's constant divided by 2π and k_B is Boltzmann's constant. The black body spectrum, L_ω, at a temperature of 600°K, is shown in Fig. 3. The abscissa is the frequency axis but the ordinate is an arbitrary scale. The area under the black body spectrum at temperature T is proportional to T^4 and the centroid frequency of the spectrum, $<\omega>$, is proportional to temperature.

A. Normal

The total normal emissivity is defined as the spectral emissivity weighted by the black body spectrum divided by the spectral emissivity of a black body weighted by the black body spectrum. By definition, the spectral emissivity of a black body is one, so

$$\varepsilon_N = \int_0^\infty \varepsilon(\omega) L_\omega \, d\omega \Bigg/ \int_0^\infty L_\omega \, d\omega . \qquad (29)$$

In general, the folding of the spectral emissivity together with the black body spectrum can only be carried out numerically; however, there are two limits where the integration can be done analytically. One of these limits is for $\omega\tau << 1$ (Eq. 24). Foote (2) derived Eq. 1 in this limit. The other limit (1) is $\omega\tau >> 1$ (Eq. 25). We must decide if either limit is appropriate for describing the total normal emissivity of good conductors.

Since the spectral emissivity is weighted by the black body spectrum, the appropriate emission frequency is that of the black body centroid:

$$<\tilde{\nu}> = 2.665 \ T(^\circ K) \ cm^{-1}. \qquad (30)$$

In Fig. 3, the centroid frequency is at 1600 cm^{-1}, which is much larger than the relaxation frequency, r, of copper shown in the same figure. At least at $600^\circ K$, the $\omega\tau >> 1$ limit should be more appropriate than the $\omega\tau << 1$ limit. What about at higher temperatures? By Eq. 30, the black body centroid is proportional to temperature, but by Eq. 27 so is the relaxation frequency of the metal. Therefore, although both quantities change, they change in such a manner that their ratio remains fixed. If the $<\omega>\tau >> 1$ limit is satisfied at room temperature then it will remain satisfied at still larger temperatures.

The high frequency limit of the spectral emissivity (Eq. 25) is independent of frequency and can be factored out of the integral given in Eq. 29. The remaining integrals cancel so the total normal emissivity becomes

$$\varepsilon_N = \frac{2}{\omega_p \tau} \qquad \text{for} \quad <\omega> \tau > 1 . \qquad (31)$$

In terms of the d.c. resistivity, this expression becomes

$$\varepsilon_N = \frac{\omega_p}{2\pi} \rho \qquad \text{for} \quad <\omega> \tau > 1 . \qquad (32)$$

This expression should give an upper limit to the total normal emissivity because the frequency dependent spectral emissivity shown in Fig. 3 has been replaced by its high frequency limit. For good conductors, such as Cu or Al, Eq. 32, not Eq. 1, gives the Drude value for the total normal emissivity.

B. Hemispherical

The total hemispherical emissivity of a metal is defined as the total power radiated by the metal into a hemisphere compared to the total power radiated by a black body into the same solid angle (3), so

$$\varepsilon_H = \int_0^\infty \int_0^{\pi/2} \varepsilon(\omega,\theta) \cos\theta \cdot \sin\theta \ d\theta \ L_\omega \ d\omega \Big/ 1/2 \int_0^\infty L_\omega \ d\omega \qquad (33)$$

where θ is the polar angle measured with respect to the metal normal, the cosine θ describes the projected area radiating at that angle and

$$\varepsilon(\omega,\theta) = 1/2[\varepsilon_{TM}(\omega,\theta) + \varepsilon_{TE}(\omega,\theta)] \qquad (34)$$

is the frequency and angular dependent emissivity for the two polarizations of the radiation. The azimuthal integral is the same in the numerator and denominator of Eq. 33 and cancels. $\varepsilon_{TE}(\omega,\theta)$ is the spectral emissivity for plane polarized radiation with the plane of polarization normal to the plane of incidence (the electric field is parallel to the metal plane, usually called transverse electric). For metals, $\varepsilon_{TE}(\omega,\theta) = \varepsilon(\omega) \cos\theta$, so by Eq. 33,

$$\frac{\varepsilon_{H(TE)}}{2} \cong <\cos^2\theta> \varepsilon_N = \frac{1}{3}\varepsilon_N.$$

For the other polarization the electric field is perpendicular to the metal plane and the emissivity is $\varepsilon_{TM}(\omega,\theta)$. This polarization has an anomalous angular dependence because at some angle, Brewsters angle, the emissivity goes from the small value characteristic of a metal to 1. For good conductors, this angle occurs near $90°$. In this limit, the emissivity is

$$\varepsilon_{TM}(\omega,\theta) \cong \varepsilon(\omega)(\cos\theta)^{-1}$$

so

$$\frac{\varepsilon_{H(TM)}}{2} \cong <1> \varepsilon_N = \varepsilon_N. \tag{35}$$

Because of the smaller effective area radiating at large angles, the total radiated power for TM waves at a given angle is independent of angle. The total hemispherical emissivity is

$$\varepsilon_H = \frac{\varepsilon_{H(TM)} + \varepsilon_{H(TE)}}{2} \cong \frac{4}{3}\varepsilon_N. \tag{36}$$

Davisson-Weeks (4) derived Eq. 2 in the limit $\omega\tau << 1$. It is now evident that $<\omega>\tau >> 1$ is more appropriate for good conductors. From Eqs. 32 and 36, the Drude value for the total hemispherical emissivity is

$$\varepsilon_H \cong \frac{2}{3\pi} \omega_p \rho .$$

(37)

This result should be compared with Eq. 2.

Whereas Eqs. 1 and 2 have only one parameter, the d.c. resistivity, Eqs. 32 and 37 have two, the d.c. resistivity and the plasma frequency or electron density. The second parameter may also be derived from a non-optical measurement by making use of the Hall constant (23):

$$R_H = - \frac{1}{(zn)ec}$$

(38)

so

$$\varepsilon_N = \frac{2.49 \times 10^{-8}}{|R_H|^{1/2}} \rho_\mu$$

(39)

and

$$\varepsilon_H = \frac{4}{3} \varepsilon_N$$

(40)

where $|R_H|$ is measured in m^3/coulomb and ρ_μ in $\mu\Omega$cm. For most metals $|R_H|$ and ρ_μ can be looked up in Tables (6).

Equations 39 and 40 have been derived from the same free electron model that was used to obtain Eqs. 1 and 2. For good conductors with $\rho \sim T$, both sets of equations predict a linear temperature dependence of the total normal and hemispherical emissivity. The main difference between the two sets of equations is that the coefficients of Eqs. 1 and 2 are much larger than those of Eqs. 39 and 40 (for copper, Eq. 1 is $\sim$ 2.6 (Eq. 39)). But from our derivation, both Eqs. 39 and 40 represent upper limits to the total emissivities, hence for good conductors Eqs. 1 and 2 should be poor approximations.

C. Model Corrections

Equations 1 and 2 do give values of the total emissivities which are larger than experimental values (10,11) but Eqs. 39 and 40 give values which are smaller. Since these latter two equations were derived as upper limits to the emissivities of the Drude model, this model as it stands does not provide an accurate description of the emissivity of a good conductor.

To correctly describe the infrared emissivity of metals one must take account of the properties of metal surfaces. Even if the electrons do not scatter in the bulk they do scatter diffusely from the surface (24) (remember, the conduction electron wavelength is a few atomic spacings). For these electrons there is one collision with the surface in the time needed by the electron to cross the electromagnetic skin depth ($\delta = \dfrac{c}{\omega_p}$ for $\omega\tau >> 1$) to the surface and back again. Thus, the surface scattering time, τ_s, is

$$\tau_s \overset{\bullet}{=} \frac{2c}{\omega_p} / v_F$$

so

$$\varepsilon_{Ns} = \frac{2}{\omega_p \tau_s} = v_F/c \; .$$

A more accurate calculation of the spectral emissivity associated with diffuse scattering gives (24,25)

$$\varepsilon_{Ns} = 3 \, v_F/4c \; . \tag{41}$$

Next we assume that bulk and surface scattering processes are independent so

$$\tau_e^{-1} = \tau_b^{-1} + \tau_s^{-1} \; .$$

The total normal surface scattering including diffuse surface
scattering is then

$$\varepsilon_N = \frac{2}{\omega_p \tau_e} = \frac{2}{\omega_p \tau} + \frac{3v_F}{4c} . \tag{42}$$

For the free electron model both the Fermi velocity and the Hall
coefficient are related to the electron density, so Eq. 42 can be
expressed in terms of ρ_μ and R_H:

$$\varepsilon_N = \frac{2.49 \times 10^{-8}}{|R_H|^{1/2}} \rho_\mu + \frac{3.55 \times 10^{-6}}{|R_H|^{1/3}} \tag{43}$$

where again

$$\varepsilon_H = \frac{4}{3} \varepsilon_N . \tag{44}$$

The values obtained from Eqs. 43 and 44 are in good agreement
with recent experimental studies on copper and aluminum (10,11).
In Fig. 4 we show the experimental temperature dependence of the
hemispherical emissivity of copper. The temperature dependence
according to Eqs. 43 and 44 is given by the solid curve. The
agreement between theory and experiment is quite satisfactory for
free electron-like metals.

For the transition elements such as tungsten $\omega <\tau> \sim 1$ and
the total emissivity can only be calculated by a numerical integration (1,26). In addition, one needs to know not only the
temperature dependence of the free electron properties, but also
the temperature dependence of the infrared properties of the
interband transitions. As this process has not been studied in
detail, we cannot incorporate it into our emissivity calculation.
By ignoring this physical process, the emissivity calculated from
Eq. 44 will be smaller than the true emissivity. As an example,
we compare the experimentally measured (27) emissivity of tungsten

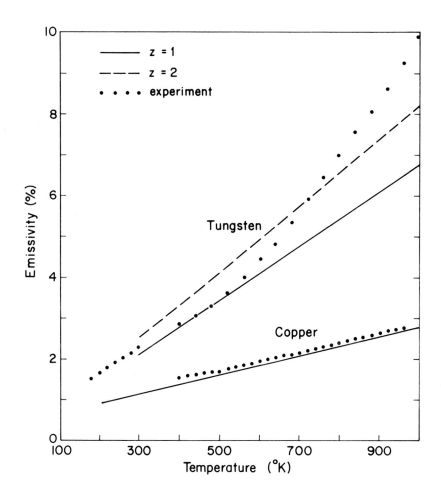

*FIGURE 4. Temperature dependence of the hemispherical emis-
sivity of copper and tungsten. The solid circles show the experimental data (10,27). The solid lines show the calculated emissivities for the Drude model with one electron per atom (z = 1)
and the dashed line, the emissivity for two electrons per atom
(z = 2).*

with the values calculated from the d.c. conductivity (6).   In

Fig. 4 the experimental data for tungsten are represented by the

solid circles. The solid line is calculated from the d.c. resis-

tivity, with an assumed one free electron per atom, while the

dashed curve shows the same calculation for two free electrons

per atom. Both for copper and for tungsten, the one free electron per atom model provides a better fit to the data. At
temperatures above 600°K, (corresponding to frequencies above
1600 cm^{-1}) interband effects become important for tungsten and
the emissivity becomes larger than that given by the Drude model.

IV. CONCLUSIONS

For metals which are not particularly good conductors, the
derivation of the emissivity must start with Eq. 21 not Eq. 22,
which we have used here and it is no longer possible to obtain a
simple analytic solution for the emissivity. To see how well Eq.
43 describes the emissivity of all metals, we have performed
exact calculations of both the total normal and hemispherical
emissivities and divided these numerical values by the values
obtained from Eq. 43. The comparison is shown in Fig. 5. For
the total normal emissivity the difference between the numerical
calculation and Eq. 43 is less than 1% for ρ_μ < 10 $\mu\Omega$cm and less
than 10% for ρ_μ < 100 $\mu\Omega$cm. Although the error is somewhat
larger for the hemispherical emissivity calculation, it is perhaps worth a ten or twenty percent error just to avoid the
numerical calculations involved.

For completeness we show in Fig. 6 for a number of elements
the numerically calculated hemispherical emissivity versus the
metal resistivity. The hemispherical emissivity ranges from 1%
for a good conductor to over 20% for the highest resistivity
metals. In every case except aluminum (3) and beryllium (2) the
metal is assumed to have one free electron per atom. The electron densities were obtained from Mott and Jones (28). This
assumption gives the smallest reasonable emissivity for the metal
and these curves represent lower bounds to the true hemispherical
emissivities.

This chapter has been devoted to the emissivity of pure
metals since the elements themselves provide the lowest thermal

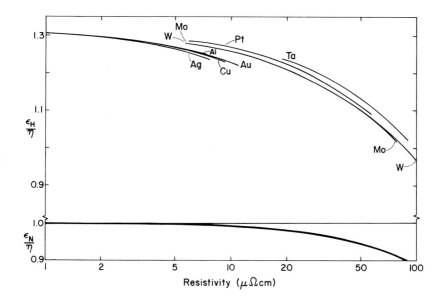

FIGURE 5. Two emissivity ratios versus d.c. resistivity.
In this figure, ε_N and ε_H are given by exact numerical calcula-
tions of the emissivity and η represents Eq. 43 in the text.
A small error discovered in Fig. 8 of Reference 1 has been
corrected here.

emissivities. Metal alloys which have large d.c. resistivities
due to disorder-induced electron scattering should have large
emissivities. If the scattering is made large enough so that
$<\omega> \tau << 1$ and the centroid of the black body spectrum is at a
frequency much less than the relaxation frequency, then the
Hagen-Rubens relation should provide a reasonable estimate of
the spectral emissivity and the Foote (2) and Davisson-Weeks (4)
equation follows. As the temperature is increased, however, the
black body centroid increases in frequency while the relaxation
time remains temperature-independent, and at some large temperature the material may still obey $<\omega> \tau > 1$ since τ is temperature
independent for disorder scattering.

An important alloy for solar applications is stainless steel
304. Recently, Roger et al. (29) have measured the temperature

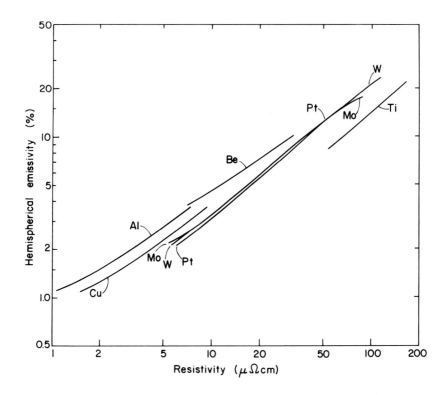

FIGURE 6. The total hemispherical emissivity as a function of d.c. resistivity for a number of elements. These curves are obtained from exact numerical calculations of the Drude model and define lower bounds for real metals. A small error discovered in Fig. 11 of Reference 1 has been corrected here.

dependence of its hemispherical emissivity from 340 to 1100°K. In Fig. 7 we compare these experimental data with calculated values obtained from various models. The Davisson-Weeks Equation (Eq. 2) gives results which are larger than the experimental values over the entire temperature range indicating that the Hagen-Rubens relation is not strictly valid in this temperature regime and indeed as expected at high temperatures the disagreement is larger than at low temperatures. By fitting the Parker-Abbott Equation (Eq. 4) to the data at 400°K, reasonably good

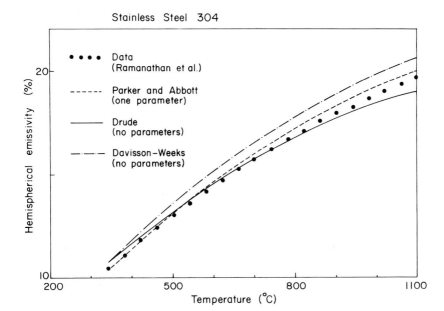

FIGURE 7.   *Temperature dependence of the hemispherical emis-*
sivity of stainless steel 304.   The solid dots show the experi-
mental data points of Ramanathan et al. (29).   The dashed curve
is calculated from Eq. 4 in the text with a one parameter fit to
the data.   The solid curve gives an exact Drude model calculation
as described in Reference 1.   The dot-dash curve is calculated
from Eq. 2 in the text.   All of these models use the same values
of the temperature dependent d.c. resistivity.

agreement is obtained over the entire range.   The solid curve in

Fig. 7 represents the Drude model as described in this paper.

The same temperature dependent d.c. resistivity is used as in

the other models but in addition the average of the electron

densities of the metal components is used here.   No free param-

eters are available and the agreement is quite satisfactory over

the entire temperature range.

To conclude this study of the emissivity of metals, let us

reexamine Fig. 1.   Data shown there on the hemispherical emis-

sivity of platinum was found to be in good agreement with the

Davisson-Weeks expression.   In this same figure, exact

calculations with the Drude model developed in the text for one
or two electrons per atom are also shown. As expected, neither
of these calculated cruves is large enough to fit the experimental data since the Drude model calculation does not include the
emissivity contribution from interband transitions and as such
represents a lower bound to the measured emissivity. Of course,
this same comment applied to the Davisson-Weeks expression. The
apparent agreement of Eq. 2 with the experimental data occurs
for two reasons: 1) because an incorrect limit was used to
describe the spectral emissivity of the metal, and 2) because
the influence of surface scattering was not included. Now that
we have obtained correct minimum values for the emissivity of
bare metals, it should be possible to extend this analysis to
find the minimum emissivity of more complex selective surfaces.

REFERENCES

1. Sievers, A.J., *J. Opt. Soc. Amer. 68,* 1505 (1978).

2. Foote, P.D., *Bull. Natl. Bur. Stand. 11,* 607 (1914-15).

3. "Thermophysical Properties of Matter" (Y.S. Touloukian, ed.),
    Vol. 7, p. 24a. Plenum Press, New York, (1970).

4. Davisson, C., and Weeks, J.R., *J. Opt. Soc. Am. 8,* 581
    (1924).

5. Abbott, G.L. in "Measurement of Thermal Radiation Properties
    of Solids", (J.C. Richmond, ed.), p. 293. NASA SP-31, (1963).

6. "American Institute of Physics Handbook", 3rd Edition,
    (D.E. Gray, ed.), p. 4-13, McGraw-Hill Book Co., New York,
    (1972).

7. Jakob, M., "Heat Transfer", p. 51. New York, Wiley, (1949).

8. Rutgers, G.A.W. in "Handbuch der Physik", (S. Flugge, ed.),
    Vol. 26, p. 154. Berlin, Springer-Verlag, (1958).

9. Parker, W.J., and Abbott, G.C., in "Symposium on Thermal
    Radiation of Solids", (S. Katzoff, ed.), p. 11, NASA SP-55,
    (1965).

10. Ramanathan, K.G., Yen, S.H. and Estalote, E.A., *Appl. Opt.*
     16, 2810 (1977).

11. Smalley, R., and Sievers, A.J., *J. Opt. Soc. Am. 68,* 1516
     (1978).

12. Bennett, H.E., and Bennett, J.M. in "Proceedings of the
     International Colloquium of Optical Properties", (F. Abeles,
     ed.), p. 175. North-Holland, Amsterdam, (1966).
     Bennett, H.E., Bennett, J.M., Ashley, E.J., and Motyka, R.J.,
     Phys. Rev. 165, 775 (1968).

13. Seraphin, B.O. in "Solar Energy Conversion--Solid State
     Physics Aspects", (B.O. Seraphin, ed.), ("Topics in Applied
     Physics", Vol. 31), Springer-Verlag, Berlin, (1979).

14. Lampert, C.M., *Solar Energy Materials 1,* 319 (1979).

15. Mattox, D.M., et al., Sandia Laboratories, Albuquerque, NM,
     SAND 75-0361 (July 1975). Also see Mattox, D.M., *J. Vac.*
     Sci. and Technol. 13, 127 (1976).

16. Seraphin, B.O., "Proc. Symp. on the Materials Science
     Aspects of Thin Film Systems for Solar Energy Conversion",
     Tucson, AZ, (July 1974). Also see Ref. 13.

17. Craighead, H.G., Bartynski, R., Buhrman, R.A., Wojcik, L.,
     and Sievers, A.J., *Solar Energy Materials 1,* 105 (1979).

18. McDonald, G.E., and Curtis, H.B., Lewis Research Center,
     Cleveland, OH, NASA TMX-71731, (May 1975).

19. Peterson, R.E., and Ramsey, J.W., *J. Vac. Sci. Tech. 12,*
     471 (1975).

20. Drude, P.K.L., "Theory of Optics", Dover, New York (1959).

21. Wooten, F., "Optical Properties of Solids", Chapter 4.
     Academic Press, New York (1972).

22. Kittel, C., "Elementary Statistical Physics", p. 207.
     New York, Wiley (1958).

23. Trotter, D.M., and Sievers, A.J., *Applied Optics,* (to be
     published 1980).

24. Holstein, T., *Phys. Rev. 88,* 1427 (1952).

25. Dingle, R.B., *Physica 19,* 311 (1953).

26. Sievers, A.J., *Solar Energy Materials 1*, 431 (1979)

27. Verret, D.P., Ramanathan, K.G., *J. Opt. Soc. Am. 68*, 1167 (1978).

28. Mott, N.F., and Jones, H., "The Theory of the Properties of Metals and Alloys", p. 268. Dover, New York, (1936).

29. Roger, C.R., Yen, S.H., and Ramanathan, K.G., *J. Opt. Soc. Amer. 69*, 1384 (1979).

CHAPTER 8

FUNDAMENTAL LIMITS TO THE SPECTRAL
SELECTIVITY OF COMPOSITE MATERIALS [1]

A.J. Sievers

Laboratory of Atomic and Solid State Physics
and
Materials Science Center
Cornell University
Ithaca, New York

I. INTRODUCTION

A variety of selective surfaces have been proposed and fabricated for solar thermal applications (1,2). Most of these
surfaces are constructed as absorber-reflector tandems. Either
the solar radiation is transmitted through a heat mirror to a
nonselective absorber or the solar radiation is absorbed by a
visibly dark mirror which has a small thermal emissivity. For
both proposed systems the end result is the same: the solar
radiation is absorbed and the thermal reradiation is suppressed.
Since the construction of these two types of spectrally selective
surfaces is very different it is important to determine which
approach has more potential. At first glance, this comparison
would appear to be a formidable task. A number of heat mirror
candidates are available (2). Each makes use of smooth or
textured metallic or semiconducting films to obtain the large

[1] This work was supported by the Solar Energy Research
Institute Contract No. XH-9-8158-1. Additional support was
received from the National Science Foundation under grant No.
DMR-76-81083 through the Cornell Materials Science Center.
MSC Report #4191.

transmissivity in the visible and the large reflectivity in the
infrared. A paramount advantage to this selective reflector-
absorber tandem is that the selective surface remains at room
temperature. If the solar transmissivity could be made large
enough, this approach would be preferred since material stability
problems would be greatly reduced.

An even larger number of candidates are available for dark
mirror applications. Composite media, interference films and
semiconducting films have all been proposed (1,2) for the solar
absorbing layer which is in direct contact with the infrared
reflecting mirror. The main advantage of this construction is
that the solar absorptivity can be made very large but the associated disadvantage is that the selective surface must run at an
elevated temperature. Because the thermal emissivity is temperature dependent, the optical properties of this absorber-reflector
tandem measured at room temperature will not provide the correct
information for a realistic assessment of the system's efficiency
at operating temperature.

The approach which is taken in this chapter in order to
assess the potential of each system is to assume in both cases
the best configuration consistent with physical constraints.
Although conducting micromeshes have been proposed to increase
the transmissivity of the heat mirror it is shown in the next
section that the intrinsic absorption produced by the texture and
finite conductivity of the mexh severely limits their usefulness.
For semiconducting meshes, it is shown that a 90% transmissivity
in the solar region is not compatible with a 90% reflectivity in
the thermal reradiation region while for aluminum and magnesium
meshes with submicron wire diameters, some potential improvement
in selectivity can be obtained over that found with metallic
films.

The procedure for obtaining the limits to the spectral selectivity of dark mirrors is somewhat more intricate since the
temperature dependent properties must be included. Starting from

the results of the previous chapter, where we determined the
smallest physically allowable emissivity for a given d.c. resistivity we go on to catalogue the different emissivity contributions as a solar absorber is constructed on the metal. By summing
all the different contributions, we find the smallest ε_H due to
intrinsic physical processes. No matter what composite film-
metal substrate combination is assembled, our results provide
definite lower limits to the hemispherical emissivity. These
limits are determined by the temperature dependent properties of
the metal, the influence of the index of refraction associated
with the composite film and the effect of thermal broadening on
the solar absorptivity profile. In the last section we compare
our findings for heat mirrors and dark mirrors and conclude that
dark mirrors have the most potential for intermediate temperature
weakly focus solar thermal systems.

II. TRANSPARENT HEAT MIRRORS

Tin oxide in the form of a thin film on glass has been used
for a number of years as a transparent electrode or a transparent
heater (4). Such films can be applied by spraying a mixture of
tin chloride with ethanol onto a glass substrate which has been
heated to 400°C. The tin chloride is transformed into the semiconducting tin oxide.

A tin oxide coating reduces the infrared emissivity of window
glass because of the large free carrier density in the semiconducting coating: however, the carrier density is still small
enough that the plasma frequency occurs below the solar spectrum.
Carrier concentrations of $10^{20}/cm^3$ have been achieved by adding
some antimony chloride or HF to the tin chloride solution before
spraying (6). The infrared emissivity is reduced to about 0.2.
An even smaller emissivity, $\varepsilon \sim 0.1$ has been obtained for indium
oxide doped with tin (7). Although these coatings may play an
important role in increasing the thermal insulation of window

glass (3,5), their usefulness for the intermediate temperature
heat mirror configuration is severely limited by the low visible
transmissivity. The coatings have a large index of refraction,
hence a large reflectivity, due to low lying interband transitions in the ultraviolet region of the spectrum. To increase the
transmissivity of these heat mirrors, several investigators (7,8)
have proposed that the continuous film be replaced by a conducting micromesh. Both metallic and semiconducting meshes have
been suggested. Such meshes already have been used successfully
as transmission filters in the far infrared and microwave region
(10).

An end view of the heat mirror selective configuration for a
tubular geometry is shown in Fig. 1. The mesh heat mirror is
attached to the inside of a glass housing enclosing a nonselective black absorber. If the black absorber is approximated by a
cylinder of radius r_a inside a mesh coated cylinder of radius r_m
then the effective emissivity is (11):

$$\varepsilon_{eff} = \left[\frac{1}{\varepsilon_a} + \frac{r_a}{r_m} \left(\frac{1}{\varepsilon_m} - 1 \right) \right]^{-1} \qquad (1)$$

where ε_a and ε_m are the emissivities of the absorber and the wire
mesh, respectively. For this equation to hold, the radiation is
assumed to scatter diffusely from either the absorber, the mesh
or both. If the absorber is much smaller than the mesh shell,
$r_a/r_m \ll 1$, ε_{eff} is controlled by ε_a; conversely, if $r_a/r_m \sim 1$,
the emissivity is determined by ε_m. Only in the second limit is
the heat mirror concept effective. Let us determine ε_m for a
mesh.

The three lengths required to define a mesh are given in
Figs. 2a and 2b. The mesh spacial period d is assumed to be
square. For the infrared, the wavelength of the radiation, λ, is
much larger than d so the mesh acts like a reflecting film. The

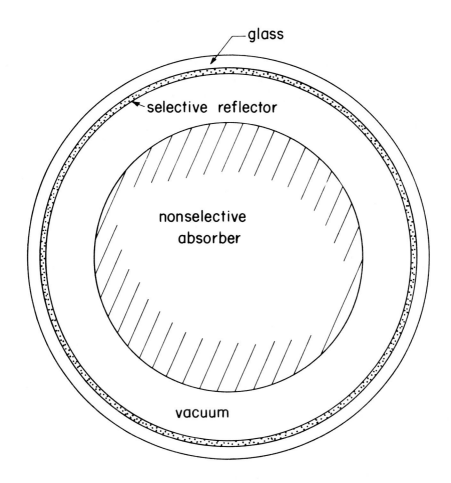

FIGURE 1.   End view of a heat mirror--non selective absorber tandem.   The assembly has cylindrical symmetry.   The conducting mesh is attached to the inside of the evacuatable glass container.

large reflectivity of both metal and semiconducting meshes is a consequence of shielding currents induced in the grid wires parallel to the incident infrared electric field.

To simplify the estimate of the emissivity of the mesh (12) in this long wavelength region, consider the problem of the parallel wire grid (not a mesh), with the infrared wave normally incident, the wire grids and infrared electric field perpendicular

to the figure and the infrared magnetic field, B_i in the plane
of the figure as shown in Fig. 2b. The grid wires are taken
thick compared with the electromagnetic skin depth, $(t > \delta)$, a
necessary requirement to obatin a reflectivity near unity.
Further, assume that the grid is characterized by transmission
and absorption coefficients, T and A, near zero and a reflection
coefficient, R, near one, since that is the only situation of
interest. The magnetic field near the grid can be characterized
by the amplitude of the incident and reflected fields, $B_i \sim B_r$,
since T = 0 and R = 1, as shown in Fig. 2b. The shielding
current is distributed within a skin depth of the surface of the
wire and for $d >> a$, the case of interest here, the current is
uniformly distributed around the wire. If we call this sheet
current density K_g, then the energy dissipation in the wire (per
unit length) around the perimeter is proportional to K_g^2 so

$$W_g \propto pK_g^2 \tag{2}$$

where p is the wire perimeter. The induced magnetic field is
proportional to the sheet current density around the perimeter so

$$B_r = B_i \propto pK_g . \tag{3}$$

For the uniform film, the surface current density, K_f, is
uniform and appears only on the incident face as shown in Fig.
2c. The energy dissipated in the surface (per unit length) is
proportional to K_f^2 so

$$W_f \propto dK_f^2 . \tag{4}$$

Again, the induced magnetic field is proportional to K_f itself
so

$$B_r = B_i \propto dK_g . \tag{5}$$

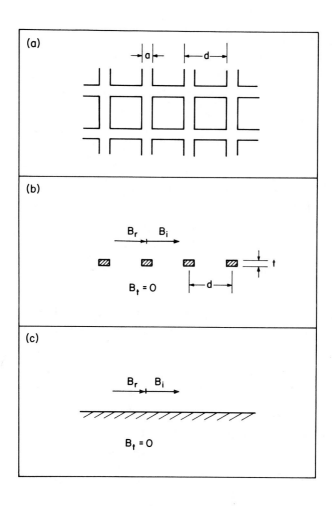

FIGURE 2.   (a) Top view of a conducting mesh.   (b) Geometrical configuration of the parallel wire grid for the long wavelength absorptivity calculation (end view).   The TM wave is
normally incident, the wire grids and the infrared electric
field are perpendicular to the figure and the infrared magnetic
field, B_i is in the plane of the figure as shown.   (c) Geometrical configuration of the uniform conducting film.   The fields
are the same as for 2(b).

The absorptivity of the grid is conveniently expressed in terms of that of the film, giving

$$\frac{A_g}{A_f} = \frac{W_g}{W_f} \sim (\frac{d}{p}) . \tag{6}$$

Taking further for convenience, $t << a$, Eq. (6) becomes

$$\frac{A_g}{A_f} \sim \frac{d}{2a} \tag{7}$$

so the grid absorptivity (or emissivity) is increased over that of the uniform film of the same material by one half of the geometric projection of the parallel wire grid or by roughly the full projection of the corresponding square mesh.

For the indium oxide doped with tin semiconducting film the homogeneous film absorptivity (3) at 10 μm wavelength is 9%. By Eq. (7) the long wavelength absorptivity is increased by a factor ≥ 1.5 when the film is converted to a particular mesh size. Experimentally, Fan (3) has found that the absorptivity of this mesh is increased by a factor of 1.9. It is clear from the theoretical calculations and from the experimental results that semiconducting meshes cannot provide the desired selectivity.

The theoretical spectral characteristics of perfectly conducting opaque meshes have already been treated in some detail (13). Although no transmission occurs for $\lambda >> d$, for $\lambda < d$ the transmission rapidly approaches that expected from geometric objects, i.e., the fraction of mesh width that is open (13). In this limit, the solar transmissivity is approximately $(1 - a/d)^2$. Thus, to obtain a 90% transmissivity through a thin aluminum mesh in the short wavelength limit, the absorptivity in the long wavelength limit (inductive shielding) limit by Eq. (7) must be 10 times the absorptivity of the uniform film. The 98.5% infrared reflectivity of an aluminum film implies only an 85% infrared

reflectivity of a thin mesh with a 90% geometrical aperture; however, for a square wire (a = t) the infrared reflectivity would
increase to 92%. In principle, the large absorptivity associated
with the mesh can be reduced still more by using wire with large
t/a ratios, although increasing t/a to much beyond unity decreases
the geometrical aperture of the system for light of non-normal
incidence. In any event, the limits are clear.

III. DARK MIRRORS

A. *Dark Metal*

An end view of the dark mirror selective configuration for a
tubular geometry is shown in Fig. 3. The dark mirror, which consists of a solar absorbing layer in direct contact with an
infrared mirror, must, because of its design, operate at elevated
temperatures. The temperature dependence of the optical properties plays an important role in determining the system performance. In this section, we indicate how one can determine the
smallest possible temperature dependent emissivity of these
selective dark mirrors.

To simulate the temperature dependent optical properties of
an ideal dark mirror, we coat a metal with a hypothetical material
which does not change the infrared emissivity of the metal but
causes the absorptivity of the composite structure to be 1
throughout the solar spectrum. We choose a steplike function
for this additional absorptivity but one which has a width of a
few $k_B T$ where k_B is Boltzmann's constant and T is the temperature
in °K. The high frequency tail of the black body thermal reradiation spectrum will overlap the low frequency end of the blurred
step function absorptivity and produce a contribution to the
emissivity of the composite structure. The total emissivity is
the sum of the total emissivity of the metal plus the total
emissivity of the coating. The temperature dependences of these
two emissivity components are shown in Fig. 4. The open circles

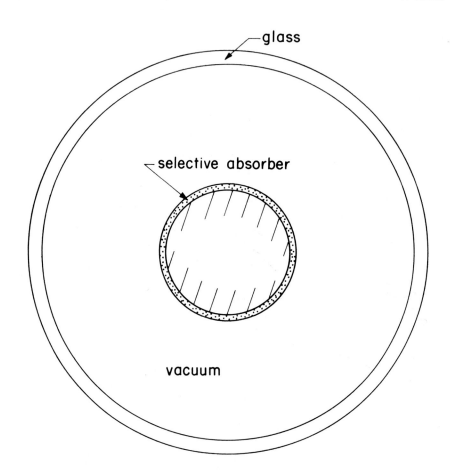

FIGURE 3. End view of a dark mirror absorber--reflector tandem. The assembly has cylindrical symmetry. The selective absorber is attached to the working fluid. The selective absorber diameter can be much smaller than the envelope diameter in contrast with the heat mirror assembly shown in Figure 1.

and the crosses give the experimental values of the total hemi-spherical emissivities of copper and tungsten, respectively. The solid lines show the calculated values for copper, tungsten and platinum, obtained from the previous chapter. The dashed curves in Fig. 4 portray the contribution to the total emissivity from the blurred absorptivity edge of the coating for different cutoff frequencies. To absorb most of the solar spectrum, the cutoff

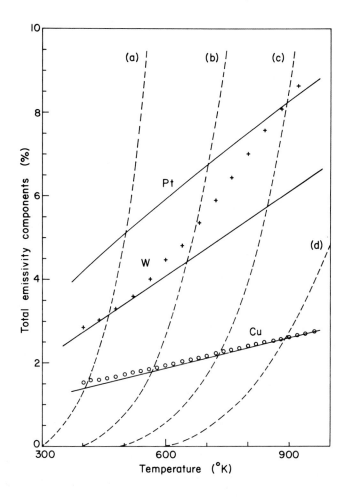

FIGURE 4. Total emissivity components for an ideal selective surface vs. temperature. The experimentally measured hemispherical emissivities are represented by discrete values: copper, ..., and for tungsten, xxx. The calculated hemispherical emissivity curves from the previous chapter are represented by solid lines. The dashed curves indicate the contribution to the emissivity from the thermally broadened absorption edge in the solar spectrum. The edge frequency is a) 3000 cm⁻¹, b) 4000 cm⁻¹, c) 5000 cm⁻¹, and d) 6000 cm⁻¹. The spectral shape of the assumed solar absorptivity is described in Reference 2.

should be near 2 microns wavelength (case c). This contribution,
when added to the Drude emissivity of the metal substrate produces a new lower bound to the hemispherical emissivity of the
selective surface. So far we have ignored the influence of the
optical properties of the selective coating on the infrared
emissivity. Including the dielectric properties increases the
hemispherical emissivity to still larger values.

B. *Metal Plus Dielectric Film*

A metal is a good reflector in the infrared because its optical constants are very different from those of free space. An
electromagnetic wave traversing the boundary between the two
media is almost completely reflected, irrespective of whether the
wave is incident from the metal or from the free space side. A
dielectric has optical constants in between those of the metal
and free space, and when inserted at the metal interface, it in
general provides a better match between the metal and free space
and the reflectivity is decreased. As an example of this matching, we consider first the interface between a half space of dielectric and a half space of metal (14). The total hemispherical
emissivity (from the metal to the dielectric) is calculated as a
function of the index of the dielectric by using the equations
developed in the previous chapter. The results for Drude copper
at three different temperatures are shown in Fig. 5. The emissivity of the interface increases almost linearly with the index,
showing that ever increasing amounts of radiation are coupled out
of the metal into the dielectric as the index is increased. The
emissivity for an index = 3 (Si) is twice as large as for an
index = 1.4 (MgF_2).

When the dielectric is made into a thin film, two additional
optical processes come into play: total internal reflection and
interference. Thermal radiation emitted from the metal into the
dielectric at angles larger than the dielectric free space critical angle remains trapped in the dielectric. This effect causes

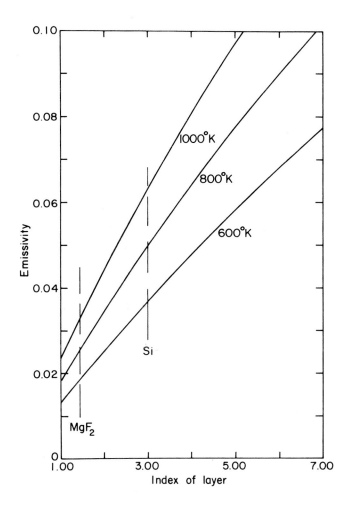

*FIGURE 5.   Hemispherical emissivity vs. index of refraction
for the interface between a half space of dielectric and a half
space of Drude copper.   The dielectric is assumed to be non-
absorbing.   (After Reference 14.)   The dashed vertical lines mark
the index of refraction of two dielectrics, MgF$_2$ and Si.*

a decrease in the emissivity. Thin film interference produces a
good match from the metal to free space for some wavelengths and
a poor match for an equal number of other wavelengths of the black
body spectrum. Since the reflectivity of the metal is typically

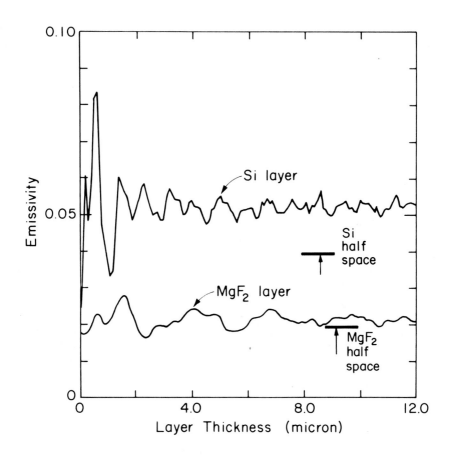

FIGURE 6. Hemispherical emissivity vs. film thickness for a composite composed of a dielectric layer on a Drude metal substrate. Upper curve: silicon on copper. Lower curve: MgF₂ or SiO₂ on copper. (After Reference 14.) The interference patterns go away when the films are made optically absorbing (see Fig. 8).

greater than 0.95 in the infrared, a good match can produce a much larger decrease in the reflectivity than a poor match can an increase. This effect causes the emissivity to increase. The net result of both effects is an increase in total emissivity, as shown in Fig. 6. The emissivity of the system as a function of film thickness is calculated for a variety of temperatures and the results are averaged to simulate operation as the solar flux

varies over the course of the day (14). The interference pattern
is due to the peak of the black body spectrum coinciding with
fractional multiples of the optical thickness of the coating.
Clearly, a dielectric film with a small index of refraction is
most desirable from the standpoint of reducing the emissivity of
the system.

C. Metal Plus Absorbing Film

We now consider the emissivity of a metal covered with an
absorbing dielectric film (15). The frequence dependence of the
absorption coefficient is illustrated in the lower part of Fig.
7. The value of α_o is determined by the film thickness d, since
α_o must be large enough so that all the solar energy is absorbed
in passing through the film.

Since the two parts of the complex refractive index of the
film are related by a Kramers-Kronig integral the contribution to
the real part of the film index, $n(\omega)$ is determined by $\alpha(\omega)$. For
$\alpha(\omega)$ given by the solid curve, the Kramers-Kronig relation yields
the $n(\omega)$ given by the top curve in Fig. 7. Note that $n(\omega)$ is
essentially constant throughout the infrared and equal to

$$n(0) \cong 1 + \frac{\alpha_o c}{\pi \omega_L} \quad .$$

(8)

Since atomic transitions in the dielectric composite at still
higher frequencies also contribute an amount, say n, to the infrared index of refraction, the total index of refraction is

$$n_t \cong n + \frac{\alpha_o c}{\pi \omega_L} \quad .$$

(9)

For films thinner than 0.4 microns, $n \sim \alpha_o c / \pi \omega_L$ and the emissivity
is increased by the appreciable increase in n_t. For films thick-
er than 1 micron, a much smaller α_o can be used to absorb almost

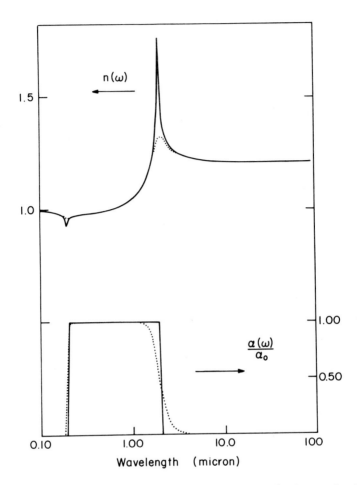

FIGURE 7. Frequency dependence of the optical constants of a
selective composite film. Lower curve, solid--the absorption coefficient, $\alpha(\omega)$ for an ideal selective surface. Lower curve,
dotted--$\alpha(\omega)$ for a real selective surface showing finite $k_B T$
width of the step in the absorptivity ($T \overset{\sim}{\sim} 800°K$). Upper curves,
solid--$n(\omega)$ for the ideal selective surface as calculated from
Kramers-Kronig. Upper curve, dotted--$n(\omega)$ for a real selective
surface. (After Reference 15.)

all the solar energy and the contribution to the effective index,
$\alpha_o c/\pi\omega_L$, no longer plays a significant role in determining the
emissivity. In fact, since the complex refractive index of the
film is usually graded over a couple of optical wavelengths to

enhance the coupling to the solar spectrum, a film thickness
d $\sim$ 2 microns is to be preferred (16).

Next, we must include the contribution to the emissivity from
the thermal blurring of the absorptivity profile in the dielectric film, which is represented by the dotted curve in Fig. 7.
Simply calculating the overlap between the absorptivity profile
and the black body distribution as we did earlier clearly under-
estimates the contribution of this term to the total hemispherical emissivity since it ignores the properties of the dielectric.
In fact, since off-thermal rays must pass through a greater
effective thickness of the absorbing dielectric, the hemispherical emissivity will be much larger than for the dark metal considered earlier.

Including front surface reflections and interference produces
a more complicated angular dependence for the emissivity, but the
tendency for both ε_{TE} and ε_{TM} to increase with increasing angle
remains.   The importance of thermal blurring can be seen by exam-
ining Fig. 8.   The hemispherical emissivity of platinum at $900^{\circ}K$
is shown for three cases (17).   The lowest curve, a, represents
the hemispherical emissivity of platinum, the next, curve b, for
platinum coated with an optically absorbing dielectric but one
which has a cutoff at 2 micron wavelength, and the last, curve c,
for an optically absorbing dielectric which has a thermally
blurred cutoff at 2 micron wavelength.   This thermally induced
absorption increases the resultant emissivity of the selective
surface by a factor 2 over that of the base metal itself.

By summing these contributions, we find the smallest hemispherical emissivity, ε_{H}, due to intrinsic physical processes.
For example, we can calculate the minimum emissivity of a dark
mirror assembly which contains a 99% optically absorbing film
(not counting front surface reflections) having a thickness of 2
microns, for a variety of metal substrates.   The calculated
results are shown in Fig. 9.   The hatched areas in the figure
identify non-physical regions of the emissivity versus temperature

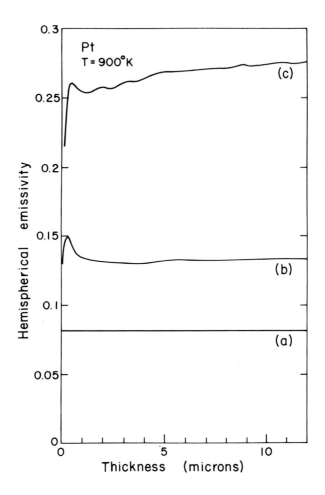

FIGURE 8. *Hemispherical emissivity of a metal coated with an ideal and a real selective surface. Curve (a) represents the ε_H of a bare metal, platinum. Curve (b) represents the ε_H of platinum coated with a selective absorber which absorbs the solar spectrum with a sharp cutoff at 2 micron wavelength. Curve (c) represents the ε_H of platinum coated with a selective absorber which has the same optical properties as (b) except thermal blurring of the 2 micron cutoff is assumed. (After Reference 17.)*

plot for different elements. The solid lines describe the smallest attainable hemispherical emissivities for a selective surface at a given operating temperature on a given substrate.

IV.   CONCLUSIONS

No matter what composite film--metal substrate combination is assembled, Fig. 9 indicates that the hemispherical emissivity will be appreciable at high temperatures. Moreover, there are physical processes which have not been included which would make the hemispherical emissivity still larger. Neither interband transitions in the metal nor lattice vibrational absorption in the composite has been included in these estimates, although both are intrinsic contributions. In addition, extrinsic processes such as those associated with surface pits, surface reconstruction, etc., have been ignored. If the coating does not have the ideal case of k_BT blurred step function behavior, so the absorptivity changes more slowly with wavelength than the step function, then the emissivity will be increased accordingly although the temperature dependence will be less rapid.

In the temperature region below $700^\circ K$ the effective emissivity is a sensitive function of the optical properties of the composite structure, both of the absorption edge and the base material. By $900^\circ K$ the difference between the effectiveness of different base metals is small. For such high temperature applications, the base metal should be chosen for its physical stability, not its hemispherical emittance, since all good conductors, when used in the selective surface mode, have about the same high temperature emissivity.

At the present there is not much measured data which can be compared with the dark mirror emissivity calculation given in Fig. 9 (see next chapter by Buhrman). The labelled points shown in the figure are taken from Table I of the previous chapter and Reference 18. Before comparing emissivity values, it should be remembered that the minimum emissivity is directly related to the solar absorptivity and the solar absorptivity used to calculate Fig. 9 corresponds to $\alpha_s \approx .94$. The appropriate limits for other values of solar absorptivity are described in Reference 16. Some

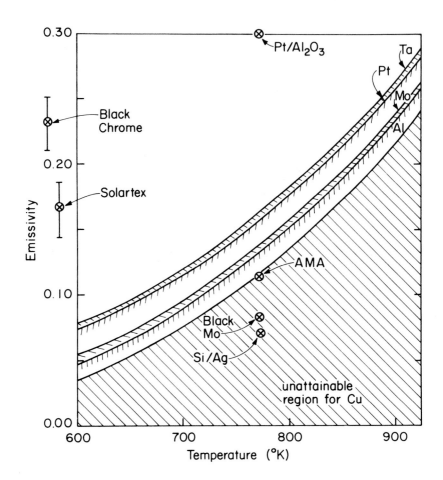

FIGURE 9. *Calculated minimum hemispherical emissivities for selective surfaces on various metal substrates. The solar absorptivity is assumed to be 0.94. These limits are determined by the temperature dependent properties of the metal, the influence of the dielectric index of refraction and the thermal broadening of the solar absorptivity profile. The hatched region of the figure identifies emissivity values which are excluded on physical grounds. The data points are taken from Table 1 in the previous chapter and also from Reference 18. The three data points in the hatched region were obtained by using room temperature optical constants to calculate the 500°C hemispherical emissivities.*

of the composites listed in this figure have smaller α_s's, hence could also have smaller ε_H's; however, all three of the data points shown in the hatched region have been obtained by using room temperature optical constants to calculate the hemispherical emissivity at 500°C.

Finally, we should compare the potential of the heat mirror and dark mirror concepts. As long as one maximizes α_s first and is constrained to weakly focused systems, a simple figure of merit of the selectivity is α_s/ε_H. For the heat mirror tandem the maximum figure of merit can be estimated by taking the best aluminum mesh with square wires and a 90% geometric transmissivity. Equation (7) then leads us to an $\alpha_s/\varepsilon_H < 11$. At low temperature, below 700°K, Fig. 9 shows that dark mirrors can have larger values of α_s/ε_H and indeed experimental values of the figure of merit at very low temperatures are already somewhat larger than this metal mesh heat mirror limit (1,18). However, because of the very temperature dependent emissivity shown in Fig. 9, should selective absorber temperatures ever rise above 800°K, the heat mirror may ultimately provide the larger figure of merit. For the immediate future, however, it appears that dark mirrors have the greater potential.

ACKNOWLEDGMENTS

The author would like to acknowledge helpful conversations with R.A. Buhrman, H.G. Craighead, R.H. Silsbee and D.M. Trotter.

REFERENCES

1.   Seraphin, B.O. in "Solar Energy Conversion--Solid State
     Physics Aspects", (Topics in Applied Physics, Vol. 31),
     (B.O. Seraphin, ed.), p. 5. Springer-Verlag, Berlin, (1979).
2.   Sievers, A.J. in "Solar Energy Conversion--Solid State
     Physics Aspects", (Topics in Applied Physics, Vol. 31),

(B.O. Seraphin, ed.), p. 57. Springer-Verlag, Berlin,(1979).

3. Fan, J.C.C. in "Solid State Chemistry of Energy Conversion and Storage", (Advances in Chemistry Series, 163), (J.B. Goodenough and M.S. Whittingham, eds.), p. 149. American Chemical Society, Washington, D.C.,(1977).

4. Rosebury, F., "Handbook of Electron Tube and Vacuum Techniques", p. 207. Addison-Wesley, Reading, MA,(1962).

5. Sievers, A.J., in "Materials Research Council Summer Conference. Vol. II-Solar Energy", (E.E. Hucke, ed.), p. 85. Dept. of Materials and Metallurgical Engineering, Univ. Michigan,(1973).

6. Groth, R., and Kauer, E., *Philips Tech. Rev. 26,* 105 (1965).

7. Groth, R., *Phys. Stat. Sol. 14,* 69 (1966).

8. Fan, J.C.C., Bachner, F.J., and Murphy, R.A., *Appl. Phys. Lett. 28,* 440 (1976).

9. Horwitz, C.M., *Optics Commun. 11,* 210 (1974).

10. Ulrich, R., *Appl. Opt. 7,* 1987 (1968).

11. Jakob, M., "Heat Transfer", Vol. 2, p. 5. Wiley, New York, (1957).

12. Pramanik, D., Sievers, A.J., and Silsbee, R.H., *Solar Energy Materials 2,* 81 (1979).

13. McPhedran, R.C., and Maystr, D., *Appl. Phys. 14,* 1 (1977).

14. Trotter, D.M., Craighead, H.G., and Sievers, A.J., *Solar Energy Materials 1,* 63 (1979).

15. Trotter, D.M., and Sievers, A.J., *Appl. Phys. Lett. 35,* 374 (1979).

16. Trotter, D.M., and Sievers, A.J., *Appl. Optics* (to be published).

17. Trotter, D.M., private communication.

18. Call, P.J., "National Program Plan for Absorber Surfaces R & D", p. 10. Solar Energy Research Institute, Golden, CO. TR-31-103, (1979).

CHAPTER 9

COMPOSITE FILM SELECTIVE-ABSORBERS[1]

R. A. Buhrman
H. G. Craighead[2]

School of Applied and Engineering Physics
Cornell University
Ithaca, New York

I. INTRODUCTION

The effectiveness of finely divided metal/dielectric composite coatings for the collection of solar radiation is well
established. Dielectric rich composites (cermets) are, in
general, strongly absorbing over much, if not all, of the solar
spectrum. In the infrared spectral range they rapidly become
highly transparent with increasing wavelength. When such a
composite coating is formed on a highly reflecting metal surface
the resulting absorber-mirror tandem can be a strongly absorbing, yet selective, photothermal collector. A schematic representation of such a system is shown in Fig. 1.

---

[1]*Research supported by Solar Energy Research Institute.*
[2]*Present address: Bell Telephone Laboratories, Holmdel,*
New Jersey.

SOLAR MATERIALS SCIENCE

277

Composite Selective Surface

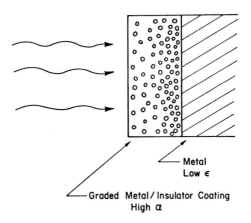

Metal
Low ε

Graded Metal / Insulator Coating
High α

FIGURE 1.   Schematic representation of a composite absorber-mirror tandem selective-surface.

The most intensively studied, if not the best understood, example of such a collector is the black chrome coating (1). This electroplated coating is basically a mixture of Cr particles and Cr oxides with particularly useful microstructure plated onto specially prepared metallic substrates. At present it is by far the most widely employed low temperature (T ≤ 300 C) selective absorber.

For demanding high temperature (T > 350 C) applications, where electro-plated coatings tend to fail, it is again dielectric/metal composites that deliver the best performance. In this case the coatings are either vacuum or sputter deposited. Such films, due to their superior integrity, tend to be considerably more stable in hostile high temperature environments. With such composite films, unexcelled high temperature selective absorber performance and stability has recently been

in many cases, including that of black chrome, the coating consists of small metallic grains dispersed within an amorphous, often porous, matrix of non-stoichiometric oxide (7).

C. *Optical Properties-Theory*

The theory that is usually invoked to explain the optical response of finely divided composite material is the so-called Maxwell-Garnett theory published in 1904 (8). This theory is a straightforward mean-field approach to the problem of accounting for the effect of particle-particle interactions in affecting the overall response of metallic particles embedded in a dielectric host. In recent years this theory has received renewed attention. The fundamental assumptions of the theory have put on a sounder theoretical footing and previously neglected structural multipole effects have been included in the theory (9).

In its simplest form the Maxwell-Garnett (M-G) theory considers the composite to consist of electrically-isolated, spherical metal particles uniformly dispersed within an ideal dielectric matrix. Generalization of the theory to non-spherical particle shape is straightforward, but it is done at the expense of introducing additional parameters (depolarization factors) into the theory. While such a generalization may be necessary to make definitive tests between theoretical predictions and the observed response of a particular composite (10) for which the particle shapes are precisely known, it is of limited value in *predicting* the optical response of composite materials.

There is an alternative mean field theory first proposed by Bruggeman (11) which is sometimes used in attempts to explain the optical response of composite materials. The basic difference between the Bruggeman approach and M-G is that the former assumes a quite different composite topology than does

the latter. Bruggeman assumes the composite to consist of
randomly intermixed particles of dielectric and metal. M-G
assumes the metal to be dispersed as particles throughout the
dielectric. While the distinction may appear subtle the result
is that the Bruggeman theory tends to predict broader absorption curves than does the M-G theory. If encountered experimentally these broader absorption curves would be somewhat
more favorable for solar energy applications.

D.   *Optical Properties-Experiment*

While composite coatings have been actively studied for
quite some time there is very little data available concerning
the intrinsic optical properties of these materials. Recently
a series of experiments have sought to establish the value of
theory in predicting the optical behavior of composites, and
to determine the suitability of various composite coatings
for selective-surface applications. Thin, semi-transparent
films of uniform composition were produced by co-evaporation
from separate electron beam heated sources. The reflectance
and transmittance of these films, which were deposited on
quartz substrates, were measured over the visible and near
infrared spectral ranges. The complex index of refraction of
the composite was then numerically determined from this data.
The results were then correlated with the topology and microstructure of the composite and compared with theoretical predictions.

A variety of vacuum deposited composites have been examin-
ed, including Ni/Al_2O_3, Au/Al_2O_3, Au/MgO and Pr/Al_2O_3 composites. The results of these and related studies can be summarized as follows: The optical behavior of a composite is in
general characterized by a broad absorption peak, centered in
the visible spectral region, with significant absorption extending into the infrared region. In the near infrared region,
this absorption typically decreases only gradually with in-

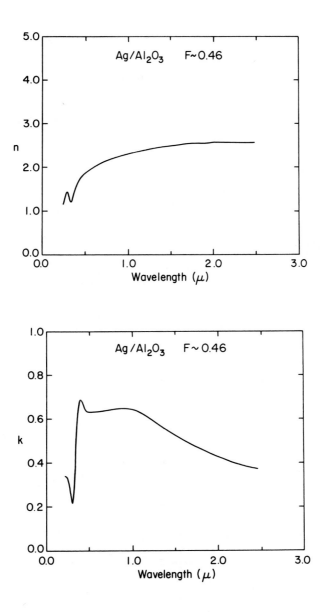

FIGURE 2.   The index of refraction and extinction coefficient
of a typical composite of relatively high (46%) metallic composition.

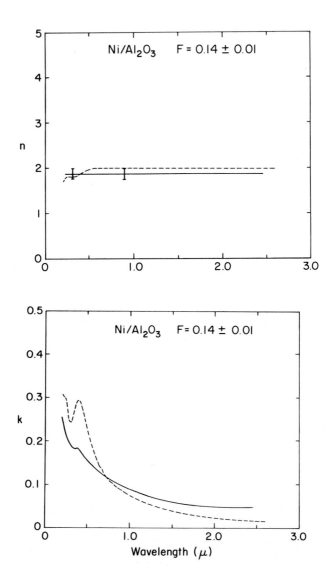

FIGURE 3.   The optical constant of a relatively low metal
fraction Ni/Al₂O₃ composite (solid line) compared to the M-G
prediction (dotted line).

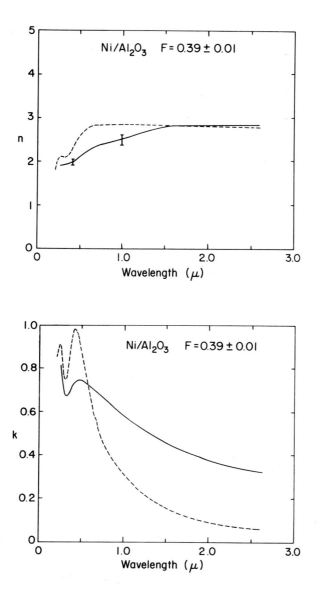

FIGURE 4.   The optical constant of a high metal fraction
Ni/Al_2O_3 composite (solid line) compared to the M-G theory
prediction (dashed lines).

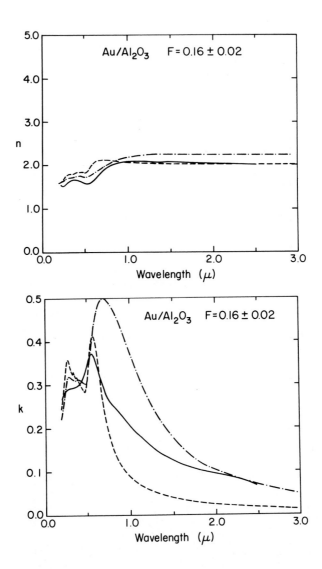

FIGURE 5. *Optical constants of Au/Al₂O₃ composites (solid line) compared to the M-G theory (dashed line) and Bruggeman theory (dot-dashed line). In making the theoretical prediction, corrections were made for short mean free path effects within the Au particles.*

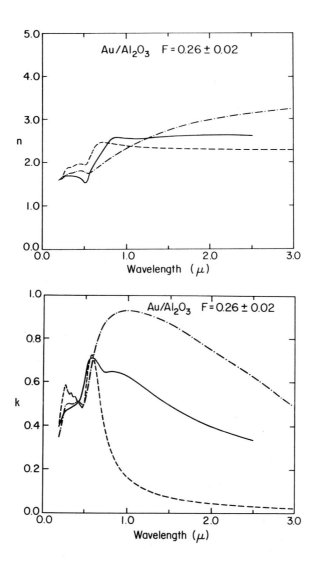

FIGURE 6. *Optical constants of a higher metal fraction composite. Again the solid line is the data, the dashed and dot-dashed lines the M-G and Bruggeman theories, respectively.*

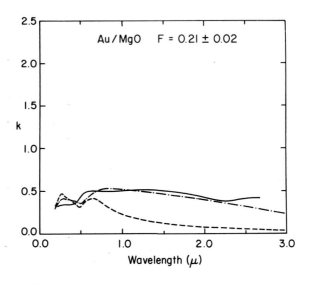

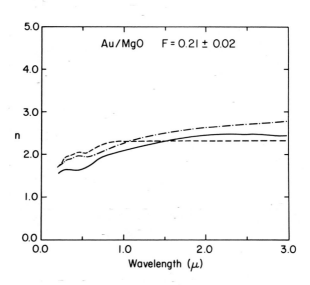

FIGURE 7. Optical constants of a Au/MgO composite (solid line) compared to the M-G (dashed) and Bruggeman (dot-dashed) theories.

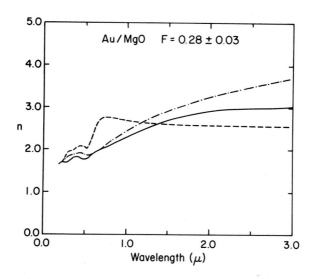

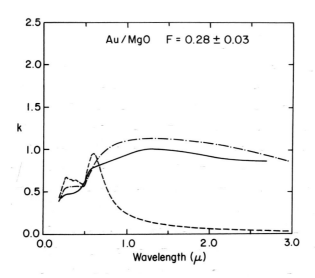

FIGURE 8. *Optical constants of a higher metal fraction Au/MgO composite.*

creasing wavelength. Examples of this behavior are shown in
Fig. 2 where the real and imaginary component of the index of
refraction of a typical dielectric-rich composite Ag/Al_2O_3 is
plotted.

If the details of the composite response is more closely
examined a strong correlation is found between composite microstructure and optical behavior. Those composites characterized
as being composed of observable metallic grains randomly dispersed in a dielectric matrix, usually amorphous, exhibit behavior in qualitative accord with the predictions of the simple
M-G theory. This is illustrated in Figs. 3 and 4 and Figs. 5
and 6 where experimental results for several Ni/Al_2O_3 and
Au/Al_2O_3 composites are compared with the predictions of the
simple M-G theory. In making these comparisons, corrections
were made to account for the shortened electron mean free path
in the metallic microcrystals. This effect acts to extensively
broaden the optical absorption over that predicted from bulk
optical data. The major disagreement between the results for
these composites and the M-G theory is an increased infrared
absorption over the predicted value. This disagreement tends
to grow with increased metal fill fraction. The cause of this
extra absorption is undetermined, although it has been specula-
ted that it is due to electron tunneling between metallic grains.
Other explorations are possible. Regardless of the explanation
the effect can be useful in improving the effectiveness of
composite absorbers.

Composites whose topology consist of randomly distributed
metal and dielectric micro-crystals have optical behavior significantly different from that discussed above. This class of
composites includes Au/MgO. The optical absorption of a composite of this type is broader than that of, for example,
Au/Al_2O_3. This is illustrated in Figs. 7 and 8, where the
extinction coefficients of several Au/MgO composites are plotted. The M-G theory does not approximate the behavior of such

composites. This is not surprising since the topology is not
that assumed by the M-G theory. The topology assumed by Bruggeman is similar to that exhibited by Au/MgO composites and useful
comparisons between theory and experiment can be made. As can
be seen in Figs. 7 and 8 the Bruggeman theory is fairly successful in predicting the optical behavior, although not surprisingly, disagreement appears as the metal fill fraction increases.

The third type of composite that is sometimes produced is
the one in which the metal is highly, even atomically dispersed,
with no microstructure detectable by electron microscopy. Such
composites are produced when the metal constituent is relatively
reactive, such as is the case of Fe/Al_2O_3 and V/SiO_2 composites.
Existing composite medium theories are not appropriate for such
materials. Experimentally these materials exhibit broad,
essentially featureless, optical extinction over the visible
and near infrared spectral ranges, as illustrated in Fig. 9.

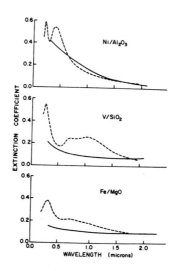

FIGURE 9. *Extinction coefficient of two amorphous and one
crystalline (Ni) composites compared to the M-G prediction
(dashed line).*

We conclude this discussion by reviewing the optical absorption expected from composite coatings. Composites with microstructure consisting of metallic grains dispersed in a dielectric matrix exhibit optical behavior in accord with the M-G theory. Composite coatings of intermixed micro-crystals of metal and dielectric exhibit broad absorption in qualitative accord with the Bruggeman theory. Composites of highly dispersed metal exhibit very broad, or flat, optical absorption through the solar spectrum. In general, if the microstructure of a particular composite system is known, reasonable predictions concerning its optical response can be made. This knowledge can then be effectively employed to engineer selective absorbers.

III. ENGINEERING OF COMPOSITE FILM SELECTIVE ABSORBERS

A.   Introduction

The basis of the selective absorber approach to solar energy collection is to develop a surface which is strongly absorbing over the solar spectrum but non-absorbing at longer wavelengths. The "ideal" absorption curve of a selective absorber is illustrated in Fig. 10. By using a solar collector surface with an absorption curve such as that in Fig. 10, the solar radiation is efficiently abosrbed. At the same time the absorptivity, and therefore the emissivity (by Kirchhoff's laws), is low at the longer wavelengths corresponding to the collector's black-body emission wavelengths. The sun's energy is thus efficiently trapped and the heat energy is not lost by reradiation. This spectrally selective absorption is a significant advantage over a flat, non-selective black-body for which the absorptivity (equal to the emissivity) is unity for all wavelengths. The highest attainable solar collector temperature for a black-body is T_o where

$$\sigma \, T_o^{\,4} = \text{incident solar flux}, \tag{1}$$

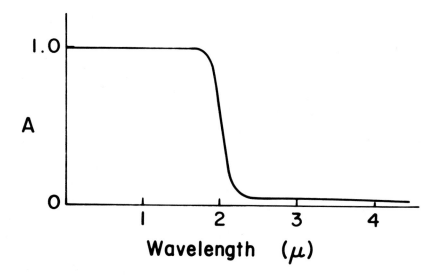

FIGURE 10. Ideal selective surface absorption curve.

and σ is the Stefan-Boltzmann constant. Taking the optimistic value of 750 Wm^{-2} for the solar flux, T_o = 66°C for a black-body. To achieve high temperatures and large Carnot and system efficiencies one must go beyond simple black-body absorbers .

An alternative way around the limit on black-body efficiency is to increase the amount of energy incident on the collector by focusing the sunlight. However, the use of a highly absorbing solar selective surface yields improved performance at all degrees of concentration compared to a non-selective surface with the same solar absorptivity.

Forming composite coatings on low emissivity surfaces, as shown schematically in Fig. 1, can yield very high performance selective absorbers. By making proper choice of components and by careful tuning of coating parameters selective absorbers can be engineered which give unexcelled performance in practically all instances. After reviewing the criteria by which selective absorbers are evaluated, we will discuss the basic guidelines for producing these composite-mirror selective absorbers.

B.   *Figures of Merit*

A quantitative measure of the performance of solar selective surfaces is necessary both for the comparison of different
surfaces and as a means of gauging the efficiency of a system
using a given selective absorber.   These relevant parameters
are the solar abosrptivity α, ant the thermal emissivity ε.

The absorptivity α is defined as the fraction of the solar
power absorbed by a film.   It is related to the reflectance of
a non-transparent film by:

$$\alpha = \frac{\int S(\lambda) \, [1 - R(\lambda)] \, d\lambda}{\int S(\lambda) \, d\Lambda} \, , \tag{2}$$

where $S(\lambda)$ is the solar power spectrum, $R(\lambda)$ is the wavelength
dependent reflectance and λ the wavelength, integrated over
the range of the solar spectrum.   The total hemispherical
emissivity of a film is the total power reradiated into a
hemisphere by the film, divided by the power radiated by a
black-body at the same temperature.   By Kirchhoff's laws the
emittance at a given angle θ and wavelength λ, is equal to the
absorptance,

$$\varepsilon(\theta,\lambda) = A(\theta,\lambda) = 1 - R(\theta,\lambda) \tag{3}$$

for a non-transparent film.   By integrating this over all
wavelengths and over the hemisphere we get the total hemispherical emissivity as

$$\varepsilon(T) = \frac{\int_0^{\pi/2} U(\lambda,T) \, \varepsilon(\theta,\lambda) \, d\lambda \; d(\sin^2\theta)}{\int U(\lambda,T) \, d\lambda} \, , \tag{4}$$

where U is the Planck distribution function.

These quantities of α and ε are intrinsic properties of
the selective surface.   However, of interest also is the total
system performance which involves both the temperature of
operation and the amount of concentration of the solar flux.
The essential quantity of interest in evaluating a selective

absorber system's performance is the fraction of the available
solar energy retained. This is expressed as the collection
efficiency η where

$$\eta = \frac{x \alpha S \div \varepsilon \sigma T^4}{xS} \tag{5}$$

or

$$\eta = \alpha - \varepsilon \sigma T^4 / xS. \tag{6}$$

Here σ is the Stefan-Boltzmann constant, T the absolute temperature, x the solar concentration factor, and S is the total solar
energy flux. The efficiency involves α and ε independently
rather than as the ratio α/ε. The collection efficiency is also
a function of both x and T. For large x, α is relatively more
important than ε. For high concentration applications, where
x is very large, the collection efficiency approaches its upper
limits of $\eta = \alpha$. For an ideal black-body α is equal to unity,
and α can be near unity for actual flat plate type absorbers.
The importance of selectivity is therefore reduced for the large
x concentration applications. It is in this high x range that
thermal stability and large α become most important. The expression for η also gives the upper limit on T for any system,
which occurs at $\eta = 0$. At this point all the energy is re-
radiated.

C. *Choice of Composite Absorber Components*

No known metal/dielectric composite has the ideal absorption properties illustrated in Fig. 10. As indicated previously,
experiments reveal that the optical absorption of composites
is broad, with only a gradual transition observed between the
strongly absorbing short wavelength regime and the weakly absorbing long wavelength regime. Indeed the behavior of most, if
not all composites, can be roughly approximated in the crucial
2 μm wavelength region by a wavelength independence extinction
coefficient k. This yields an absorptivity, $\alpha = 2\pi k/\lambda$, that

decreases only as $1/\lambda$.

Theoretically, neither the M-G nor the Bruggeman theory predicts the possibility of a sharp 2 µm cut-off for any elemental metal composite for which the bulk optical constants are known. The only exception to this statement is that if very non-spherical particles of particular orientation are assumed a sharp absorption edge at 2 µm can be predicted (10). Achieving such control over particle shapes in composites seems highly unlikely. The Bruggeman theory does indicate that composites of the appropriate topology can have a somewhat sharper cut-off than composites with the M-G topology. The differences however are not very strong.

While the ideal composite material is yet to be discovered, the range of choice of constituents for high performance, composite selective absorbers is quite large. Consequently the choice for a particular application can and should be made largely on the basis of cost and stability considerations. For high temperature applications stability is the prime consideration. Thus it is here that vacuum deposition offers the most options, since the choice of the dielectric is basically independent of the choice of the metal. Therefore good candidates for high temperature systems are the relatively non-reactive refractory metals such as Ni, Pt and Mo, and dielectrics such as Al_2O_3, due to its effectiveness as a passivating material, or other stable oxides and nitrides. For lower temperature applications, wet chemical techniques are advantageous due to lower cost even though the dielectric typically must be an oxide or sub-oxide of the metal constituent. The metal choice is made on the basis of its stability and wet chemical deposition characteristics.

The choice of the metal mirror that underlies the composite coating in a dark-mirror selective absorber is also governed by considerations of cost, stability and performance. However since available composite absorbers do not have the ideal ab-

sorption cut-off in the 2 μm region, the near infrared reflectivity of the metal is not as critical as it would be for a more
ideal composite absorber. In currently realizable dark-mirror
systems, the predominant source of emissivity losses at high
temperature (T > 300 C) is the composite film and not the metal
mirror. Consequently nearly any relatively high reflectance
metal mirror will serve, with good quality Mo and Pt films being
particularly good choices due to their high temperature stability. For low temperature operations, cost and compatibility with
a particular wet chemical process tends to dictate the mirror
choice.

D.   Graded Composition Coatings

Conceptually the simplest approach to the formation of a
dark-mirror tandem is to produce a *uniform* composite coating on
the metal mirror surface. The necessary thickness d of the
coating is determined by the requirement that the absorption
of the composite remain strong out to 1.5 to 2.5 μm wavelengths.
Since the attenuation of radiation upon passing through an absorbing medium of thickness x and extinction coefficient k is

$$I = I_o \, e^{-2\pi kx/\lambda} \tag{7}$$

the requirement is that kd ~ .1 - .3 μm. Depending upon the
metallic content, the extinction coefficient of most composites
in the 2 μm wavelength region varies from < 1., for low metallic
content, to $\gtrsim$ .5 for fractional metallic content in the .2 to
.5 range. (Typically for metal volume fraction above .5 the
composite becomes electrically conducting which is undesirable
for selective absorption since the extinction coefficient of
the composite then increases with increasing wavelength.)

It is usually advantageous to employ relatively thin
(d $\lesssim$ .5 μm) composite coatings, as this tends to minimize problems with film adhesion and serves to eliminate undesirable
interference effects in the infrared. Thus a .5 μm composite

coating with metal fill fraction of the order of .3 to .4 is
typical, although thicker layers with less metal per unit
thickness will also serve.

A detrimental aspect of a composite absorber is that the
index of refraction of composites tends to be large, increasing
directly with metallic content. The result is significant
front surface reflection losses and hence lowered absorber
performance. The most effective way to counter this effect is
produce the absorber with a compositional gradient (11-14) such
that the metallic content is zero at the front surface and
maximum at the composite-mirror interface. If properly executed
such a compositional gradient can effectively eliminate reflection losses due to the metallic content of the composite,
leaving only front surface losses due to the dielectric mismatch
between the air and the pure dielectric. Ideally the composite
should be graded to a front surface index of refraction of one,
to also avoid this loss mechanism. In practice this usually
requires the composite to be porous, which is apparently the
case for some electroplated black chrome coatings. For vacuum
deposited coatings this can only be achieved by additional
processing.

The precise nature of the compositional grade is not critical to achieving the desired effect. This is illustrated in
Fig. 11 where the results of some model calculations for the
reflectivity of composite films deposited on copper surfaces
are shown. (The optical constants assumed in these calculations
is typical of composite materials.) One of the model films
(No. 1) is of uniform composition with a metal content of
$F = .25$ and a thickness of 4000 $\overset{\circ}{A}$. The other three films graded
are with the same average metal content and thickness as the
uniform film; so the only parameter varied is the shape of the
composition grade. The shape of these three composition grades
along with the calculated absorptivity α and thermal emissivity

CALCULATED NORMAL REFLECTANCE

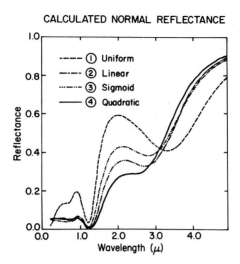

FIGURE 11. The reflectance curves of a uniform composite film and three graded composition films, all with the same thickness and total metal content.

ε at 300°C are shown in Fig. 12. All three graded films, although graded in quite different ways, have essentially identical α and ε values. Distinct advantages in both α and ε can be seen for the graded films compared to the uniform film. More subtle changes in the optical response among the graded films can be seen in Fig. 11. But these differences are insignificant in terms of practical solar performance. The reflectance in the 0.2 - 1.0 μ range was reduced to the lowest possible value for all three grades. At longer wavelengths the quadratic grade formed a slightly sharper absorption edge.

The primary benefit of a graded composition can be seen as the increase of α from 0.84 for the uniform film to 0.94 for the graded films. A secondary benefit of the graded composition is a reduction in ε of about 0.03 for the graded films.

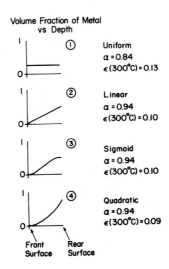

FIGURE 12. The shape of the composition-depth profile used for calculating the reflectance curves of Fig. 12. Also shown are the calculated absorptivity and emissivity at 300°C.

This reduction in ε was explained by Sievers (15) as being due to a decreased optical phase shift for the graded film, yielding a better reflection coating at the long thermal wavelengths. The temperature dependent emissivity for one of these graded films is shown in Fig. 13a. This can be compared with the higher emissivity for the uniform film shown in Fig. 13b.

The solar performance calculated for these typical composite selective absorbers is well within the limits for graded composite absorbers calculated by Trotter and Sievers (16). They have calculated the maximum attainable α/ε value for a graded composite film on Cu to be about 33 at 300°C. For the above calculation for practical graded composite materials, α/εwas 9.4 at 300°C. This indicates the advances which remain to be made by obtaining the ideal selective absorber material, i.e. one with a more abrupt absorption cut-off.

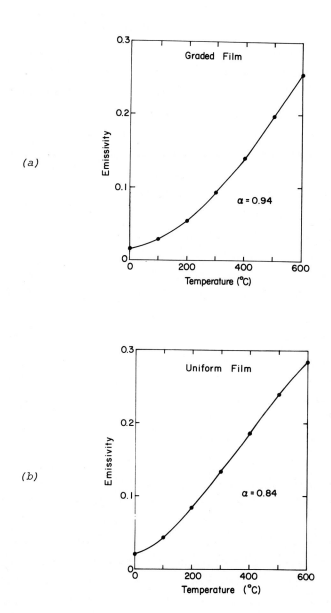

(a)

(b)

FIGURE 13.   *Calculated emissivity of a graded and of a uniform composition film.*

E. *Production of High Temperature Selective Absorbers*

With a knowledge of the optical behavior of composite materials and with the means available to achieve compositional gradients, it is a straightforward matter to produce selectively absorbing coatings. Since a composite with ideal optical constants is as yet not available, the performance of such coatings is not perfect; it can be quite good. Moreover, by adjusting the parameters of the coating it is a straightforward matter to engineer the response of a coating to optimally match the requirements of nearly any selective absorber application.

To illustrate this point we describe the behavior of several Ni/Al$_2$O$_3$ and Pt/Al$_2$O$_3$ composite coatings. These coatings were produced by electron beam deposition in the course of a research program designed to develop selective absorbers for a high temperature applications. The composite materials were chosen mainly on the basis of physical stability considerations. In this work graded and ungraded composite films were deposited on metal coated quartz, stainless steel and copper substrates. The metal coating was either a Ni or Pt film. Hemispherical reflectivity and emittance calorimetry measurements were made on these films and from this data the solar performance of each film was determined. Series of films, each with different thickness but all with maximum back surface metal concentration of approximate 40%. In Fig. 14 we show the reflectance spectrum of one such Ni/Al$_2$O$_3$ film deposited on a copper substrate. The extrema in the curve are the result of thin film interference effects rather than intrinsic sharp absorption lines of the composite. The overall absorption edge of this film is approximately at the desired wavelength of ~ 2 μ. The solar absorptivity for this film is a fairly high 0.93. The price for this fairly high α value is a rather high ε value as well which we will discuss below.

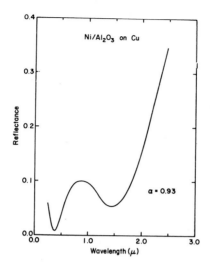

FIGURE 14.   Reflectance curve of a typical graded Ni/Al$_2$O$_3$ film on Cu.

Reflectance spectra for three members of this family of Ni composite films on Cu are shown in Fig. 15. (This figure includes the spectrum of the film in Fig. 14. - here it is number 5). The numbering of the films is in increasing order of thickness ranging from ~ 1500 Å with film number one the thinnest. The α values for this group of films is listed in Table 1 A. The list of α values in this table shows that the absorptivity clearly increases with film thickness. The whole point of solar selective surfaces, is, of course, to have both high α and low ε. Figure 16 shows the measured ε for this family of surfaces. The emissivity of the surfaces is seen to generally increase with the thickness with α. The rather strong increase in ε, with increasing temperature, is the result of the broad absorption edge of these composites. Table 1 a lists the ε values for the films at selected temperatures for comparison. As the temperature increases and the

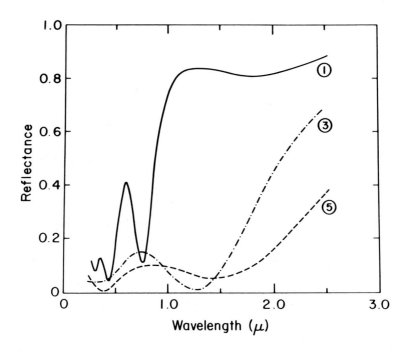

FIGURE 15. Reflectance curves for a series of Ni/Al₂O₃
composite coatings on Cu. Films 1, 3 and 5 are of progressively
greater thickness.

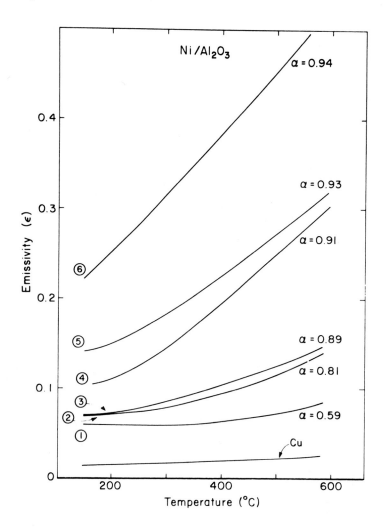

FIGURE 16. The curves are the measured temperature dependent emissivity of six Ni/Al₂O₃ films on Cu and the curve for pure Cu. The curves also are labeled with the solar absorptivity values α, calculated from the measured reflectance.

TABLE 1 a   Performance Parameters of Ni/Al$_2$O$_3$ Films

Film #	α	ε(150°C)	ε (300°C)	ε (500°C)
1	0.59	0.06	0.06	0.07
2	0.81	0.07	0.08	0.12
3	0.89	0.07	0.09	0.13
4	0.91	0.10	0.15	0.26
5	0.93	0.14	0.19	0.27
6	0.94	0.22	0.32	0.45

TABLE 1 b   Performance Parameters of Pt/Al$_2$O$_3$ Films

Film #	α	ε(150°C)	ε (300°C)	ε (500°C)
1	0.78	0.06	0.07	0.09
2	0.80	0.08	0.10	0.16
3	0.89	0.12	0.16	0.25
4	0.94	0.19	0.24	0.33

TABLE 1.   Performance parameters for the two series of Ni and Pt composite films. The data are organized in order of increasing film thickness.

black-body spectrum begins to overlap this absorption edge the emissivity begins to increase rapidly. An ideal composite with a much sharper absorption edge would not exhibit this effect nearly so strongly.

One important feature should be noted in the family of curves shown in Fig. 16. In this group of films the only uniform film, formed without a graded composition, is film number two. Films two and three have nearly identical emissivity curves but film number three has a larger α. This shows experimentally the enhanced performance derived from a composition and refractive index grade.

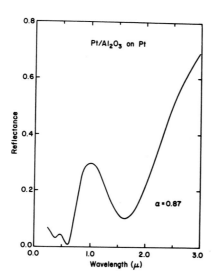

FIGURE 17.   Reflectance of a typical Pt/Al$_2$O$_3$ composite
on Pt.

The second composite system to be discussed here is
Pt/Al$_2$O$_3$. To compensate for its expensive components the
Pt/Al$_2$O$_3$ composite demonstrates nearly unsurpassed thermal stability. A typical reflectance curve for a Pt/Al$_2$O$_3$ film on Pt
is shown in Fig. 17. The general features are similar to those
seen with the Ni composite films. The oscillatory nature of
R(λ) again is due basically to thin film interference effects.

In Fig. 18, the total hemispherical reflectance curves
are shown for three members of a family of Pt/Al$_2$O$_3$ films on
polished Cu. Here again the numbering of the films is in the
order of increasing thickness. Table 1 b lists the solar absorptivities for this family of films. Again the trade-off
between α and ε can be seen in Fig. 19, where increases in ε
accompany increases in α. ε remains low, ~ 0.05 - 0.10 over
the temperature range for the α = 0.78 film but for the highly
absorbing α = 0.94 film ε ranges from ~ 0.18 to 0.36 over the

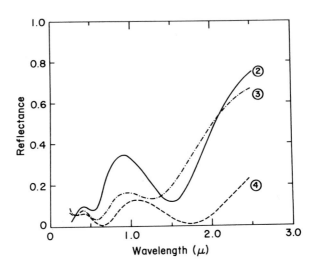

FIGURE 18. Reflectance curves for three members of a series of Pt/Al_2O_3 coatings on Cu. The coatings are numbered in order of progressively greater thickness.

temperature range. This ε information also is tabulated in Table 1 b.

An imperfect measure of the value of a selective surface is the ratio of α/ε. Figures 20 and 21 show the measured α/ε as a function of temperature for these Ni and Pt composite absorber films. The quantity of α/ε is often quoted in the literature as a constant for a film. This is clearly incorrect since ε is a strong function of T. Figures 20 and 21 give α/ε over the temperature range of 100°C – 500°C. The ratio α/ε becomes lower, less favorable, at higher temperatures as the wavelength range of the emission spectra overlaps more of the film's absorbing range. The value of α/ε > 13, obtained for both Ni and Pt composites, is quite large.

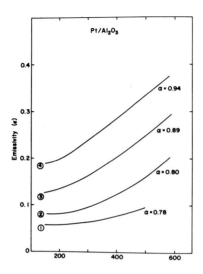

FIGURE 19.   *Emissivity as a function of temperature for four*
Pt/Al$_2$O$_3$ *films on Cu.*

However, α/ε alone does not fully characterize a selective
surface. Both α and ε are independently important in characterizing a film's performance. The collection efficiency defined previously is the proper quantity to consider.

The collection efficiency η as a function of solar concentration x has been calculated for the composite films of Figs.
16 and 19, using the measured values of α and ε. For the incident solar radiation the air mass 2 value was taken as given
in reference 16. Figs. 22 and 23 show η vs. x at 150°C for
several of the Ni and Pt composites. Figs. 24 and 25 show
η vs. x for the same films at 500°C. The advantage of selectivity was seen by comparing these curves to the efficiency of a
non-selective black-body ($\alpha = \varepsilon = 1$). As pointed out by
Seraphin (17), even at large concentration factors, x > 100,
the lower emissivity of the selective surface leads to efficiencies significantly better than for a black-body. It is

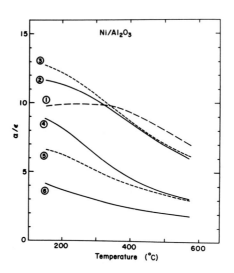

FIGURE 20. Measured α/ε for Ni/Al₂O₃ selective surfaces.

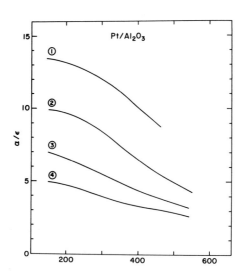

FIGURE 21. Measured α/ε for Pt/Al₂O₃ selective surfaces.

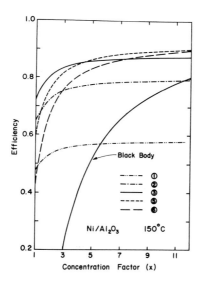

FIGURE 22. Collection efficiency η as a function of the solar concentration x for Ni/Al$_2$O$_3$ composite 150°C.

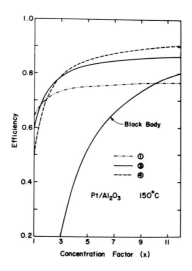

FIGURE 23. Collection efficiency for Pt/Al$_2$O$_3$ composites of 150°C.

readily seen that different films have the largest η at different values of x and T. The essential point is that with reasonable control of composite film parameter one can easily create
a film with the best combination of α and ϵ for a given application within the constraints imposed by the basic optical response of these composite materials. For large x the high
α (α = 0.94) samples are the best choice. For lower x the best
choice is $\alpha \approx .9$ with ϵ much lower than that of the α = .94
films.

F.   High Temperature Selective Absorber Stability

The ability of a selective absorber to survive in a hostile
high temperature environment is essential if that absorber is to
be of practical value. Fortunately composite absorbers, at
least those produced by vacuum deposition techniques, are distinguished by their high temperature stability.

For example to test the inherent stability of Ni/Al_2O_3
and Pt/Al_2O_3 selective absorbers, a number of samples were
produced by depositing graded composite layers on fused quartz
substrates. This deposition occurred immediately after the
substrate had been coated with either a Ni or Pt metal film
mirror. Figure 26 shows the specular reflectance spectrum of
a Ni/Al_2O_3 composite on a Ni reflecting layer on quartz before
and after extended tests. There was an initial change of α
from 0.96 to 0.95 in the first 70 hours at 400°C in air.
Within the accuracy of the measurement the absorptivity then
remained at this value for the length of the test. The triangles in Fig. 26 show the reflectance after 260 hours at 400°C
plus 68 hours at 500°C. Problems with film adherence eventually
developed at and above 500°C but the tests reveal the Ni/Al_2O_3
composites to be quite rugged.

Perhaps not too surprisingly, the Pt/Al_2O_3-Pt film selective absorbers exhibit excellent stability. For example the

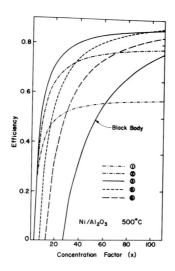

FIGURE 24.   Collection efficiency for the same Ni/Al$_2$O$_3$ composited at 500°C.

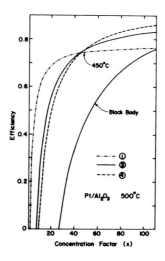

FIGURE 25.   Collection efficiency for the same Pt/Al$_2$O$_3$ composites at 500°C.

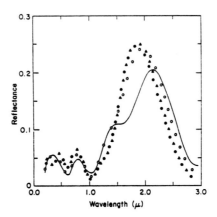

FIGURE 26. Reflectance of a Ni/Al$_2$O$_3$ composite on a Ni layer on quartz: as deposited - solid curve, after 70 hours at 400°C in air - open circles, after 260 hours at 400°C - closed circles, 68 hours at 500°C - triangles.

reflectance of Pt/Al$_2$O$_3$ composite deposited on a Pt layer on fused quartz showed no significant change in reflectance after 50 hours at 500°C or an additional 300 hours at 600°C. The reflectance spectra are shown in Fig. 27. The Pt/Al$_2$O$_3$ composite appears to be very stable, better than any other known metal-insulator composite film, and is at least as stable as any other presently available selective absorber. It is quite promising as a selective absorber material for application in high temperature oxidizing environments.

To show the use of the Pt composite films as a practical surface, coatings have been deposited on bulk pieces of 316 stainless steel. This temperature resistant steel is one of the candidate materials for the high temperature solar central receiver. The total hemispherical reflectance curve for one coating Pt/Al$_2$O$_3$ on stainless steel is shown in Fig. 28. The solid curve in this figure is the reflectance measured immediately after the film deposition. Also plotted with open circles

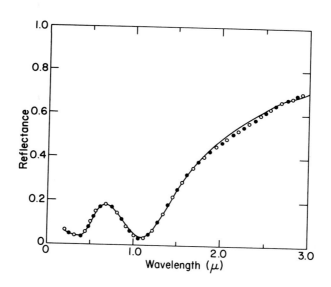

FIGURE 27. Reflectance of a Pt/Al_2O_3 composite on Pt:
as deposited - solid curve, after 50 hours at 500°C in air -
open circles, after 300 hours at 600°C - solid circles.

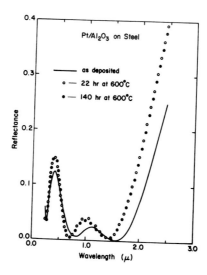

FIGURE 28. Reflectance of Pt/Al_2O_3 composite deposited on
316 stainless steel before and after heating in air.

is the reflectance after heating for 22 hours at 600°C in air.
Failure due to detachment of the film from the substrate occurred after heating to 700°C but more sophisticated surface pretreatment could possibly alleviate this problem. Obviously
these results are quite encouraging.

IV. SUMMARY

We have descirbed the general nature of the optical properties of metal-insulator composite films and have discussed
the engineering of these films into high performance selective
absorbers. For demanding high temperature applications vacuum
deposited composites, particularly Pt/Al_2O_3, have been shown
to have both excellent solar selective properties and the
necessary thermal stability.

Theorectically, selective absorber performance superior to
that achieved so far is possible. With present composites
some improvement can be realized by additional processing, at
additional cost, such as, for example, depositing anti-reflection coatings on the composite. A major advance in performance
is dependent on the elusive goal of developing an absorber with
a more ideal absorption curve. Barring such a breakthrough,
in the coming years we can expect the development of a number
of alternative composite absorbers with similar optical performances. While all alternative composites should be considered,
the eventual development of improved selective absorbers is
most dependent on a research program that deals effectively
with the problems of coating cost and durability.

REFERENCES

1.   McDonald, G. E., *Solar Energy, 17,* 119 (1975).

2.   Craighead, H. G., Bartynski, R., Buhrman, R. A., Wojcik, L.
     and Sievers, A. J., *Solar Energy Materials, 1,* 105 (1979).

3. Granqvist, C. G., Anderson, A. and Hunderi, O., *Appl. Phys. Lett., 35,* 268 (1979).

4. Dickson, J., Proceedings DOE/DST Thermal Power Systems Workshop on Selective Absorber Coatings, SERI, Golden, Colo., p. 347 (1977).

5. Gittleman, J. I., Abeles, B., Zanzucchi, P. and Arie, Y., *Thin Solid Films, 45,* 9 (1977).

6. Blickensderfer, R., Deardorff, D. K. and Lincoln, R. L., *Solar Energy, 19,* 429 (1977).

7. Pettit, R. B. and Sowell, R. R., *J. Vac. Sci. Technol., 13,* 596 (1976).

8. Garnett, J. C. M., *Phil. Trans. Roy. Soc. London 203,* 385 (1904).

9. Lamb, W., Wood, D. M. and Ashcroft, N. W., *Phys. Rev. B, 21,* 2248 (1980).

10. Granqvist, C. G. and Hunderi, O., *Phys. Rev. B, 17,* 3513 (1977).

11. Bruggeman, D. A. G., *Ann. Phys. (Leipzig) 24,* 636 (1935).

12. Lord Rayleigh, *Proc. Roy. Soc. London 86,* 207 (1912).

13. Strong, J., *J. de Physique, 11,* 441 (1950).

14. Sievers, A. J., *J. Opt. Soc. Am., 68,* 1505 (1978).

15. Moon, P., *J. Franklin Inst., 230,* 583 (1940).

16. Trotter, D. M. and Sievers, A. J., *Appl. Phys. Lett., 35,* 374 (1979); Trotter, D. M. and Sievers, A. J., *J. Opt. Soc. Am.,* in press.

17. Seraphin, B. O., *Thin Solid Films, 39,* 87 (1976).

CHAPTER 10

CORROSION SCIENCE AND ITS APPLICATION TO
SOLAR THERMAL ENERGY MATERIAL PROBLEMS[1]

Steven L. Pohlman

Solar Energy Research Institute[2]
Golden, Colorado

I. INTRODUCTION

In Chapter One, a basic overview of the role that materials
play in the development of solar energy was presented. The
importance of understanding the limitations imposed upon a system
by the materials employed was emphasized. As with other energy
technologies, it is becoming clear that basic material problems
must be addressed and resolved before the successful advancement
of solar energy can be anticipated. The purpose of this chapter
is to explore the material question once again, but this time
concentrate on problems that may limit the development of solar
thermal systems due to material failure. Failure in this sense
includes chemical, mechanical, or combined modes of degradation.
It is important that, during the early stages of solar thermal
development, basic material problems be identified so that a
proper research and development program can be established to
address the major issues.

[1]*This work is supported by the Division of Solar Technology,
U.S. Department of Energy (DOE), under contract EG-77-C-01-4042.*
[2]*U.S. DOE facility*

The objective of the first section of this chapter is to review each solar thermal technology and discuss the anticipated
material problems, and to present to the reader a ranking of
their currently perceived relative importance. The second section
of this chapter will present the basic theories used in corrosion
and electrochemical research and show their application to laboratory studies. Weight loss, direct current (dc), and digital alternating current (ac) methods of studying the kinetics of corrosion will be described and an example of how weight loss data
were employed to determine the corrosion kinetics of a thermal
receiver will be presented.

Solar thermal conversion, like other developing technologies,
has experienced its share of material problems. For example,
shortly after the start-up of the linear concentrator and freon
turbine loop at the Willard, New Mexico solar irrigation site,
the system developed excessive pressure and blew a rupture disc.
An investigation revealed that build-up of corrosion product in
the working fluid created by the corrosion of the steel piping
caused the problem. Also the freon working fluid contained water,
and at operating temperatures, hydrochloric acid was being generated. Fortunately, the problem was discovered and solved before
excessive material damage had occurred (1).

In another case, a medium temperature concentrating solar
thermal system was forced to shut down shortly after initial testing because corrosion product caused extensive erosion and eventual destruction of the high pressure circulation pump. Unfortunately, the corrosion problem was not quickly solved and thousands
of dollars of damage occurred (2). There are many other examples
of material failures of short-lived solar thermal systems which
show that these problems cannot be ignored and that they are life
limiting.

Material degradation in one form or another appears to be a
common problem facing the development of solar thermal power
(3-5). The corrosion damage anticipated in solar is similar to
that found in other energy technologies and, therefore, knowledge
and experience that has been gained earlier should be incorporated
into any comprehensive research plan. The different forms of
corrosion that will be discussed can be divided into categories
of uniform, local, galvanic and stress assisted attack. Uniform
attack includes all forms of surface degradation that do not appear to be selective in nature. This type of attack is generally
the easiest to prevent by closely controlling the chemistry of
the working fluid or by the use of corrosion resistant coatings.
Local attack includes pitting, crevice corrosion, erosion, and
selective leaching and is more difficult to control than uniform
attack. By proper system design and proper selection of construction materials and working fluids, the damage occurring from this
form of attack can be mitigated. Galvanic attack includes corrosion caused by surface potential differences arising from thermal
transients, stray currents, cementation of metal ions, and multi-
metal contact. This form of corrosion can be controlled by
proper design that insures the elimination of galvanic cells.
Stress assisted corrosion/cracking includes stress corrosion
cracking, corrosion fatigue, creep/creep-fatigue, and hydrogen
damage. This form of material degradation is the most insidious
of all forms because frequently no overt signs of damage will
appear prior to catastrophic failure. Proper selection of material and control of working conditions can help diminish the like-
lihood of this type of attack   (6-8).

A complete list of definitions of the terms used throughout
this chapter is included in Appendix A.

II. MATERIAL PROBLEMS AND SOLAR THERMAL CONVERSION

Due to the importance of examining specific material problems
associated with the proposed solar thermal systems, a workshop
was organized to identify and assess anticipated material problems
which could limit their performance or feasibility. Of those
material problems identified, priorities were assigned according
to their potential for limiting solar thermal development. A
lot of the information generated at the workshop will be presented
in this chapter and procedings are available from the Solar Energy
Research Institute information division   (9).

The highest priority was given to the understanding of creep,
creep-fatigue, thermal and corrosion fatigue, and their interaction in high temperature thermal conversion systems. Thermal
energy conversion and transport systems are exposed to high temperature working fluids and experience unique diurnal temperature
cycles that result in severe creep and creep/fatigue interactions.

The development and/or optimization of materials for unique
solar applications was identified as the second highest priority
issue. The data base of relevant existing information must be
expanded to cover the solar thermal needs. For example, high-
temperature receivers that are inexpensive and can withstand
thermal stresses and diurnal cycles must be developed. Flat-
plate collector systems must not only be inexpensive, but must
employ compatible materials. Optical concentrators must be made
from the materials which resist rapid degradation due to outdoor
exposure. New mirroring processes and mirroring materials must
be developed to provide the solar industry with high quality,
durable mirrors. Other solar energy conversion technologies,
such as thermal storage and ocean thermal energy conversion
(OTEC) also present unique material requirements.

The development of accelerated life tests used to predict long-term reliability received the next highest priority. It was also concluded that inexpensive, reliable, easily employed, and maintained corrosion monitoring devices must be developed.

The development of industrial standards and design rules especially for solar application was given the next highest priority. For example, solar thermal conversion differs substantially from nuclear conversion in terms of safety and failure consequences, yet the container materials of both systems are subjected to the same pressure vessel code. New design rules and standards must be developed that are specific for the actual needs and requirements of solar thermal energy.

Reviewing the major material concerns anticipated for the various solar thermal systems it should be obvious to the reader that many of the same material problems occur in the various solar thermal technologies but differ slightly in degree of severity due to different operating conditions. The operating condition for each thermal conversion system that will be discussed is given in Table I. It is the objective of the following portion of this chapter to discuss the specific material degradation problems such as creep, fatigue, stress corrosion cracking, oxidation, etc. for each of the solar energy technologies and list these in order of importance.

Material problems of thermal concentrating systems that employ gas, liquid metal, molten salt, water or organic fluids as a primary heat transfer fluid will be discussed first. Degradation of the reflective surfaces used to concentrate thermal energy will be followed by a discussion of material problems associated with thermal storage. Flat plate collector systems will then be reviewed and finally the material problems facing ocean thermal conversion systems will be presented.

TABLE 1. *Operating Conditions for Solar Thermal Systems*

Maximum System[b]	Operating Temperature	Working Media	Special Working Conditions
High Temperature			
Gas systems	800° C to 1400° C	air, helium	controlled purity He
Molten salt[b]	250° C to 600° C	alkali nitrates and nitrities	chlorides, carbonates, sulfates as impurities
H_2O/steam[a]	500° C to 570° C	H_2O	deionized and treated (NH_3 and hydrazine) water, 100 to 200 kg/cm^2
Liquid metal[b]	570° C to 700° C	Na, NaK	controlled O_2, C, N
Organic fluid	250° C to 400° C	petroleum and silicon fluids	
Storage:			
Sensible heat	Ambient to 150° C	MgO, rocks, Al_2O_3	
Thermo-chemical and latent heat	moderate temperature to 300° C	CH_4, H_2O, products $Ca(OH)_2$, CaO, H_2O, SO_3, SO_2, O_2	
Medium Temperature			
H_2O	200° C to 350° C	H_2O	treated water (coordinated phosphate), 200 kg/cm^2
Oil	150° C to 300° C	synthetic oils, distillate cuts	oxidized in air
Low Temperature			
H_2O	Ambient to 90° C	H_2O	200° C stagnation temperatures are possible
H_2O/alcohol	Ambient to 90° C	H_2O/ethanol	freezing conditions
Chemical storage Ammoniated/ methanolated salts reactions		$MgCl_2 \cdot 6\ NH_3$ $CaCl_2 \cdot 4\ NH_3$ $CaCl_2 \cdot X\ MeOH$ $CaCl_2 \cdot H_2O$	
Phase change storage Glaubers salt	30° C to 90° C	$Na_2SO_4 \cdot 10\ H_2O$	
OTEC	20° C to 40° C	sea water	periodic cleaning required due to biofouling

[a]Single-pass steam generator receiver tubes experience departure from nucleate boiling (DNB) within the receiver tubes and are therefore subject to high cycle fatigue.

[b]All systems are subject to diurnal cycling plus more rapid thermal cycling from intermittent cloud cover. The liquid metal and molten salt systems will experience thermal stripping (Thermal fatigue induced by fluid flow).

A.   Thermal Concentrator Systems

Advanced central receiver concepts being studied by the U. S. Department of Energy (DOE) are pushing the limit of receiver material technology. (5). Several types of gas, sodium, salt, organic, and water cooled prototypes are being investigated. Materials research is being performed on ceramics, metals, coatings, and their interaction with the various heat transfer fluids. Although some conditions in solar-thermal power systems are similar to those in conventional nuclear and fossil fired systems, the thermal fluctuations caused by diurnal cycling and by intermittent cloud cover represent a thermal and cyclic stress environment not common to the conventional technologies. The major material failure problems anticipated for each solar thermal technology as dictated by the various heat transfer fluids and operating conditions will be discussed. Chapter one briefly describes each of these systems and is recommended for readers who are not familiar with the various solar thermal systems.

1.   *Gas-Cooled Systems*

a.   *Creep, Fatigue, and Their Combination in Solar Receiver Material.* Characterization of the behavior of both metallic and ceramic receiver materials under creep, fatigue, and creep-fatigue conditions, that result from steady state loads, mechanical vibrations, and thermal cycling conditions is required. The characterization should include both developing and validating predictive methods to permit extrapolation of relatively short time test data to desired design lifetimes. Materials under current design consideration include nickel and cobalt base alloys (e.g., Inconel 617, Haynes 188, Incoloy 601) and ceramics, including silicon carbide and silicon nitride. Metallic receivers should be reliable up to about $950^{\circ}C$ and ceramic receivers to about $1400^{\circ}C$.

b. Oxidation of High-Temperature Alloys and Nonoxide Refractories. The long term durability of super alloys and silicon carbide as solar receiver materials must be demonstrated in the total solar receiver environment. The particular problem is one of structural stability under oxidizing conditions at temperatures in the range of 600 to 1400°C. Breakaway oxidation reactions will occur at some point in time. This could result in failure by an accelerated uniform surface recession or localized attack and penetration that would weaken the material. Currently available data on materials have not been obtained under the complete solar environment and may not accurately represent actual behavior. More importantly, it is difficult to extrapolate data with respect to time of exposure due to the nature of oxide breakaway and, therefore, long-term exposure tests are required.

c. Slow Crack Propagation in Ceramic Receivers. Slow crack growth can occur as a result of localized creep or by stress corrosion cracking in the ceramic and metallic components of the receiver tubes and structural supports. Such crack growth can be driven by residual stresses and by operating stresses caused by thermal gradients generated under operating conditions.

d. Fretting and Wear of Ceramic and Alloy Components. Fretting and wear of metals and ceramics is a potential problem associated with the joining and sealing of metals and ceramics particularly when sliding or line contact compression joints are used to accommodate effects of thermal cycling. Metal-to-metal interactions at elevated temperatures can lead to welding of the metals and subsequent mechanical failure. Likewise metal-to-ceramic and ceramic-to-ceramic contacts can result in unacceptable wear and/or high stress concentration due to frictional forces.

e. Alternative Materials for Very High Temperature Solar Receivers. Silicon carbide has been selected as the material of construction for all current receiver designs that will operate

above 900°C. Although results of preliminary efforts promise
success for SiC as a material of construction for this application,
there is no experience that demonstrates suitability for long-
term application. It is possible that work during the next 2-5
years may reveal that SiC is deficient in one or more areas. If
this is the case, no alternate candidate material has been identified, and programs to develop high temperature solar receivers
will encounter serious delays.

 f. Microstructural Instability in High-Temperature Alloys.
Many candidate alloys for use in high-temperature, gas-cooled
systems undergo significant microstructural change during elevated
temperature service. These changes usually alter the mechanical
properties significantly. For example, thermal aging may reduce
ductility and toughness. Therefore, detailed information on
operating parameters is required to permit creation of reliable
designs for high temperature systems. Candidate materials for
high temperature structures, such as Inconel 617, Incoloy 601,
Haynes 188, etc., must be evaluated. Exposure temperatures should
cover the range of interest and exposure times should be long
enough to permit extrapolation of the predicted lifetime of the
alloy with fair confidence.

 g. Carburization/Decarburization of High-Temperature Alloys.
Certain of the central receiver concepts utilize high-pressure
helium to remove the solar heat from the receiver. Such systems
typically employ a Brayton cycle gas turbine to generate electricity. Pure helium gas is chemically inert; however, it is
difficult to assure a complete absence of impurity gases. These
impurities can arise from a variety of sources, including leakage
of air or of oil from rotating equipment. The contamination of
the container walls with these impurities can alter the chemical
potential and physical characteristics of the surface sufficiently
to allow the onset of carburization or decarburization. This

problem is particularly important when helium coolant gas is employed, since the oxygen potential of the gas is generally so
low that protective oxides do not form on metal surfaces.

   *h. Degradation of Insulation and Changes of Emissivity
Caused by Sintering and/or Devitrification.* Ceramic insulation
materials are used in high-temperature, gas-cooled heat exchangers primarily to protect structural metals. Fibrous ceramics
are the present candidate materials since low weight and maximum
insulating value are principal requirements and strength requirements are minimal. In a cavity receiver, fibrous insulation can
aid in redistributing the radiant thermal energy by acting as a
diffuse reflector. Thus, a fibrous insulator with a very low
absorptivity will function as a good diffuse reflector and,
when exposed to direct radiation, provide increased thermal protection. By improving thermal distribution within a cavity,
fibrous ceramic insulation aids in reducing thermal gradients in
the heat exchanger tubes within the cavity. Problems with fibrous insulation are related to changes in insulating value and,
to a lesser degree, emissivity.

   2. Liquid Metal and Molten Salt Systems

   *a. Creep/Fatigue Behavior of Container Materials in Molten
Sodium or Molten Salt.* Material behavior under creep-fatigue
conditions is extremely important in determining the lifetime
performance of solar receiver tubes. The absorber tubes will
be subjected to daily cyclic loads from shut-down and start-up
operations, with additional cycles imposed upon the system from
intermittent cloud cover and focusing and defocusing cycles.
Thus, the absorber tubes will be expected to undergo hundreds of
thousands of strain cycles over a 30-year service life.

   *b. Physical and Chemical Data (400-600°C) on Alkali
Nitrate Salt Mixtures.* Mixtures of molten alkali metal nitrates

and nitrites (i.e., HITEC) have been used extensively as heat
transfer agents. Both these and mixtures of alkali metal nitrates (e.g., draw salt) are being considered for use as heat
storage media. For economic reasons, interest is mainly in the
sodium and potassium salts. While the chemical and physical
properties of the above systems are fairly well known at temperatures below 400°C, information on their properties at temperatures up to 600°C is scarce.

   c. Sodium Corrosion and Mass Transport Studies. Present
solar system designs specify the use of austenitic steels and
nickel-rich alloys in the receiver, austenitic and ferritic
steels in the piping, ferritic steels in the steam generators,
and carbon steel in the low-temperature storage tanks. Metallic
mass transfer and carbon transfer, which will occur during operation and be complicated by the presence of the various metals,
may be enhanced by the potentially higher oxygen content of the
sodium employed for solar use. Future high temperature designs
will probably require specifying new receiver materials with
unknown sodium corrosion/carbon transport behavior. Therefore,
corrosion and transport studies under simulated or actual environments are required.

   d. High-Strength Ferritic Alloy Development. Alternative
structural materials to the 316 and 304 austenitic stainless
steels and Incolloy 800 are needed for use in sodium-heated
steam generators. These materials are prone to chloride and
caustic stress corrosion cracking, and are sensitized at
500-600°C in weld heat affected zones. They have high expansion
coefficients and lower thermal conductivity than ferritic steels,
and in proposed systems require transition weld joints. The
austenitic stainless steels are already ASME code approved,
and it would be advantageous to develop and obtain ASME code
acceptance of high strength ferritic steel that could withstand

temperatures up to 650°C.

   e. Thermal Striping and Thermal Fatigue Damage. Material
evaluation under thermal cycling conditions of the proposed
sodium central receiver system is extremely important. In addition to diurnal and intermittent thermal cycling, thermal striping, a phenomenon seen in the nuclear steam generators, will
occur in the receiver and piping. Both of these thermal cycling
phenomena can cause failure due to a thermal fatigue mechanism
and should be investigated.

   f. Forced Convection Loop Studies of Molten Nitrate Salt.
The Martin Marietta Company conceptual design for a central receiver incorporates molten sodium nitrate as the primary heat
transfer medium in both the receiver and thermal storage subsystems. The major concern in the qualification of this salt
for the intended application is the need for quantitive corrosion
information above 400°C.

   *g. Development and/or Qualification of Coolant Containment
Alloys for Receiver Operation to 700°C for Second-Generation
Plant Designs.* Present design strategies for both sodium-cooled
and molten salt-cooled solar thermal conversion systems are based
on present steam generator temperatures and technology. This is
the proper technology base for the short term, but both the
problems and opportunities in the solar thermal technology are
different from those in nuclear technology and, over the long
term, many benefits may result from using higher receiver operating temperatures.

   h. Caustic Cracking of Steam Generator Tubes. The possibility of caustic attack or stress corrosion cracking of the water/
steam circuit (steam generator) of a sodium-cooled central receiver is great. Changes in coolant chemistry may happen as a
result of unplanned intrusion of sodium or sodium hydroxide.

These intrusions result from ineffective control of chemical
constituents in the feed water, as well as from leaks between
the sodium and water/steam circuits. The resistance of container
materials exposed to typical solar environments and to caustic
cracking needs to be explored.

   i. Stress Corrosion Cracking of Alloys by Molten Salts.
Identification of the temperature ranges and stress levels which
cause stress corrosion cracking when molten salts are in contact
with container materials and their weldments is needed. It is
evident that considerable knowledge and expertise are already
available from the nuclear energy industry. However, some interesting and important differences between the two energy-generat-
ing approaches also became apparent. For example, temperature
cycling, resulting from changes in the cloud coverage or the
day-night cycle is likely to lead to unusual high-temperature
creep/fatigue effects, and these will be advanced by the presence
of the chemically active liquid metal/salt environments. Very
little technical data, needed for safe design, is available on
such effects, and this information will have to be generated as
soon as possible. Another difference between solar and nuclear
systems is the absence of radioactive substances. Hopefully,
this will allow use of construction under less-restrictive standards, hence reducing the cost of the construction, maintenance,
and operation of solar liquid metal and salt loops. Another
difference will be the presence in solar loops of large storage
tanks of molten metal or salt, these being needed to damp out
the temperature oscillations noted above. These will be expensive to construct and maintain, but the number required in any
given loop could be reduced if the maximum operating temperature
in the loop could be increased from 600° up to $700^{\circ}C$. This
seems possible with sodium loops, but will require the qualification of alloys not presently in use for these applications. For

salt loops, it will be necessary to obtain more basic chemical
operating data in mixed alkali nitrate salts on the mechanical
reliability of the construction materials in contact with them.

    j. Secondary Problems. The following problems are not
considered as significant as those previously discussed but do
warrant consideration: degradation of external containment surfaces by oxidation, pitting, sensitization; control of the
chemical constituents in the sodium coolant; control of sodium
fires; improved nondestructive testing of flaws or cracking;
wear and galling of parts due to clean up by sodium and molten
salt; on-line monitoring of the coolant; heat flux and asymmetric
thermal stress effects on corrosion and oxide scale spalling in
the salt coolants; establishment of the mechanics of fracture
mechanics of receiver materials in a sodium or salt environment;
corrosion and degradation of internal (storage tank) insulation
materials in sodium and salt; and development or modification
of material standards and codes.

    3. Water/Steam Cooled Systems

    a. Methodology and Basis for Receiver Design. The mechan-
ical reliability of the receiver, particularly as a function of
the thermal and environmental variables encountered in this application, is of primary importance. These are the two major issues
in this area. First, the applicability of the existing ASME
high-temperature design code (Code Case 1592) needs to be examined. This code often requires detailed elastic-plastic-creep
analysis, which is both expensive and time consuming to apply,
and may not be appropriate for the solar receiver. For example,
during operation the hot side of a receiver tube is loaded in
axial compression which is not appropriately reflected in the
code. The code would also demand ultraconservative designs which
would economically penalize solar applications. Development and
validation of a simpler design basis compared to that required

of a nuclear system would be a very useful step in advancing solar technology.

Secondly, data are lacking in a number of areas that would facilitate developing reliable, economical receiver designs. High-temperature design data are available on relatively few alloys (i.e., 304 and 316 stainless steel, Incoloy 800, and 2 1/4 Cr-1 Mo). Even with these relatively well known materials, information is lacking on the creep-fatigue interaction, particularly in the multiaxial aspects; on corrosion fatigue in the relevant aqueous and steam environments; on the effects of low-frequency cycling stress superimposed on crack growth rate; about long term creep, particularly with respect to temperature transients; about creep-fatigue and corrosion-fatigue on weldments and bimetallic joints; on the fundamental, time-dependent inelastic, deformation behavior under multiaxial conditions and solar environments; and on the potential importance and role of residual stresses.

   b. *Coolant Chemistry.* There are five general concerns regarding the maintenance of coolant chemistry and the interaction of the working fluid with container materials. First, the selection of feedtrain and condenser materials with respect to coolant technology and systems design must be considered in order to minimize corrosion product generation during operation and the diurnal lay-up cycle. This selection process will decrease the probability of receiver-tube corrosion or burn-out. Secondly, the frequency and difficulty of cleaning receivers which is required to maintain proper chemical resistance and thermal transfer properties, needs to be determined. Thirdly, the extent of thermally shock-induced spalling of the natural oxides of heat exchanger tubes needs to be determined. Fourthly, the effect of soluble contaminants in superheated steam needs to be measured

for their effect on promoting localized attack based on the
diurnal wet/dry cycle. Finally, the long-term compatibility of
the selective coating on the outside diameter of metallic receiver tubes needs to be studied.

4. *Aqueous and Oil Systems*

a. *System Design--Choice of Coolants/Container Materials.*
The choice of coolant and the selection of container materials
should be integrated decisions. For example, the choice of an
organic coolant will reduce system operating pressures and make
possible the use of thinner wall sections than necessary for an
aqueous coolant, and may reduce the necessity for corrosion control. On the other hand, a substantial inventory of organic
coolant may be necessary, and continuous clean-up capability may
need to be installed. The decisions about the coolant and container material should be made with joint input and designers
must be advised that changing from one coolant to another will
have ramifications with regard to overall design and material
selection.

b. *Thermal and Thermal-Mechanical Stability of Selective
Coatings.* Selective coatings are essential to the technical
feasibility of mid-temperature (150-350°C) thermal systems. These
coatings must have high absorptivity to solar radiation and low
emissivity at the operating temperature. They must not be degraded by either thermal history or environmental interactions.
They must be produced economically and have consistent properties.
While the technology for black chrome seems adequate to meet
current needs, its degradation mechanism is not well understood.
Perhaps, the currently used black chrome process could be improved, or a new coating technology could be developed to further
the design options.

c.   Thermal and Corrosion Fatigue of Receiver Tubes. The primary piping and receiver tubing will be subjected to a non-uniform load history resulting from thermal cycling of the concentrator heat source, to impulse overloads due to wind and movement of coupled structures, to high-cycle, low-amplitude loads due to vapor locks in aqueous cooling systems, and to differential thermal expansion at seals and couplings at the receiver. These combined loadings are likely to be more variable and extreme than in process-heat boilers due to the unique thermal cycling, longer piping runs, varying degrees of insulation, and greater use of dissimilar materials. Therefore, thermal and corrosion fatigue of the receiver tubes and primary heat transfer systems needs to be investigated.

d.   Control of Critical Parameters. Success of solar thermal designs depends on awareness of the critical parameters, experience of existing technology, and careful and continual control of variables. For example, for a water-cooled carbon steel system to be reliable, there must be a control of water chemistry (pH, redox potential, dissolved oxygen and solids) to avoid operating under corrosive conditions. Similar controls are required independent of the alloy selection. The nature of solar conversion systems will compound the problems compared with conventional steam generators operating under the same temperature and pressure boundaries.

e.   Standardized Information on Organic Coolants. Physical and chemical data on organic heat transfer fluids and systems are not readily accessible to designers and builders of solar systems. Lack of uniform service simulation tests has resulted in conflicting and misleading performance data. Potential users of organic heat transfer fluids often lack bases for evaluating comparative performances of different fluids and therefore, find it difficult to determine how different fluids will perform in their systems.

Solar system designers/builders are much less familiar with
organic heat transfer technology than with water/steam technology.
Therefore, they express preference for water/steam systems· even
when there are technical advantages favoring organic coolant
working fluids.

5. *Solar Concentrators*

a. *Development of Industry Standards and Test Procedures.*
At the present time there are few standards or test procedures
that are generally accepted by the industry for many materials
used for solar concentrators. This situation exists with respect
to metals, glasses, and polymers, and is due in part to the rapid
growth of the industry. Accordingly, manufacturers and test
laboratories are often unable to compare data on the properties
and performance of materials. For example, when a mirror vendor
states that his product has a weathering resistance of six years,
a potential user has no idea what this means, since many times
no standard test conditions are specified. Specific tests should
be developed for evaluating weatherability of the mirrors that
can be related to loss of reflectance and estimated life.

b. *Degradation of Metallic Reflective Coatings.* Reflective
surfaces normally lose their reflective properties as a result of
chemical degradation of the metallic surface. The chemical degradation is probably caused by corrosion of the reflective surface resulting from the combination of water and at least one
other contaminant. A loss in reflectance results in a decrease
in performance so either more concentrator are required or the
mirrors have to be replaced. In existing systems, the corrosion
of the silvered surface has been the major reason for loss of
reflectance. Extensive degradation has been observed within a
few months of atmospheric exposure, as discussed in more detail
in the chapter by Czanderna.

   c. *Degradation of Reflective Support Materials.* Substrate
or superstrate materials support and/or protect the reflective
surface. Degradation or failure of supporting materials may
shorten the useful life of the concentrator.

   d. *Interfacial Adhesion of Materials.* Good interfacial
adhesion between materials, i.e., metal/glass, metal/polymer,
and metal/metal, is of primary importance for long term reliability of solar concentrator systems. Reflective metals on glass
or polymers must maintain good adhesion to have a cost-effective
lifetime. Weakly bonded silver, for example, will flake or peel
away from the glass resulting not only in a loss in reflectance
but will provide further reactive sites for moisture and other
aggressive contaminants. Protective copper coatings over silver
reflective coatings must adhere to the silver to impede the in-
gress of corrosive species. Protective organic coatings over
metallized surfaces must adhere to them in order to provide protection from moisture, atmospheric pollutants, etc.   Structural
adhesives used to join the structural back-up panel to the reflective surface must adhere in order to maintain the structural
integrity and dimensional stability of the subsystem. Sealants
must adhere intimately to panel edges (glass, plastic, metal,
wood, etc.) to stop moisture or airborne pollutants from inward
migration and eventual attack of the metallized reflective surface. As indicated, good adhesion is essential for all concentrator components and the effect of thermal cycling and long-term
environmental exposure can create bond degradation at an interface even though initial adhesion was good.

   e. *Surface Contamination by Environment.* Solar collector
systems inevitably accumulate dust or dirt on surfaces exposed
to the environment. The rate of accumulation depends on a number
of parameters including system design, mode of operation, materials of construction, weather, and location. Nevertheless,

degradation of system performance can be expected due to both
absorption and scattering of light by the contaminants.

   f. Dimensionally Stable Material and Systems. The long
required lifetimes of solar energy components, specifically concentrators, require dimensionally stable materials. This specific problem deals not only with the need for a long-term stable
structure, but also with the dimensional stability of bulk materials used in the construction of these concentrator systems.

   g. Development of Improved Reflective Materials. Mirrors
currently available commercially, were not designed for application to solar systems. Degradation by erosion and/or corrosion
will significantly reduce system performance. The future of the
solar energy program depends on being able to generate power
reliably or convert its thermal energy to a useful output at
competitive costs. An improved corrosion-resistant mirror assembly is needed to guarantee that both objectives are met.

   6. Thermal Storage

   a. Sensible Heat-Dual Media Systems. The underlying materials reliability concern of sensible heat storage is compatibility
of the dual media with each other and with the containment vessel.
The primary motivation for designing storage systems to incorporate a second material other than the working fluid is economic
(i.e., to reduce the amount of the expensive working fluid). A
secondary issue is that a steep thermocline cannot be maintained
in a single medium. Present and anticipated dual-phase storage
concepts include hydrocarbon oils with granite, molten salts with
taconite (iron ore), air or gas with refractories, and air with
rock.

   Inevitably, cyclic-elevated temperatures will degrade both
the solids, and may promote detrimental interactions between the
two. For example, hydrocarbon oils will readily crack at high
temperatures in the presence of oxygen, react with the granite,

and form organic acids which could then corrode the storage
vessel. Also, the liquid or gas may cause fragmentation of the
rock by combined action of chemical attack and thermal stress.
Thermal-stress fracture and corrosion of rock are probably coupled phenomena, as they are in most ceramics. It would be useful
to search for ceramics or rocks which are more resistant to one
or the other aspects of this phenonemon. For example, taconite
is more resistant to chemical attack than is granite; hence,
the former is used in molten salt systems. Ability of the rocks
to arrest cracks should be separated from their absolute resistance to fracture. In other words, crack propagation may be quite
tolerable if it ultimately ceases before the rock fractures.
Candidate solid storage media should be ranked on the basis of
appropriate thermophysical properties. Representative tests are
needed to guide the development of economic systems with adequate
thermomechanical and chemical stability for both near-term demonstrations and long-range application.

Representative tests should be developed, and conducted on
proposed systems. These tests must accurately duplicate operating conditions, (i.e., be capable of scale-up to full size). The
tests must incorporate mass transport, heat transfer, and cyclic
operation. For example, thermomechanical and chemical stability
of rock and brick in the presence of working fluids and under
cyclic temperature conditions is of specific concern. Contamination of the liquid by the solid should be addressed because it
may affect the heat transfer characteristics of the fluid, or
induce chemical attack on the containment vessel. Potential
solid media (rocks, bricks, ceramics) should be rank-ordered with
respect to critical thermophysical properties, and then with
respect to chemical stability. Increase in storage vessel diameter by ratchetting may be induced by the cyclic thermal stresses

and, therefore, should be investigated. Analytical predictions are conflicting, and some well-designed tests to resolve the issue definitely would be useful. Environmental considerations including the effects of working fluid vapors on local atmosphere, potential spillage of nitrate salts, and disposition of the products of chemical degradation within the storage vessel, should be addressed.

b. Latent Heat Thermal Storage Systems. The highest priority problem area is a requirement to establish compatibilities of the inorganic hydrates, organics (waxes), eutectics, and nitrate compounds identified for storage use with containment materials. Molten salts are currently the leading candidates for the high-temperature thermal storage in large scale solar systems. The normally corrosive characteristics of molten salts are aggravated by the cyclic temperature conditions encountered in solar systems. Containment is the single most important concern, since large storage components, once in place, will be difficult and expensive to repair.

Compatibility tests including both localized and general corrosion between working fluids and containment materials should be conducted and the effects of varying the operating parameters such as flow velocities and fluid chemistry should be studied. In all tests, cyclic temperature control must be featured to reflect solar needs. Working fluids should include inorganic hydrates, organics, urea-based eutectics, and alkali nitrates/ hydroxide systems. Container materials should include stainless, ferritic, and carbon steels, as well as common building materials.

A continuing basic research effort to obtain fundamental data relevant to predicting the stability of candidate container materials in contact with their proposed high-temperature storage media is required to increase the cost effectiveness of development programs. Research investigation employing redox potentials, polarization curves, Pourbaix diagrams, and reaction kinetics is essential.

    c. Thermochemical Energy Storage Systems. Thermochemical energy storage utilizing reversible chemical reactions have several advantages over sensible and latent heat process. Among these, higher energy density, long-term and ambient temperature storage capabilities and lower energy capacity cost are the most important.

    Work to date has identified several promising types of reversible reactions such as thermal decomposition reactions (e.g., $CaO/Ca(OH)_2$), and solution/dissolution reactions (e.g., SO_3/SO_2). A system based on sulfuric acid concentration has been tested successfully in a small pilot-type unit. Cost considerations and materials selection appear to be the major unresolved issues.

    Important issues affecting reliability of several promising processes vary according to system type. For example, in the thermal decomposition reactions, the major outstanding problem is lack of knowledge and understanding of the long-term dynamic behavior of reactant/product particles. For thermal decomposition reactions, such as ammonium hydrogen sulfate decomposition, the main reliability issue is lack of understanding and knowledge of the long-term kinetic reversibility of the reaction steps involved.

    The lack of long-term data on catalyst performance under temperature cycling conditions is an important unresolved problem in the catalyzed type of reactions. The corrosion and selection of construction are issues of lower priority at this stage of development. Based on the present state of development of process conditions in these systems, processes involving sulfur and oxygen species at elevated temperature are judged as providing the most severe conditions for material containment. An additional complication in some of the designs considered is the presence of thermal cycling of the system components.

7. Flat-Plate Collectors

The most conventional thermal conversion system is the flat-plate collector used for space and water heating. Although solar heating and cooling is technically feasible, it is not yet economically competitive with conventional heating and cooling methods. The main barriers to commercialization are the high initial cost and low reliability associated with liquid-cooled, flat-plate collectors. The annual costs of solar systems are primarily dependent upon investment recovery and thus, the expected lifetime of the system is an extremely important factor.

High costs and corrosion of the absorber plate are the major problems with flat-plate collectors. The most widely used absorber materials are copper, aluminum, carbon steel, and stainless steel. Except for its high cost, copper is the most attractive metal because of its excellent corrosion resistance in aqueous media and its good thermal properties. Aluminum is less expensive and lightweight, but is susceptible to severe galvanic and localized attack. Although carbon steel is the least expensive in dollars per pound, it's general corrosion characteristics makes the use of thicker absorber plates necessary. Certain stainless steels exhibit excellent corrosion resistance and look very promising. Plastics have also been considered, but they generally cannot withstand stagnation temperatures (150-230°C) or pressures (about 1.73 mpa) of high-performance flat-plate systems.

Protection against freezing is necessary for most systems and this complicates corrosion problems for all the candidate materials. Presently, ethylene or propylene glycol/water solutions are most frequently used. Unfortunately, glycols degrade at temperatures near their boiling point to form corrosive organic acids. Since stagnation and high temperatures are likely to occur, inhibition, buffering, monitoring, and periodic replenishing of the solution are necessary to prevent corrosion by such acids. The use of glycols with proper treatment is effective, but the complicated water chemistry and maintenance make it a complicated procedure for typical homeowners. Organic fluids, such as paraffin oils and silicone oils, are possible alternatives to aqueous antifreeze solutions. They are noncorrosive, but their expense, high viscosity, and poor thermal properties make their use less practical at this time.

Another approach to the freezing problem is the use of water (an excellent heat transfer fluid) along with automatic drainage of the collector at low temperatures or at night. Untreated water could possibly be used with copper or stainless steel absorber plates, but distilled or deionized water and some inhibition must be used for aluminum or carbon steel collectors. The main disadvantage of the drainage method is that the corrosivity of water increases dramatically upon exposure to air, and frequent drainage would create corrosion problems unless some protective scheme, such as, vapor phase inhibition, inert gas blankets, or chemically adsorbed coatings was used.

The reliability of flat-plate collectors must be improved before space heating and cooling of buildings becomes competitive with traditional heating and cooling methods. Protection against freezing and corrosion must be developed that is less complicated, inexpensive, and more reliable than current methods. The enthusiasm with which the public greets the solar energy program will be strongly influenced by the success of the most visible component of solar energy systems, the flat-plate solar collector.

a. *Materials Selection*. Generally materials of construction have not been chosen on the basis of adequate supporting data. All too often the choice of materials has been left to low-technology fabricating organizations with no appreciation of

the potential dangers from corrosion, etc. It is important that
sufficient long-term data be assembled to permit optimization of
material selection for both air- and liquid-type flat-plate solar
collector systems. The following material problems have been
identified and could be mitigated by the proper material selection. Corrosion of metallic materials by interaction with the
working fluid or by weathering of external surfaces; degradation
of non-metallic materials under conditions of stagnation, ultraviolet exposure and thermal cycling; degradation of coatings
(applied to external surfaces) caused by weathering, thermal
cycling and ultra violet radiation; degradation of the insulation
materials; and degradation of the heat transfer fluids.

   b. Monitoring Performance of Materials. Corrosion damage
which proceeds undetected may lead to unexpected catastrophic
failure of solar collectors. The potential for such failures is
especially pronounced in systems with recirculated liquids.
Localized corrosion (pitting, crevice corrosion) is particularly
insidious since perforations can occur with relatively little
total loss of metal. The corrosiveness of recirculated fluids
may change with time under operating conditions as the result of
breakdown of organic chemicals, exhaustion of inhibitors, etc.
Thus, initially unreactive fluids may become corrosive. A low-
cost, reliable, on-line monitoring system is needed to detect
changes that show fluids have become corrosive. If changes are
detected early enough, timely corrections can be made.

   c. Cleaning and Maintenance. To maintain good thermal
efficiency, surfaces of flat-plate collector systems must be
cleaned without surface degradation. Cleaning is normally performed without disassembly of the apparatus, and thus chemical
cleaners must be compatible not only with the specific part to be
cleaned, but also with all other adjacent parts. This includes
all metallic and nonmetallic collector materials (metals, plastic,

glass, sealants, etc.).

A greater appreciation is needed concerning the importance
of cleaning procedures to the long-term reliability of solar
conversion systems. Corrosion of the solar panels may result
from inadequate cleaning and maintenance, use of improper cleaning methods, or the use of corrosive cleaning agents. Procedures
should be developed and cleaning agent identified which may be
used by nontechnical personnel with minimum supervision in the
cleaning of solar apparatus.

8. Ocean Thermal Energy Conversion (OTEC)

Another solar thermal technology that is capable of generating large outputs of energy is ocean thermal energy conversion.
(10).

The problem of biofouling and corrosion in heat exchanger
tubes exposed to a primary heat transfer fluid seawater is a
major technical obstacle to making OTEC power plants economically
feasible. The most crucial problem is the loss of heat transfer
efficiency associated with biofouling and corrosion of the heat
exchanger surface on the seawater side. Because of the small
temperature differences involved in the heat transfer process
in OTEC systems, heat transfer surfaces must be maintained at
optimum efficiency. Since the cost of the heat exchanger comprises nearly 50% of the cost of an OTEC plant, the efficiency of
the heat transfer process is an important economic consideration.
The loss of heat transfer efficiency can be associated with all
types of corrosion and biofouling, but of most importance is the
formation of primary organic films. These films are the results
of adsorption of organic materials and living organixms on the
exposed surfaces. They begin to form almost immediately upon
exposure to marine invironments. A film 0.25 mm thick can reduce
heat transfer efficiency by a minimum of 50%. Corrosion products,

macrofouling (attachment of algae, barnacles, fungus), and scale
(chemical precipitates from seawater) also contribute to and complicate the problem   (11).

Biofouling can also accelerate the corrosion process. Micro-
organisms (bacteria) can directly accelerate anodic dissolution
(corrosion) of the surface or they can create a corrosive environment by producing localized concentrations of acid. Macroorgan-
isms can ititiate crevice attack and produce organic acids which
create corrosive conditions   (12).

Presently, titanium is the preferred material for heat exchanger tubes because it shows good corrosion resistance and is
compatible with ammonia (the preferred working fluid) and aqueous
mixtures of ammonia. Other materials being considered are aluminum, 90/10 copper/nickel, high nickel/chromium/molybdenum iron-
base alloys, and plastics. The cupronickel alloy is an excellent
material for marine use because the toxic effect of copper on
marine organisms makes it the least susceptible to fouling. However, using copper is questionable because of severe stress-corr-
osion compatibility problems with ammonia, the proposed working
fluid. Aluminum is very attractive because of cost advantages,
but its predicted lifetime is much shorter than that of titanium.
The predicted lifetime of a plastic heat exchanger is even
shorter.

Research is required to characterize all candidate materials
with respect to localized attack and the effect of cleaning on
the long-term compatibility of the heat exchanger material   (13).

III. CORROSION SCIENCE

The identification of the critical material issues that are
confronting the expeditious development of solar thermal energy
is only the first step. Once the anticipated material problems

have been identified initiating and completing the research required to address and solve the problems become the next objectives. Since the material problems have been identified it is
appropriate to discuss the various scientific techniques by which
these material problems may be examined. First, a brief discussion of the basic theories of corrosion science will be presented
and then a description of some of the laboratory techniques which
apply these theories will be given. It is not practical at this
time to discuss all the laboratory techniques available to examine the interaction of a material with its environment, so an
emphasis will be placed upon describing the electrochemical techniques employed to evaluate nonmechanical electrolytic degradation.

A. Corrosion Kinetics

The investigation of corrosion kinetic processes is normally
pursued by either measuring weight change or monitoring electrochemical reactions. The weight change method gives the total
magnitude of the corrosion reaction(s), integrated over both
time and area of the corroding surface. An increased understanding of the processes involved and how they change with time may
be obtained by using electrochemically induced polarization techniques which continuously measure the rate of a reaction(s) as
an electrical current. Polarization curves are therefore a
graphical representation of the current to an electrode surface
as a function of an applied potential. The current is a measure
of the rate of a reaction or reactions on the electrode, and it
is usually dependent on the environment in which the measurements
are carried out. The measurement of polarization curves in different environments thus provides a rapid and sensitive method
for the evaluation of corrosion rates and aids in the selection
of the material or the environment which may give superior
service performance. Such techniques have been employed in the

fields of alloy development, inhibitor research and application, localized corrosion, and industrial corrosion monitoring and control   (14-16).

The direct current electrochemical methods (i.e. linear polarization, and Tafel polarization) may be used to determine corrosion rates and the change in rate with time if the change is slow enough. Alternating current techniques have also been used to monitor corrosion rates but the variation with time is usually very tedious to follow. For determining rates, the ac measurements must be made at several different frequencies, and if the corrosion rate changes before the necessary frequency range can be swept, errors will result   (17). In principle, however, the ac measurement techniques are very powerful and provide more information than dc methods. Both of these methods will be described in greater detail.

B. *Corrosion Thermodynamics*

The useful energy that may be derived from a chemical reaction is the Gibbs free energy (ΔG) of the reaction. The free energy can be obtained as electrical energy in an electrochemical cell and the two are related by the expression $\Delta G = - n F E$, where ΔG is the free energy change per mole of reactant, n is the number of electrons involved in the reaction per mole of reactant, F is Faraday's constant, and E is the cell voltage   (18). The free energy provides the driving force for the reaction to occur, and a larger (more negative) driving force produces a larger cell voltage.

Consider the reaction $Zn(metal) + 2H^+ = Zn^{+2} + H_2(g)$, where the reaction is carried out in an electrochemical cell. In the cell the reaction can be considered as half-reactions, one involving the zinc oxidation, $Zn = Zn^{+2} + 2e^-$ and the other the

hydrogen reduction, $2H^+ + 2e^- = H_2$ (g). The separation of the
two reactions causes electrons to build up at the anodic electrode and to be depleted at the hydrogen electrode. This creates
a cell voltage where the zinc electrode is negative with respect
to the hydrogen electrode. Useful work may then be derived by
allowing current to flow in the external circuit. Different
metals will produce different voltages with respect to the hydrogen electrode, allowing the metals to be ranked in the electromotive force series by these voltage differences   (19).

There is a relationship between the free energy change and
the cell potential of an electrochemical reaction. If the free
energy change for a given reaction is negative, the reaction is
thermodynamically possible. If the sign is positive, the reaction will not proceed unless the conditions are altered. Therefore, a measureable electrochemical variable, the cell potential,
can be related directly to a thermodynamic quantity.

The application of thermodynamics to corrosion processes has
been greatly enhanced by the efforts of Pourbaix who implemented
the use of potential -pH diagrams to predict the direction of
reactions, estimate corrosion products by measuring solution pH
and corrosion potential, and to predict environmental conditions
that would be benign to various materials   (20).

The Pourbaix diagrams are based upon the conditions of the
Nernst equation:

$$E = E^O + \frac{RT}{\eta F} \ln Q$$

where E^O is the standard half cell potential, R is the gas constant, T is the absolute temperature, η is the number of electron
transfered, F is the Faraday constant, and Q is the activity of
the oxidized species divided by the activity of the reduced
species   (21). An example of a Pourbaix diagram is shown in

Figure 1. One must be careful in using the Pourbaix diagrams because they are subject to the same limitations as any thermodynamic calculation, but they can be extremely valuable if used properly   (22). Once it has been determined that thermodynamically the reaction may occur, it then becomes desirable to use electrochemical equations to calculate the rate at which the reaction will take place.

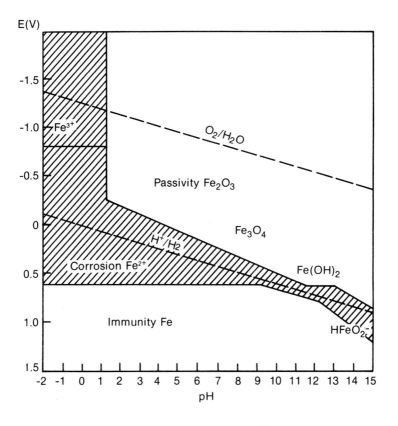

FIGURE 1.   Potential Versus pH Diagram for Iron.

C. *Electrochemical Relationships*

The rate of an electrochemical reaction is measured by the current flow to the electrode surface. The current is related to the driving force for the reaction by the expression

$$i = i_o \left[\exp \left(\frac{\alpha_a F}{RT} (E - E_o) \right) - \exp \left(- \frac{\alpha_c F}{RT} (E - E_o) \right) \right] \quad (1)$$

where i_o is the exchange current density, α_a and α_c are kinetic parameters, and $(E - E_o)$ is the driving force in volts   (23). The driving force is the actual measured potential of the electrode minus its equilibrium potential, and is termed the "over-potential." The overpotential is also denoted by the symbol η in the electrochemical literature. At equilibrium, there is zero net current since the anodic reaction rate is balanced by the reverse cathodic reaction rate. The magnitude of this (balanced) current at equilibrium is i_o, the exchange current density. The exchange current density, i_o, is a function of the concentration of the oxidation and reduction species in solution, and is affected by the physical state and condition of the surface such as the roughness of the surface or the presence of films or contaminants on the surface   (24).

For most corroding systems, multiple chemical reactions occur simultaneously on the corroding surface. The simplest way to treat multiple electrode reactions is to take the net current as the sum of the partial currents for each reaction. This practice is useful only if the characteristics of the individual reactions are known. Multiple reactions are encountered, for example, in the corrosion of a zinc electrode in an acid solution. The zinc electrode now has two reactions which occur simultaneously, the zinc dissolution reaction and the hydrogen discharge reaction. The zinc electrode will adopt a potential which is

intermediate between the reversible zinc potential and the
reversible hydrogen potential. This "mixed" potential is the
potential at which the total zinc anodic dissolution current is
balanced by the total hydrogen ion reduction current. Each
reaction drives the other until the balance is achieved. The
net current density, i_{total} is equal to i_a (the anodic current)
minus i_c (the cathodic current) and is equal to zero at the
mixed or corrosion potential   (25).

D. Corrosion Rate Calculations

Wagner and Traud's mixed-potential theory, as described above,
forms the basis for two electrochemical methods used to determine
corrosion rate. These are the Tafel extrapolation and linear-
polarization techniques   (26).

1. *Tafel Extrapolation Method*. The Tafel extrapolation
method uses data obtained from cathodic and anodic polarization
measurements and is based on the current to potential relationship given in equation (1). The exponential behavior of current
density with potential allows the anodic contribution to the
total current density to be ignored if the specimen potential is
50 mV more negative than the open circuit value, while the cathodic contribution can be ignored at potentials 50 mV more positive than the open circuit value. Using this principle of electrode polarization corrosion currents can be calculated   (27).

By polarizing a working electrode at least 50 mV negative of
the open circuit potential and then polarizing the same electrode
at least 50 mV positive of the open circuit electrode and plotting the data on a semilogarithmic plot, the corrosion current
density value can be obtained. This is shown in Figure 2.

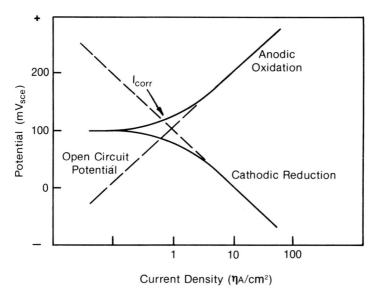

Current Density (ηA/cm^2)

FIGURE 2.   Tafel Extrapolation Method for Estimating Corrosion Rate.

Under controlled conditions, the results of the Tafel ex-
tropolation method has correlated well with rates calculated
using conventional weight-change methods. Using this technique
it is possible to measure extremely low corrosion rates, and to
perform the measurements in a short period of time. Although
these measurements can be performed rapidly with high accuracy,
there are numerous restrictions which must be met before this
method can be used successfully. To ensure reasonable accuracy,
the linear region of the semilogarithmic plot (Tafel region)
must extend over a current range of at least one order of magnitude. In many systems this cannot be achieved because of the
interference from concentration polarization and other extraneous
effects. Furthermore, the method is difficult to use with systems containing more than one reduction process, since the Tafel
region is usually distorted and a linear plot cannot be obtained
(28).

2. *Linear Polarization Method.* The linear polarization technique is based on the theory that the current is activation controlled (i.e., not diffusion controlled) and that the relationship between current and potential is linear when the overpotential is very small. The mathematical basis of this theory, as developed by Stern and Geary, extracts the corrosion rate from the slope of the polarization curve, called the polarization resistance, as it passes through the corrosion potential.

Stern and Geary derived the relationship (29) between the polarization resistance and corrosion rate;

$$R_P = \frac{\Delta E}{\Delta I} = \frac{1}{2.3 I_{corr}} \left(\frac{B_a B_c}{B_a + B_c} \right) \qquad (2)$$

where R_P is the polarization resistance, B_a and B_c are the Tafel constants.

The polarization resistance can then be measured experimentally as the slope of the linear polarization plot at the corrosion potential as shown in Figure 3. The corrosion rate (I_{corr}) can then be calculated by substituting values into equation (2).

The Tafel constants, B_a and B_c, may be approximated for a quick, semiquantitative estimate of corrosion rate or they may be measured experimentally using semilogarithmic Tafel plots. (30). The latter will give more accurate corrosion rate but requires additional experimental work.

3. *Polarization Characteristics.* Besides aiding in the calculation of corrosion rates, the shape of polarization curves can be used to determine the relative susceptibility of an alloy to localized attack (31). By examining the polarization curves of a material susceptible to pitting, the pitting and protection potentials can be determined. The potential at which the anodic current density begins to dramatically increase as the potential reaches higher values in the passive region is called the pitting

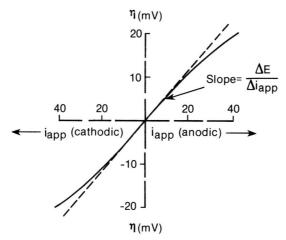

FIGURE 3. *Linear Polarization Method of Estimating Corrosion Rate.*

potential. The potential at which the current returns to a passive current density upon its return scan is called the protection potential   (32).

In very simplified terms, the pitting and protection potentials divide the corrosion characteristics of a metal into three regions with regard to electrode potential. At potentials below the protection potential, the metal will not pit or crevice corrode; between the protection and pitting potentials, the metal may be susceptible to both pitting and crevice corrosion; and at potentials above the pitting potential, the alloy or metal will undergo very rapid failure. Therefore, after the pitting and protection potentials are determined, the corrosion potential or natural electrode potential of the material will show whether the system is residing in a region of general corrosion, possible pitting and crevice corrosion, or extremely severe localized corrosion. Figure 4 indicates the position of these various potentials on an idealized curve. Potential and polarization measurements can also be effectively employed to predict and

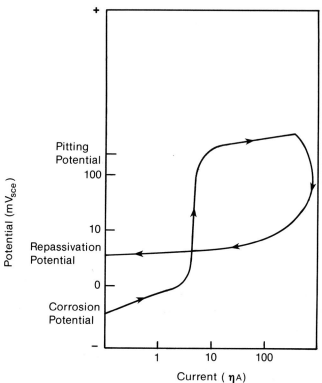

FIGURE 4. *Polarization Curve Showing Corrosion, Pitting and Repassivation Potentials.*

measure galvanic corrosion   (33).

4. *Digital AC Impedance Measurements.* Direct current elect-rochemical techniques are extremely powerful in analyzing corrosion processes. The advance of modern digital analysis has also made it possible to consider digital alternating current (ac) impedance measurements as a viable technique for corrosion rate termination   (34). The important characteristic of the digital system is that of high speed computation, and in partic-ular, the use of the Fast Fourier Transform (FFT). This permits the transformation of the data from the time domain into frequen-cy domain, and as described below, permits the immediate calcu-lation of the ac impedance as a function of frequency. Thus a

multiple frequency spectrum input which covers the necessary
frequency range can be used to determine the corrosion rate
rapidly and accurately.

In discussing the digital FFT technique, it is convenient to
use the idealized equivalent electrical circuit for the metal-
solution interface   (35). In this approximation, the interface
behaves as a "leaky" capacitor as shown in Figure 5. The electrical double layer capacitance, C_{dl}, is in parallel with the
Faradaic reaction, R_p. The Warburg impedance, W, represents the
effect of diffusion in solution on the electrochemical reaction.
The solution resistance term R_s, which is taken to be in series
with the interfacial terms, is normally considered the magnitude
of the ohmic drop in solution caused by current flow between
the reference electrode and the working electrode.

In systems where the diffusion impedance is not important,
the total impedance of the circuit as a function of the frequency
is written in complex notation as (36).

$$Z(\omega) = Z_r(\omega) - iZ_i(\omega) = R_s + \frac{R_P}{1 + \omega^2 C_{dl}^2 R_P^2} - i\frac{C_{dl}R_P^2}{1 + \omega^2 C_{dl}^2 R_F^2} \quad (3)$$

where i is the $\sqrt{-1}$, ω is the frequency, R_p is the Faradaic
corrosion resistance, R_F is the solution resistance, and C_{dl} is
the double layer capacitance.

The use of the above relationship to describe a corroding
metal has other limitations in addition to that imposed by ignoring the effect of the Warburg diffusion factor. The most important limitation is that the system be "linear," and for the
Faradaic corrosion reaction to be linear, the ac signal must be
of small amplitude (i.e., 1-2 millivolts). Also, the system must
be in a steady-state over the time period of the measurement.

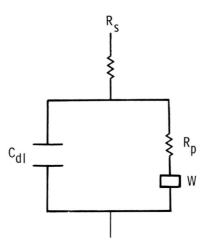

FIGURE 5. Equivalent Circuit of the Impedance of an Electrochemical Cell.

The use of digital data acquisition and signal analysis has made it possible to automate the measurements and to greatly reduce the time required to obtain the data. The key elements in digital signal analysis are the sampling of a time varying signal to obtain N samples with an interval spacing of t, so that the total sampling period is $N\Delta t$; the transformation of this discrete set of samples to a discrete frequency set, usually performed by some variation of the Fast Fourier Transform computing algorithm; (37-38) and the calculation of the impedance as a function of frequency from the voltage and current spectra.

Once the voltage and current data have been obtained, and the Fourier Transform calculated, the impedance may be calculated. The impedance "transfer function" may be shown to be

$$Z(\omega) = \frac{V\omega}{I(\omega)} = \frac{\text{Fourier Transform of Voltage Spectrum}}{\text{Fourier Transform of Current Spectrum}}$$

at each frequency, ω, of the discrete set. Essentially, one applies noise spectra of known frequency content, such as Gaussian

"white noise," and the impedance function is known immediately
at these frequencies.   Imposing the multiple frequencies simultaneously then avoids both the tedium involved in sweeping the
measurement frequency and the uncertainty involved with variation
of the interfacial processes in time   (39-41).   In the treatment
of data, (42-45), the most useful plot is a plot of the imaginary
part of the impedance versus the real part, as a function of
frequency   (46).   From the intercept and slope of the
platted linear data the values for C_{dl}, R_p and R_s can be
determined.

The experimental system to be described here is based on a
Hewlett Packard 9830 Programmable Calculator.   This calculator
with 16 K words of memory controls the repeated acquisition of
digital data on a Biomation 1015 Transient Waveform Recorder.
The calculator initiates the data-recording cycle and controls
the readout of data into the calculator memory for further
manipulation.   The transient recorded operates as a buffer,

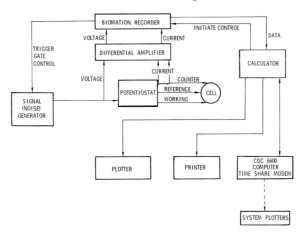

*FIGURE 6.   Block Diagram of the Digital Signal Analysis
System.*

wherein data is stored on its memory (1024) words/channel, with
4 channels) until it is called back into the calculator.

Data are recorded as the voltage input impedance to a poten-
tiostat, and the current output from the same instrument. The
voltage input signal (Gaussian white noise) is supplied by a
Hewlett Packard 3722A Noise Generator, which provides repeated
sequences of pseudo-random noise. The sequence length is an
integral multiple of the periods of the frequencies of the signal,
which thus minimizes round-off errors. The noise generator is
operated so that 102 frequencies are used in the input spectra.
A summary of the various elements of the digital system are
shown on the block diagram (Figure 6). The data transmission
lines and the logic control line are Hewlett Packard 11202A
Interface Cables.

The technique is rapid and experimental results show it to be
quite reliable   (47). It may be applied to systems with a wide
range of corrosion parameters, and solution conductivities.

5. *Corrosion Science Application.* It should be concluded
that both dc and ac electrochemical techniques are extremely
useful and can be employed to investigate corrosion processes
with proper limitations   (48). All of the ac and dc techniques
described have been used to investigate corrosion processes related to solar thermal energy ranging from the investigation of
the corrosion of mirrored surfaces to the characterization of
aluminum alloys for use as OTEC heat exchangers. Weight-change
measurements, the simplest corrosion rate determination technique,
can also be used to develop the mathematical models that can be
used to predict the life expectancy of a thermal conversion
system. As an example, a recent investigation of a parabolic
dish thermal conversion system will be outlined   (49).

A solar concentrator purchased for the evaluation of its

thermal and electrical performance used a cavity-type receiver
mounted at the focus of a 6 m parabolic reflector. Figure 7 is
a schematic diagram showing the receiver and reflector mounted
on the tracking base. The manufacturer had recommended that the
receiver be operated at temperatures up to 980°C in order to
deliver high-pressure steam to an experimental turboelectric
generator. The cavity was fabricated from Inconel 600 alloy and
the steam coil and receiver walls were constructed of ASI-type
316 stainless steel. In operation, solar radiation is focused
into the cavity by the mirrors which heats the aluminum-filled
interior of the receiver. The aluminum then conducts heat from
the cavity interior to the steam coil. The aluminum melts at
temperatures above 660°C, making the energy of fusion thus stored
available to generate a continuous supply of steam even if the
solar flux is briefly interrupted by occasional cloudiness.

Aluminum is a good heat transfer and storage medium because
of its high heat of fusion, high thermal conductivity, relatively
low melting point, and availability. However, because iron,
Inconel, and ferrous alloys are extremely soluble in aluminum,
(50-52), the reliability of the receiver at temperatures above
the melting point of aluminum was questioned. Concern was expressed that a dissolution-induced weakening of the receiver
wall might cause a failure resulting in the release of molten
aluminum or dissolution and failure of the steam coil might cause
a failure that would release highly-pressurized wet steam into
the interior of the receiver vessel, which is not designed to
withstand high pressures. The resultant explosion could scatter
molten aluminum over a wide area. Also, momentary confinement
of the steam in the receiver vessel could lead to the generation
of gaseous hydrogen in amounts sufficient to ignite upon release,
(53), greatly increasing the violence of the explosion.

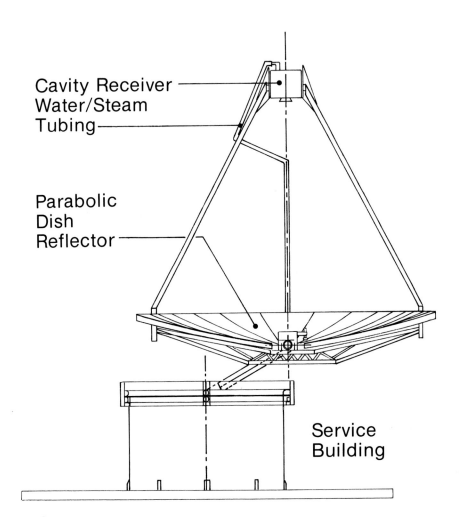

Cavity Receiver
Water/Steam
Tubing

Parabolic
Dish
Reflector

Service
Building

FIGURE 7. *Point-Focus Solar Concentrator.*

With these concerns in mind, a study was performed to determine the effect of molten aluminum, similar to that employed in
the receiver, on Type 316 stainless steel. The goals of this
effort were to assess the rate of attack of aluminum on 316
stainless steel; to predict a useful lifetime for the steel components of the receiver under exposure to molten aluminum through
study of the temperature dependence of the attack rate; and to
compare these predictions to actual data gathered from tested
receivers.

To accomplish these goals, samples of Type 316 stainless
steel were immersed for varying lengths of time in crucibles
containing molten aluminum. The crucibles were maintained at
several temperatures ranging from $700^{\circ}C$ to $930^{\circ}C$ in order to span
the range of operating temperatures proposed for the receiver.
After immersion, the steel samples were cleaned and weighed to
determine the amount of steel lost by dissolution. Using these
data, a dissolution rate model was constructed which enabled
lifetime predictions for steel receiver components to be made.
These predictions were then checked against operating data gathered from actual receiver tests. To provide additional information, the interior portion of another receiver containing the
aluminum filling was tested in a furnace at $820^{\circ}C$.

Laboratory tests on samples of stainless steel of the type
used in the receiver were conducted to determine the mechanism
and rate of attack after it was determined that dissolution of
steel components of the receiver was a primary failure mode.
Steel samples immersed in molten aluminum in a laboratory furnace
showed a rapid growth of hemispherical pits accompanied by an
exponential increase in weight loss with time. A simple geometric model for the pitting process was developed which correlated
well with the weight loss data (See Figure 8).

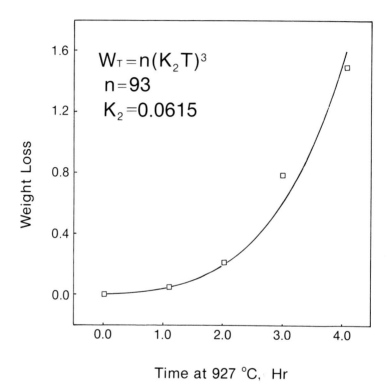

FIGURE 8. Fit of Pitting Rate Model to Steel Coupon
Weight Loss Data. Correlation Coefficient of Linearized
Data = 0.99.

The model equation was

$$W_T^{1/3} = n^{1/3} K_2 t, \qquad (4)$$

where W_T is total sample weight loss, n is number of pit initiation sites, and K_2 is a rate constant. Using this relation, values of K_2 for the dissolution process were determined at $700^{\circ}C$, $760^{\circ}C$, $820^{\circ}C$, $870^{\circ}C$, and $930^{\circ}C$, and an activation energy of 17.5 kcal/mole was calculated. Validity of the correlation was established using standard statistical analyses of variance.

The general expression for the rate constant K_2 enabled use of the pitting model to predict failure time envelopes for the receiver steam coil and container walls. The maximum (upper limit) lifetimes of these components were predicted to be 94 and 172 hr at $700^{\circ}C$, respectively. The lower and upper limits of the envelopes were determined at the 95% level of confidence.

Mean predicted lifetimes and lifetimes predicted at higher temperatures were much shorter. For example, at $820^{\circ}C$, a mean lifetime of only 17 hr was predicted for receiver steam coils immersed in molten aluminum. A receiver module, consisting of the steel receiver interior, underwent a violent steam coil failure after 12 hr of testing in a diffusion furnace at $820^{\circ}C$. In conclusion, the materials used in the design were not compatible and results showed that there was a need to involve material scientists at an early stage of the system design. The use of weight changes or electrochemical methods to predict life performance is extremely important and is becoming even more important for solar applications, due to the high capital investment costs of solar installation and long life requirements.

IV.   CONCLUSION

By the application of thermodynamic data, environmental
observations and electrochemical measurements, chemical and
mathematical models of corrosion processes can be developed and
used to effectively predict corrosion failure and can be employed
to develop accurate and meaningful accelerated testing techniques.
The future of solar energy is not solely dependent upon the
capability of material scientists to determine means to prevent
material failure but the advance of solar energy can be severely
hindered by a lack of preventative procedures to decrease the
frequency of failure. Earlier in this chapter, most of the major
material problems facing the solar thermal industry were identified, now it is necessary that a sound analytical research
approach be employed to address the problems and increase the
reliability of the materials and subsequent reliability of the
various solar thermal systems.

REFERENCES

1. Pohlman, S.L., and Dees, D., Sandia Laboratory Internal
    Memo, New Mexico (1977).

2. Pohlman, S.L., and Dees, D., Sandia Laboratory Internal
    Memo, New Mexico (1977).

3. Pohlman, S.L., "Corrosion of Solar Energy Systems: An
    Overview", NACE Publication No. 198, (March, 1979).

4. Pohlman, S.L., "Materials and Corrosion in Solar Energy
    Systems, Materials and Corrosion Problems in Energy
    System", p. 10-1, NACE, New Orleans, (February 1980).

5. Pohlman, S.L., and Swearengen, J. (eds.), "Containment
    Materials for Solar Thermal Energy Transport Systems: An
    Overview of Research and Development Needs. SERI/MR
    334-347.

6.   Uhlig, H.H., "Corrosion and Corrosion Science", John Wiley
     & Sons, N.Y., (1971).

7.   Fontana, M.G., and Breene, N.D., "Corrosion Engineering",
     McGraw-Hill, N.Y., (1978).

8.   Shreir, L.C., "Corrosion", Newnes-Butterworths, London,
     (1978).

9.   Staehle, R., and Pohlman, S.L. (eds.), "Reliability of
     Materials for Solar Energy". Proceedings Vols. 1 and 2,
     SERI/TP-31-248, (1978).

10. The National Research Council, "Selected Issues of the
     Ocean Thermal Energy Conversion Program," CON-EX-76-C-01-2,
     (1977).

11. Fetkovich, J., et al., "Fouling and Corrosion Studies in
     OTEC-Related Heat Exchanger Tubes," TD27028, (1976).

12. Gray, R.H. (ed.), "Proceedings of the OTEC Biofouling and
     Corrosion Symposium," PNL-S-A-7115, (1977).

13. Draley, J. (ed.), "Proceedings of the OTEC Biofouling,
     Corrosion, and Materials Workshop," ANL/OTEC-BCM-002,
     (1978).

14. Bockris, J., and Reddy, A., "Modern Electrochemistry",
     Vol. 1 and 2, Plenum, N.Y. (1977).

15. Smyrl, W., and Pohlman, S.L., "Experimental Application of
     Design Principles in Corrosion Research," *Corrosion,*
     (April, 1979).

16. Jones, D.A., "Ind. Eng. Chem. Prod. Res. Develop.,", Vol.
     II, No. 1, (1972).

17. Pohlman, S.L., "Determination of Corrosion Parameters by
     Electrochemical Techniques," NACE (October, 1978).

18. Moore, W., "Physical Chemistry," Prentice Hall, (1962).

19. Uhlig, H.H. (ed.), "Corrosion Handbook", Wiley, N.Y.
     (1948).

20. Pourbaix, M., "Atlas of Electrochemical Equilibrium in
     Aqueous Solutions," Pergamon Press, London, (1966).

21. Scully, J.C., "Fundamentals of Corrosion," Pergamon, (1975).

22. LaQue, F., "Shortcuts to Prediction of Corrosion Rates," *Materials Performance*, (June, 1979).

23. Robinson and Stokes, "Electrolyte Solutions," Academic Press, N.Y., (1955).

24. Oldham, K.B., and Mansfield, F., *Corrosion Science, 13,* No. 70, (1973).

25. Wagner, C., and Traud, W., *Electrochem., 44,* 391 (1938).

26. Tafel, J., *Phys. Chem., 50,* 641 (1905).

27. Mansfield, F., *Corrosion, 29,* 10 (1972).

28. Greene, N.D., *Corrosion, 18,* 136t-42t (1962).

29. Stern, M., and Beary, A.C., *J. Electrochem. Soc., 104,* 56-63 (1957).

30. Brown, R.H., and Means, R.B., *Trans. Electrochemical Soc.,* 74 (1938).

31. Pourbaix, M., *Corrosion, 26,* 431 (October 1970).

32. Verink, Jr., E.D., and Pourbaix, M., *Corrosion, 27,* 495 (December 1971).

33. Pohlman, S.L., "Corrosion and Electrochemical Behavior of Boron/Aluminum Composites," NACE No. 30, San Francisco, (1977).

34. Smyrl, W.H., and Pohlman, S.L., "Determination of Corrosion Parameters by Digital Impedance Analysis," NACE, Pub. No. 166, (March, 1978).

35. Randles, J.B., *Faraday Soc. Discussions, 1,* 11, (1947).

36. Sluyters, J.H., *Rec. Trav. Chem., 79,* 1092, (1960).

37. Stearns, S.D., "Digital-Signal Analysis," Hayden, Rochelle Park, New Jersey, (1976).

38. Smith, D.E., "Electroanalytical Chemistry," A.J. Bard, (ed.), M. Dekker, Inc., N.Y., Vol. 1, (1966).

39. Smith, D.E., "Computers in Chemistry and Instrumentation," Vol. 2, M. Dekker, Inc., N.Y., (1966).

40. Smith, D.E., *Anal. Chem., 48,* 221A, (1976).

41. Creason, S.C., and Smith, D.E., *J. Electroanal. Chem., 36,* App. 1, (1972).

42. Rehbach, M., and Sluyters, J.H., *Rec. Trav. Chem., 80,* 469, (1961).

43. Rehbach, M., and Sluyters, J.H., *Rec. Trav. Chem., 82,* 553, (1963).

44. Rehbach, M., and Sluyters, J.H., *Rec. Trav. Chem., 82,* 535, (1963).

45. Rehbach, M., and Sluyters, J.H., *Rec. Trav. Chem., 82,* 525, (1963).

46. Cole, K.S., and Cole, R.H., *J. Chem. Phys., 9,* 341, (1941).

47. Smyrl, W., and Pohlman, S.L., NACE No. 166, (March, 1978).

48. Sheldon, D., "Electrochemical Methods of Corrosion Testing," NACE No. 113, (March, 1976).

49. Webb, J., and Pohlman, S.L., SERI/TR-31-338, (September, 1979).

50. Lelby, C.C., and Ryan, T.C., - ARCRL-TR-73-p421, Air Force Cambridge Research Laboratories, (1973).

51. Ecully, E., *Mem. Sci. Rev. Metall,* Vol. 72 (No. 11), pp. 815-822, (November 1975).

52. Fegel, O., - Farnholz, et al., *Aluminum,* Vol. 5, pp. 496-498, (August 1969).

53. Hess, P.D., and Brondyke, K.J., "Met. Prog.," pp. 93-100, (April 1969).

APPENDIX

GLOSSARY OF TERMS (6, 7, 8)

Breakaway Oxidation - a type of corrosion in which a finely-
    subdivided porous oxide layer forms at the metal-oxide
    interface.

Carburization/Decarburization - the process of providing a metal
    surface with a carbon layer to provide a harder, more wear-
    resistant surface. Consists of heating the surface in an
    atmosphere rich in Co or hydrocarbon gases at a temperature
    of 900-1000°C. Decarburization is the reduction of the
    carbon level by heating in air or other oxidizing or re-
    ducing gases.

Cementation - the process by which one substance is caused to
    penetrate and change the character of another, by interfaci-
    al interactions, at temperatures below their melting points.

Corrosion Fatigue - the reduction of fatigue resistance due to
    the presence of a corrosive medium. Strongly influenced by
    environmental factors. Prevented by reducing stress on the
    component or introducing corrosion inhibitors. Probably a
    special case of stress-corrosion cracking.

Creep - the continous plastic elongation of a metal under an
    applied stress. Increasingly important at high temperatures.
    Creep rate or resistance is usually expressed as percent
    plastic deformation for a given time period at a constant
    applied load.

Creep-Fatigue - stressing caused by repeated cyclic deformations
    in conjunction with long-term continuous deformation.

Crevice Corrosion - pitting corrosion occurring in metal crevices
    or other areas where a corrosive medium becomes stagnant.
    Can be combated by keeping corrosive medium in motion over
    metal surface.

Erosion - the wearing away of a material by friction processes
    of another material or liquid.

Erosion Corrosion - corrosion of metal when subjected to high-
    velocity liquids, characterized by pitting. Also called
    impingement attack. Examples of attacked materials are
    copper and brass condenser tubes.

Fatigue - the tendency of a metal to fracture under repeated
    cyclic stressing; failures usually occur at stress levels
    below the yield point after many cyclic stress applications.

Fretting - Damage occuring at the interface of two contacting
    surfaces (one or both metal), subject to slight relative
    slip, usually oscillatory or vibrational. Also called wear
    corrosion and friction oxidation. Characterized by dis-
    coloration of metal surface and pitting if oscillatory.

Galvanic Corrosion - if two dissimilar metals are placed in a
    conductive solution and are placed in electrical contact, a
    potential difference will exist which will produce an
    electron flow. This flow will be due to two reactions; an
    oxidation occuring at the anode, and a reduction reaction
    occuring at the cathode.

Intergranular Corrosion - localized attack at and immediately
    adjacent to grain boundaries with relatively no attack on
    the grains themselves.

Inhibition - substance which when added in relatively small
    concentrations to an environment decreases the corrosion
    rate.

Hydrogen Embrittlement - severe loss of ductillity in some
    stressed alloys on metals due to the absorption of atomic
    hydrogen into the metal lattice.

Linear Polarization - a method of electrochemical corrosion-rate
    measurement. Within 10mV more passive or more active than
    a material's corrosion potential, the applied current density
    is a linear function of the electrode potential. The slope
    of this Function, $\dfrac{\Delta E}{\Delta i_{applied}}$, is approximately equal to
    $.026/i_{corr}$ where i_{corr} is the current density for corrosion
    of the material .

Passivity - loss of chemical reactivity of a material due to the
    presence of a self-produced, self-healing film (oxide salt,
    etc.) which seperates the active metal surface from the
    environment.

Pitting - a form of extremely localized attack which results in
    cavities in the metals. By convention the surface diameter
    to depth ratio is less than one.

Polarization - the deviation from equilibrium potential between
    two metal electrodes in a cell, defined as the displacement
    of electrode potential resulting from a net current.
    Frequently measured in terms of over voltage, η.

Polarization curve - the graphic representation of a material's
    potential, taken by plotting the potential of an electrode
    of the material against the logarithm of applied current.
    The curve becomes linear at higher currents on a semi-
    logrithmic plot.

Pourbaix Diagram - a graphic interpretation of thermodynamic
    data relating to the electrochemical and corrosion behavior
    of a metal in water. Using potential vs. pH, the diagram
    indicates specific conditions in which the metal either does
    not react (immunity) or can react to form specific oxides
    on complex ions.

Selective Leaching - the removal of one component on element of
    a solid alloy by corrosion processes.

Sensitization - making an alloyed metal surface susceptible to
    intergranular corrosion by heating. The heating causes
    the alloy to disintegrate (grains of material fall out)
    and/or lose its strength.

Slow Crack Growth - crack extension caused by stress less than
    that of the materials critical stress.

Spallation - the cracking, breaking, or splintering of materials
    due to heat or impact stress.

Stress Corrosion - corrosion caused by the simultaneous presence
    of tensile stress and a specific corrosive medium.
    The metal is basically unattached over most of its surface,
    while fine cracks progress through it.

Stress Corrosion Cracking - if a metal, subject to a *constant*
    tensil stress and exposed simultaneously to a *specific*
    environment, cracks immediately or after a given time,
    that failure is called stress corrosion cracking (SCC).

Tafel Constant - $2.3\frac{RT}{\alpha nF}$, where R = gas constant, T = absolute
    temperature, n = number of electrons transferred between
    two electrodes, F = Faraday constant, α = symmetry coeffic-
    ient describing the shape of the rate-controlling energy

barrier. This constant is used in the equation describing
the relationship between reaction rate and overvoltage,
$\eta = B \log \frac{i}{i_o}$, where η = overvoltage, B = Tafel constant,
and i = rate of oxidation or reduction in terms of current
density.

Thermal Striping - thermal fatigue caused by thermally gradiated
   fluid flow.

Transpassivity - potential region in which the protective/passive
   film is no longer stable causing a rapid increase in reactivity of the substrate metal.

Uniform Attack - Removal of material by chemical or electriochem-
   ical reaction over an entire exposed surface is called
   uniform attack.

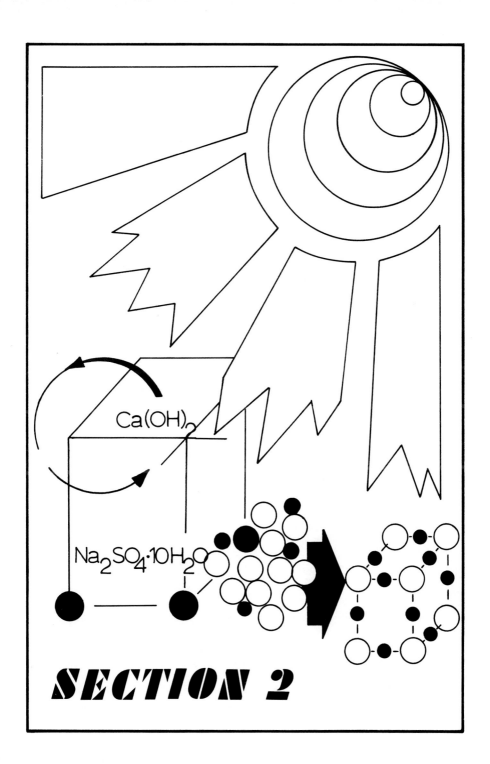

Ca(OH)₂

Na₂SO₄·10H₂O

SECTION 2

2. SOLAR STORAGE AND THERMOCHEMICAL MATERIALS

Many successful solar applications are dependent and will continue to be dependent upon the ability to store heat energy, or to be able to utilize collected solar heat energy at some future time. This can be achieved by a number of means, but the most economical and practical involve simple thermodynamic principles. To a large extent, solar storage systems and system economics are dictated by storage materials cost and the necessary storage volume. Two storage principles characterizing two categories of materials are described in this section. These involve the heat of fusion and the storage of solar heat in salts, salt hydrates and other materials as the heat of fusion, and reversible solar thermal reactions of various kinds in a variety of materials systems. The so-called heat of fusion materials generally require 15-18 times smaller volumes than rock storage systems and 8-10 times less volume than that required for water storage of solar thermal energy. Such materials are, with the addition of nucleating and thickening agents, able to undergo unlimited cycling, and this concept can be utilized in heating as well as cooling cycles. Heat of fusion salt storage is ideally suited to home heating and other low volume systems with space limitations. At higher temperatures and in systems involving large solar thermal power fluxes, photochemical or more appropriately thermochemical materials and reactions constitute an attractive and practical scheme for solar energy storage as well as transport. Thermochemical materials science is concerned not only with the materials reactions but also materials compatibility and related system problems involved in both heating and cooling.

The first two chapters of this section describe practical salt-hydrate storage systems, the heat of fusion concept, the thermodynamic basis for heat storage materials selection; and provide an overview of related materials problems. The second two chapters present an overview of thermochemical reaction applications to solar energy storage and finally an outline of associated materials science activities and problem areas, including effects of thermal cycling, heat transfer, corrosion and structural degradation, and the like. Examples of reaction control by product separation, energy transport in distributed systems, and the chemical heat pump concept are also presented. This section, together with the presentations in the previous section, should provide the reader with a practical and fundamental basis for assessing and understanding the materials science problems related to contemporary and future solar energy collection and storage.

CHAPTER 11

THERMAL STORAGE IN SALT - HYDRATES

Maria Telkes

Solar Thermal Storage Development
AEC Research Institute
Killeen, Texas

I. INTRODUCTION

Salt hydrates may be regarded as "alloys" of anhydrous salts
with a definite number of moles of water forming typical crystalline solids. The melting point of salt-hydrates is usually,
(but not necessarily), below $100^{\circ}C$, even if the melting point
of the anhydrous salt is much higher. Water itself has a rather
high entropy of fusion per unit weight, and when combined with
anhydrous salt of high entropy of fusion, a salt hydrate of
conveniently lower melting point may result, usually with the
combined high entropies of fusion of the components.

II. MELTING AND RECRYSTALLIZATION

Salt hydrates of the type $AB.nH_2O$, when heated, usually
change to another salt-hydrate $AB.mH_2O$ containing fewer moles
of water, while some part, or the entire amount of $AB.mH_2O$
dissolves in (n-m) moles of water. Several salt hydrates change
to the anhydrous form when heated, changing from $AB.nH_2O$ to
$AB + nH_2O$, as the anhydrous salt dissolves partly or completely
in mH_2O. The amount of water forming the salt-hydrate is the

SOLAR MATERIALS SCIENCE
377

"water of crystallization" and the traditional expression:
"melting in its water of crystallization" aptly describes the
process of transition at the melting point.

The solubility of the salt determines the condition above
the melting point. The composition of the salt-hydrate is
usually ex pressed in percent anhydrous salt AB,in the salt-
hydrate $(AB/AB.nH_2O)$. In Tables 1A and 1B, this is compared
with the solubility of the anhydrous salt above its melting
point. Congruent melting occurs when the solubility is sufficiently high. Partly incongruent melting occurs when some
salt remains undissolved above the melting point, or even at
some higher temperature. In such melts, the solid residue is
usually of greater density than the melt and it settles to the
bottom of the container.

When the partly incongruently melting salt-hydrate is subsequently cooled, while the mixture is stirred, recrystallization occurs without difficulty. When cooled in closed containers, without stirring or mixing, the solid residue at the
bottom of the container is "frozen in", being surrounded by
solid salt-hydrate crystals. Some of the residue, therefore,
cannot re-combine with its water of crystallization and some
saturated solution remains, after the mixture has cooled below
its melting point. In such cases the conditions are not completely reversible, that is, only part of the heat required
to melt the salt-hydrate can be recovered when the system is
cooled. Several methods have been developed to prevent the
settling of residual solids in partly incongruently melting
salt-hydrates, with the result that the conditions can become
completely reversible.

Table 1A.   Properties of Salt-Hydrates

Compound	H_2O Change	Melting Point $^\circ C$	Observed Heat of Fusion $\frac{cal}{g}$	$\frac{Btu}{lb}$	Entropy of Fusion Obs	Calc	$\frac{Calc}{Obs}$
$Ba(OH)_2$	8-0	78	72	130	64.5	53.3	0.83
$CaCl_2$	6-2	29.5	40.7	73	29.4	31.6	1.07
$Ca(NO_3)_2$ α β	4-2	42.6 39.7	33.5	60	25.0	25.5	1.02
$Cd(NO_3)_2$	4-3	59.5	25.3	45	23.5	30.7	1.30
$Co(NO_3)_2$	6-4	57	30.4	55	26.7	36.0	1.34
$CoSO_4$	7-1	96	40.7	73	31.5	40.1	1.27
$Cu(NO_3)_2$	6-4	24	29.4	53	29.5	36.0	1.22
$FeCl_3$	6-0	37	54	97	47.1	46.8	0.99
$LiNO_3$	3-0	30	71	128	28.7	29.0	1.01
$Mg(NO_3)_2$	6-4	90	38.2	68	27.0	36.0	1.33
$MgSO_4$	7-1	48	48.2	87	37.5	40.1	1.07
$MgCl_2$	6-4	117	41.2	74	21.6	21.7	1.01
$MnCl_2$	4-2	58	42.5	76	25.5	21.7	0.85
$Mn(NO_3)_2$	3-2	35.5	28.8	52	21.0	30.7	1.46
$Mn(NO_3)_2$	6-4	26	33.5	60	32.1	36.0	1.12
$NaC_2H_3O_2$	3-0	58	43.0	77	21.0	26.1	1.24
$NaOH$	1-0	64	65	117	11.2	11.4	1.02
Na_2CO_3	10-1	34	60	108	56.6	64.5	1.14
Na_2CrO_4	10-4	20	39	270	46	53.1	1.15
Na_2HPO_4	12-2	36.5	63	114	66	71.3	1.08
Na_3PO_4	12-2	70	52.5	94	58	71.9	1.24
$Na_2S_2O_3$	5-0	49			37	42.3	1.14
Na_2SO_4	10-0	32.4	60	108	63.5	62.3	0.98
$Ni(NO_3)_2$	6-4	57	36.4	65	33.3	36.0	1.08
$Zn(NO_3)_2$	6-4	36.4	31	56	29.7	36.4	1.23
PROP.$-136^\circ F$		58.0	63	113	25.6	26.1	1.02
PROP.$-190^\circ F$		94	62	112	77	84.6	1.10

Table 1B. Properties of Salt-Hydrates

Compound	H_2O Change	Density	Anhyd. Salt %	Solub. > mp %	Specific Heat Solid	Liquid
$Ba(OH)_2$	8-0	2.18	54.5	50	0.28	
$CaCl_2$	6-2	1.68	50.5	56	0.345	0.55
$Ca(NO_3)_2$	4-2	1.82	69.5	74	0.35	
$Cd(NO_3)_2$	4-3	2.45	76.5	73	0.26	0.38
$Co(NO_3)_2$	6-4	1,87	63	64	0.37	0.50
$CoSO_4$	7-1	1.95	55	45		
$Cu(NO_3)_2$	6-4	2.07	63.5	62	0.33	0.48
$FeCl_3$	6-0		60	76		
$LiNO_3$	3-0		56	61		
$Mg(NO_3)_2$	6-4	1.46	58	65	0.54	0.88
$MgSO_4$	7-1	1.64	49	39	0.36	
$MgCl_2$	6-4	1.56	47	50	0.38	0.68
$MnCl_2$	4-2	2.01	63.5	52		
$Mn(NO_3)_2$	3-2		77	v.s.	0.34	0.41
$Mn(NO_3)_2$	6-4	1.82	62.5	67		
$NaC_2H_3O_2$	3-0	1.45	60.5	60	0.47	0.80
$NaOH$	1-0		69	75		
Na_2CO_3	10-1	1.44	37	34	0.45	0.80
Na_2CrO_4	10-4	1.48	47	48		
Na_2HPO_4	12-2	1.52	40	44	0.37	0.76
Na_3PO_4	12	1.64	43	44		
$Na_2S_2O_3$	5-0	1.69 s 1.66 l	64	67	0.35	0.57
Na_2SO_4	10-0	1.46	44	34	0.425	0.79
$Ni(NO_3)_2$	6-4	2.05	63	64	0.38	0.74
$Zn(NO_3)_2$	6-4	2.07	64	67	0.32	0.54
PROP.-136°F		1.30	62	62	0.48	0.77
PROP.-190°F		1.64		78	0.36	0.80

III. SUPERCOOLING

Salt-hydrates heated in closed containers above their melting points, when subsequently cooled, may supercool considerably below their melting points, before they recrystallize. When melts are cooled below a definite temperature, (the temperature of labile supersaturation), crystal formation occurs spontaneously, but the difference between the real melting point and the temperature of labile supersaturation may be as high as 30 to 40°C (54 to 72°F). This is usually too large and impractical for heat storage systems. The typical example of water-ice is well known, supercooling to 40°C (-40°F) may occur before supercooled water freezes spontaneously. In supercooled melts, solidification or crystallization can be induced by nucleation or "crystal seeding".

IV. NUCLEATION OR CRYSTAL SEEDING

The literature of the problems of crystal formation and nucleation is very extensive, involving crystallography, phase equilibria, the physical-chemistry of solid-liquid transitions, and finally the fascinating subject of crystal growth and the formation of large, hyper-pure crystals. Reviews by Mullin (1), Van Hook (2), and Tipson (3), summarize most of the history, theory, and practical applications of these subjects.

For practical heat storage applications, the basic problem is the prevention of supercooling when the material is permanently sealed into its container. Supercooling must be prevented and reversible melting and recrystallization must be attained at the melting point, or within a few degrees of it, without opening the container. External influences must be avoided, such as stirring, shaking, or applying other mechanical or physical means, except the delivery or withdrawal of thermal energy.

The following methods can be used to attain nucleation of salt-hydrate melts, when sealed into containers and cooled slightly below their melting points:

a) Introducing solid crystals of the original salt-hydrate forming the melt. (Self-nucleation.)

b) Providing solid crystals which are isomorphous, epitaxial, or partly isomorphous, with the salt-hydrate crystals. (Heterogeneous nucleation.) These nucleating agents should not appreciably influence the melting point, or the heat of fusion of the salt-hydrate. Telkes (4) studied epitaxy, that is, nucleation of inorganic salt-hydrates by partly isomorphous crystals. Heterogeneous, nucleating crystals must be at least partly similar in crystal form, lattice spacing and atomic arrangement, within a range of 15 percent. This was originally found to be the condition of nucleation in metallic alloys (Hume-Rothery (5)).

c) Nucleation can occur in melts which are heated only slightly above their melting points. Crystals grow from the walls of the container, or from the surface of heterogeneous materials within the container. According to Turnbull (6), this observation can be explained by assuming that submicroscopic crevices are present in solid walls, and that minute crystals of salt-hydrate are lodged there. Interfacial tension increases the melting point of the salt-hydrate in the crevices slightly above the true melting point ("superheating"). The surviving solid crystals act as self-nucleating agents.

Additional methods suggested in the past included the use of nucleating devices in the form of tubular or other external attachments that are kept cooler than the melt and, in this way, retain solid crystals for subsequent nucleation. From the shape of these external devices, the name "cold finger"

describes them rather aptly. The performance of these devices
is best if they are near the top level of the container, but
in good contact with the melt, so that growing crystals can
protrude, break off, and provide numerous additional nuclei.
Crystals grow more slowly if they have to surmount gravity.
This occurs when growth starts from a tube or other small container located near the bottom of the melt that should be nucleated.

The patent literature is replete with various methods including supersonic devices, or thermal expansion disks, which
have not been successfully used due to high cost or other drawbacks. Relatively low cost proprietary devices exist which
have been used successfully (7).

V.   RATE OF CRYSTAL GROWTH

Extensive literature covers the subject of nucleus formation, supercooling and viscosity of the melt, or solution.
The classical work of Tamman (8) (9), proved that in supercooled
melts the rate of nucleus formation is very low near the melting
point, but increases rapidly at lower temperatures, reaching
a maximum and then decreasing again. The velocity of crystallization is usually expressed as the linear velocity of the
movement of the solid-liquid boundary in a tube of small diameter, containing the supercooled melt. The ice-forming velocity
of water, supercooled to $-9^{\circ}C$, is 6 cm/sec. (nearly 12 foot/min-
ute) while in salt solutions, at the same temperature, the
velocity is 1.0 to 4.0 cm/sec. (2 to 8 foot/minute). It is
probable, therefore, that a container of one foot length, filled
with supercooled salt-hydrate and nucleated at one end, should
be filled with a network of growing crystals in a matter of
minutes. During this short time, the temperature of the salt-

hydrate increases abruptly to its melting point. Crystallization, of course, may continue for a long time, depending upon
heat removal from the container.

The conditions differ considerably with melts that are already nucleated, or where supercooling has been prevented by
using a nucleating device. In the vicinity of the melting point
T_f and at the temperature of the liquid T_1, the rate of crystal
growth C is related to the heat of fusion H_f, the density D,
and the specific heat conductivity k in the following equation:

$$C = \frac{(T_f - T_1)k}{H_f D}$$

For typical salt-hydrates such as those listed in Tables 1A,
1B, and 2, average values are the following:

$k = 0.002$ to 0.008 cal/$(cm^2 sec \,^oC)$

$H_f = 40$ to 65 cal/gram

$D = 1.5$ to 1.7 gram/cm^3

Using these values in eq. 7., the range of the linear velocity
of crystal growth is:

$C = 0.12 \, (T_f - T_1)$ to $0.48 \, (T_f - T_1)$ cm/hour

For an acceptable temperature difference of $3\,^oC$

C for $3\,^oC = 0.36$ to 1.5 cm/hour (about 0.14 to 0.6 inch/hour)

VI.   THE RATE OF HEAT REMOVAL

The rate of heat removal from a crystallizing melt, encased
into a container of high heat conductivity, depends upon the
specific heat conductivity of the salt-hydrate.

The corresponding equation is:

$$Q = \frac{(T_f - T_w) k \cdot A}{d} \qquad \text{Btu/hour} \qquad \ldots \text{eq. 5)}$$

A = area of container wall in contact with salt-hydrate, assuming that the container has parallel walls for heat removal, (ft^2).

d = the thickness of the salt-hydrate layer (inch) between parallel wall.

T_f = melting point of salt-hydrate $(^\circ F)$.

T_w = wall temperature $(^\circ F)$.

Q = heat removal from A area in Btu/hour.

k = thermal conductivity of the salt hydrate $\frac{Btu}{hr.ft.\,^\circ F}$

The equation can be modified for containers of cylindrical or other shape, or for internally finned containers.

In actual experiments, the value of k was found to be as much as k = 1 to 2 Btu/(hr.ft $^\circ F$). The reason for this increased heat transfer is due to the fact that the interface between solid and liquid is not perpendicular to the distance between plates, but it is more complex following the contours of the growing crystals.

VII.   CALCULATION OF THE HEAT OF FUSION FROM HEATS OF SOLUTION

The heat of solution of salts and salt-hydrates can be used to calculate the heat of fusion (H_f) of the salt-hydrate, provided that the heat of solution is known for the hydrates involved in the phase change.

H_s is the heat of solution (cal/mole) of an anhydrous salt. H'_{sh}, H''_{sh} of its salt-hydrates.

$$H_f = H_s - H_{sh} \qquad \text{or} \qquad H_f = H'_{sh} - H''_{sh}$$

Summary data of heat of solution have been published by Rossini
(10) and Bichowsky and Rossini (11). As an example:

$$H_s \quad Na_2SO_4 \qquad\qquad 280 \text{ cal/mole}$$

$$H'_{sh} \quad Na_2SO_4 \cdot 10H_2O \qquad -18740 \qquad "$$

$$H_s - H'_{sh} = 280 - (-18740) = 19020 \text{ cal/mole} = 59.1 \text{ cal/gram}$$

This is in good agreement with the observed value (60.5) and
the value calculated from the entropies of fusion. The importance of selecting the correct heat of solution, relating to
the number of moles of water involved in the phase change, is
shown in the following example:

$$MgCl_2 \cdot 6H_2O \longrightarrow MgCl_2 \cdot 4H_2O + 2H_2O$$

$$H_s \quad MgCl_2 \qquad\qquad 36300 \text{ cal/mole}$$

$$H'_{sh} MgCl_2 \cdot 4H_2O \qquad 12000 \qquad "$$

$$H''_{sh} MgCl_2 \cdot 6H_2O \qquad 3400 \qquad "$$

$$H'_{sh} - H''_{sh} = 12000 - 3400 = 8600 \text{ or } 42.3 \text{ cal/g (observed 41.2)}$$

If the difference 36300 - 1200 = 24300 is used, a much higher
value (120 cal/g) is obtained, but this does not agree with
the facts, as the phase change does not involve the formation
of anhydrous $MgCl_2$, but only that of the tetrahydrate.

VIII. CALCULATION OF THE HEAT OF FUSION FROM ENTROPIES OF FUSION

As previously outlined, the entropy of fusion (S_f) of a
salt-hydrate is calculated from the sum of the entropy of
fusion of the anhydrous compound and the entropy of fusion of
water, ($S_f = 5.26$). In those salt-hydrates which change from
$AB \cdot nH_2O$ to $AB \cdot mH_2O$ the number (n - m) is used for the number
of water moles, as listed in column two of Table 1A. The calculated S_f is then compared with the observed S_f, when this

is known. The method can be used to calculate the unknown heat
of fusion of any salt-hydrate, provided its melting point is
known and also the number of moles of water that are involved
in the phase change.

As an example, using data from Table 1A:

$Na_2SO_4 \cdot 10H_2O \quad Na_2SO_4 + 10\ H_2O$	S_f Calc	S_f Obs	S_f Calc/Obs
Na_2SO_4 (transition + fusion)	9.7	8.38	1.16
$Na_2SO_4 \cdot 10H_2O \quad 9.7\ +\ 52.6\ =$	62.3	63.5	0.98

Using S_f Calc and the melting point of $Na_2SO_4 \cdot 10H_2O$ (305.6°K),
we obtain for the heat of fusion:

H_f = 62.3 x 305.6 = 19,100 cal/mole or 61.5 cal/gram

This is an excellent agreement with the observed value. The
same type of calculation can be applied to another example:

$MgCl_2 \cdot 6H_2O \quad MgCl_2 \cdot 4H_2O + 2H_2O$	S_f Calc	S_f Obs	S_f Calc/Obs
$MgCl_2$	11.2	9.4	1.07
$MgCl_2 \cdot 6H_2O \quad 11.2 + (2 \times 5.26) =$	21.7	21.6	1.01

The melting (transition) point is 390°K, giving:

H_f = 21.7 x 390 = 8450 cal/mole or 41.5 cal/gram.

This value is in excellent agreement with experimental observations.

IX. CALORIMETRIC MEASUREMENTS AND DATA

Several books describe calorimetric methods, based on the pioneering work of Thomsen (12), and others. "The Modern Calorimeter" by White (13), and the publications of Kubaschewski (14) (15) describe most of the methods in use.

The heat of fusion of salt-hydrates can be measured by direct calorimetry. A somewhat modified method has been used for salt-hydrates which can be supercooled below their melting points. The method consists of placing the supercooled salt-hydrate, (encased in its container) into the calorimeter. After temperature equilibrium is reached, at a temperature below the melting point, the melt is nucleated with small crystals of the same salt-hydrate. The amount of heat evolved during crystallization is measured in the calorimeter. This amount of heat must be corrected for the heat content of the salt, between its melting point and the temperature of the calorimeter. The correction can be obtained by measuring the specific heat of the salt-hydrate.

This method has been applied by Perreu (16) (17) to a number of salt-hydrates and to their saturated solutions. The results are incorrect when the saturated solution contains excess water. In such cases, salt remains in solution and, as a result, only part of the heat of fusion can be measured. Perreu's measurements were at least 10 percent lower than other data for congruently melting salt-hydrates and up to 60 percent lower for partly incongruent melts.

Some of the relatively few direct calorimetric measurements of the 19th century are still quoted in literature summaries, but could be confirmed by exact measurements. Person, in 1949 (18), may have been the first to measure the heat of fusion of di-sodium phosphate dodekahydrate. Berthelot, in 1878 (19), and Thomsen, in 1883 (12), quoted most of the earlier results.

Koppel, in 1905 (20), published results on sulfates, Leenhardt, in 1912 and 1913 (21) (22), on several salt-hydrates, Riesenfeld, in 1914 (23), on nitrates. Muller used the supercooled nucleation method on a number of organic chemicals. Lithium salt-hydrates were measured by Slonim (24), sodium thiosulfate pentahydrate and sodium acetate trihydrate by Sturley (25), $MgCl_2 \cdot 6H_2O$ by Auzhbekovich (26), $CaCl_2 \cdot 6H_2O$ by Lannung (27).

Special calorimeters for precision measurements have been described by Johnston (28), Ticknor and Bever (29), Leake (30), Oelsen (31), Nachtrieb (32), Schottky and Bever (33), and Bever (34).

Tables 1A, 1B, and 2 contain available data, listing salt-hydrates, change in the number of moles of water, density, salt content in weight percent based on total weight of salt-hydrate, solubility above the melting point expressed as percent anhydrous salt in 100 gram salt-hydrate, specific heat of solid and liquid. Tables 1A and 1B list the melting point in $^{\circ}C$ and $^{\circ}F$, the observed heat of fusion in calories per gram, and Btu/lb, the entropy of fusion S_f observed (Obs.) and calculated (Calc.), and finally, the ratio S_f Calc/Obs, serving the purpose of establishing the validity of calculations.

X.   PROPERTIES OF SELECTED SALT-HYDRATES

A.   Barium Hydroxide Octahydrate

This salt-hydrate has many advantages. In addition to its high heat of fusion, 72 cal/gram = 130 Btu/lb, its density is high (2.18) and the heat of fusion per unit volume is consequently also high (157 cal/cm^3 = 17,700 Btu/ft^3, based on the volume of solid salt-hydrate). The heat of fusion calculated from the heat of solution is 82 cal/g (de Forcrand (35)). The volume change during melting is small.

The solubility of $Ba(OH)_2$ in water has been determined by
Cabot (36). The salt melts congruently and recrystallizes with
negligible supercooling. One of the disadvantages is due to
the fact that $Ba(OH)_2$ absorbs CO_2 from the air, but this can
be prevented in sealed containers. The major difficulty with
this material is its corrosiveness and toxicity. The melting
point $78°C = 172°F$ may be decreased to some extent by additives.

Table 2. Properties of Hydrates of Bromides and Iodides

Compound	Mols H_2O	Melting p. °C	Solid %	Soluble at Mp.%	Heat of Fusion cal/g	Btu/lb	$S_f$calc
LiI	3	73	71	67	41	74	22.2
NaBr	3	51	74	54	36	65	15.5
MgI_2	8	41	66	48	38	68	52
$CaBr_2$	3	80	79	95	38	68	27.5
$CaBr_2$	6	38	65	92	43	78	43.3
CaI_2	6	42	73	95	34	61	43.6
$FeBr_3$	6	27	73	v.s.	36	65	48.2
FeI_2	4	90	81	v.s.	32	58	33
$CoBr_2$	6	47	66	60	43	78	43.4
$SrBr_2$	6	20		v.s.	35	63	43.4
SrI_2	6		76	v.s.			43.6
$CdBr_2$	4	36	79	v.s.	30	54	32.8
$BaBr_2$	2	75	90	67	23	41	22.3

v.s. = very soluble

B.   *Sodium Hydroxide - Monohydrate*

The properties of $NaOH.H_2O$ and $LiOH.H_2O$ are listed in Tables
1A and 1B.   In addition to the monohydrate, the following hydrates are known.

			Melting p. ^{o}C	^{o}F	Heat of fusion cal/gram	Calc Btu/lb
NaOH.	*1*	H_2O	*64.3*	*148*	*65*	*117*
NaOH.	*2*	"	*12.3*	*54*	*62*	*112*
NaOH.	*3.5*	"	*5.0*	*41*	*65*	*117*

The calorimetrically determined heat of fusion of $NaOH.H_2O$
agrees well with the theoretical value of 65 cal/gram, or 113
cal/cm^3.

NaOH.H_2O is highly corrosive and hygroscopic, but it has
several advantages.   It melts congruently and recrystallizes
with only slight supercooling.   Its crystal form is orthorhombic,
group P_{cab}, a = 6.21, b = 11.72, c = 6.05, Z = 8.   Density is
1.72.   The specific heat of the melt is 0.69, of the solid 0.38.
The mono-hydrate contains 62.5 percent anhydrous salt.   The
phase diagram and solubility have been determined by Staniford
(37).   The thermal conductivity is advantageously high, according to measurements by Riedel (38), and Boehm (39).   The volume
contracts slightly when the salt solidifies.

NaOH.H_2O can be handled in sealed polyethylene or polypropylene containers.   For special applications, corrosion resistant silver containers can be used.

C.   *Sodium Acetate Trihydrate*

Sodium acetate trihydrate melts at $58.3^{o}C$, $(137^{o}F)$.   The
density of the solid is 1.45.   The trihydrate melts partly
incongruently but dissolves entirely in its water of crystallization at $77^{o}C$ $(170^{o}F)$.

The heat of fusion reported in the literature was 39 to
43 cal/gram; Leenhardt (22), Sturley (25). Recent measurements
made by the writer with melts of the trihydrate (supersaturated
solution) give a higher value of 63 cal/gram = 114 Btu/lb in
excellent agreement with calculations based on the entropy of
fusion. The heat of fusion per unit volume is 81 cal/cm^3 or
9100 Btu/lb. The heats of solution of anhydrous sodium acetate
and its trihydrate have been determined. Their difference is
8490 cal/mole = 63 cal/gram, in perfect agreement with experimental results and calculations.

The solubility of sodium acetate decreases in the presence
of acetic acid; Heubel (40).

D. Lithium Acetate

Lithium acetate forms a dihydrate, $LiC_2H_3O_2 \cdot 2H_2O$, which
at its reported melting point 58°C (136°F), dissolves congruently. This salt could provide an interesting confirmation
of the calculations based on the entropy of fusion. Calculations based on acetic acid give 67 cal/gram, but when based
on the sum of entropies of fusion of the elements, the heat
of fusion should be 94 cal/gram.

According to Nesterova (41), a ternary eutectic, melting
at 73°C (163°F), can be obtained in the anhydrous system:
sodium acetate + potassium acetate + urea. Although this material is not a salt-hydrate, it is listed here for comparison.

E. Ammonium Aluminum Sulfate Dodekahydrate

This salt-hydrate has the highest heat of fusion of "alums"
that have been studied. Its heat of fusion is 62 cal/gram or
112 Btu/lb. The density is 1.64, with negligible volume change
at the melting point. Its heat of fusion per unit volume is
102 cal/gram^3 or 11,500 Btu/ft^3.

Ammonium aluminum sulfate or "ammonium-alum" forms cubic crystals of the T_h^6 group, with unit cell edge of 12.18 A; Z = 4. It is isomorphous with an extensive group of alums, including those of sodium and potassium; Klug (42). It supercools, but when combined with a nucleating device, the effect of supercooling can be eliminated completely. This salt-hydrate has been used in one of the heat storage containers developed commercially. After extensive tests in an ambient of minus $65^{o}F$, the salt-hydrate continued to give trouble-free performance.

Based on solubility studies, the system Li_2SO_4 + Na_2SO_4 + water shows the formation of a double salt $Li_2SO_4 \cdot 3Na_2SO_4 \cdot 12H_2O$ with transition temperature at $48.5^{o}C$ $(119^{o}F)$ according to Lepeshkov (43), Bodaleva (44), and Khu (45).

The heat of fusion of the related $AlK(SO_4)_2 12H_2O$ has been measured. It is only 44 cal/gram = 79 Btu/lb. The salt hydrate melts at $91^{o}C = 196^{o}F$.

F. Hydrated Sulfates

Tables 1A and 1B list three sulfates, all melting partly incongruently. Their pertinent data are the following:

	Melting p.		*Heat of fusion*	
	^{o}C	^{o}F	*cal/g*	*Btu/lb*
$Na_2SO_4 \cdot 10H_2O$	32.4	90	60.5	108
$MgSO_4 \cdot 7H_2O$	48.4	118	48.2	87
$CoSO_4 \cdot 7H_2O$	40.7	105	40.7	74

In this list, $Na_2SO_4 \cdot 10H_2O$ has the highest heat of fusion, measured repeatedly by Cohen (46), Perman (47), Kobe (48), Pitzer (49). The density is 1.46, giving 89 cal/cm^3 or 10,000 Btu/ft^3 as the heat of fusion per unit volume.

The writer has used $Na_2SO_4.10H_2O$ for the storage of heat
derived from solar energy and has evolved a nucleating mixture,
consisting of $Na_2SO_4.10H_2O$ with 3 to 5 percent of the isomorphous borax $Na_2B_4O_7.10H_2O$; Telkes (50). This mixture melts
at 89°F and invariably nucleates when the temperature drops
to 82°F. The melting is partly incongruent; the salt-hydrate
contains 44 percent anhydrous salt, but only 34 percent is
soluble at its melting point. The balance, 15 percent of the
original salt-hydrate, settles to the bottom of the container
(it "stratifies") and on subsequent cooling it is "frozen in",
being unable to recombine with its water of crystallization.

Some experimentors, doubtless unaware of the requirements
for nucleation and the possibilities of the prevention of settling, have reported pessimistically about the heat storage
capacility of this salt-hydrate; Whillier (51), and Hodgins
(52). The writer has conducted a number of alternate heating
and cooling cycles with this salt-hydrate, proving definitely
that settling can be prevented with thickening agents, Telkes
(7) (53) (54). Recent applications of this material have been
described (55) (56) (57) (58).

G. *Disodium Phosphate Dodekahydrate*

Disodium phosphate Na_2HPO_4 and water form several salt-
hydrates:

		Density	Crystal form
$Na_2HPO_4.12H_2O$	dodekahydrate	1.52	monoclinic
$Na_2HPO_4.7H_2O$	heptahydrate	1.68	monoclinic
$Na_2HPO_4.2H_2O$	dihydrate	2.06	rhombic
$Na_2HPO_4.H_2O$	monohydrate		

Transition of hydrates	Temperature °C	°F	Heat of fusion cal/g	Btu/lb
12 to 7	36	97	53	60
7 to 2	48	119	30.0	54
2 to 1	60	140		
1 to 0	95	203		
12 to 2			63	114
12 to 0			67	120

The heat of fusion has been measured by Person (59), Leenhardt (22), and Menzel (60). The 12 to 7 transition has a partly incongruent melting point, that is, the salt with $12H_2O$ does not melt completely if the $7H_2O$ form is present. Partington (61), measured the heats of solution.

The solubility of the salt has been measured by Menzel (62), Ingerson (63), and Hammick (64), indicating that $Na_2HPO_4 \cdot 12H_2O$ dissolves completely in its water of crystallization, if it is heated above 50°C (122°F). According to the writer's observations, it is possible to melt the $12H_2O$ salt completely at its melting point (97°F) if there are no crystal nuclei of the $7H_2O$ salt in a closed container.

The specific heat is 0.37 Btu/(lb.°F) solid
                     0.65 " " " liquid

A nucleating process has been described by the writer (7).

H. Sodium Thiosulfate Pentahydrate

This salt is extensively used in photographic development, (named "hypo"). It melts slightly incongruently at 49°C = 120°F, but dissolves completely at 52°C = 126°F. It changes to the dihydrate at 48°C and to anhydrous at 52°C. Its heat of fusion has been determined by Leenhardt (22), Muller (61), Sturley (25), and Perreu (16). The most probable value of the heat of fusion is 47.9 cal/gram or 86 Btu/lb., although calculations

based on the entropies of fusion lead to a somewhat higher value
of 55 cal/gram = 99 Btu/lb. The density of the solid salt is
1.69, of the liquid 1.66, indicating negligible volume cahnge
at the melting point. The heat of fusion per unit volume of
the melt is 80 cal/cm^3 or 10,200 Btu/ft^3. The salt-hydrate
supercools easily but with a nucleating device, supercooling
is completely eliminated.

I. Magnesium Thiosulfate Hexahydrate

Magnesium thiosulfate hexahydrate melts at $100^\circ C$, changing
to the trihydrate, which in turn transits to the anhydrous state
at $120^\circ C$; Okabe and Hori (66). The calculated entropy of fusion,
involving 3 mole water is S_f = 33.0 and the heat of fusion is
50 cal/gram = 90 Btu/lb.

J. Ferric Chloride Hexahydrate

The rather complex solubility relations and the formation
of the hydrates of $FeCl_3$ have been studied by Bakhius Roozeboom
(67). The heat of fusion of $FeCl_3 \cdot 6H_2O$ is moderately high,
54 cal/gram = 97 Btu/lb., its melting point is $98^\circ F$.

K. Magnesium Chloride Hexahydrate

$MgCl_2 \cdot 6H_2O$ melts congruently at $117^\circ C$ ($242^\circ F$); its boiling
point is $159^\circ C$ ($318^\circ F$), density is 1.56.

The heat of fusion has been determined by Riesenfeld (23),
as 41.2 cal/gram and by Auzhbekovich (26), who found 40.0
cal/gram, or 72 to 74 Btu/lb. At the melting point, the transition is from 6 to 4 H_2O and if these values are used, the calculated value agrees perfectly with the experimentally found 41.2
cal/gram. Based on unit volume, the heat of fusion is 64 cal/
cm^3 or 7300 Btu/ft^3.

L. Other Hydrated Chlorides

Hydrated chlorides listed in Tables 1A and 1B include the following:

	Melting Point ^oC	^oF	Heat of Fusion cal/g	Btu/lb
$CaCl_2 \cdot 6H_2O \longrightarrow 2H_2O$	29.4	85	40.7	73.5
$MnCl_2 \cdot 4H_2O \longrightarrow 2H_2O$	58	136	42.5	76.5

Data by Leenhardt (22), Lannung (27), and Perreu (16).

Additional melting, or transition temperatures, have been reported for the following chlorides:

	Melting Point ^oC	^oF	Calculated Heat of Fusion Cal/g	Btu/lb
$CoCl_2 6H_2O \longrightarrow 2H_2O$ (*)	54.4	130	47	85
$SrCl_2 \cdot 6H_2O \longrightarrow 2H_2O$ (*)	61.6	143	34.5	62
$BaCl_2 \cdot 2H_2O \longrightarrow H_2O$ (*)	101.9	213	25	45
$NiCl_2 \cdot 6H_2O \longrightarrow 2H_2O$ (**)	60	140	50	90
$FeCl_2 \cdot 4H_2O \longrightarrow 2H_2O$ (**)	37	99	34	61

*   Borchardt (1957)

** Mellor (1928)

M. Hydrated Nitrates

The salt-hydrates of nitrates listed in Tables 1A and 1B show rather low heat of fusion values, with the exception of $LiNO_3 \cdot 3H_2O$ (71 cal/gram = 128 Btu/lb., melting at $30^oC = 86^oF$). These salts are highly soluble and melt congruently. None of the hydrated nitrates can be expected to show unusually high heat of fusion values, with the exception of $Al(NO_3)_3 \cdot 9H_2O$, but according to the literature, it decomposes at $70^oC = 158^oF$.

N. Other Salt-Hydrates

Compounds formed of light-weight, low atomic number elements
and anions, have potentially higher heat of fusion than compounds
formed of heavier elements. The compounds and their salt-hydrates have already been described, forming acetates, hydroxides, carbonates, nitrates, oxides, fluorides, sulfates, chlorides and phosphates. Additional salt-hydrates can be found
in the group of bromides, iodides (Table 2), and others, but
their heats of fusion are rather low, in the range of 34-44
cal/g, (60 to 80 Btu/lb) and they cannot be considered as suitable heat storage materials. Additional salt-hydrates exist
in the group of borates, perchlorates, chlorates, fluorides
and fluosilicates, listed in the following table. The melting
points of some of them are not known, but where possible, the
expected heat of fusion has been calculated from the entropy
of fusion of their component elements, assuming that the salt-
hydrate changes to its anhydrous form at the melting point.

Material	*Mp* ^{o}*C*	*Calc. Heat of Fusion cal/g*	
Metaborates:			
$NaBO_2 \cdot 4H_2O$	49	56	*Melts incongruently*
$LiBO_2 \cdot 8H_2O$	47	87	*Melts partly incongruently*
Tetraborate:			
$Na_2B_4O_7 \cdot 10H_2O$	60	79	*Melts incongruently*
Perchlorate:			
$LiClO_4 \cdot 3H_2O \ H_2O$	93	69	*Ref. Ropp (68)*

Material	Mp °C	Calc. Heat of Fusion cal/g	Solubility Percent at 20°C
Chlorates:			
$Al(ClO_3)_3.6H_2O$ d.	60	68	v.s.
$NaClO_3.10H_2O$			v.s.
$Sr(ClO_3)_2.8H_2O$			v.s.
$Zn(ClO_3)_2.H_2O$	60	47	v.s.
Fluorides:			
$BF_3.2H_2O$	6	64	
$KF.4H_2O$	18	34	v.s.
$KF.4H_2O$	41	52	v.s.
$MnF_2.4H_2O$	24	58	v.s.
$CrF_3.9H_2O$ $4H_2O$			v.s.
$CoF_2.6H_2O$			v.s.
Fluosilicates:			
$Li_2SiF_6.2H_2O$	100	70	43
$MgSiF_6.6H_2O$	60 ?	75	39
$FeSiF_6.6H_2O$	70 ?	73	56
$CuSiF_6.6H_2O$	80 ?	65	70

In this group of salt-hydrates, $LiBO_2.8H_2O$ has the highest heat of fusion although it melts partly incongruently. Sodium tetraborate (borate) melts incongruently and according to tests already made, it does not crystallize reversibly. Magnesium fluosilicate $.6H_2O$ and aluminum chlorate $.6H_2O$ appear to merit further study.

REFERENCES

1. Mullin, J.W., "Crystallization", CRC Press, Cleveland, (1972).

2. Van Hook, A., "Crystallization, Theory and Practice", 325 pp. Reinhold Publishing Co., New York, (1961).

3. Tipson, R.S., "Crystallization and Recrystallization", in "Technic of Organic Chemistry", Vol. 3, (A. Weissberger, ed). Interscience Publishers, New York, (1956).

4. Telkes, M., *Ind. and Eng. Chem. 44,* 1308-10 (1952). Nucleation of supersaturated inorganic salt-solutions.

5. Hume, Rothery W., "Atomic Theory for Students of Metallurgy", Institute of Metals, London, (1946).

6. Turnbull, D., "Solid State Physics", Vol. 3, p. 283, Academic Press, New York, (1956).

7. Telkes, M., Unpublished results.

8. Tamman, G., "Kristallisieren und Schmelzen", Barth, Leipzig, (1903).

9. Tamman, G., "The States of Aggregation", trans. R.F. Mehl, Van Nostrand, New York, (1925).

10. Rossini, F.D., Wagman, D.D., Evans, W.H., Levine, S., and Jaffe, I., National Bureau of Standards (U.S.) Circular 500 (1952).

11. Rossini, F.D., Wagman, D.D., Evans, W.H., Levine, S., and Jaffe, I., "Selected Values of Chemical Thermodynamic Properties", National Bureau of Standards, Washington, Circular 500 (1952).

12. Thomsen, J., "Thermochemische Untersuchungen", Vol. 3, p. 179 (1883). Heats of hydration of numerous salts.

13. White, W.P., "The Modern Calorimeter", New York, (1928).

14. Kubaschewski, O., Catterall, J.A., "Thermochemical Data of Alloys", Pergamon Press, (1956).

15. Kubaschewski, O., Dench, W.A., "Calorimetry", p. 171, in
     "Physico-chemical Measurements at High Temperatures" (J.
     Brockris, ed). Butterworths, London, (1959).

16. Perreu, J., C.R. *199*, 48-51, 1934. Heat of crystallization
     of several hydrates.

17. Perreu, J., C.R. *212*, 701-3, 1941. Heats of solution of
     $FeCl_2 \cdot 4H_2O$, $Ni(NO_3)_2 \cdot 6H_2O$.

18. Person, C.C., C.R. *23*, 162, 1946, and *29*, 300, 1949.
     Heat of fusion of $Na_2HPO_4 \cdot 12H_2O$.

19. Berthelot, C.R. *87*, 573, 1978, Heat of fusion of Na_2CrO_4
     $\cdot 10H_2O$.

20. Koppel, J., Z. Anorg. Allgem. Chem. *52*, 385-436 (1905).
     Sollubility and formation of salt hydrates.

21. Leenhardt, C., Boutaric, A., C.R. *154*, 113-4, 1912,
     $Na_2S_2O_3 \cdot 5H_2O$ crysoscopy.

22. Leenhardt, C., Boutaric, A., Bull. Soc. Chim. de France *13*,
     651-7 (1913). Heat of fusion by calorimetry.

23. Riesenfeld, E.H., Milchsack, C., Z. Anorg. Chem. *85*,
     401-21 (1914).

24. Slonim, C., Huttig, G.F., Z. Anorg. Chem. *181*, 55-64 (1929).
     Heats of hydration of Li-halogen-hydrates.

25. Sturley, L.R., J. Soc. Chem. Ind. *51*, 271-37 (1932).
     Specific heat and heat of fusion of $Na_2S_2O_3 \cdot 5H_2O$, Na
     acetate $\cdot 3H_2O$, and K-alum.

26. Auzhbekovich, A.E., J. Appl. Chem. USSR, *9*, 594-598 (1936).
     Heat capacity and hydrates of MgCl.

27. Lannung, A., Z. Anorg. Chem. *228*, 1-18 (1936). Heat of
     hydration of $CaCl_2$.

28. Johnston, H.L., Kerr, E.C., J.A.C.S. *72*, 4733-7 (1950).
     Calorimetry (description).

29. Ticknor, L.B., Bever, M.B., Trans. AIME *194*, 941 (1952).
     Calorimetry using tin as liquid.

30. Leake, L.E., J. Sci. Instr. *31*, 447-9 (1954). Calorimeter.

31. Oelsen, W., *Arch. Eisen hutten w. Bergakademie, Clausthal, German 26,* 253-66 and 519-22 (1955). New calorimetric method and specific heats and heats of fusion of metals.

32. Nachtrieb, N.H., Clement, N., *J. Phys. Chem. 62,* 876 (1958). Heat of fusion of InSb.

33. Schottky, W.F., Bever, M.B., *Acta Met. 6,* 320 (1958). Heat of fusion of GaSb and InSb.

34. Bever, M.B., Calorimetric Investigations of III-V Compounds in "Compound Semiconductor Series", Vol. I (Willardson and Goering, eds) (1961).

35. de Forcrand, R., *C.R. 147,* 165-168, 1908. Heat of solution of $Ba(OH)_2$.

36. Cabot, G.L., *J. Soc. Chem. Ind. 16,* 417 (1897). Solubility of $Ba(OH)_2$.

37. Staniford, F.C., Badger, W.L., *Ind. Eng. Chem. 46,* 2400-3 (1954). Solubility and phase diagram of NaOH • H_2O system.

38. Riedel, L., *Chem. Engn. Techn. 22,* 54-6 (1950). Thermal conductivity of NaOH solutions.

39. Boehm, J., *Arch. ges. Warmetchen. 1,* 209-14 (1950. Thermal conductivity of NaOH solutions.

40. Heubel, J., Coup, D., *C.R. 250,* 1058-60, 1960, *C.A. 54,* 14876h, 1960. The system Na acetate - acetic acid - H_2O.

41. Nesterova, A.K., Bergman, A.G., *Zhur. Obschei Khim 30,* 317-20 (1960), *C.A. 54,* 19138f, 1960. The system sodium acetate - potas sium acetate-urea.

42. Klug, J., Alexander, L., *J.A.C.S. 62,* 2992-5 (1940). Crystal-chemical studies of alums.

43. Lepeshkov, I.N., Romashova, N.N., *Zhur. Neorg. Khim. 4,* 2812-5 (1959), *C.A. 54,* 23690g, 1960. Double salt $LiSO_4$ • $3Na_2SO_4$ • $12H_2O$.

44. Bodaleva, N.V, Khu, K.Y., *Zhur. Niorg. Khim. 4,* 2816-9 (1959). Solubility in the system Li_2SO_4-Na_2SO_4-H_2O.

45. Khu, K.Y., *Zhur. Neorg. Khim. 4,* 1910-18 (1959), *C.A. 54,* 11677e, 1960. The system Na_2SO_4-Li_2SO_4-H_2O.

46. Cohen, E., *Z. Phys. Chem. 14*, 86 (1894). Heat of fusion of
     $Na_2SO_4 \cdot 10H_2O$.

47. Perman, E.P., Urry, W.D., *Trans. Far. Soc. 24*, 337 (1928).
     Heat of hydration of $Na_2SO_4 \cdot 10H_2O$.

48. Kobe, K.A., Anderson, C.H., *J. Phys. Chem. 40*, 429-33 (1936).
     Heat capacity of sa. Na_2SO_4 solutions.

49. Pitzer, K.S., Coulter, L.V., *J.A.C.S. 60*, 1310-3 (1938).
     Heat capacity, entropy and heat of solution of Na_2SO_4.

50. Telkes, M., "Solar Heat Storage", in "Solar Energy Research",
     (F. Daniels and J.A. Duffie, eds). University of Wisconsin
     Press (1955).

51. Whillier, A., *Sun at Work 2*, 2 June 1957 (letter to the
     editor), Stored heat of $Na_2SO_4 \cdot 10H_2O$.

52. Hodgins, J.W., Hoffman, T.V., *Canadian J. Techn. 33*, 293,
     July (1955). Glauber's salt as a heat storage material.

53. Telkes, M., U.S. Patent No. 3,986,969, "Thiotropic Mixture
     and Method of Making Same", issued October 19, 1976.

54. Telkes, M., U.S. Patent No. 4,187,189, "Heat Exchange Bodies
     Utilizing Heat of Fusion Effects and Method of Making Same",
     to issue February 5, 1980.

55. Telkes, M., "Heat of Fusion Systems for Solar Heating and
     Cooling", Solar Engineering Magazine, September (1977).

56. Telkes, M., "Trombe Wall with Phase Change Storage Material",
     Second National Passive Solar Conference, Philadelphia,
     Pennsylvania, pp. 283-287, March (1978).

57. Telkes, M., "Latent Heat Storage Techniques", Institute of
     Gas Technology Symposium on ENERGY FROM THE SUN, April 3-7,
     1978, Chicago, Illinois.

58. Telkes, M., "Thermal Storage in Salt-Hydrate Eutectics",
     1978 Annual AS/ISES Solar Conference: SOLAR DIVERSIFICA-
     TION", August 28-31, 1978, Denver, Colorado.

59. Person, C.C., *Ann. de. Chim. Phys. 27*, 250 (1849). Heat of
     fusion of $Na_2HPO_4 \cdot 12H_2O$.

60. Menzel, H., Sieg, L., *Z. f. Elektrochem. 38,* 283, 1932. Solubilities of phosphates.

61. Partington, J.R., Winterton, R.J., *J. Chem. Soc. (London),* 635-43. Heat of hydration of Na_2HPO_4.

62. Menzel, H., von Sahr, *Z. f. Elektrochem. 43,* 104-19 (1937). Alkali phosphates and Na_3PO_3.

63. Ingerson, E., *Amer. Mineral 28,* 448 (1943). Solubility of Na_2HPO_4.

64. Hammick, D.L., *J. Chem. Soc. 117,* 1589 (1920). Solubility of Na_2HPO_4.

65. Muller, A.H.R., *Z. Phys. Chem. 86,* 177-242 (1914). Heat of fusion of $Na_2S_2O_3$, also many organic chemicals.

66. Okabe, T., Hori, S., *Technol. Repts. Tohoku Univ. 23,* No. 2, 85-9 (1959). *C.A. 54,* 20449g, 1960. Entropy of fusion of $MgSO_3 \cdot 6H_2O$ and $MgS_2O_3 \cdot 6H_2O$.

67. Bakhuis, Roozeboom, H.W., *Z. Phys. Chem. 10,* 479 (1912). Solubility of $FeCl_3$.

68. Ropp, C.D.L., *J.A.C.S. 50,* 1650-3 (1928). Hydrates of $LiClO_4$.

CHAPTER 12

THERMODYNAMIC BASIS FOR
SELECTING HEAT STORAGE MATERIALS

Maria Telkes

Solar Thermal Storage Development
AEC Research Institute
Killeen, Texas

I. INTRODUCTION

A. *Theoretical Basis of Estimating Heat of Fusion Values*

Search for materials with high heat of fusion values in
any designated temperature range, can follow two paths. The
empirical way is to examine all compounds which melt in the
selected ranges of temperatures and to search for heat of fusion
data in the literature. Such work could involve thousands of
materials and although many could be bypassed by analogy, the
necessary work could involve many man-years (1).

The other approach is to consider the theoretical basis
of heat of fusion and its correlation with other known physical
properties, with the aim of evolving a selection-method, pre-
ferably corroborated with data obtained for wider temperature
ranges.

The effect of heat on materials is one of the primary factors in our consummate knowledge. The melting point T_f is used
for the identification of chemical compounds; therefore, it
is a determining factor.

SOLAR MATERIALS SCIENCE

Materials in the solid state contain atoms arranged in
lattices with three vibrational degrees of freedom. Theory
predicts that the *specific heat* per unit volume C_v = 3R = 6
cal/degree per gram atom. The specific heat at constant pressure
C_p is higher in most solids and increases with temperature in
a well known manner. *Kopp's rule* postulates that the specific
heat of any compound is equal to the sum of specific heats of
the elements forming the compound. This rule has been confirmed
for a large number of materials. The usual value of C_p is
between 7 to 7.25 cal/degree per gram atom. It is well known
that the specific heat of some solids of very high melting points
follow this rule only at higher temperatures. The specific
heat of any compound is obtained by adding the specific heats
of all atoms forming the compound.

   The heat of evaporation of a liquid Q_e, according to *Trouton's*
rule may be calculated from the entropy of evaporation (S_e)
and the boiling point $(T_e$ in $K^O)$.

$$S_e = Q_e/T_e = 22 \text{ cal/}^O\text{K per mole. (e.u.)}$$

The fact that the molar entropy of evaporation is approximately
22 (e.u.) has been confirmed for many compounds. It does not
apply to low boiling point materials, or to substances which
change in chemical composition during evaporation.

   Hildebrand, (2) modified Trouton's rule, using the heats
of evaporation at equal vapor pressures (Q_e') and the corresponding temperature (T_e'). This modified entropy of evaporation
(S_e') is around 27 entropy units (e.u.) for most normal liquids
and around 32 e.u. for polar liquids (including water).

   The change from liquid to vapor phase involves the larger
change in bonding energy between atoms and the corresponding
entropy of evaporation is nearly ten times greater than the
entropy of fusion.

Schinke and Sauerwald, (14) (15), measured the $V_1 - V_s$ molar-volume change during melting and compared it with the entropy of fusion per molar volume (S_f/V_s) and the effect of the applied pressure (P) on the melting point. They used a modified Clausius-Clapeyron equation:

$$\frac{V_s - V_1}{V_s} = \frac{dT_m}{dP} \frac{S_f}{V_s}$$

Vogel, Schinke and Sauerwald, (16), found that the volume change, expressed as per cent of the solid volume, was proportional to the lattice energy and increased with the entropy of fusion. It was not possible, however, to calculate heats or entropies of fusion from their data.

Ubbelohde, (17) (18) (19) (20), studied the entropy of fusion in relation to volume changes during melting and changes in lattice parameters, viscosity of melt and electrical conductivity.

The physical aspects of "Changes of State" have been treated by Temperley, (21) (22) who has summarized various theories of the melting process.

The relationship of the heat of fusion, melting point, lattice energy and other physical properties have been studied by Vorob'ev, (23) (24).

Rather extensive work is being carried out on phase equilibria in binary, ternary and even more complex systems, including measurements of heats of fusion.

This intentionally brief survey of the theoretical approach indicates a deficiency in our present knowledge in using known physical constants for the purpose of calculating the entropy or heat of fusion of elements.

The conclusion can be reached that Kubaschewsky's rule can
be applied to calculate the heat of fusion of inorganic compounds, by using definite values for the entropy of fusion per
gram atom. These values are approximately 3.0 to 3.5 e.u. for
ionic compounds, 2.6 e.u. for layer-type lattices and less than
2.0 e.u. for molecular crystals of inorganic compounds.

B. Heat of Fusion and Entropy of Fusion of Elements

Most of the theoretical work of the past has been hampered
by lack of data on the heats of fusion of elements. Using
recently published results a nearly complete table has been
prepared, listing melting points, transition points, heats of
fusion and transition, also entropies of fusion and transition.
Table 1 shows these data arranged according to the atomic numbers of the elements. Combined data of entropies of transition
and fusion in relation to atomic numbers are shown in Figure 1.
There is an obvious correlation with the periodic table
of the elements as this is demonstrated by Figure 1. High
entropy of fusion values are shown in the following tabulation
with values in excess of 3.3 e.u. being outlined heavily.

III A	*IV A*	*V A*	*VI A*	*VII A*
B	C	N	O	F
2.48	(6.1)	2.13	3.47	3.37
Al	Si	P	S	Cl
2.76	7.2	0.47	1.12	4.45
Ga	Ge	As	Se	Br
4.42	6.7	(1.17)	2.65	4.76
In	Sn	Sb	Te	I
1.82	3.4	5.25	5.77	4.85
Tl	Pb	Bi	Po	At
1.86	1.86	4.77	(1.76)	(5.0)

Table 1. Heat of Fusion and Entropy of Fusion of Elements

Atomic No. Element			Melting Point $^{\circ}K$	Heat of Fusion $\dfrac{cal}{mole}$	$\dfrac{cal}{8}$	Entropy of Fusion
1	H		14	14	14	1.00
2	He		3.5	5	1.2	1.4
3	Li		454	717	104	1.58
4	Be		1556	3520	390	2.26
5	B		2310	5720	525	2.48
6	C		(4100)	(25000)	(2100)	(6.1)
7	N	tr	36	27	2.0	0.77 ⎤
		f	63	86	6.1	1.36 ⎦ 2.13
8	O	tr	23.7	11.2	0.7	0.47 ⎤
		tr	43.8	88.8	5.5	2.02 ⎬ 3.47
		f	54.4	53.1	3.3	0.98 ⎦
9	F		54	186	10	3.37
10	Ne		25	80	4	3.26
11	Na		371	622	27	1.68
12	Mg		923	2127	88	2.30
13	Al		933	2570	95	2.76
14	Si		1683	12100	432	7.20
15	P		317	150	5	0.47
16	S	tr	369	85	2.7	0.27 ⎤
		f	392	336	10.5	0.85 ⎦ 1.12
17	Cl		172	766	21.5	4.45
18	A		84	281	7	3.35
19	K		337	570	15	1.69
20	Ca	tr	713	220	7	0.38 ⎤
		f	1123	2070	52	1.85 ⎦ 2.23
21	Sc		1845	3700	84	2.05
22	Ti	tr	1155	950	20	0.82 ⎤
		f	1950	3700	77	1.90 ⎦ 2.72
23	V		2120	(3830)	(75)	(1.76)
24	Cr		2120	3470	67	1.62

Table 1. Heat of Fusion and Entropy of Fusion of Elements
(cont'd.)

Atomic No. Element			Melting Point °K	Heat of Fusion cal/mole	cal/8	Entropy of Fusion
25	Mn	tr	1000	535	10	0.53
		tr	1374	545	10	0.40 ⎤ 3.53
		tr	1410	430	8	0.30
		f	1517	3500	64	2.30
26	Fe	tr	1183	215	4	0.18
		tr	1673	165	3	0.10 ⎤ 2.30
		f	1812	3670	66	2.02
27	Co		1766	3700	69	2.10
28	Ni		1728	4220	72	2.44
29	Cu		1356	3120	49	2.3
30	Zn		693	1765	27	2.55
31	Ga		303	1335	19	4.42
32	Ge		1210	8100	112	6.70
33	As		1087	(6600)	(83)	(6.1)
34	Se		490	1300	16	2.65
35	Br		266	1263	8	4.76
36	Kr		116	391	5	3.47
37	Rb		312	560	7	1.8
38	Sr		1043	2400	28	2.3
39	Y		1820	2730	30	
		tr	1135	1040	0.91	
40	Zr		2130	5000	53	2.36 3.27
41	Cb		2770	(4800)	(52)	1.76
42	Mo		2890	6650	69	2.3
				6100		
43	Tc		(2400)	(5500)	(56)	(2.3)
44	Ru		2700	(6200)	(61)	(2.3)
45	Rh		2239	5200	50	2.3
46	Pd		1823	4200	40	2.3
47	Ag		1234	2780	26	2.25

Table 1.   Heat of Fusion and Entropy of Fusion of Elements
(cont'd.)

Atomic No. Element		Melting Point °K	Heat of Fusion		Entropy of Fusion
			cal/mole	cal/8	
48	Cd	594	1450	13	2.44
49	In	429	780	7	1.82
50	Sn	505	1720	15	3.4
51	Sb	903	4740	40	5.25
52	Te	723	4180	33	5.77
53	I	387	1885	7	4.85
54	Xe	161	549	4	3.42
55	Cs	302	510	4	1.69
56	Ba tr	643	150	1	0.23 ⎤
	f	983	1830	13	1.86 ⎦ —2.09
57	La	1193	(2750)	(20)	(2.3)
58	Ce	1077	3100	22	2.32
59	Pr	1208	3800	27	3.15
60	Nd	1297	2990	21	2.3
61	Pm	(1300)	(3000)	(21)	(2.3)
62	Sm	1325	3050	20	2.3
63	Eu	(1100)	(2500)	(17)	(2.3)
64	Gd	(1600)	(3700)	(24)	(2.3)
65	Tb	(1700)	(4100)	(25)	(2.3)
66	Dy	1773	(4100)	25	2.3
67	Ho	1733	(4600)	25	2.3
68	Er	1800	(4100)	(26)	(2.3)
69	Tm	(1900)	(4400)	(26)	(2.3)
70	Yb	1097	(2200)	(13)	(2.3)
71	Lu	(2000)	(4600)	(26)	(2.3)
72	Hf	2495	5790	32	2.3
73	Ta	3250	(7500)	(41)	(2.3)
74	W	3650	(8400)	(46)	(2.3)

Table 1. Heat of Fusion and Entropy of Fusion of Elements (cont'd.)

Atomic No. Element		Melting Point $^{\circ}K$	Heat of Fusion $\frac{cal}{mole}$	$\frac{cal}{8}$	Entropy of Fusion
75	Re	3400	8000	43	2.3
76	Os	3000	(6900)	(36)	(2.3)
77	Ir	2730	(6300)	(33)	(2.3)
78	Pt	2043	4700	24	2.3
79	Au	1336	2955	15	2.2
80	Hg	234	549	3	2.3
81	Tl	577	1020	5	1.86
82	Pb	601	1140	6	1.86
83	Bi	545	2600	12	4.77
84	Po	527	(3000)	(14)	(5.7)
85	At	(575)	(2850)	(14)	(5.0)
86	Rn	202	693	3	3.43
87	Fr	(300)	(495)	(2)	(1.65)
88	Ra	973	(2250)	(10)	(2.3)
89	Ac	(1470)	(3420)	(15)	(2.3)
90	Th tr f	1650 1980	860 3300	19	2.3
91	Pa	(1500)	(3500)	(15)	(2.3)
92	U	1406	4750	20	3.38

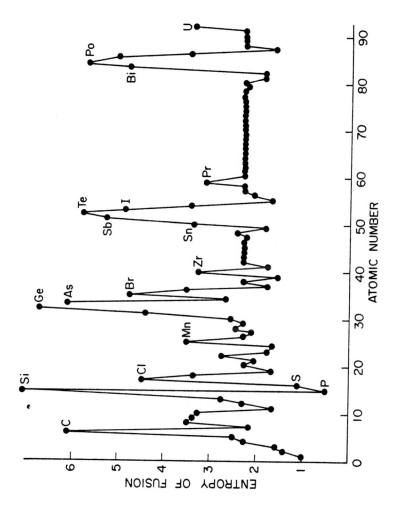

FIGURE 1.

The highest heat of fusion values in cal/gram are listed in decreasing order as follows:

Atomic No.	Element	Melting Point $^{\circ}K$	$^{\circ}C$	Heat of fusion cal/gram
6	C	(4100)	(3800)	(2100)
5	B	2310	2040	525
14	Si	1683	1410	432
4	Be	1556	1283	390
32	He	1210	940	112
3	Li	454	181	104
13	Al	933	660	95
12	Mg	923	650	88

It is of interest to note that high heat of fusion values (cal/g) are found only in elements of low atomic numbers. Values higher than 50 cal/gram cannot be found in elements with atomic numbers in excess of 47.

The purpose of preparing an exact list was to test the applicability of a proposed method of calculating heat of fusion, which is now outlined.

C. *Proposed Method of Calculating the Heat of Fusion*

The proposed method consists of calculating the heat of fusion of compounds, from their entropies of fusion, obtained by adding the entropies of fusion of their component elements. We designate the entropy of fusion of elements A and B as S_{fA} and S_{fB}. The elements form a compound of the type A_2B_3 (other numbers may be applicable). The proposed method of calculation is:

$$S_f \text{ of } A_2B_3 = 2yS_{fA} + 3zS_{fB}$$

y and z are to be evaluated later.

For compound A_2B_3: $S_f \cdot T_f = Q_f$ cal/mole

II. THERMOPHYSICAL PROPERTIES

To test the validity of the proposed method of calculating
the heat of fusion of any compound, a survey has been made of
the most extensive data that can be found for groups of compounds. The most prevalent data occurs in the groups of
chlorides, fluorides and oxides and for this reason the calculations have been applied to these groups.

CHLORIDES. Data is listed in Table 2 in the order of the
atomic number of the atom forming the chloride. The table lists
72 chlorides with known compositions and melting points, values
of heats of fusion, and entropy of fusion have been reported
in the literature for more than 50 chlorides, (S_f Obs.). The
entropy of fusion of all listed chlorides has been calculated
from equation 3, (S_fCalc.). The ratio of S_f Calc/Obs is shown
in the last column of Table 2. Agreement for calc/Obs = 1.00
cannot be expected in view of the obvious difficulties in measuring heats of fusion, especially at higher temperatures.
Ratios between 0.7 to 1.3 were therefore accepted as "reasonable
agreement", of 1.0$\pm$0.3. The following table summarizes results
of Table 2. Similar categories are shown for 1.65 + 0.35 or
1.31 to 2.0 and for higher deviations.

Elements forming chlorides	Number of Compounds	S_f Cal/Obs.
H Li F Na Mg Al K Mn Fe Co Ni Zn Ga Rb Zr Pd Cd Cs Ta Tl Th	21	0.7 to 1.0
B C Ca Cr Cu Sr Cb Mo Ag I Ba La Ce Pr Nd Gd Hg Pb Bi	19	1.31 to 2.0
Be Si P Ti As Sn Sb Te Ho Er W	11	2.01 to 4.0

Table 2. Heat of Fusion and Entropy of Fusion of Chlorides

Compound		Melting Point °C	Heat of Fusion cal/g	Btu/lb	Entropy of Fusion Obs	Calc	Calc/Obs
1 HCl	tr	−175	8	14	5.9	5.45	0.93
	f	−115	13	23			
3 $LiCl$		607	112	202	5.43	6.03	1.11
4 $BeCl_2$		440	25	45	2.8	10.7	3.8
5 B_2Cl_4		−93	16	29	14.31	19.96	1.39
6 CCl_4	tr	−48	7	13	7.27	13.35	1.83
	f	−23	−9	7			
7 NCl_3		−43	(30	54)	−	15.48	−
NH_4Cl		185	18	32	2.05		
8 Cl_2O_7		−92	(33	59)	−	33.2	−
9 F_3Cl		−76	24	43	11.14	14.56	1.40
11 $NaCl$		800	114	205	6.20	6.13	0.99
12 $MgCl_2$		714	108	194	10.44	11.2	1.07
13 $AlCl_3$		193	65	117	18.4	16.11	0.88
14 $SiCl_4$		−70	11	20	9.0	25.0	2.8
15 PCl_3		−87	8	14	6.0	13.82	2.3
$POCl_3$		1	20	36	11.4		
16 S_2Cl_2		−76	(16	29)	−	11.1	
17 Cl_2		−101	22	40	8.9	8.9	1.00
19 KCl		770	86	155	6.10	6.14	1.00
20 Cl_2		772	61	110	6.44	11.13	1.73
21 $ScCl_3$		939	(127	229)	−	15.85	−
22 $TiCl_4$		−25	12	22	9.04	19.6	2.17
23 VCl_4		−26	(26	47)	−	20.1	−
24 $CrCl_2$		815	62	112	7.1	11.2	1.57
25 $MnCl_2$		650	71	128	9.8	11.2	1.14
26 $FeCl_2$		677	81	146	11.9	11.2	0.94
$FeCl_3$		304 307	55	99	15.4	15.35	1.00

Table 2.   Heat of Fusion and Entropy of Fusion of Chlorides (cont'd.)

Compound	Melting Point °C	Heat of Fusion		Entropy of Fusion		
		cal / g	Btu / lb	Obs	Calc	Calc / Obs
27 $CoCl_2$	724	79	142	10.1	11.2	1.10
28 $NiCl_2$	1030	142	256	14.15	11.34	0.80
29 $CuCl$	430	26	47	3.72	6.75	1.81
$CuCl_2$	498	(64	115)	-	11.2	-
30 $ZnCl_2$	275	40	72	10.0	11.45	1.14
31 $GaCl_3$	78	37	67	17.8	17.8	1.00
32 $GeCl_4$	-49	(38	68)	-	24.5	-
33 $AsCl_3$	-16	13	23	9.35	19.08	2.0
34 $SeCl_4$	305			-	20.5	-
35 $BrCl$	-23			-	9.2	-
37 $RbCl$	722	47	85	5.70	6.25	1.09
38 $SrCl_2$ tr	730	25	45	3.38	6.75	1.3
f	873	10	18	1.65		
39 YCl_3	677	(77	139)	-	15.77	-
40 $ZrCl_4$	437	61	110	17.0	20.1	1.18
41 $CbCl_5$	205	26	47	14.4	24.45	1.7
42 $MoCl_5$	188	29	52	17.2	24.5	1.4
44 $RuCl_3$	d.					
45 $RhCl_3$	d.					
46 $PdCl_2$	678	55	99	10.2	11.2	1.1
47 $AgCl$	455	21	38	4.25	6.76	1.6
48 $CdCl_2$	568	41	74	9.0	11.34	1.25
49 $InCl_3$	586	(41	74)	-	10.7	-
50 $SnCl_2$	247	23	41	6.3	12.3	1.5
$SnCl_4$	-33	8	14	9.1	21.2	2.3
51 $SbCl_3$	73	13	23	8.82	18.6	2.1
52 $TeCl_4$	224	17	31	9.05	23.5	2.6
53 ICl	27	11	20	6.1	9.3	1.5

Table 2. Heat of Fusion and Entropy of Fusion of Chlorides
          (cont'd.)

Compound	Melting Point °C	Heat of Fusion cal/g	Btu/lb	Entropy of Fusion Obs	Calc	Calc/Obs
55 CsCl	645	30	54	5.27	6.14	1.16
56 $BaCl_2$	920 960	38	68	6.6	11.0	1.65
57 $LaCl_3$	857	53	95	13.0	15.6	1.20
58 $CeCl_3$	822	52	94	11.8	16.23	1.3
59 $PrCl_3$	827	49	85	12.1	16.5	1.35
60 $NdCl_3$	784	48	86	12.0	15.6	1.3
62 $SmCl_3$	678	(57	103)	–	15.6	–
63 $EuCl_3$	627	(54	97)	–	15.6	–
64 $GdCl_3$	632	36	65	10.6	15.6	1.48
65 $TbCl_3$	588	(50	90)	–	15.6	–
66 $DyCl_3$	657	(54	97)	–	15.6	–
67 $HoCl_3$	718	26	47	7.0	15.6	2.2
68 $ErCl_3$	773	29	52	7.4	15.6	2.1
69 $TmCl_3$	845	(63	113)	–	15.6	–
80 $HgCl_2$	277	15	27	7.6	11.2	1.5
Hg_2Cl_2	539	2.5	4.5	1.5	6.8	4.5
81 TlCl	429	16	29	5.4	6.1	1.13
82 $PbCl_2$	495	20	36	7.53	10.8	1.5
83 $BiCl_3$	234	18	32	11.2	18.0	1.6

The table indicates that agreement between calculated and observed values was found in about half of the compounds. Largest
deviations were found in groups IV A, and V A of the periodic
table of elements.

FLOURIDES. Table 3, shows data for fluorides; containing
data for 41 compounds, with observed heat of fusion values
listed for 31 compounds, with the following results:

Elements forming fluorides	Number of Compounds	S_f Calc/Obs.
H Li Be Na Mg P Cl K Rb Sr Zr Cs Ce Re	14	0.7 to 1.3
B N Al Si Ca As Br Cb Mo Cd I U	12	1.31 to 2.0
C Br Ba Pb Bi	5	2.01 to 3.1

The deviation factor is 1.0 in half of the compounds. Some
of the data may be in error due to difficulties in determining
the heat of fusion of fluorides, which melt at higher temperatures than the corresponding chlorides. Repeated measurements
have been made with LiF, MgF_2 and NaF - which show rather high
heat of fusion values. Here the ratio of S_f Calc./Obs. is near
1.0.

OXIDES offer an excellent opportunity for testing the applicability of the proposed calculation. They are mostly stable
compounds and usually do not disassociate or decompose during
melting. Most of the measurements have been performed with
precision calorimetry. (Table 4)

During calculation of S_f the recently determined combined
entropy of transition and fusion of oxygen has been used
(S_f=3.47), while the previously available value for fusion alone
(1.0) was not suitable for calculating the entropy of fusion
of water. The first two lines of Table 4 show excellent agreement (within 4 percent) between the calculated and observed

Table 3. Heat of Fusion and Entropy of Fusion of Fluorides

Compound Mol. W.		Melting Point °C	Heat of Fusion cal/g	Heat of Fusion Btu/lb	Entropy of Fusion Obs	Calc	Calc/Obs
F					3.37		
1	HF	-83	47	85	4.93	4.	0.90
3	LiF	848	249	448	5.77	5.0	0.87
4	BeF$_2$	543	139	250	8.0	8.5	1.06
5	BF$_3$	-128	16	29	7.6	12.4	1.6
6	CF$_4$ tr / f	-197 / -184	6	11	6.52	17.0	2.6
7	NF$_3$ tr / f	-216 / -207	6	11	7.78	12.2	1.6
11	NaF	995	192	346	6.20	5.1	0.8
12	MgF$_2$	1263	224	403	9.15	9.1	1.0
13	AlF$_3$	1040	96	173	6.25	12.9	2.06
14	SiF$_4$	-87	22	40	12.0	20.7	1.7
15	PF$_5$	6	22	40	10.0	10.0	1.0
17	ClF$_3$ tr / f	-82 / -76	23	41	11.1	14.5	1.3
19	KF	856	106	191	6.0	5.1	0.85
20	CaF$_2$ tr / f	1151 / 1418	14 / 92	25 / 166	0.79 / 4.2	8.9	1.8
24	CrF$_2$	1097	(111	200)	-	9.0	-
25	MnF$_2$	856	(113	203)	-	10.2	-
27	CoF$_2$	1127	(111	200)	-	12.4	-
28	NiF$_2$	1450			-	9.2	-
29	CuF	908	(82	148)	-	5.7	-
30	ZnF$_2$	872	(77	139)	-	9.2	-
31	GaF$_3$	1027	(150	270)	-	14.5	-
32	GeF$_4$				-	20.2	-
33	AsF$_3$	-6	19	34	9.3	15.8	1.7
35	BrF$_3$	9	21	38	10.2	14.9	1.45
35	BrF$_5$	-60	9	16	7.05	21.61	3.1

Table 3. Heat of Fusion and Entropy of Fusion of Fluorides (cont'd.)

Compound Mol. W.		Melting Point °C	Heat of Fusion		Entropy of Fusion		
			cal g	Btu lb	Obs	Calc	Calc Obs
37 RbF		795	59	106	5.76	5.17	0.9
38 SrF_2		1397	94	169	8.1	9.0	1.1
39 YF_3		1367	98	176	8.7	12.5	1.45
40 ZrF_4		932	92	166	12.7	15.8	1.2
41 CbF_3		77	20	36	8.3	12.4	1.5
42 MoF_6	tr	-10	9	16	7.5	22.5	2.0
	f	18	5	9	3.6		
48 CdF_2		1047	55	99	6.2	9.2	1.50
53 IF_5		9	18	32	13.9	21.70	1.56
55 CsF		713	34	61	5.32	5.06	0.95
56 BaF_2		1277	29	52	3.2	8.83	2.8
58 CeF_6		1459	67	121	7.62	9.6	1.26
75 ReF_6		19	17	31	17.1	22.5	1.32
82 PbF_2		818	18	32	3.81	8.6	2.25
83 BiF_3		649	20	36	5.6	14.9	2.65
92 UF_6		64	13	23	13.6	22.4	1.7
94 PuF_3		1425	51	92	9.0	11.4	1.27

Table 4. Heat of Fusion and Entropy of Fusion of Oxides

Compound	Melting Point °C	Heat of Fusion			Entropy of Fusion		
		$\frac{cal}{mole}$	$\frac{cal}{g}$	$\frac{Btu}{lb}$	Obs	Calc	$\frac{Calc}{Obs}$
O					3.47		
H_2O	0	1437	80	144	5.26	5.47	1.04
H_2O_2	2	2560	76	137	9.35	8.94	0.96
Li_2O	1700	(13000	440	790)	-	6.63	-
BeO	1500	17000	680	1220	6.10	5.27	0.9
B_2O_3	450	5500	79	142	7.6	15.0	2
CO tr+f	-200	351	13	23	5.4	7.0	1.3
CO_2	-56	1990	46	83	9.15	10.44	1.14
NO	-163	550	18	32	5.1	5.6	1.09
N_2O_5	57	8300	77	140	27.5	21.6	0.78
N_2O	-93	1563	36	65	8.6	7.7	0.89
O_2	-219	153	10	18	6.94	6.94	1.00
Na_2O	920	11200	180	320	9.35	6.83	0.73
MgO	2800	18500	462	830	6.0	5.77	1.0
Al_2O_3	2000	26000	258	460	11.4	15.9	1.4
SiO_2	1610	3600	34	62	1.91	14.1	7.4
P_2O_5	572	5750	40	72	6.8	18.3	2.7
SO_2	-75	1769	28	50	8.95	7.06	0.78
$(SO_3)_2$	62	6090	38	68	18.95	23.0	1.21
CaO	2600	17000	305	550	5.9	5.7	0.97
TiO	2020	14000	219	395	6.1	6.19	1.01
TiO_2	1940	15500	194	350	7.3	9.66	1.3
VO_2	1545	14000	168	303	7.7	9.24	1.2
V_2O_5	670	15500	86	155	16.4	20.9	1.25
MnO	1750	13000	183	330	6.5	7.0	1.08
FeO	1380	8000	111	200	4.8	5.77	1.20
Fe_3O_4	1600	33000	143	260	17.6	20.8	1.17
CoO	1800	(12000	160	290)	-	5.81	-

Table 4. Heat of Fusion and Entropy of Fusion of Oxides (cont'd.)

Compound		Melting Point °C	Heat of Fusion			Entropy of Fusion		
			cal mole	cal g	Btu lb	Obs	Calc	Calc Obs
NiO		1950	(13100	164	295)	-	5.9	-
Cu$_2$O		1230	13400	94	170	8.95	8.1	0.91
ZnO		1800	(12600	155	280)	-	6.0	-
GeO$_2$	tr	1030	5050 tr	48	86	3.9	13.6	2.0
	f	1115	4100 f	39	70	2.95		
ZrO$_2$		2680	20800	169	305	7.17	9.24	1.28
Cb$_2$O$_5$		1510	24590	93	188	13.8	21.9	1.6
MoO$_3$		795	12500	87	156	11.6	12.7	1.09
BaO		1920	14700	90	164	6.7	6.86	1.03
PbO		875	6100	28	50	5.3	5.33	1.00
Bi$_2$O$_3$		820	7400	16	29	6.8	8.24	1.21
UO$_2$		2840	25300	94	170	8.1	10.3	1.28

entropy of fusion for water and hydrogen peroxide. Confirmation
of the calculations as applied to water indicate that these
could be extended to salt-hydrates, as this will be shown later,
reaching rather excellent agreement with known measurements.

HYDROXIDES are listed in Table 5, containing the data for
9 anhydrous compounds; and 4 hydrates. Deviations occur only
with the hydroxides of Ca, Sr, and Ba.

BROMIDES AND IODIDES, listed in Table 6 and 7, contain most
of the observed values of heats of transition, fusion and entropy of fusion. An analysis of Bromides and Iodides follows:

$$S_f \text{ Calc/Obs} = 1+0.3 \qquad S_f \text{ Calc/Obs} = \quad 1.3$$

BROMIDES

Li Na Mg K Fe   *C   N   Al Ca Ti Cu Zn As Sr Ag*
RB CD Cs Ta Tl   *Sn Sb Ba La Pr Nd Gd Hg Pb U*

IODIDES

Li Na K Ca Fe   *N Al Ti Cu Sr Cd Sn Ba Ce Pr Pb*
Rb Ag Cs Nd Hg

OTHER COMPOUNDS are listed in Table 8.

NITRATES. In the group of nitrates, those of NH_4, Na, K,
and Ag have average deviation factors of 1.8, Nitric acid, its
hydrates and $LiNO_3$ do not deviate from the additive rule.

SULFATES. In the group of sulfates, sulfuric acid shows
a deviation factor of 2. With this correction the entropies
of fusion of the sulfates of Li Na K Ba and Pb, as well as the
acid sulfates of NH_4 Na and K follow the additive rule. The
sulfates of Ca and Tl show higher ratios. It is, of course,
possible that some of the data is inaccurate, or that decomposition occurs during melting. Later data may provide better
proof of the additive rule, which has been reasonably accurate
in giving heat of fusion values for more than half of the studied compounds.

Table 5.   Heat of Fusion and Entropy of Fusion of Hydroxides

Compound	Melting Point °C	Heat of Fusion cal/mole	Heat of Fusion cal/g	Heat of Fusion Btu/lb	Entropy of Fusion Obs	Entropy of Fusion Calc	Calc/Obs
(OH)					4.45	4.45	1.0
H_2O	0	1437	80	144	5.26	5.43	1.03
LiOH	471	5290	220	395	7.1	6.03	0.85
$B(OH)_3$	d 185	(7250	118	212)	-	15.8	-
NH_4OH	-179	1568	45	81	8.1	10.5	1.10
NaOH tr	300	1720	43	77	3.00 ⎤		
f	320	1520	38	69	2.56 ⎦	6.15	1.10
$NaOH.H_2O$	64	3850	66	119	11.41	11.41	1.00
$NaOH.2H_2O$	13.5	3859	51	92	13.5	15.7	1.16
$NaOH.3.5H_2O$	15.5	5380	52	94	18.5	21.0	1.13
KOH tr	249	1520	27	49	2.93 ⎤		
f	400	2050	37	67	3.20 ⎦	6.14	1.00
NaOH.KOH	177	(5100	53	95)	-	12.29	-
$Ca(OH)_2$	835	6900	93	165	6.25	11.1	1.8
$Sr(OH)_2$	510	5230	43	77	6.7	11.2	1.67
$Ba(OH)_2$	417	4590	27	49	6.65	10.7	1.60
$Ba(OH)_2 8H_2O$	78	22700	72	130	64.6	53.1	0.82
RbOH tr	225	1290	13	23	2.58 ⎤		
f	385	2120	21	37	3.22 ⎦	5.25	0.90
CsOH tr	220	1450	10	18	2.95 ⎤		
f	315	1090	7	13	1.85 ⎦	5.05	1.05

Table 6. Bromides

Atom No.			Melting Point °C	Heat of Fusion $\frac{cal}{g}$	$\frac{Btu}{lb}$	Entropy of Fusion Obs	Calc	$\frac{Calc}{Obs}$
3	LiBr		550	51	92	5.1	6.3	1.2
6	CBr_4	tr	47	4	7	4.4	—22.5	3.1
		f	93	3	5	2.6		
7	NH_4Br		542	51	92	6.1	10.9	1.8
11	NaBr		750	61	110	6.1	6.4	1.05
12	$MgBr_2$		711	48	86	9.0	11.8	1.3
13	$AlBr_3$		98	25	45	7.3	17.0	2.3
19	KBr		735	51	92	6.1	6.4	1.05
20	$CaBr_2$		743	35	63	6.9	11.8	1.7
22	$TiBr_4$		39	9	16	9.9	21.7	2.2
29	CuBr	tr	382	10	18	2.1	— 7.0	1.4
		f	487	16	29	3.0		
30	$ZnBr_2$		394	17	31	5.6	12.1	2.15
33	$AsBr_3$		31	9	16	9.2	20.4	2.2
37	RbBr		692	34	61	5.8	6.6	1.15
38	SrBr	tr	645	12	22	2.7	— 11.8	2.0
		f	657	10	18	3.2		
47	AgBr		430	12	22	3.1	7.0	2.3
48	$CdBr_2$		568	29	52	9.5	11.9	1.25
50	$SnBr_2$		232	6	11	3.4	12.9	3.8
51	$SbBr_3$		97	10	18	9.5	19.5	2.0
55	CsBr		636	27	49	6.2	6.4	1.0
56	$BaBr_2$		850	26	47	6.7	11.6	1.7
57	$LaBr_3$		788	53	95	11.6	16.6	1.4
59	$PrBr_3$		699	30	54	11.7	17.4	1.5
60	$NdBr_3$		682	28	50	11.4	16.5	1.45
64	$GdBr_3$		785	28	50	8.3	16.5	2.0
73	$TaBr_5$		267	19	34	19.8	26.1	1.3
80	$HgBr_2$		238	12	22	8.4	11.8	1.40

Table 7. Iodides

Atom No.			Melting Point °C	Heat of Fusion $\frac{cal}{g}$	$\frac{Btu}{lb}$	Entropy of Fusion Obs	Calc	$\frac{Calc}{Obs}$
3	LiI		469	26	47	4.7	6.4	1.36
7	NH_4I		551	35	63	6.1	11.0	1.8
11	NaI		660	38	68	6.04	6.5	1.08
13	AlI_3		191	10	18	8.6	12.5	1.45
19	KI		681	35	63	6.0	6.4	1.09
20	CaI_2		779	34	61	9.5	9.9	1.04
22	TiI_4		150	8	14	9.9	22.1	2.23
26	FeI_2		594	46	83	16.1	12.0	0.86
29	CuI		590	10	18	2.2	7.15	3.2
37	RbI		647	25	45	5.7	6.65	1.16
38	SrI_2		538	17	25	5.8	12.0	2.06
47	AgI	tr	150	6	11	3.5 ⎤	⎫ 7.15	1.16
		f	558	10	18	2.7 ⎦	⎭	
48	CdI_2		387	14	25	7.6	12.1	1.6
50	SnI_4		145	7	13	11	22.8	2.1
55	CsI		626	22	40	6.27	6.5	1.04
56	BaI_2		711	16	29	6.4	11.7	1.84
58	CeI_3		760	24	43	12.0	17.4	1.45
59	PrI_3		738	24	43	12.6	17.7	1.40
60	NdI_3	tr	574	6	11	4.0 ⎤	⎫ 12.0	0.90
		f	787	19	34	9.2 ⎦	⎭	
80	HgI_2	tr	129	1	1	1.6 ⎤	⎫ 12.0	0.90
		f	254	10	18	8.60 ⎦	⎭	
82	PbI_2		412	13	23	8.8	11.5	1.3

The conclusion can be reached that the proposed rule can definitely establish the maximum possible value of the heat of fusion of any compound.

A. *Entropy of Fusion Per Gram Weight*

In Tables 1 and 2 the entropy of fusion is shown in calories/mole^{o}K. It would be desirable to compare the entropy of fusion of elements and anions forming the compounds on the basis of equal weights.

In Table 9, the cationic elements are arranged according to atomic numbers and the anions Cl F O Br I NO_3, according to their atomic numbers, showing deviation factors. The second column lists the combined heats of transition S_{tr} and fusion S_f, while the rest of the columns give the deviations from the additive rule. For the anions the following S_{tr} and S_f values were used: Cl = 4.45, F = 3.37, 0 = 3.47, Br = 4.76, I = 4.85, NO_3 = 11.6. The elements and anions with the lowest atomic weight(atomic No) show the highest entropy of fusion.

B. *Entropy of Fusion and the Periodic System*

It is advantageous to correlate the ratios S_fCalc/Obs and the deviation factors from the additive rule, with the periodic system of the elements by listing them according to atomic numbers. This is shown in Table 8, where the entropy of fusion of elements in included for comparison. There are striking similarities between the peaks of the curve for the elements and the deviations of chlorides, fluorides and oxides. In many groups the ratios are close to unity (within the 0.7 to 1.3 range). This group includes the compounds of H Li Ma K Rb Mg Mn Fe Al, where data was available for at least two compounds. It is probable that several other elements belong to this group but data was not available to prove this.

Table 8. Heat of Fusion and Entropy of Fusion of Compounds

Nitrates	Melting Point °C	Heat of Fusion cal/mole	Heat of Fusion cal/g	Heat of Fusion Btu/lb	Entropy of Fusion Obs	Entropy of Fusion Calc	Calc/Obs
(NO_3)						11.6	
HNO_3	-42	2503	40	72	10.81	12.6	1.15
$HNO_3 \cdot H_2O$	-37	4184	52	93	17.76	17.9	1.0
$HNO_3 \cdot 3H_2O$	-18	6954	54	107	27.30	28.4	1.0
$LiNO_3$	251	6390	93	167	12.1	13.2	1.1
NH_4NO_3 tr	32	1710	21	38	4.66 ⎤		
tr	125				⎬ 16.6		2.2
f	170	2900	16	29	2.94 ⎦		
$NaNO_3$ tr	276	944	11	20	1.26 ⎤ 13.3		1.7
f	310	3800	45	82	6.50 ⎦		
Kno_3 tr	128	1400	14	25	3.5 ⎤		
f	337	2800	28	51	4.6 ⎬ 13.3		3.3
tr	160	930	9	16	2.15 ⎦		
$Ca(NO_3)_2$	561	5700	35	63	6.8	15.2	2.25
$Cd(NO_3)_2$	300	4350	14	25	7.6	14.0	1.85
$AgNO_3$ tr	159	560	3	6	1.30 ⎤ 13.9		1.85
f	208	3020	18	32	6.25 ⎦		
$LiNO_3 2KNO_3$	129	(16000	59	107)	-	39.8	
		(11200	42	75)	-	27.9	
$LiNO_3 NaNO_3$	204	(12700	82	148)	-	26.5	
$Sr(NO_3)_2$	645	10650	50	90	11.6	25.3	2.2
$Ba(NO_3)_2$	592	9950	38	69	11.5	25.3	2.2
$RbNO_3$	308+u	2480	17	31	6.0	13.3	2.2
$CsNO_3$	406	3370	17	31	5.0	13.3	1.85
$KNO_3 NaNO_3$	222	(7900	42	76)	-	15.9	-
$LiNO_3 \cdot 2NH_4NO_3$	97	(10000	43	78)	-	26.9	-
$LiNO_3 \cdot 4KNO_3$	112	(17500	37	67)	-	45.6	-

Table 8.   Heat of Fusion and Entropy of Fusion of Compounds
(cont'd.)

| Sulfates | Melting Point °C | Heat of Fusion | | | Entropy of Fusion | | |
		cal mole	cal g	Btu lb	Obs	Calc	Calc Obs
SO_2	-75	1769	28	50	8.95	8.06	0.90
H_2SO_4	10	2360	24	44	8.32	17.00	2.0
$H_2SO_4 \cdot H_2O$	9	4630	40	72	16.44	22.26	1.3
SO_4 (basis)					6.32	15.00	2.35
Li_2SO_4 tr	575	7000	64	115	8.25 ⎤ 9.48		0.85
f	859	3350	30	54	2.95 ⎦		
Na_2SO_4 tr	241	2068	15	27	3.48 ⎤ 9.7		1.5
f	884	5800	41	74	4.90 ⎦		
$Na_2SO_4 \cdot 10H_2O$	32	19400	60	108	63.5	62.3	1.0
$NaHSO_4$	182	2480	21	38	5.45	5.84	1.5
NH_4HSO_4	144	4300	43	77	8.18	10.29	1.5
K_2SO_4 tr	583	2570	43	27	3.0 ⎤ 9.7		1.0
f	1070	9060	52	93	6.75 ⎦		
$KHSO_4$ tr	164	590	4	7	1.33 ⎤ 5.85		1.0
f	219	2220	16	29	4.52 ⎦		
$CaSO_4$ f	1300	6700	49	88	4.26	8.55	2.0
$BaSO_4$ tr	1149 ⎤ 9700	42	76	6.00	8.41	1.4	
f	1350 ⎦						
Ti_2SO_4	632	5500	11	20	6.1	10.0	1.5
$PbSO_4$ tr	866	4060	13	23	3.56 ⎤ 8.18		0.8
f	1087	9600	32	58	7.1 ⎦		
$Na_2S_2O_7$ tr		1620	7	13 ⎤			
f	403	9760	44	80 ⎦	14.4	29.1	2.0
$MgSO_4$	1130	3500	29	52	2.5	8.6	3.45

The majority of elements of higher valence form compounds where the ratio S_fCalc/Obs is greater that $1 + 0.3$. At least two observations of this type are available for B C Si P As Cb Mo, forming deviations. With elements of lower valence higher deviations occur in Ca Sr Ba Cd Sn Pb.

According to Table 9 the ratios for the chlorides, fluorides and oxides of the same element deviate from $1 + 0.3$ to the same extent. We can, therefore, assign a factory to each element as a multiplier to change its ratio to unity.

$$y(S_f\text{Calc/Obs}) = 1 + 0.3$$

With this simplification the previous equation is changed to:

$$S_f\text{Calc/Obs for } A_2B_3 = y(2S_{fA} + 3S_{fB}) = 1 + 0.3$$

The factory y is shown in the following Table:

Element	Valence	Average deviation	1/y		y
Al	3	1.5	3/2	=	1.5
C	4	2.0	4/2	=	2.0
B	3	1.7	7/4	=	1.75
Si	4	2.0 (except SiO_2)	4/2	=	2.
P	5	2.5	5/2	=	2.5
Ca	2	1.8 (except CaO)	4/2	=	2
As	3	1.7	3.2	=	1.5
Cd	2	2	4.2	=	2
Ba	2	2	4/2	=	2
Pb	2	2	4/2	=	2

This comparison shows that the y factors are common fractions where the denominator corresponds to the valence of the element forming the compounds. In monovalent elements the factor should be y=1 and this explains the fact that the entropies of fusion of such compounds can be obtained directly by adding the s_f values of the elements. This rule is applicable to the bromides and iodides of Na and K.

According to Table 9 the y factor is within the 1+0.3 range for the compounds of several bivalent and tribalent elements, including Mg, Mn, Fe and possible several others. This could be explained only by introducing an integer multiplier for these elements, equal to their valence.

Until further data will be available, we may conclude, therefore, that it is possible to calculate the entropy of fusion by adding those of the elements and using the correction factors y as shown in the table.

C. Materials of High Heat of Fusion

The highest heat of fusion is shown by LiH, a compound of the lowest molecular weight, (7.95), storing 1120 cal/gram. The material melts at 1267°F, but it decomposes near its melting point. It has been tested for space applications (25). The next highest heat of fusion values are shown by lithium compounds, by fluorides, the refractory carbides, nitrides, borides, silicides and some oxides. Some of these have heat of fusion values up to 700 cal/gram, 1260 BTU/lb as shown in Table 8.

The high melting point materials are of interest because some of them can be used in binary or ternary systems, forming compounds or eutectics with considerably lower melting points. The most extensive change in melting point can be achieved in salt-hydrates. some of the high melting point materials with high heats of fusion form salt-hydrates which melt below 100°C.

III.   CONCLUSIONS

Collection of extensive data are presented, giving heats
of transition and fusion of elements and their compounds.
Analysis of data, based on theoretical considerations, reveal
that it is possible to calculate heat of fusion values of compounds by using the entropies of fusion of elements, forming
the compounds, correcting these data with correlation factors
to compensate for deviations from the simple additive rule.
In this way entropy of fusion and heat of fusion data can be
obtained for eutectics, or compounds, provided their melting
point is known, in addition to molar composition. Calculated
values agree with observed values within ±30% and often within
± 10%, which can be considered satisfactory, in view of experimental difficulties in determining heat of fusion values, especially at high temperatures and with materials that may decompose near their melting points.

The search for the "highest" heat of fusion material in
a selected melting point range can be simplified greatly, by
calculating the probable entropy and heat of fusion before em-
barking on highly time consuming experiments, because such experiments can be restricted to those compounds, or eutectics,
that have been found to be "promising", by calculations.

Tables and Figures indicate that only those compounds have
higher heat of fusion values, based on unit weights, that are
formed of elements of low atomic weights. These consideration
eliminates from the group of candidates a large group of materials of higher atomic weights.

Materials formed of low cost and low atomic weight compounds
include H, B, C, N, O, Na, Mg, Al, Si, P, S, Cl, K, Ca, Ti and
Fe. Some of these elements can combine to form groups, such
as borates, carbonates, nitrates, phosphates, including the
addition of water in the form of hydrates.

Other compounds that include Li and fluorides of some of
the groups listed above are available only at a higher cost
level.

REFERENCES

1. Lane, George A., et al., "Solar Energy Subsystems Employing
    Isothermal Heat Storage Materials", ERDA-117, UC-94a, May,
    (1975).
2. Hilderbrand, J. Anal. Chem. 24, 720 (1952).
    Hilderbrand, J. Chem. Phys. 20, 1520 (1957).
3. Tamman, G., "Kristallisieren und Schmelzen", 1903, Barth,
    Leipzig.
4. Tamman, G., "The States of Aggregation", trans. R.F. Mehl,
    1925, Van Nostranc, New York.
5. Kubaschewski, O., Dench, W.A., "Calorimetry", p. 171, in
    "Physico-chemical Measurements at High Temperatures" (J.
    Brockris, ed), Butterworths, London, (1959).
6. Kubaschewski, O., et al., "Metallurgical Thermochemistry",
    Pergamon Press, 1967 edition.
7. Kubaschewski, O., Trans. Faraday Soc. 45, 931 (1949).
8. Blanc, M., C.R. 246, 570-1, 1958. Heat of fusion of Li, Na,
    K bromides.
9. Blanc, M., C.R. 249, 2012-3, 1959. Heat of fusion of KF
    + K_2SO_4 and LiCl + $SrCl_2$.
10. Blanc, Mme M., C.R. 247, 273, 1958. Heat of fusion
    (Calculated) of A, Cl, AgBr.
11. Blanc, M., Compt. Rend. Acad. Sci. Paris 247, 273 (1958).
12. Blanc, M., and Petit, G., Compt. Rend. Acad. Sci. Paris
    248, 1305 (1959).
13. Blanc, Mme M., C.R. 248, 1305, 1959. Heat of fusion of Pb
    salts.
14. Schinke, H., Sauerwald, F., Z. Anorg. Chem. 287, 313-24
    (1956). Volume change during melting.

15.   Schinke, H., Sauerwald, F., Z. *Anorg. Chem. 304,* 25-36
      (1960). Volume change during melting.

16.   Vogel, E., Schinke, H., Sauerwald, F., Z. *Anorg. Chem. 284,*
      131-41 (1956). Volume change during melting.

17.   Ubbelohde, A.R., *Quart. Rev. Chem. Soc. 4,* 356 (1950).
      Entropy of fusion of inert gases.

18.   Ubbelohde, A.R., *Quart. Rev. (London) XI,* 246 (1957).
      Thermal transformations in soldis.

19.   Ubbelohde, A.R., Heat of fusion calculations.

20.   Ubbelohde, A.R., Melting and Crystal Structure, Clarendon
      Press, Oxford, (1965).

21.   Temperly, H.M.V.,"Changes of State", (1956).

22.   Temperly, H.M.V., "Changes of State, A Mathematical-
      Physical Assessment". Interscience Publishers, New York,
      (1956).

23.   Vorob'ev, A.A., *Izvest. Vysshikh Ucheb. Zavedenii Fiz No. 1,*
      160-2 (1958), *C.A. 54,* 9418d (1960). Properties of ionic
      crystals and lattice energy, heat of fusion theory.

24.   Vorob'ev, A.A., et al., *Izvest. Vysshikh Ucheb. Zavedenii*
      Fiz No. 6, 162-5 (1959), *C.A. 54,* 9416f (1960). Heat of
      fusion theory.

25.   Caldwell, R.T., McDonald, J.W., and Pietsch, A., "A Solar
      Energy Receiver with Lithium-Hydride Heat Storage", *Solar*
      Energy J. 9(1), 48-60 (1965).

CHAPTER 13

THE APPLICATION OF REVERSIBLE CHEMICAL REACTIONS
TO SOLAR THERMAL ENERGY SYSTEMS

Raymond Mar

Sandia Laboratories
Livermore, California

I. INTRODUCTION

Numerous investigators have proposed reversible thermochemical
reactions as a means for storing thermal energy in solar energy
systems. In this application, thermal energy is stored in the
form of chemicals created by endothermic reactions. The thermal
energy is recovered by recombining the chemicals, which releases
a quantity of energy equivalent to the heat of reaction. In addition to the storage applications, there is interest in applying
reversible reactions to solar thermal energy transport and solar
thermal heat pumping for space heating and cooling systems.

This paper reviews all three of these applications in order to
introduce the reader to the use of reversible chemical reactions
in solar energy systems. The characteristic features of reversible chemical reaction systems are described and compared to
sensible and latent heat systems. The different uses of reversible chemical reactions in solar energy applications are reviewed
and their potential impact (i.e., economic viability) is
appraised. The following chapter discusses materials science
issues relevant to the development of thermochemical schemes.

Only those reversible chemical reactions which are thermal in
nature are discussed; electrochemical, photochemical, and

SOLAR MATERIALS SCIENCE

439

radiochemical reactions are specifically excluded. The term
"thermochemical" is used based on the fact that thermal energy is
stored and released as a consequence of net changes in chemical
bonding.

II. GENERAL CHARACTERISTICS OF THERMOCHEMICAL SYSTEMS

Solar thermal energy can be stored in three forms: 1) as
sensible heat, 2) as latent heat of transitions, and 3) as chemical bond energy, i.e., the use of thermochemical reactions.
Because enthalpies of reaction are generally much greater than
either enthalpies of transition or sensible heat changes over
reasonable temperature intervals, storage densities (based on
either mass or volume) are greater for thermochemical storage
systems than for sensible or latent heat systems. Typical
energy density differences are illustrated on Figure 1, where the

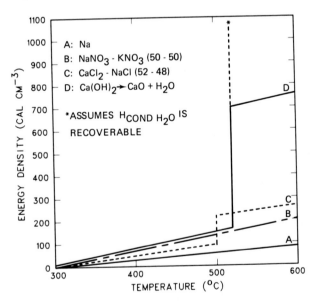

*FIGURE 1. A comparison of energy storage densities for sensi-
ble heat (curves A and B), latent heat of fusion (curve C) and
reversible chemical reaction (curve D) storage media. Energy
densities are based on chemical properties only.*

requirements of thermal energy storage systems in solar central receiver systems (i.e., a nominal temperature swing of 300°C - 600°C) are met by four different systems: 1) a sensible heat liquid sodium system, which is the baseline approach adopted for systems designed by General Electric (1) and Rockwell International (2); 2) a sensible heat nitrate draw salt system, adopted by the Martin Marietta design (3); 3) a heat of fusion system based on the $NaCl-CaCl_2$ eutectic; and 4) the use of the $Ca(OH)_2 = CaO + H_2O$ reaction. The energy densities were calculated based on the chemical properties and are not totally reflective of the density in actual practice. For example, a dual tank approach to sensible heat storage would require the storage volume to double, thereby decreasing the effective storage density by a half. Nevertheless, the implications drawn by Figure 1 are reasonably accurate: storage densities associated with thermochemical systems are typically greater than sensible and latent heat systems by a factor ranging from two to more than an order of magnitude.

One important criterion for selecting thermochemical reactions is that one must be able to prevent the products of the endothermic reaction from reversing. This is typically accomplished by selecting reactions whose products are easily separated physically, or reactions which do not proceed in the absence of a suitable catalyst. Control over the reaction enables one to store energy at ambient temperatures by cooling and storing the products of the endothermic reaction. The ability to store energy at ambient temperatures has associated attributes. First, the potential for corrosion is reduced because large quantities of material are not kept at elevated temperatures. Second, one is not faced with potential heat losses and insulation needs. However, it should be noted that storing energy at ambient temperatures requires careful process design and engineering to use the sensible heat associated with cooling to ambient. If these sensible heat losses are not utilized in the operational cycle, the overall system efficiency suffers. The ability to store at ambient temperatures provides operational design flexibility in that long-term storage options may be considered.

Because chemicals at ambient temperature can be easily transported over long distances, thermochemical reactions can be used to transport thermal energy. Chemical reactions involving liquids or gases are the most amenable to energy transport applications, since pipeline transport is simple and straightforward. However, the transport of solids has also been proposed (4,5).

The cost of an energy storage system is divided into power related costs and energy related costs (6). The former are costs associated with transferring heat to and from the storage unit, and includes costs for heat exchangers, chemical reactors, and pumps. Energy related costs are associated with the quantity of energy stored, and include costs for the storage media, storage tanks and insulation. Thermochemical systems are characterized by the ability to physically separate and power and energy components; therefore, the sizes of these components can be varied independently. Since the energy related costs are generally low for thermochemical systems, scenarios calling for added storage capacity (e.g. long-term storage) at fixed power delivery can be explored. This situation can be contrasted with latent heat systems where the heat exchanger is an integral part of the storage tank. Any increase in storage tank volume must necessarily be accompanied by a corresponding increase in the heat exchanger surface area.

Thermochemical systems have several drawbacks. Round trip efficiencies, defined as the ratio of amount of energy out of storage to that entering, are low because of the numerous energy loss steps (such as heat exchanging and gas compression) required for a complete operational cycle. Also, the operational and maintenance requirements may be more demanding and costly due to the complexity of thermochemical systems.

III.   THERMAL ENERGY STORAGE

The intermittant nature of solar energy requires that there be a means to store thermal energy to cover short term transients such as daily cloud coverage, and longer term transients, such as diurnal cycling and days of poor insolation. A simplified flow schematic for the reaction $Ca(OH)_2 = CaO + H_2O$ is used to illustrate the basic features of a thermochemical energy storage scheme (Figure 2). To store thermal energy, $Ca(OH)_2$ is transported to the endothermic chemical reactor and decomposed into CaO and H_2O. H_2O is condensed and the products of reaction, CaO and H_2O, are stored until there is a need for the energy. The system is discharged by first vaporizing the water and then running the exothermic reaction between steam and CaO in the exothermic reactor. $Ca(OH)_2$ exiting from the exothermic reactor is cooled and stored until used in the charging cycle.

It is possible to design the system whereby only one heat exchanger is used for both the exothermic and endothermic reaction. Further, the chemical reactor may be used for chemical

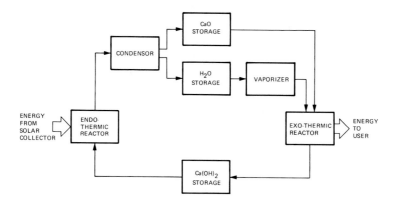

FIGURE 2.   The use of the reaction $Ca(OH)_2 = CaO + H_2O$ for thermal energy storage in a two reactor/heat exchanger system.

storage; such a system is schematically represented in Figure 3. In this approach, the energy and power related components are no longer decoupled. While the overall system is less complex than the moving solids approach of Figure 2, the fixed bed concept suffers from heat transfer and mass transport concerns.

Figure 4 gives an example of a reaction which is catalytically controlled, as opposed to being controlled by physically separating the products. In this case, SO_3 is decomposed into SO_2, which is condensed and stored as a liquid, and O_2, which is compressed and stored as a gas. Energy is recovered by catalytically recombining SO_2 and O_2 in the exothermic reactor; the product of this exothermic reaction is then condensed and stored.

The schematics in Figures 2 through 4 are very simplistic and do not show all of the heat recovery steps necessary for developing an efficient process. More realistic process schemes have been developed and analyzed (7-9) but their discussion is outside of the scope of this paper.

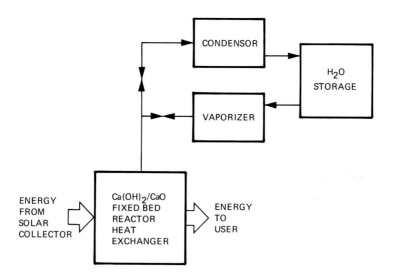

FIGURE 3. The use of the reaction $Ca(OH)_2 = CaO + H_2O$ for thermal energy storage in a one reactor/heat exchanger system.

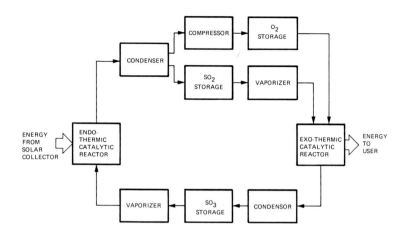

FIGURE 4.   A simplified schematic for a thermal energy storage system based on the reaction $2SO_3 = 2SO_2 + O_2$.

IV.   TRANSPORT OF THERMAL ENERGY

Thermochemical reactions can be used in distributed solar power systems, where solar energy is collected by numerous collectors and transmitted to an energy storage and central power generating facility.   Figure 5 shows a specific example, the Solchem process proposed by Chubb (10), which uses the reaction $2SO_3 = 2SO_2 + O_2$ to transmit thermal energy from a field of high concentration parabolic collectors to a molten salt heat of fusion energy storage system.   The ammonia dissociation reaction has also been proposed for this application by Carden (11).   In principle, thermochemical reactions can also be used as the primary heat transfer fluid in central receiver thermal electric large power systems of the type discussed in Reference 12.

High concentration solar central receiver systems have been proposed where the end product is thermal energy rather than electricity (13,14).   In this application, thermochemical re- actions are used to transport the thermal energy from the solar collection site to the user.   An example of this application is

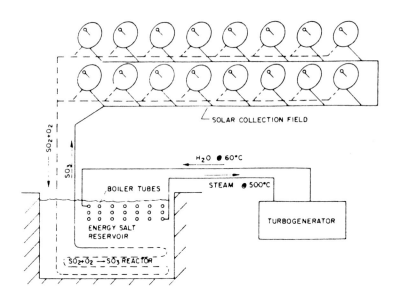

*FIGURE 5. The use of the reaction $2SO_3 = 2SO_2 + O_2$ to trans-
port thermal energy in a distributed solar power plant (Reference
10).*

given in Figure 6, where the reforming/methanation reaction
$CH_4 + H_2O = CO + 3H_2$ is used to transport energy from a solar
collection device to an end user, e.g. industrial process heat.

A slight variation of this process has been studied by Baker,
et al., (12) in which the thermochemical reaction is not used in a
reversible manner, but rather as a step in the generation and
transmission of synthetic natural gas. The concept shown in
Figure 7 is identical to the system studied in detail by Baker,
et al., (15) only the ash agglomerating gasifier plant is replaced
by a solar powered coal gasifier. Pulverized coal and steam are
heated to generate a mixture of H_2 and CO, which is then piped
over long distances to a methanator where the methanation
reaction is catalytically carried out. The end result is the
production of substitute natural gas, with the added feature that
the heat of methanation is now generated at a site selected such
that the heat can be put to good use (e.g. industrial process
heat).

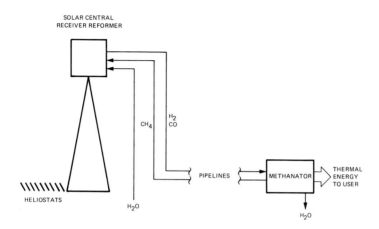

FIGURE 6.   The transport of thermal energy from a solar central receiver facility, using the reaction $CH_4 + H_2 = CO + 3H_2$.

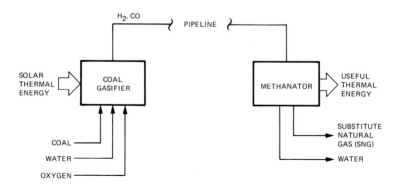

FIGURE 7.   The use of the methanation reaction in an open loop energy transmission system.

V. SPACE HEATING AND COOLING

Solar chemical heat pump systems capable of space heating and cooling have been proposed by many different investigators (16-18). The essential elements of the chemical heat pump system are illustrated in Figure 8. Two thermochemical reactions which have a common vapor species are needed. In the example, $NH_{3(g)}$ is the transferring gas species, and the two reactions used are:

$$MgCl_2 \cdot 6NH_3 = MgCl_2 \cdot 2NH_3 + 4NH_3 \qquad (1)$$
$$CaCl_2 \cdot 8NH_3 = CaCl_2 \cdot 4NH_3 + 4NH_3 \ . \qquad (2)$$

These two reactions are selected such that Reaction (1) takes place at a higher temperature than Reaction (2). During the charging cycle, solar energy is used to decompose the salt in the high temperature reactor (HTR); this may be done by direct

CHARGING CYCLE

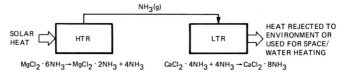

DISCHARGE CYCLE FOR SPACE HEATING

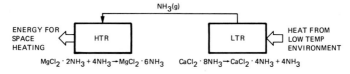

DISCHARGE CYCLE FOR SPACE COOLING

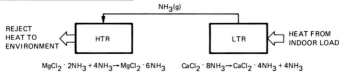

FIGURE 8. The operation of a chemical heat pump system based on carrying out the reaction $CaCl_2 \cdot 8NH_3 = CaCl_2 \cdot 4NH_3 + 4NH_3$ in the low temperature reactor (LTR) and $MgCl_2 \cdot 6NH_3 = MgCl_2 \cdot NH_3 + 4NH_3$ in the high temperature reactor (HTR).

absorption (18), or indirectly by means of a heat transfer fluid (17). The ammonia generated is transferred to the low-temperature reactor (LTR) where an exothermic reaction forms $CaCl_2 \cdot 8NH_3$. This heat of reaction may be used for space or water heating, stored, or simply rejected to the environment. When the heat pump system is discharged for space heating, energy is absorbed by the low-temperature unit at outdoor ambient temperatures; the $CaCl_2 \cdot 8NH_3$ salt in the low-temperature reactor is decomposed, and the ammonia transfers to and reacts with the $MgCl_2 \cdot 2NH_3$ in the high-temperature unit. The resulting heat of reaction is used for space heating. Space cooling is achieved by using the low-temperature reactor to absorb heat from the indoor environment, and the exothermic heat of reaction generated in the high-temperature reactor is rejected outdoors. The two reactions must be selected such that their pressure-temperature curves are consistent with temperature and pressure differential requirements for spontaneous gas transfer.

The example given in Figure 8 is for two solid dissociation reactions. If the vapor species were easily condensible (such as in the case of hydrated salts and methanolated salts), one could replace the low-temperature reactor with a condenser unit (16,18). One further modification is worth mentioning; the high-temperature reaction may be a liquid phase solution/dissolution reaction such as that found for sulfuric acid systems (19).

VI.   ASSESSMENT OF POTENTIAL

Before discussing the materials science issues associated with each of the thermochemical concepts, it is important to place the applications in the proper perspective. Therefore, their economic viabilities as determined by cost/benefit studies are discussed in this section. It should be noted that definitive conclusions cannot be drawn in many cases because the

solar and thermochemical technologies have not yet been developed
to a stage where reliable data are available. In addition,
systems studies generally tend to be system specific, and unique
characteristics of different systems are often neglected. Therefore, one is cautioned against universally applying the results
of one study. However, preliminary assessments have been made
and their implications must be seriously considered.

A. *Thermal Energy Storage*

The capital cost of an energy storage system can, as a first
approximation, be divided into two components: a power related
component and an energy capacity term. Therefore,

$$C = C_p + C_E \cdot t \tag{3}$$

where C is the cost of the storage system ($/kW), C_p is the power
related cost ($/kW), C_E is the energy capacity related cost
($/kW-hr), and t is the time of storage (hr). In general, C_p for
thermochemical systems are large relative to sensible and latent
heat systems, and C_E is small. Therefore, it is not cost
effective to use thermochemical systems in lieu of sensible and
latent heat systems for short storage times. However, at long
storage times, the costs of thermochemical systems become low
relative to sensible and latent heat systems, which is simply a
manifestation of the relatively small energy capacity related
costs for thermochemical systems. This suggests that extended
storage scenarios which take advantage of the low energy
capacity costs associated with thermochemical storage systems may
be attractive. However, it is the cost of energy to the consumer
and not the cost of the energy storage system that is of interest.
Therefore, the impact of thermochemical systems must be determined by considering the total solar energy system and deriving
a delivered energy cost.

Detailed cost/performance analyses have recently been
completed at Sandia (20-23) and Rocket Research (23-24). In these
studies, computer simulations of solar thermal electric

conversion facilities were carried out in which the solar energy
system facility was parametrically modeled as a collection of
subsystems (collector field, receiver, turbogenerator, and energy
storage) which process energy with various efficiencies and
capacities. Using real time insolation data and assuming a
constant baseload output, the minimum energy cost configuration
was identified. The results of these studies, summarized recently by Iannucci, Fish, and Bramlette(25), show that long duration
storage scenarios which take advantage of the low energy capacity
costs for thermochemical systems will probably be unattractive.
The basic problem is the low thermal efficiencies characteristic
of thermochemical systems. These low efficiencies are levered
back through the most expensive component of the solar facility,
the heliostat, and the end result is a greatly increased total
system cost. When very long duration storage scenarios are
considered (where the majority of storage capacity is used for
infrequent periods when isolation is poor), the use of this additional storage capacity and associated collector area is too low
to justify the initial capital investment. Therefore, sensible
heat storage systems (perhaps hybridized with a small quantity
of fossil fuel (24-25)) and not thermochemical systems seem to be
best suited for use in solar thermal electric power generating
systems.

B. *Transport of Thermal Energy*

Cost/benefit studies of reversible chemical reactions used to
transport thermal energy from a point of collection to a central
point (for storage, power generation, or thermal use) are just
now evolving. Turner (26) and Williams (27) have calculated energy
transport costs using reversible chemical reactions, and they
show significant improvements over those projected for sensible
heat systems. However, Iannucci (28) has pointed out that the
cost of energy transport is itself not the most meaningful figure

of merit. Rather, the total system cost (or cost of energy delivered to the consumer) must be calculated. As such, the costs associated with the required chemical reactors and process inefficiencies must be factored in. On the other hand, it may be that for some high-temperature applications, thermochemical techniques are the only option available due to the thermal instability of sensible heat transfer fluids. To date, the potential role of thermochemical reactions in distributed solar systems is as yet undetermined.

The use of the reforming/methanation reaction $CH_4 + H_2O = CO + 3 H_2$ to transport solar generated energy for industrial process heat has been the subject of several studies (13-15). Vakil et al. (14) concluded they could not unambiguously determine the economic viability because of cost and performance uncertainties associated with the solar energy collection system. Assuming all technical/materials problems can be overcome, their study concludes that a solar based industrial process heat distribution system can provide energy at an incremental cost of $2-4/MBTU above that of the solar energy collection system. For comparison, current costs for process heat generated from synthetic natural gas boilers are approximately $5/MBTU.

In another study, Richardson and Wendlandt developed and analyzed a process cycle called SOLTHERM (13). The estimated cost of process steam delivered 100 miles from a solar energy collection facility was $3.5/MBTU; this cost is incremental to the solar collection plant. Comparative costs for process steam generated by other energy sources were $3.75/MBTU, $6.00/MBTU, and $6.50/MBTU for natural gas, synthetic natural gas, and electrically generated steam respectively. Given projected escalation rates of fossil fuels, Richardson and Wendlandt concluded their SOLTHERM process would be competitive by the early 1990's. It should be noted that Richardson and Wendlandt's conclusions, like Vakil's, are subject to uncertainties associated with the cost of the solar collection system.

Baker et al. (15) have carried out cost/performance analyses for several thermal energy transport concepts.   They concluded the production of process steam and substitute natural gas with a concept consisting of a coal gasification plant to produce hydrogen and carbon monoxide was economically promising. A 100-mile long pipeline that would transmit the reformed gas to a methanator which would produce heat for process steam and substitute natural gas may be currently competitive and would almost certainly be attractive in the immediate future (15). Solar-powered coal gasifiers are a possibility, but are in a very early stage of conceptualization and evaluation; therefore, the potential impact of a solar-powered, open-loop energy transport system reamins undetermined.

C. *Solar Heating and Cooling*

A recent study (29) has suggested chemical heat pump systems for space heating cannot possibly compete with the more conventional solar heating systems. However, several critical oversights were made in this study. The fact that chemical heat pumps can operate with a coefficient of performance greater than unity was not duly accounted for; the direct effect of a greater than unity COP could be a sizeable reduction in the collector area. Since collectors are a major cost item in the system, a significant reduction in the overall system cost can potentially be realized. Although the heat pump system itself is expensive, the total system cost may be lower since reduced collector costs may compensate for the heat system costs.

The same chemical heat pump device can be used for both heating and cooling, which is not possible with conventional solar heating devices. When the cooling capability is included in cost/benefit analyses, chemical heat pumps become even more attractive. Preliminary cost/benefit studies have been carried out by several investigators (30-33), and all results show the chemical heat pump device has potential merit.

In summary, of the three applications discussed, preliminary cost/benefit studies have shown solar chemical heat pump systems for space heating and cooling applications to be the most attractive. Several innovative schemes have been proposed for the use of reactions for transporting thermal energy for industrial process heat, but a definitive evaluation of the economic potential has not yet been made. Furthermore, solar collection technologies must be advanced before reliable data are generated for use in systems studies. The use of thermochemical reactions strictly for thermal energy storage in solar thermal electric power plants does not appear to be attractive; high efficiency, less costly sensible heat systems are difficult to displace for this application.

VII. SUMMARY

Reversible thermochemical reactions can be used for various solar energy applications. Based upon studies to date, their use in chemical heat pump systems for space heating and cooling appears to be very promising. Several innovative schemes to use thermochemical reactions to transport industrial process heat from a solar energy source have been proposed; preliminary cost/ benefit studies have concluded the application has potential. The most obvious use of thermochemical reactions is thermal energy storage; however, detailed cost/benefit analysis have shown this application to be of marginal interest. It is doubtful that thermal energy storage by thermochemical reactions will ever be competitive with less complex, less expensive, higher efficiency sensible heat systems.

REFERENCES

1. Brower, A.S., Doyle, J.F., Gerels, E.E., Ku, A.C., Pomeroy, B.D., Roberts, J.M., and Salemme, R.M., "Conceptual Design

of Advanced Central Receiver Power Systems", General Electric Corporate Research and Development report No. SED-79-035 (DOE Report No. SAN/20500-1), June, 1979.

2. "Conceptual Design of Advanced Central Receiver Power Systems Sodium-Cooled Receiver Concept", Rockwell International report ESG-79-2 (DOE report No. SAN/1483-1/1), June, 1979.

3. Tracey, T.R., "Conceptual Design of Advanced Central Receiver Power System, Phase I", Martin Marietta Corporation, final report for contract EG77-C-03-1724, September, 1978.

4. Taube, M., Furrer, M., Frick, E., and Chevalley, B., "Thermochemical System for the Management of Heat from LWR's, "American Nuclear Society 1978 Annual Meeting, San Diego, CA, June 18-23, 1978.

5. Taube, M., *Nuclear Technology,* April, 1978.

6. Kalhammer, F.R., "Energy Storage: Applications, Benefits and Candidate Technologies", in Proc. Symposium on Energy Storage, Berkowitz, J.B. and Silverman, H.P. (eds), The Electrochemical Society, Inc., 1976.

7. Bhaktar, M.L., "Chemical Storage of Thermal Energy Using the SO_3-SO_2O_2 System", M.S. Thesis, University of California, Berkeley, 1976.

8. Dayan, J., Foss, A.S., and Lynn, S., "Evaluation of a Chemical Heat Storage System for a Solar Steam Power Plant", Proc. 12th IECEC Conf., Washington, D.C., 1977.

9. Springer, T., "Solar Energy Storage by Reversible Chemical Processes Final Report", Rockwell International report under contract FAO-92-7671, 1979.

10. Chubb, T.A., *Solar Energy, 17,* 129 (1975).

11. Carden, P.O., *Solar Energy, 19,* 365 (1977).

12. Tallerico, L.N., "A Description and Assessment of Large Solar Power Systems Technology", Sandia Laboratories Report SAND79-8015, August, 1979.

13. Richardson, J.T., and Wendlandt, W.W., "Cyclic Catalytic Storage Systems", University of Houston final report for

DOE Contract #EG-77-C-04-3974, May, 1979.

14. Vakil, H.B., and Flock, J.W., "Closed Loop Chemical Systems for Energy Storage and Transmission", General Electric Company, Power Systems Laboratory Report COO-2676-1, 1978.

15. Baker, N., "Transmission of Energy by Open-Loop Chemical Energy Lines", Institute of Gas Technology, final report for Sandia Contract #87-9181, February, 1978.

16. Offenhartz, P. O'D, Turner, M.J., Brown, F.C., Warren, R.B., Pemsler, J.P., and Brummer, S.B., "Methanol-Based Heat Pumps for Storage of Solar Thermal Energy, Phase I", issued as Sandia report SAND79-8188, September, 1979.

17. Jaeger, F.A., Howerton, M.T., Podlaseck, S.E., Myers, J.E., Beshore, D.G., and Haas, W.R., "Development of Ammoniated Salts Thermochemical Energy Storage Systems", Martin Marietta Corporation report SAN/12294, May, 1978.

18. Greiner, L., "The Chemical Heat Pump: A Simple Means to Conserve Energy", Chemical Energy Specialists, final report for DOE Contract #EY-76-C-03-1332, February, 1977.

19. Hiller, C.C., and Clark, E.C., "Development and Testing of the Sulfuric Acid-Water Chemical Heat Pump/Chemical Energy Storage System", Sandia Laboratories report SAND78-8824, July, 1979.

20. Iannucci, J.J., "The Value of Seasonal Storage of Solar Energy", Jet Propulsion Laboratory Thermal Storage Applications Workshop, Volume II, p. 213, February 14-15, 1978.

21. Iannucci, J.J., and Eicker, P.J., "Central Solar/Fossil Hybrid Electrical Generation: Storage Impacts", Proceedings of the 1978 Annual Meeting, Americal Section of International Solar Energy Society, Inc., Denver, Colorado, p. 904, August 28-31, 1978.

22. Iannucci, J.J., "Goals of Chemical Energy Storage", Focus on Solar Technology: A Review of Advanced Solar Thermal Power Systems, Report DOE/JPL-1060-78/5, Department of Energy, p. 145, November 15-17, 1978.

23. Iannucci, J.J., Smith, R.D., and Swet, C.J., "Energy Storage
     Requirements for Autonomous and Hybrid Solar Thermal Elec-
     tric Power Plants", proceedings, International Solar Energy
     Congress, New Delhi, India, January, 1978.

24. Smith, R.D., Poole, D.R., Li, C.H., Carlson, D.K., and
     Peterson, D.R., "Chemical Energy Storage for Solar Thermal
     Conversion", Rocket Research Company, final report for
     Sandia Contract #18-2563, April, 1979.

25. Iannucci, J.J., Fish, J.D., and Bramlette, T.B., "Review and
     Assessment of Thermal Energy Storage Systems Based Upon Re-
     versible Chemical Reactions", Sandia Laboratories report
     SAND79-8239, August, 1979.

26. Turner, R.H., "Economic Optimization of the Energy Transport
     Component of a Large Distributed Collector Solar Power
     Plant", Proceedings 11th Intersociety Energy Conversion
     Engineering Conference, Volume II, p. 1239, September, 1976.

27. Williams, O.M., *Solar Energy, 20,* 333 (1978).

28. Iannucci, J.J., Sandia Laboratories, Livermore, CA, private
     communication.

29. Lawrence, T., "Economic Assessment of Thermal Energy Storage
     Technologies", Little, A.D., Inc. final report under con-
     tract #31-109-38-3944, 1979.

30. Offenhartz, P. O'D, EIC Corporation, Newton, Massachusetts,
     private communication.

31. Clark, C., Rocket Research Company, Redmond, Washington,
     private communication.

32. Hiller, C.C., Sandia Laboratories Livermore, Livermore,
     CA, private communication.

33. Gorman, R., and Moritz, P.S., "Metal Hydride Solar Heat Pump
     and Power System (HYCSOS)", Argonne National Laboratory
     report CONF-781133-5, 1978.

CHAPTER 14

MATERIALS SCIENCE ISSUES ENCOUNTERED
DURING THE DEVELOPMENT OF THERMOCHEMICAL CONCEPTS

Raymond Mar

Sandia Laboratories
Livermore, California

I. INTRODUCTION

This chapter introduces the reader to materials problems
that have been encountered during development of thermochemical
concepts and schemes. Because space does not allow an in-depth,
detailed discussion of all problems, an overview approach is
used with a few selected topics treated in greater detail. The
thermodynamics of reactions is discussed first, emphasizing
those aspects which directly affect the selection of candidate
reactions. Materials problems are then discussed in generic
fashion; this discussion is broken down with the class of
reaction as a common element. Then, a more detailed discussion
of corrosion and catalyst development activities follows.

II. THERMODYNAMICS OF REACTIONS

Thermodynamic considerations are useful for preliminary
screening of reactions for solar energy applications; reactions
which do not meet thermodynamic requirements are dismissed from
further consideration. Reactions which meet the thermodynamic

Copyright © 1980 by Academic Press, Inc.
All rights of reproduction in any form reserved.
ISBN 0-12-511160-6

requirements must be evaluated further with regard to parameters such as rate of reaction, kinetic reversibility, long term cyclability, cost, and toxicity.

Any reaction that occurs with increasing temperature on an equilibrium temperature-composition diagram will necessarily occur with the absorption of heat ($\Delta H > 0$) and a positive entropy change ($\Delta S > 0$). In other words, any reaction of potential interest for solar energy applications will necessarily have $\Delta H > 0$ and $\Delta S > 0$ when written in the energy charge direction. Also, the exothermic discharge reaction will take place at a temperature less than the endothermic reaction, unless one adds energy to the system, for example as pressure/volume work. These statements are various corallaries to the "principle of successive entropy states" (1), which is derived directly from the First and Second Laws of Thermodynamics.

A. *Reaction Temperature*

For any reaction the free energy change can be written:

$$\Delta F_T = \Delta F_T^\circ + RT\ln K \tag{1}$$

where ΔF_T is the free energy change at temperature T, ΔF_T° is the free energy change for the reaction when all reactants and products are in their standard states, and K is the equilibrium constant. Free energies are generally calculated from enthalpy, entropy, and heat capacity data; the useful relationships are:

$$\Delta F_T^\circ = \Delta H_T^\circ - T\Delta S_T^\circ \tag{2}$$

$$\Delta H_T^\circ = \Delta H_{298}^\circ + \int_{298}^{T} \Delta C_p \, dT \tag{3}$$

$$\Delta S_T^\circ = \Delta S_{298}^\circ + \int_{298}^{T} \frac{\Delta C_p}{T} \, dT. \tag{4}$$

Also, the equilibrium constant can be written as

$$K = \frac{a_Q^q \, a_R^r \cdots}{a_L^l \, a_M^m \cdots} \tag{5}$$

for the general reaction

$$lL + mM + \cdots = qQ + rR + \cdots , \tag{6}$$

where a_I is the activity of species I. Substituting Equations (2) through (5) into (1) gives

$$\Delta F_T = \Delta H_{298}^\circ + \int_{298}^T \Delta C_p \, dT - T\Delta S_{298}^\circ - \int_{298}^T \frac{\Delta C_p}{T} \, dT$$

$$+ RT\ln \frac{a_Q^q \, a_R^r \cdots}{a_L^l \, a_M^m \cdots} \, . \tag{7}$$

When considering reactions for solar applications, one is interested in reversible conditions. Under such conditions, Equation (7) can be set equal to zero and then solved for the temperature T^*. The significance of this T^* is that at $T > T^*$ the endothermic reaction proceeds; at $T < T^*$, the exothermic reaction takes place. This T^* serves as the critical thermodynamic parameter which is compared to the requirements of the application. The application requires energy to be supplied and extracted at a certain temperature(s), and these application needs must be consistent with T^*.

Several problems exist with regard to solving Equation (7) for T^*: 1) enthalpy, entropy, and heat capacity data are not always available; 2) the solution is often tedious and time consuming, and therefore is not amenable to quick screening uses; and 3) a unique solution for T^* does not always exist (an example is given below when the reaction $2SO_3 = 2SO_2 + O_2$ is discussed). A useful simplification has been proposed and used by Wentworth and Chen (2), in which all reactants and products are assumed to be in their standard states and ΔC_p is taken as 0. In this special case, Equation (7) becomes

$$\Delta F_T = \Delta H^\circ_{298} - T\Delta S^\circ_{298} \tag{8}$$

and at equilibrium,

$$T^* = \Delta H^\circ_{298}/\Delta S^\circ_{298}. \tag{9}$$

The standard state equilibrium temperatures calculated by Equation (9) are compared with realistic endothermic and exothermic reaction temperatures on Table I. These latter temperatures were determined by preliminary design studies of thermochemical plants based on the reaction in question (3,4). It is seen that the standard state equilibrium temperature is, in general, a useful screening tool; however, the actual temperatures of operation (endothermic and exothermic) can vary significantly from T*.

Simple thermal decomposition reactions can be treated more precisely by a simple extension of Equation (9). Consider the general reaction

$$\text{Solid I} = \text{Solid II} + \sum_i^i \nu_i G_i \tag{10}$$

where G_i is the gas species formed upon reaction and ν_i is a numerical coefficient. In this case,

$$\Delta F_T = \Delta F^\circ_T + RT \sum_i \ln P_{G_i}^{\nu_i} \tag{11}$$

where P_{G_i} is the partial pressure of species G_i. Assuming equilibrium conditions and $\Delta C_p = 0$,

$$O = \Delta H^\circ_{298} - T\Delta S^\circ_{298} + RT \sum_i \ln P_{G_i}^{\nu_i}. \tag{12}$$

Solving for T gives

$$T^* = \frac{\Delta H^\circ_{298}}{\Delta S^\circ_{298} - R \sum_i \ln P_{G_i}^{\nu_i}} \tag{13}$$

If P_T represents the total pressure, the reaction stoichiometry requires that

TABLE I. A Comparison Between Standard State Equilibrium Temperatures
and Operational System Temperatures[a]

Reaction	ΔH°_{298} (KCAL)	ΔS°_{298} (CAL/DEG)	$T^* = \dfrac{\Delta H^{\circ}_{298}}{\Delta S^{\circ}_{298}}$ (K)	T endo (K)	T exo (K)
$C_2H_6(g) = C_2H_4(g) + H_2(g)$	32.7	28.8	1136	1200	1000
$CaCO_3(s) = CaO(s) + CO_2(g)$	42.6	38.4	1110	1125	1000
$2SO_3(g) = 2SO_2(g) = O_2(g)$	47.0	45.4	1037	1100	800
$CH_4(g) + H_2O(g) = CO(g) + 3H_2(g)$	49.0	51	960	1100	700
$Ca(OH)_2(s) = CaO(s) + H_2O(g)$	26.1	34.7	752	800	675
$NH_4HSO_4(l) = NH_3(g) + H_2O(g) + SO_3(g)$	80.4	108.9	740	1200	700
$MgCO_3(s) = MgO(s) + CO_2(g)$	28.0	41.8	620	700	600
$C_6H_{12}(g) = C_6H_6(g) + 3H_2(g)$	49.2	86.7	568	590	670
$Mg(OH)_2(s) = MgO(s) + H_2O(g)$	19.4	36.5	531	550	450

a - Taken from preliminary design studies in Reference (3).

$$P_{G_i} = \frac{\upsilon_i}{\sum_i \upsilon_i} P_T. \tag{14}$$

Finally, from Equations (13) and (14), one can derive the expression

$$T^* = \frac{\Delta H^\circ_{298}}{\Delta S^\circ_{298} - R \sum_i \left[\upsilon_i \ln P_T - \upsilon_i \ln \dfrac{\upsilon_i}{\sum \upsilon_i} \right]}. \tag{15}$$

In most cases of interest (see Table I), only one gas species is produced upon reaction, in which case, Equation (15) reduces to

$$T^* = \frac{\Delta H^\circ_{298}}{\Delta S^\circ_{298} - R \ln P_T} \tag{16}$$

when the reaction is written for one mole of gas generated.

The most complex situation to analyze is one whereby gaseous species are present both as reactants and products. In this case, there is not a single unique temperature above which the endothermic reaction proceeds and below which the exothermic reaction is favored. Rather, the reaction reverses itself over a range of temperatures. To illustrate the point, consider the reaction $2SO_3 = 2SO_2 + O_2$. If one initially has n moles of SO_3, from which x moles of O_2 form, the amounts of SO_3 and SO_2 present are (n − 2x) and 2x moles respectively, as determined by the reaction stoichiometry. If the total system pressure is P_T, the partial pressures of each chemical constituent in the reactor are

$$P_{SO_3} = \frac{(n - 2x)}{(n + x)} P_T \tag{17}$$

$$P_{SO_2} = \frac{(2x)}{(n + x)} P_t \tag{18}$$

$$P_{O_2} = \frac{(x)}{(n + x)} P_T. \tag{19}$$

The free energy change for the reaction is given as

$$\Delta F_T = \Delta F_T^\circ + RT\ln \frac{P_{SO_2}^2 P_{O_2}}{P_{SO_3}^2}.$$

(20)

Under reversible conditions, one sets Equation (20) equal to zero; then substituting the pressure expressions given in Equations (17), (18), and (19), one has a relationship established between temperature, pressure, and the relative amounts of SO_2, SO_3, and O_2:

$$\Delta F_T^\circ = -RT\ln P_T + RT\ln \frac{(n+x)(n-2X)^2}{4x^3}.$$

(21)

Equation (21) can be solved to give the amount of SO_3 as a function of temperature and pressure; the solution is plotted in Figure 1 for several different values of P_T. It is clear from Figure 1 that there is not a singular temperature below which the exothermic reaction proceeds to completion (i.e., the SO_3 is the only stable chemical species) and above which the endothermic reaction is complete (i.e., complete conversion of SO_3 to SO_2 and O_2). Rather, the reaction reverses itself over a range of temperatures, and a design decision must be made with regard to the degree of reaction completion to be achieved during both the endothermic and exothermic process.

Williams and Carden [5] have developed a singular analytical expression for a T* applicable to the type of reaction just discussed. For a generalized gaseous dissociation reaction,

$$\sum_i \upsilon_{ri} G_{ri} \rightarrow \sum_j \upsilon_{rj} G_{rj}$$

they have shown that T* as defined by

$$T^* = \frac{\Delta H_{298}^\circ}{\Delta S_{298}^\circ - (m-1)\, R\ln P_T + \Delta S_{mix}}$$

(23)

where

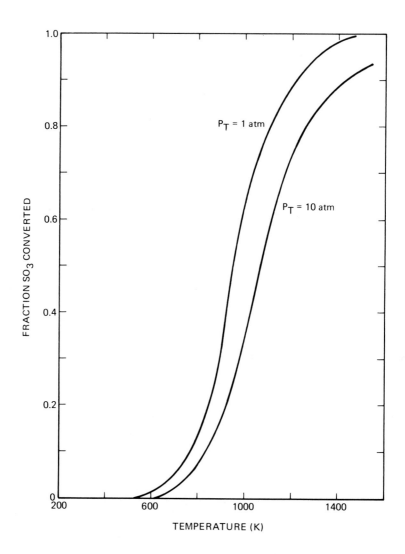

FIGURE 1.   The equilibrium conversion of SO₃ to SO₂ and O₂ as a function of temperature at two different pressures.

$$\Delta S_{mix} = -R \left[\sum v_{pj} \ln \frac{v_{pj}}{m} - \sum v_{ri} \ln v_{ri} \right] \tag{24}$$

$$m = \sum_{j} v_{rj} / \sum_{i} v_{ri} \tag{25}$$

is a useful approximation for screening these types of reactions.

B. *Coupled Reaction Considerations*

The discussions above pertain to the use of one thermo-chemical reaction. Additional constraints are introduced when two reactions must work in concert, as in the chemical heat pump concept. For example, consider a system based upon the two reactions:

$$MgCl_2 \cdot 6NH_3 = MgCl_2 \cdot 2NH_3 + 4NH_3 \tag{26}$$

$$CaCl_2 \cdot 8NH_3 = CaCl_2 \cdot 4NH_3 + 4NH_3. \tag{27}$$

The charge and discharge reactions of the chemical heat pump cycle are illustrated in Figure 8 of the preceding chapter. The charge and discharge reaction can also be shown on a ln P vs 1/T plot (see Figure 2). Three key temperatures are shown: 1) T_H - the solar charging temperature, 2) T_M - the outlet temperature of pumped heat, and 3) T_L - the temperature from which heat is pumped. During the charge cycle, reaction 26 occurs at T_H, and the reverse of reaction 27 at T_M. The free energy changes for each reaction are given by

$$\Delta F_{26} = \Delta H_{26} - T_H \Delta S_{26} \tag{28}$$

$$\Delta F_{27} = \Delta H_{27} - T_M \Delta S_{27} \tag{29}$$

where the subscripts 26 and 27 refer to reactions 26 and 27. Assuming equilibrium conditions, $\Delta F_i = 0$ and therefore

$$\Delta H_{26} = T_H \Delta S_{26} \tag{30}$$

$$\Delta H_{27} = T_M \Delta S_{27}. \tag{31}$$

It is assumed that entropies for similar reactions are equal

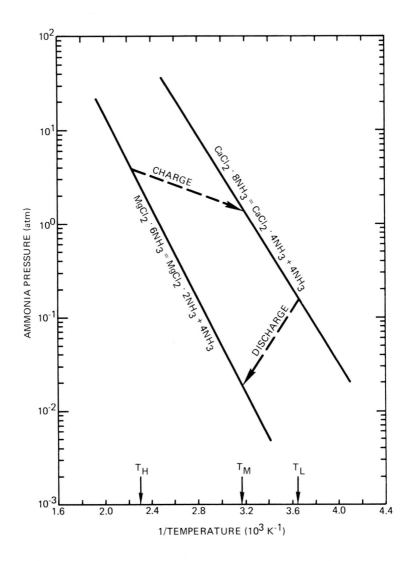

FIGURE 2. Chemical heat pump operation depicted on a pressure-temperature plot of the dissociation of $MgCl_2 \cdot 6NH_3$ and $CaCl_2 \cdot 8NH_3$. The solar collector is at temperature T_{H1} heat is pumped from T_2 to T_M.

(discussed in the next section); therefore

$$\Delta S_{26} = \Delta S_{27} \tag{32}$$

and from Equations 30 and 31

$$\frac{\Delta H_{26}}{\Delta H_{27}} = \frac{T_{H.}}{T_M} \tag{33}$$

Similarly, for the discharge reaction where the reverse of reaction 26 takes place at T_M and reaction 27 at T_L, one can show that

$$\frac{\Delta H_{26}}{\Delta H_{27}} = \frac{T_{M.}}{T_L} \tag{34}$$

Chemical heat pump reactions must be selected with these relationships (Equations 33 and 34) in mind. Studies of factors influencing the enthalpies of formation of chemical compounds are useful for "tailoring" compounds for heat pump applications. For example, it is well known that the dissociation pressures of AB_5 hydrides can be altered by the substitution of elements for B (6,7,8). Gruen et al. have studied the $LaNi_{5-x}AL_x$ system (9) with the objective of optimizing the hydrides used in the HYCSOS chemical heat pump and conversion system (10,11). The enthalpy dependence on the amount of Al is illustrated in Figure 3.

C. *Entropy Considerations*

Positive entropy changes are associated with net increases in the following: number of molecules, number of gaseous species, molecular complexity, and number of liquid species. The magnitude of entropy changes of reactions are predictable, and "rules of thumb" have been developed and are often used for estimation purposes. Familar examples are Troutons rules (the entropy of vaporization of metals is circa 92 J/K/mole) and Richards rule (the entropy of fusion of metals is circa 9.2 J/K/mole). Reactions involving the net generation of one or more gas species are generally of interest for solar applications for several

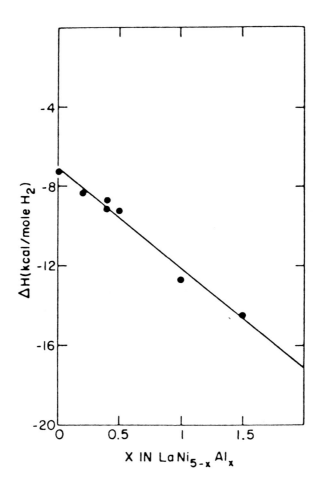

FIGURE 3. The enthalpy of hydrating $LaNi_{5-x}Al_x$ as a function of χ (Reference 9).

reasons: 1) back reactions can be guarded against by simply separating the products of reaction, and 2) the large entropy changes are associated with high energy densities. Entropies of reactions typical of various classes of reaction are given in Table II. As shown, the generation of one mole of polyatomic gas is associated with an entropy change of approximately 40 cal/ mole deg. To maximize the energy density, one must select reactions with large values of ΔH; based on Equation (9), this means selecting reactions with large entropy changes. A corollary is that if a system is designed to use a class of reaction (for example the decomposition of carbonates), a limit to the energy density is established once the application temperature is set, since both T and ΔS will have been established. Further, it is

TABLE II. *Entropies of Thermal Decomposition Reactions*

Reaction		S_{298}° [a] (cal/deg)
$M(OH)_2(s)$	$= MO(s) + H_2O(g)$	35.6
$MCO_3(s)$	$= MO(s) + CO_2(g)$	40.5
$M_2CO_3(s)$	$= M_2O(s) + CO_2(g)$	36.7
$MSO_4(s)$	$= MO(s) + SO_3(g)$	45.5
$M_2SO_4(s)$	$= M_2O(s) + SO_3(g)$	42.7
$2/n\ M_mN_n(s)$	$= 2m/n\ M + N_2(g)$	45.2
$2/n\ M_mO_n(s)$	$= 2m/n\ M + O_2(g)$	44.4
$2/n\ M_nS_n(s)$	$= 2m/n\ M + S_2(g)$	43.0
$2/n\ MF_n(s)$	$= 2/n\ M + F_2(g)$	39.3
$2/n\ MCl_n(s)$	$= 2/n\ M + Cl_2(g)$	36.5
$2/n\ MBr_n(s)$	$= 2/n\ M + Br_2(g)$	35.5
$2/n\ MI_n(s)$	$= 2/n\ M + I_2(g)$	37.3

a - *Taken from References (1) and (2)*

clear from Equation (9) that for analogous reactions (i.e., constant ΔS), higher energy densities are achieved for higher temperature systems.

III. MATERIALS SCIENCE ACTIVITIES

A. Generic Materials Problems

In this section an overview of the materials science issues faced during the development of thermochemical systems for solar applications is presented. It is convenient to organize the discussion into three areas according to the three classes of thermochemical reactions which are of interest: catalyzed reactions, thermal decomposition reactions, and dissolution/solution reactions.

1. *Catalyzed Reactions.* Catalyzed reactions which have stimulated the greatest interest in the thermal energy storage and transport community are

$$2SO_3 = 2SO_2 + O_2 \tag{35}$$

$$CH_4 + H_2O = CO + 3H_2 \tag{36}$$

$$C_6H_{12} = C_6H_6 + 3H_2. \tag{37}$$

Major materials concerns arise from the fact that catalytic reactor technologies, which are generally well established, must now be interfaced with solar energy, an energy source which is intermittant in nature. Chemical reactor technologies have been developed for constant isothermal heat loads, and very little emphasis has been placed on thermally cyclic conditions. Therefore the major unresolved issues are the effects of thermal cycling on reactor performance, catalyst performance, and structural materials. Catalyst pellets typically used in chemical reactors will be subject to break-up and crushing on repeated thermal cycling. The response of structural metals to the cycling thermal environment are also of concern. These metals behavior problems are of particular concern for the high temperature reactions such as the reforming/methanation reaction. The metals problems are by no means unique to thermochemical systems; the same concerns exist for any high temperature solar energy system. The unique element added by thermochemical reactions is the specific chemical environment.

The nature of catalyst availability problems varies depending upon the specific reaction. The endothermic and exothermic reactions of the SO_3/SO_2 system are effectively catalyzed by platinum based catalysts; however they are prohibitively expensive. Low-cost V_2O_5-based catalysts have been developed for the catalysis of the low-temperature exothermic reaction. However, the endothermic reaction takes place at temperatures where V_2O_5 is not stable; therefore, alternate catalyst compositions are needed. Transition metal oxides that melt at high temperature should be studied with regard to the catalytic effectiveness in decomposing SO_3.

Hydrogenation/dehydrogenation reactions have been shown to be potentially attractive for energy transport applications (12) but here again current catalyst technologies fall short. Selective catalysts have been developed in the petroleum industry for these reactions, but with the emphasis on once-through reactors where 99% efficiency is more than adequate. However, in cyclic closed loop operation, the generation of small amounts of non-reversibles on each cycle is unacceptable.

Suitable catalysts are available for the reforming/methanation reaction, where Ni catalysts are well developed.

A concern with all catalyzed reactions is that one must develop a thorough understanding of poisoning and degradation mechanisms. Solar energy systems are generally designed to meet 20-30 year lifetimes, and our understanding of long-term catalyst performance must be consistent with these lifetime requirements.

2. *Thermal Decomposition Reactions.* Thermal decomposition reactions can be used over a large temperature range; for example, the decomposition of carbonates, sulfates, and oxides

requires high temperatures, hydroxides require intermediate
temperatures, and lower temperature processes involve coordination compounds such as hydrates and ammoniates. Despite the wide
range of decomposition temperatures, the materials problems
encountered tend to be generic.

Very little data exists on reaction kinetics. Previous
studies have tended to concentrate on the endothermic decomposition reaction; very few studies of the exothermic recombination
reaction have been carried out, and even fewer studies of the
effects of reaction cycling have been reported. Evidence to date
suggests potential cycling problems. For example, selected
results from Barker's study (13,14) of the reaction $CaCO_3 = CaO +$
CO_2 are shown in Figure 4. It is seen that there is a continual
decrease of reaction rate upon cycling. Barker was able to find
a sample which showed a constant reaction rate; he concluded the
particle size distribution was the key parameter to achieving
constant rates. Bowry and Jutsen (15) in their studies observed

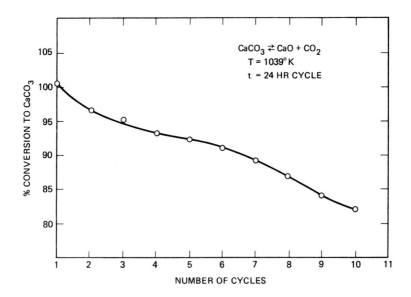

FIGURE 4. *The effect of reaction cycling on the reversibility of the $CaCO_3 = CaO + O_2$ reaction (Reference 13).*

variations in rates for the reaction $2BaO_2 = 2BaO + O_2$ which was attributed to crusting of the powders; they concluded that crusting was caused by high heating rates and high temperatures. These studies have identified several parameters which correlate with reaction rates, but indepth experimentation has not been carried out to convincingly prove causation.

An important consideration when studying reaction kinetics and designing component hardware is the volume change occurring during reaction. Volume changes for several reactions are given in Table III, where it is seen that changes of 100-200% are not uncommon. These large volume changes can have grave consequences on the design and operation of energy systems. If expansion takes place in a restricted volume, densification and sintering can occur that result in the formation of a dense, sintered mass of material. This mass may be an impermeable barrier to reactants, and hence render the entire chemical bed inactive. Volume expansion upon reaction has also been observed to cause large stresses on structural members, and structural metal failures have resulted (16,17). One approach to solving these problems is to use fluidized or moving beds. Dynamic systems also offer

TABLE III.

Volume Changes for Some Thermal Decomposition Reactions

Reaction	Volume Expansion (Percent)
$CaO + CO_2 = CaCO_3$	101
$CaCl_2 + 8NH_3 = CaCl_2 \cdot 8NH_3$	195
$CaO + H_2O = Ca(OH)_2$	95
$MgCl_2 + 6H_2O = MgCl_2 \cdot 6H_2O$	230
$ZnO + SO_3 = ZnSO_4$	214

for the possibility of improving the poor heat transfer associated with fixed-bed systems. However, the behavior of materials
and chemicals in a dynamic environment would have to be thoroughly researched if long life systems are to be designed.

While dynamic bed systems are attractive for several reasons,
they are generally not applicable to small, low-power applications such as residential heating and cooling devices. Here,
fixed-bed approaches seem to make the most sense. The mass
transport of reactants (gases) through the fixed bed has been
found to be a critical problem (16). The use of pelletized
chemicals eliminates the mass transport problem, but introduces
other potential concerns such as the performance and integrity
of the pellets during reaction and temperature cycling.

Thermal decomposition reactions generate gaseous species that
must be stored. The cost for the compression and high-pressure
storage of gases is extremely high and is the one major cause for
poor storage system efficiency and economics. Many candidate
reactions involve the generation of O_2 and CO_2; there is a clear
need for new materials which are capable of solid-state storage
of these gases in the same manner in which hydrogen can be stored
in AB_5 compounds at densities greater than liquid hydrogen (18).

   3. *Solution-Dissolution Reactions.* Numerous solution/disso-
lution reactions exist with high heats of solution. Perhaps the
most promising reaction, and certainly the one that has been most
thoroughly studied, is the sulfuric acid reaction (19-21):

   H_2SO_4(dilute) $= H_2SO_4$(conc.) $+ H_2O$(g).

This reaction is very well understood, and its use in thermochemical schemes lends itself to high-confidence designing since
the heat exchanger/reactors will be liquid phase systems. The
primary concerns are material corrosion/compatibility issues.
Sulfuric acid can be handled and contained in glass lined tanks
and glass piping. However, the development of a sulfuric acid
compatible metal alloy has obvious merit as a heat exchanger
material and as a non-breakable material of construction.

B. *Materials Corrosion and Compatibility*

Corrosion is a problem common to all thermochemical systems. The nature of the problem tends to be specific to the reaction and application; but the impetus is the same, namely the desire to predict and guarantee long-term performance without failure due to materials corrosion.

The operating conditions (e.g. temperature) and operational lifetimes of energy systems are generally limited by corrosion and compatibility phenomena. General corrosion, stress corrosion, pitting corrosion, and fatigue cracking are concerns which must be addressed for every system. Compounding the problem in solar energy systems is the cyclic nature of solar thermal energy which cause cyclic stress states. These concerns are of course not unique to thermochemical systems, but apply to all solar energy systems.

Of particular concern to the operation of thermochemical systems is the generation of noncondensable gases as a result of corrosion reactions. The thermochemical system's performance will degrade with time since the transport of gaseous chemical species from one portion of the system to another will be impeded (vapor flow blockage). The corrosion necessary to cause vapor blockage is generally much less than that required to compromise the structural integrity of containment and component alloys.

Experiments have shown that if water vapor is present, as found in systems using hydroxides and hydrated salts, the potential for a serious problem exists. The reaction between water vapor and metals to form the metal oxide and hydrogen is highly favored thermodynamically. Normally, oxidation resistant alloys rely upon the formation of protective oxides on the surface. However, in closed systems the formation of protective oxides may still generate an unacceptable amount of hydrogen. Furthermore, it is uncertain that the oxide layers formed in an operating thermochemical system will indeed be protective against continued

oxidation and hence continued hydrogen generation, since the
protective qualities of oxide layers are strongly dependent upon
the conditions of formation.

Pittinato (22) has studied the effect of hydrogen generation
on the performance of water heat pipes. His results can be
directly related to materials problems anticipated in thermochemical energy storage and transport systems. In fact, one may
argue that evaporation/condensation heat pipes are a special case
of thermochemical system. Pittinato's data have been re-cast in
terms of the number of moles of hydrogen present in the heat pipe
as a function of time on Figure 5. The hydrogen concentration
was measured indirectly by monitoring the performance of the heat
pipe; the temperature difference between the hot and cold ends of
the pipe was directly related to the concentration of hydrogen in
the pipe. The general behavior is identical for all alloys.
There is a rapid initial increase in the hydrogen concentration,

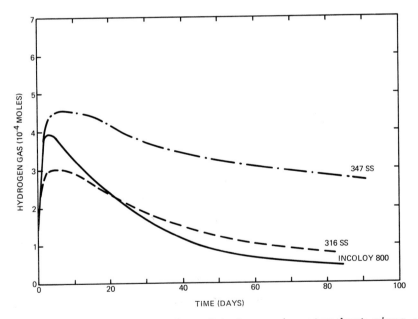

FIGURE 5. *The generation of hydrogen in water heat pipes as
a function of time. The heat pipes were operated at 110°C
(Reference 22).*

followed by a slow but steady decrease. The hydrogen generation portions of the curves on Figure 5 undoubtedly result from the oxidation of the metal by H_2O, with metal oxide and hydrogen as the products of reaction. The decrease in hydrogen concentration has been attributed primarily to the loss of hydrogen by permeation through the oxide and metal wall.

The generation of non-condensables in an operating thermochemical system would most probably be fatal to the operation of a commercial thermochemical system. It may not be practical to remove the non-condensables by periodic maintenance. However, surface pre-treatment techniques might be all that is required to eliminate the problem (22).

C. *Catalyst Development Activities*

Considerable interest has been expressed in the reaction $2SO_2 = 2SO_2 + O_2$ for solar energy applications (23,24), and catalyst development activities have been encourage in support of these interests. The requirement of continuous operation at temperatures of $\sim$ 1100 K places severe requirements on the performance of the catalyst support material. Support materials, marketed under various trade names, are typically composed of stable ceramics such as alumina, silicate, and aluminosilicate. Studies have shown these support materials are not inert to changes in surface area upon heating.

The surface areas of several commercial catalyst carriers, as determined by N_2 adsorption BET, are shown in Figure 6 to vary with time when held at 1155 K. In general, there is an immediate drop in surface area, followed by a slower decrease with time. Scanning electron photomicrographs of γ-Al_2O_3 carrier supplied by Reynolds are shown in Figure 7, where it is seen that the reduction in surface area appears to be correlated with the reduction in fines and the enhanced definition of crystallites. The consequences of the decrease in surface area seen on Figure 6 are not yet known, since the ideal surface areas

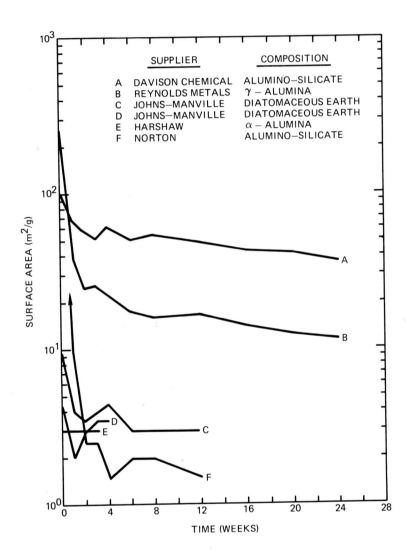

FIGURE 6. The loss of effective surface area of commercially available catalyst support structures when held at 1155 K (data courtesy of E. Schmidt, The Rocket Research Company, Division of Rockcor, Inc.).

(a)   (b)

FIGURE 7.   Scanning electron photomicrographs at γ-Al_2O_3 catalyst support materials before (a) and after (b) exposure to 1155 K for 20 weeks.

have not yet been established.

In general, one looks to the transition metals and rare earths for catalytic activity. In the SO_3/SO_2 application, the environment is highly oxidizing and, except for the noble elements, the active catalyst element will be in the form of an oxide. In addition to being catalytically active, the ability to achieve the chemical conversions at high space velocities is highly desirable in order that smaller, more compact reactors can be designed.

An extensive screening study to identify active catalysts has been conducted by Schmidt [15]; selected results are summarized on Figure 7. A differential catalytic reactor was used for these studies, and the effectiveness of the catalyst in catalyzing the dissociation of SO_3 at 1100 K was measured by observing the extent of SO_3 conversion in and out of the reactor. The experiments were carried out at space velocities ranging from 10,000 to 90,000 hr^{-1}. The catalysts shown in Figure 8 were all loaded to 10% on γ-Al_2O_3 support materials.

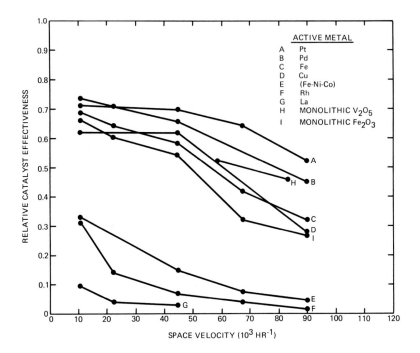

FIGURE 8. The relative effectiveness of various active
elements in catalyzing the dissociation of SO_3 to SO_2 and O_2 at
1100 K (data courtesy of E. Schmidt, The Rocket Research
Company, Division of Rockcor, Inc.).

The results in Figure 8 show that the top performing supported catalyst elements are Pd and Pt; of the lower cost
alternatives, Fe and Cu look promising. Monolithic V_2O_5 and
Fe_2O_3 are seen to be reasonably effective. Evidence for
significant mass transfer, due evidently to vaporization and
condensation processes, has been observed experimentally for
V_2O_5 and CuO (25,26). Therefore, Fe_2O_3 emerges as the leading
low-cost alternative to the noble metal catalysts. This conclusion is consistent with the results by Norman (27), but
counters the experience reported by Farbman (26). Long-term
performance and de-activiation studies are required before one
can comment further on the potential of Fe_2O_3 catalysts.

IV. SUMMARY

Materials science plays an obvious role in energy systems which have long intended lifetimes; long-term materials performance and degradation issues must be well understood. For thermochemical systems, materials science issues are also at the immediate focal point; that is, problems must be solved in order to prove a thermochemical system is technically feasible. Concerns with catalyzed reaction systems include effects of thermal cycling on reactor materials, catalyst availability and lifetime, and undesirable side reactions. Problems have arisen in thermal decomposition systems due to poor or variable kinetics and volume expansion effects. Reactions which make use of the heat of solution suffer least from materials problems; the major concerns are with the corrosive nature of the chemicals involved (e.g., sulfuric acid, sodium hydroxide).

ACKNOWLEDGMENTS

The author acknowledges the support given by the Sandia staff members who have been active in the planning and conduct of the Thermochemical Energy Storage and Transport Program for the Department of Energy. Discussions and comments with Taz Bramlette and Carl Hiller regarding the operational aspects of systems, and with Joe Iannucci regarding cost/performance studies were particularly helpful. The editorial assistance provided by Peter Dean is gratefully acknowledged.

REFERENCES

1. Searcy, A.W., "High-Temperature Reactions" in Survey of Progress in Chemistry, Vol. 1 (Academic Press, Inc., New York) 1963.

2.   Wentworth, W.E., and Chen, E., *Solar Energy, 18,* 205 (1976).

3.   "Reversible Chemical Reactions for Electric Utility Energy Applications", Rocket Research Company report RRC-77-R-559, April, 1977.

4.   Smith, R.D., Poole, D.R., Li, C.H., Carlson, D.K., and Peterson, D.R., "Chemical Energy Storage for Solar Thermal Conversion", Rocket Research Company, final report for Sandia contract 18-2563, April, 1979.

5.   Williams, O.M., and Carden, P.O., *Solar Energy, 22,* 191 (1979).

6.   Lundin, C.E., Lynch, F.E., and Magee, C.B., *J. Less Common Metals, 56,* 19 (1977).

7.   Mendelsohn, M.H., Gruen, D.M., and Dwight, A.E., *Nature, 269,* 45 (1977).

8.   Takeshita, T., Malik, S.K., and Wallace, W.E., *J. Solid State Chem., 23,* 271 (1978).

9.   Mendelsohn, M.H., Gruen, D.M., and Dwight, A.E., *J. Less Common Metals, 63,* 193 (1979).

10. Gruen, D.M., Schreiner, F., and Sheft, I., Proc. 1st World Hydrogen Energy Conference, Vol. II, 8B-73-77, Miami Beach, FL, March 1-3, 1977.

11. Gruen, D.M., Sheft, I., Lamich, G., and Mendelsohn, M., "HYCSOS: a Chemical Heat Pump and Energy Conversion System based on Metal Hydrides", Argonne National Laboratory report ANL-77-39, June, 1977.

12. Vakil, H.B., and Flock, J.W., "Closed Loop Chemical Systems for Energy Storage and Transmission", General Electric Corporate Research and Development final report for DOE contract #EY-76-C-02-26-76, February, 1978.

13. Barker, R., *J. Appl. Chem. Biotechnol., 23,* 733 (1973).

14. Barker, R., *J. Appl. Chem. Biotechnol., 24,* 221 (1974).

15. Bowrey, R.G., and Jutsen, J., *Solar Energy, 21,* 503 (1978).

16. Jaeger, F.A., Howerton, M.T., Podlaseck, S.E., Myers, J.E., Beshore, D.G., and Haas, W.P., "Development of Ammoniated

Salts Thermochemical Energy Storage Systems", Martin
Marietta Corporation report SAN/12294, May, 1978.

17. "Solar Energy Storage by Reversible Chemical Processes",
     Rockwell International final report for contract FAO-92-7671,
     January, 1979.

18. Juipers, F.A., and von Mal, H.H., J. *Less Common Metals, 23,*
     395 (1971).

19. "Sulfuric Acid-Water Energy Storage System", Rocket Research
     Corp. report RRC-76-R-530, August, 1976.

20. Huxtable, D.D., and Poole, D.R., "Thermal Energy Storage by
     the Sulfuric Acid-Water System", Internat. Solar Energy
     Society Meeting, Winnepeg, Canada, Vol. 8, p. 178 (1976).

21. Hiller, C.C., and Clark, E.C., "Development and Testing of
     the Sulfuric Acid-Water Chemical Heat Pump/Chemical Energy
     Storage System", Sandia Laboratories report SAND78-8824,
     July, 1979.

22. Pittinato, G.F., J. *Engr. Materials and Technology, 100,*
     313 (1978).

23. Gintz, J.R., "Technical and Economic Assessment of Phase
     Change and Thermochemical Advanced Thermal Energy Storage
     Systems", Boling Engineering and Construction Company report
     EPRI EM-256, December, 1976.

24. Chubb, T.A., *Solar Energy, 17,* 129 (1975).

25. Schmidt, E., Rocket Research Company, Redmond, Washington,
     private communication.

26. Farbman, G.H., "The Westinghouse Sulfer Cycle Hydrogen
     Production Process", in Proc. Hydrogen for Energy Distribu-
     tion Symposium, p. 317, Chicago, IL, July 24-28, 1978.

27. Norman, J., General Atomics, La Jolla, CA, private communi-
     cation.

SECTION 3

3. SOLAR CONVERSION (PHOTOVOLTAIC) MATERIALS

Solar cells or semiconductor junction devices can produce electricity in the form of a DC voltage and current directly from the sun's radiation by the photovoltaic effect. Silicon solar cells have already proven their utility in the powering os satellites and the like, but for large-scale power production, their efficiency and cost of production are currently major obstacles to their use in modules, arrays, and system design and fabrication. To a large extent, photovoltaic materials represent perhaps the broadest ranges of solar materials science and materials and design problems because there are uncertainties involved in materials preparation, materials performance, and materials utilization. The approaches and techniques involved in photovoltaic materials preparation are discussed in this section along with basic junction physics, the role of crystal defects, interfacial phenomena, and the combined role of these features upon device efficiency, array and system design, and specific materials or system utilization. Many of these features involving the physics, materials aspects, and technology of photovoltaics along with analytical methods for junction characterization (applying concepts originally described in Chapter 3) are described in a broad introductory fashion in the first two chapters in this section, while the third and fourth chapters elucidate some of the fundamentals associated with junction thin-film devices, including electronic properties of junctions, the role of energy gaps, direct and indirect gap semiconductors, Schottky barrier concepts, buried junction behavior, and other related phenomena.
In the fourth chapter of this section, several photochemical devices are described and the role of grain boundaries and related grain boundary effects and other device performance are described. The enhancement of solar cell efficiencies seems to be a consequence of passivation while the nature of passivation of grain boundaries is not completely understood. Passivation is, however, discussed in the context of contemporary materials science. In the last chapter in this section amorphous semiconductor photovoltaics are described. The challenges presented in this section focus on the manipulation of materials properties and fabrication technologies to achieve efficient photoconversion at a reasonable cost. To a large extent, these features are the ultimate goal of all solar programs underlined by a basic understanding utilizing principles of materials science. This section also emphasizes the role of modern analytical techniques in the elucidation of surface and interfacial phenomena involved in solar energy conversion and at the same time provides for a linking of the initial or introductory section of this book with the very last chapters. It is to a large extent this emphasis of the application of basic materials science principles and techniques to the elucidation of solar materials-related problems which forms one of the principal thrusts of this book.

488

CHAPTER 15

INTRODUCTION TO PHOTOVOLTAICS:
PHYSICS, MATERIALS AND TECHNOLOGY

Lawrence Kazmerski

Photovoltaics Branch
Solar Energy Research Institute
Golden, Colorado

I. INTRODUCTION

Before 1978, the word "photovoltaics" would have scarcely been recognized outside the scientific community. With the onset of the energy crisis and the subsequent interest in obtaining independent and inexhaustible sources of energy, the term has become increasingly common in the vocabularies of legislators, media representatives, voters, and even school children. The term "photovoltaics" is now commonly applied to those devices that provide energy from sunlight. The *solar cell* or *photovoltaic device* (or solar battery, which it was descriptively called in its early stages of development) is usually a solid-state (semiconductor) device that produces useful electricity in the form of a DC voltage and current *directly* from the sun's radiation via the photovoltaic effect. The *photovoltaic effect,* discovered over 100 years ago, involves the creation of an electromotive force (voltage) by the absorption of light (or any ionizing radiation) in an inhomogeneous solid or materials system. Three processes are necessary for the photovoltaic effect:

489

- Negative (electrons) and positive (holes) carrier pairs (in excess of thermal equilibrium conditions) must be generated by the radiation.
- The excess charges of opposite charge sign must be separated at some electrostatic inhomogeneity (e.g., pn-junction, metal-semiconductor contact, etc.)
- The carriers generated must be mobile and must continue in their separated state for a time that is long compared with the time they require to travel to the localized charge-separating inhomogeneity.

Therefore, the first and third of these requirements are associated with the *current generation* aspect of the photovoltaic effect. The second is necessary for producing the *voltage*.

A. *Device Makeup and Hierarchy of Terms*

The generic solar cell is illustrated in Fig. 1. The major elements of the junction-type device presented here include: (a) the top contact, which must either be transparent (e.g., a conducting glass) or partially open (e.g., a metal grid or finger pattern) in order to allow the light to reach (b) the top semiconductor layer or emitter, here shown as n-type. The junction is made to the (c) base semiconductor, having the opposite carrier type of (b). An ohmic contact is made to the base, and the top and bottom contacts are connected to the external load. Usually, an antireflection coating (abbreviated ARC) is deposited on the top semiconductor to minimize loss of radiation by reflection. In its basic form, the solar cell is composed of these five layers.

For even the most complex and largest photovoltaic energy-generation installation, the single-component *solar cell* is the basic building block. These individual devices are connected into series (to provide increased voltage) and parallel (to provide increased current) combinations in a *module* that encapsulates the cells, protecting them from the environment. The connection of

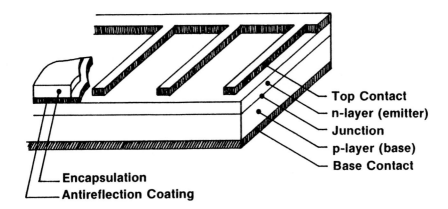

FIGURE 1. *The generic solar cell.*

modules into a single structure is called an *array*. An *array field* is the grouping of such arrays with support structures to provide the needed electricity. This hierarchy of photovoltaic components is presented in Fig. 2. The full *photovoltaic system* is composed of the array field, any necessary power conditioning (inverters to provide AC, voltage regulators, etc.) and any needed storage components (e.g., batteries).

The history of photovoltaics is not as recent as most people presume, although most of the more significant advances have occurred since the mid-1950's. The first report of the photovoltaic effect was that of Becquerel in 1839 (1). He studied metal electrodes emersed in a liquid electrolyte and noticed two phenomena: (1) the current between two electrodes changed significantly when exposed to light; and, (2) the measured photocurrent depended upon the spectral content of the light source. Working on selenium with platinum contacts, Adams and Day provided in 1877 the first report of the photovoltaic effect (although they did not identify it with that term) in an all-solid materials system (2). From that time through the early 1950's, a variety of devices

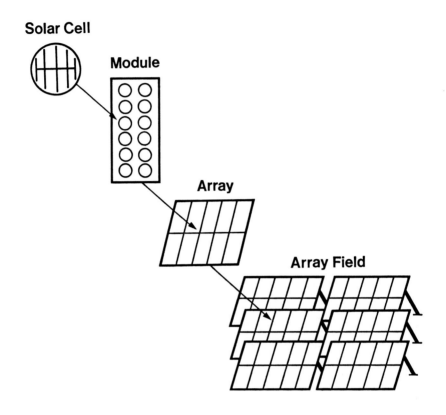

FIGURE 2.   Hierarchy of photovoltaic terms.

and material systems were studied, with the best efficiencies
(converting sunlight into electricity) usually less than 4%. Research progressed on a variety of materials. Notably, in 1954
Reynolds *et al.* observed the photovoltaic effect at metal (including Cu)/single crystal CdS junctions (3). This work led to
the development of the Cu_2S/CdS thin-film solar cell during the
early 1960's (4-6). However, in 1954 a significant step was the
fabrication of a ~6% efficient, p-n junction device on single-
crystal Si by Chapin, Fuller, and Pearson (7). This work led to
the commercialization the next year (8) of the device that still
remains the only type available in the present market - the

single-crystal silicon solar cell.

B. *Device Categories*

There are several systems used to classify photovoltaic devices. Two of these are employed in this chapter.

The *first* classification system focuses on the macro- and microstructure or degree-of-perfection of the materials used to make the solar cell. In this classification scheme, cells are:

(1) *Single crystal:* primarily bulk (e.g., silicon) materials, although some single-crystal film devices (9) are included. These types are the highest efficiency solar cells due to the crystal perfection (low-loss situation) of the materials. These are also the most expensive, primarily due to the time and energy-intensive methods needed to provide such quality material. Single-crystal solar cells are presently used in both non-concentrator (flat-plate) and concentrator (using focusing devices such as lenses, parabolic troughs, etc.) applications.

(2) *Polycrystalline:* primarily thin-film devices, but some bulk types also. It is less energy intensive and time-consuming to produce polycrystalline materials. Bulk materials can be cast (10), sintered (11), directionally-solidified (12), screen-printed (13), and grown by simple melting (14). Thin films can be evaporated (15), sputtered (16), ion-beam deposited (17), and sprayed (18). These and other fabrication techniques are discussed in detail elsewhere (19). Although less-expensive to produce, these polycrystalline device types sacrifice efficiency and possibly stability (20) due to inherent imperfections, primarily the grain boundaries. The complexity is illustrated for the case of a Cu_2S/CdS thin-film device in Fig. 3.

(3) *Amorphous:* thin-film almost exclusively. This material has no short-range order, and electronically and optically

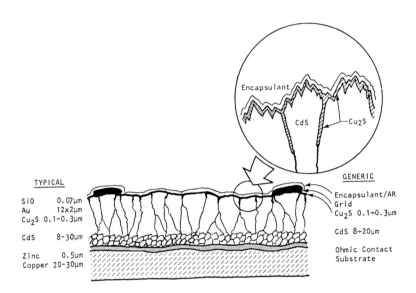

TYPICAL			GENERIC
SiO	0.07μm		Encapsulant/AR
Au	12x2μm		Grid
Cu$_2$S	0.1-0.3μm		Cu$_2$S 0.1→0.3μm
CdS	8-30μm		CdS 8→20μm
Zinc	0.5μm		Ohmic Contact
Copper	20-30μm		Substrate

FIGURE 3. Thin-film solar cell, illustrating complexity due to microstructure and non-uniformities (from Rothwarf, et al, (55).

different in behavior from either the polycrystalline or single-crystal material. Research on these devices is still quite new, and much more work on basic material properties remains to be done before efficient, stable devices appear on the commercial market. However, amorphous materials presently hold the greatest promise for an inexpensive solar cell - *if* certain performance and materials problems can be overcome.

The *second* classification is according to the method of junction formation. The major methods are illustrated in the energy-band diagrams shown in Fig. 4. The *homojunction* is formed by metallurgically joining two semiconductors of the same species (e.g., Si) but doped (e.g., with B and P) differently to provide different majority carrier types (p- and n-type). If the junction is formed between p-type and n-type semiconductors of different species (e.g., Cu$_2$S and CdS) a *heterojunction* is formed. In each

(a) Homojunction

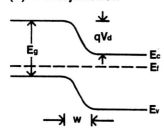

(b) Heterojunction

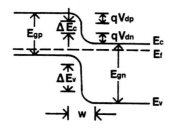

(c) Metal-Semiconductor

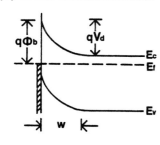

**(d) Metal-Insulator-
        Semiconductor**

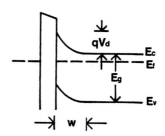

**(e) Semiconductor-Insulator-
        Semiconductor**

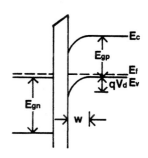

(f) Electrolyte-Semiconductor

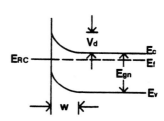

FIGURE 4. Energy-band diagrams for (a) homojunction,
(b) heterojunction, (c) Schottky barrier, (d) MIS, (e) SIS
and (f) electrochemical cells.

of these cases, an ohmic contact must be made to each side of the
junction for connection to an external circuit. A rectifying contact may also be made between a metal and semiconductor, as shown
in Fig. 4c, forming a *metal-semiconductor* or *Schottky barrier*
solar cell. A thin insulating layer can also be used to separate
the metal from the semiconductor, forming a *metal-insulator-semi-*
conductor, or MIS, structure. The purpose of the insulator (usually an oxide form of the semiconductor) is to increase the voltage output of the device. It must be thin (usually 10-30Å), however, in order to keep reasonable current levels. The metal in
the MIS structure can be replaced by a degenerate semiconductor
(e.g., indium-tin-oxide, or ITO) to form the *semiconductor-*
insulator-semiconductor, or SIS, device. Finally, a liquid electrolyte can be used to contact the semiconductor and provide the
necessary inhomogeneity. This *electrolyte-semiconductor,* or *elec-*
trochemical cell, is shown in Fig. 4f. The structures and devices
4b-f are discussed in detail in later chapters of this volume.
For the purpose of emphasizing the major materials properties
necessary to fabricate an efficient device, the homojunction will
be used as an example in the following sections. The devices and
the materials used in these devices are varied and numerous. A
summary and status of the major types are included in Appendix I
attached to this chapter.

II. AN ARGUMENT FOR MATERIALS/DEVICE RESEARCH

Currently, the only commercially-available solar cell is the
single-crystal Si solar cell. This device owes its development
status to two sources. First, Si forms the basis of the electronics industry and more is known about Si than any other element.
The data base is extensive. Second, during the space efforts in
the 1960's, the silicon solar cell was developed and used almost
exclusively, primarily due to its reliability as a power source
for that application. The argument for Si can continue. It is an

abundant element (21) and could only suffer shortages if *produc-*
tion could not meet the demand for pure, single-crystal material
(22). It can be processed in several device types and configurations. It has provided reliable, efficient, terrestrial solar
cells. With these attributes, why is more R&D needed, let alone
research on the development of alternative materials for photovoltaic applications? The major reason is cost: unless several
breakthroughs or major developments are realized in materials
production, device processing, contacting, interconnection, and
encapsulation, the Si solar cell *may* remain too expensive for all
but special remote, rural, and federal/military applications (23).

Future research directions and the status of photovoltaic
commercialization efforts can be assessed using Fig. 5. This anal-

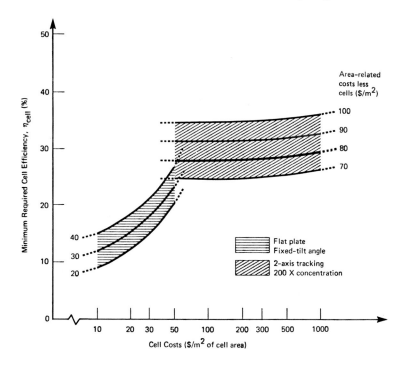

*FIGURE 5.   Necessary cell efficiency as a function of cell
cost (in 1975 dollars) for (a) flat-plate and (b) concentrator
array types. Range of area-related costs are indicated (from
DeMeo (24)).*

ysis was provided by DeMeo (24). The minimum required cell efficiency is shown as a function of cell cost per unit area (in 1975
$/m^2), for various area-related (land, structure, wiring) costs.
The shaded regions represent cost-effectiveness for two technologies. The shaded area on the left of Fig. 5 represents the analysis for a flat-plate array. For even a very inexpensive solar
cell - a thin-film device, for example - with a $10-12/m^2 cell
cost, a greater than 10% AM1 conversion efficiency would be required for reasonable area-related costs of $25-30/m^2. In the
case of a concentrator system (indicated in the Fig. 5 analysis
by the shaded region on the right of the diagram), a much higher
cell cost can be tolerated, since even the more expensive single-
crystal cell costs can be minimal with respect to concentrator-
system (lenses, or focusing mechanisms, tracking, support, etc.)
costs. However, for reasonable area-related costs ($80-90/m^2) for
a concentrator), a greater than 30% efficient concentrator solar
cell would be required. Neither the 10% efficient thin-film nor
the 30% single-crystal concentrator photovoltaic device has been
demonstrated - not even a laboratory or research version. The
present costs of silicon solar cells are about $1000/m^2, with
best efficiencies of selected devices nearing 20% (25). Even pro-
jecting costs to $80-100/m^2 for these devices would require an
efficiency near the theoretical limit of the device (26). Clearly,
much R&D remains to be done both on materials and device development if the solar cell is to become cost-competitive and viable
as an alternative energy source.

III. DEVICE PHYSICS

For the purpose of discussion, the photovoltaic homojunction
(see Fig. 4a) is used as an example in this chapter. The results
for other device types (Fig. 4b-f) are similarly obtainable, and
are discussed in other chapters of this book. In this section,
the current-voltage (or J-V, where J is the current density)

characteristics are derived in a form that highlights the major
materials properties and requirements necessary for a photovoltaic device. Finally, efficiency and equivalent circuits are discussed.

A. Homojunction Solar Cell

The solar cell is basically a diode that is configured and
designed to provide a voltage and current when exposed to light.
From basic pn junction theory, the dark, current-voltage characteristic for a diode (shown in Fig. 6) is (27):

$$J = J_s \left[\exp \left(qV/AkT \right) - 1 \right] , \tag{1}$$

where J_s is the reverse saturation current; q, electronic charge;
V, the voltage; A is the diode ideality factor, usually > 1 (28);

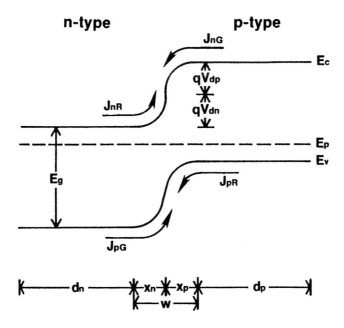

FIGURE 6.   Basic energy-band diagram for homojunction.

k, Boltzmann's constant; and, T is temperature. The reverse saturation current can be expressed in basic materials parameters
(27),

$$J_o = q \left[(D_n/L_n)/n_{po} + (D_p/L_p) \, p_{no} \right], \qquad (2)$$

where $D_{n,p}$ is the diffusion coefficient of electrons or holes;
$L_{n,p}$, the diffusion length of electrons or holes; and n_{po} and p_{no}
are the equilibrium minority carrier concentrators (i.e., n_p
refers to the density of electrons in the p-type material). The
simple diode equation given by Eq. 1 must be modified for the
case of the solar cell to account for (1) absorption of radiation,
inherent to device operation, and (2) width of the p and n regions, which can be quite thin.

For the present discussion, two assumptions are made:

- $h\nu \geq E_g$ (i.e., the incident photon energy exceeds the
  energy bandgap value),

- $d_p < L_n$; $d_n < L_p$ (i.e., the widths of the p and n layers
  are small with respect to the minority carrier diffusion
  lengths).

The first assumption ensures that carriers gain sufficient energy
to be excited from valence to conduction bands or vice-versa.
The second one enables the minority carriers to diffuse across the
n or p layer to reach the junction.

The starting point in deriving the J-V characteristics is to
find the time rate of change of minority carriers in terms of the
fields (thermal, electric, radiation) to which the carriers are
exposed. This is known as the *continuity equation*, and for holes
(on the n-side) it is given by:

$$\frac{dp_n}{dt} = \underbrace{G(T) - R}_{(i)} + \underbrace{G(X)}_{(ii)} + \underbrace{D_p \frac{d^2 p_n}{dx^2}}_{(iii)} \qquad (3)$$

where term (i) is the net recombination rate; (ii) the carrier
generation rate due to photon energy $h\nu$ at a distance X from the

surface; and (iii) is the diffusion term. In equilibrium, Eq. 3 becomes

$$(p_{no} - p_n)/\tau_p + G(x) + D_p(d^2p_n/dx^2) = 0, \qquad (4)$$

where $p_{no}-p_n$ represents the excess carrier density, and τ_p the effective minority carrier lifetime. A similar expression can be written for electrons on the p-side. Since

$$J = J_p + J_n \qquad (5)$$

and

$$J_{p(or\ n)} = qD_{p(or\ n)}(dp(or\ n)/dx)\big|_{edge\ of\ SCL} \qquad (6)$$

one must solve the differential Eq. 4 and its complement for dp/dx (and dn/dx), substitute this result into Eq. 6, and sum the electron and hole currents via Eq. 5. However, the generation rate $G(x)$ remains to be expressed in more basic materials parameters.

In general, the generation term can be written

$$G(x) = \int_{E_g}^{\infty} \phi(E)\ \alpha(E)\ \exp[-\alpha(E)\ x]\ dE, \qquad (7)$$

where $E=h\nu$, the incident photon energy; $\phi(E)$, the incident light flux; and $\alpha(E)$, the absorption coefficient of the semiconductor. The incident light flux depends upon the solar spectrum to which the device is exposed. Fig. 7 presents three such spectra for air mass zero (AM0), or extraterrestrial conditions; air mass one (AM1), terrestrial condition with one atmosphere between the device and sun (this is essentially at high noon); and air mass two (AM2) conditions (29).

The expression for the absorption coefficient, OC, depends upon the material and the wavelength of the incident radiation. For *direct bandgap* semiconductors (e.g., GaAs or InP),

$$\alpha(E) \propto (h\nu-E_g)^{1/2} \qquad (8)$$

For *indirect bandgap* materials (e.g., Si), the need to conserve momentum in the transition requires the participation of a phonon

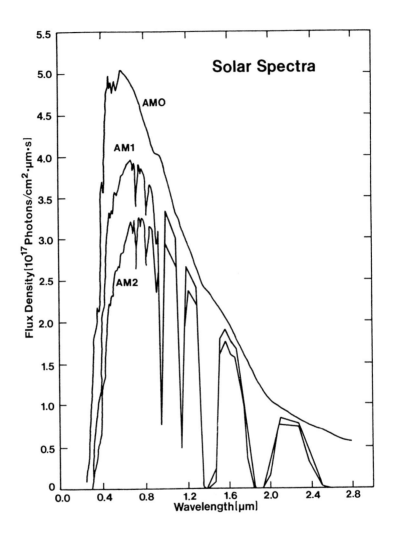

FIGURE 7.   Solar spectra for Golden, Colorado, under air-
mass-one (AM1) and air-mass-two (AM2) conditions. Air-mass-
zero (AM0) spectrum is included for reference (from Ireland,
et al (29)).

or a scattering center. In a pure material, the phonon mechanism dominates (30), and

$$\alpha(E) \propto (h\nu - E_g)^2.$$ (9)

This general functional dependence also holds for a material that has a large concentration of defects. Comparing Eqs. 8 and 9, it can be observed that the rise in α with photon energy for a direct bandgap material is more abrupt.

For this solution, assume that $\alpha(E)$ is a *constant*, representative of the energy range of interest. For this less exact situation,

$$G(x) = g(\alpha, \Phi_0),$$ (10)

where

$$\Phi_0 = \int_{Eg}^{\infty} \phi(E)\, dE,$$

the total flux density.

Substituting $g(\alpha, \Phi_0)$ into Eq. 4 and solving that differential equation (27) yields the expression for dp_n/dx (or dn_p/dx). Using Eqs. 6 and 5, the current density can be expressed:

$$J = [q\, g\, (\alpha, \Phi_0)\, \tau_p\, (D_p/L_p)\, \tanh\, (d_n/L_p)] +$$
$$[q\, p_{no}\, (D_p/L_p)\, \tanh(d_n/L_p) + q n_{po}\, (D_n/L_n)\, \tanh(d_p/L_n)].$$
$$[\exp(qV/AkT) - 1]$$ (11)

or,

$$J = J_L + J_S[\exp(qV/AkT) - 1],$$ (12)

where J_L is the current due to the incident radiation. The total current expressed in Eqs. 11 and 12 is seen to be a combination of the normal diode (dark) term and the light-generated term. The important light-generated current is a function of both external factors (i.e., the light flux) and many materials-determining parameters (e.g., absorption coefficient, minority carrier lifetime, diffusion coefficient, diffusion length).

B. Equivalent Circuit

From Eq. 12, the solar cell can be represented electrically as a light-dependent current generator in parallel with a diode. Additional resistance arises from the bulk n- and p-layers and the contacts giving rise to a series resistance, R_s. Losses occur from junction leakage and alternate current paths providing a parallel or shunt resistance across the diode. The resulting equivalent circuit is presented in Fig. 8a. R_L represents the load resistance. The effects of changing R_s and R_{sh} on the light J-V characteristic are shown in Fig. 8b. The best case is with $R_s=0$

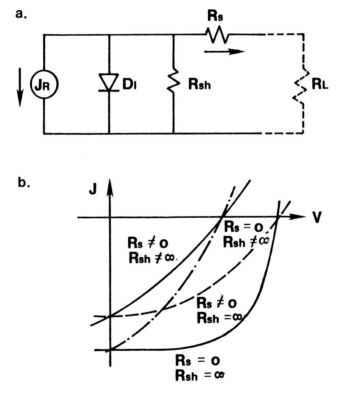

FIGURE 8.   (a) Equivalent electrical circuit for solar cells, connected to load resistance, R_L; (b) J-V characteristic of solar cell under illumination indicating effects of changing series and shunt resistances.

and $R_{sh} = \infty$. Changing either of these resistances leads to the lowering of useful voltages and currents, and the limiting of the overall device performance.

C. Efficiency

Fig. 9 shows the generic dark and light characteristics for a solar cell. Under illumination, the characteristics shift downwards, and the intersection of the curve with the current and voltage axes are the short-circuit and open-circuit voltage, respectively. The maximum of the current-voltage product along this

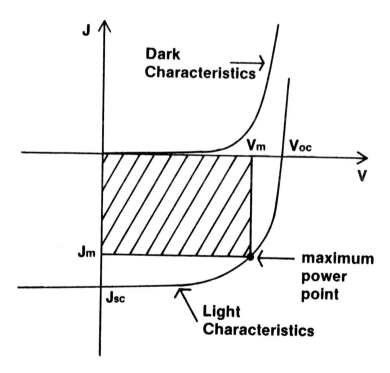

FIGURE 9. *Typical and light J-V characteristics for solar cell, defining open-circuit voltage (V_{oc}), short-circuit current density (J_{sc}), and maximum power point (J_m, V_m). Shaded region is maximum power rectangle.*

curve gives the maximum power point J_m, V_m. The maximum area rectangle inscribed within this characteristic is shown in Fig. 9
and represents the maximum power output from the device.

The efficiency is defined

$$\eta = P_{out}/P_{in}, \tag{13}$$

where $P_{out} = J_m V_m$, the maximum power output density, and P_{in}, the
input power density, usually expressed in mW/cm^2. The output
power density is

$$P_{out} = J_m V_m$$

$$= J_{sc} V_{oc} \cdot FF, \tag{14}$$

where FF $(=J_m V_m / J_{sc} V_{oc})$ is called the fill-factor. For a perfect
device, FF = 1 (i.e., the maximum power rectangle would coincide
with the light characteristic). This is never attained in real
devices. For good devices, 0.7<FF<0.9. Thus, the device efficiency can be determined using the J-V characteristics from

$$\eta = J_{sc} V_{oc} FF / P_{in} . \tag{15}$$

The parameters (temperature, load, air mass conditions, etc.) are
critical in the determination of device efficiency. Standards
have been established (31) and should be adhered to when report-
ing device performance. In this way, a uniformity in measurement
provides a basis for comparison both among and between device
types.

IV. PHOTOVOLTAIC MATERIALS: PROPERTIES AND REQUIREMENTS

Many different materials and materials' interfaces are encountered in fabricating the photovoltaic array. Support structures, encapsulation materials, contacts and interconnections,
and antireflection coatings are major and important components
of such a system. However, this section will emphasize the

basic requirements for the photovoltaic materials - i.e., those
that provide the essential ingredients for the photovoltaic effect discussed in Section I. Seven such properties, some of which
are interrelated, are presented herein.

A. *Energy Gap*

The energy gap of the absorbing layer is considered the first
indication of performance potential for a photovoltaic material.
An examination of these layers in Fig. 4 indicates that photons
with energies less than E_g cannot contribute to the light-generated current (Eq. 11) by creating electron-hole pairs. Absorbed
photons with energies exceeding E_g can provide the desired electron-hole pairs, but will also provide excess energy (losses) in
terms of lattice heating. If the solar spectrum were constant at
all wavelengths, the choice of potential absorbing materials would
be broad indeed. However, the spectrum under any air mass condition has some maximum region and a great deal of structure (see
Fig. 7) due to various atmosphere absorption bands (29). The
matching of the energy gap to the solar spectra has led to a number of calculations of the optimum energy gap for a photovoltaic
material (26, 32-34). These calculations involve modeling the
basic solar cell performance, such as the approach used in Section III-A. All involve some assumptions but lead to basically
the same results. Fig. 10 presents one set of results for a homojunction device, under AMO conditions. Curve (a) represents the
solar cell with ideal behavior (i.e., A=1 in Eq. 13), and curve
(b), A=2, the case in which generation-recombination mechanisms
dominate (28). Curve (a) is usually cited to predict an optimum
energy gap of 1.5-1.6 eV for a semiconductor. However, three factors are usually not considered in these references:

(1) Curve (a) is for AMO, not terrestrial conditions. A slightly
lower optimum E_g would be expected for AM1 or AM2, primarily

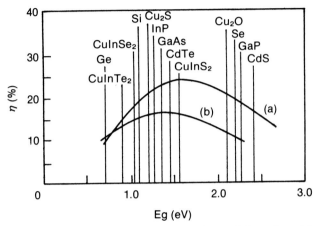

FIGURE 10. *Theoretical efficiencies of homojunction solar cells as a function of energy gap for (a) ideal diode with A=1 in Eq. 12, and (b) non-ideal diode with A=2 under AMO conditions. (From Loferski (26)).*

due to the presence of the absorption bands observable in Fig. 7 (29).

(2) Many of the solar cells reported in the literature have non-ideal (A>1) dark characteristics. Therefore, curve (b) is more proper to use. In this case, the peak is not as pronounced, with the maximum near 1.3-1.4 eV.

(3) These curves are for homojunctions.

Recently, corresponding calculations for heterojunctions have been reported (32). These are shown in Fig. 11, with the maximum efficiency as a function of absorber energy gap for various window layers. The window is a large bandgap semiconductor that provides the junction with the active absorber but does not absorb any significant quantity of photons itself. These curves are shown for AM1 conditions. The calculations do not account for refinements such as grain size (33-35) or lattice matching between the absorber and window (36-37), which can further affect performance. It should be stressed that calculations, such as those presented in Figs. 10 and 11, should be considered as guides for selecting ranges of optimum bandgaps in solar cell materials.

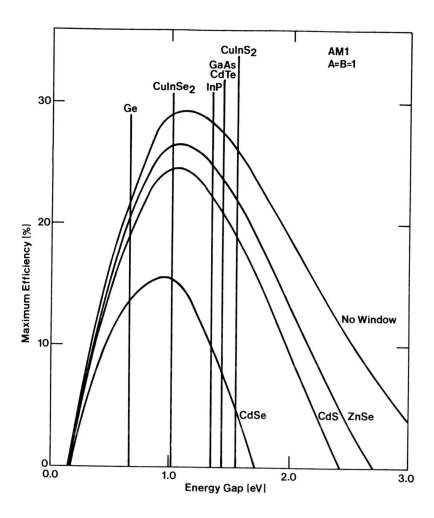

FIGURE 11. *Theoretical efficiencies of heterojunction solar cells as a function of energy gap. Results are calculated for various absorber levels and window materials.*

Optimum bandgap is a necessary condition, but not sufficient since many other requirements must be met.

B. Absorption Coefficient

Photon absorption is a very important property for a photo-voltaic material for two reasons: (1) for satisfying the device performance criteria discussed in Section III-A and specifically in Eqs. 7-10, and (2) for device economics. In general, absorber materials should have large absorption coefficients, which result from interband transitions (38). Since the necessary absorber thickness equals $1/\alpha$, large absorption coefficients permit the absorber layer to be thin (i.e., photons are absorbed nearer the surface). As a result, less material is necessary for solar cell fabrication, which has positive implications for device cost.

Fig. 12 presents the absorption coefficient as a function of photon energy for various photovoltaic candidate materials. In general, α increases with energy greater than E_g. Two types of materials are evident: *direct bandgap* (e.g., GaAs), in which increase to high α (10^4/cm) is abrupt; and *indirect bandgap* (e.g., Si), in which the rise is more gradual. A direct bandgap material would be more desirable, since the absorption takes place within a smaller depth, requiring less material. As an example, all photons would be absorbed within ~2 μm of GaAs, but about 100 μm is required for Si. Fig. 13 compares the effect of absorption co-efficient in determining the generated photocurrents for the direct (GaAs) and indirect (Si) bandgap materials. It can be seen that under the three air mass conditions, the photocurrents for the GaAs device reach their saturation values at lower thicknesses than the corresponding Si cases. In addition, the GaAs photocur-rents are higher than those for Si until the saturation region is reached.

C. Diffusion Length

The photogenerated carriers in the absorber semiconductor must be able to move across that region to the junction, space charge, or depletion layer of the solar cell. Carriers that

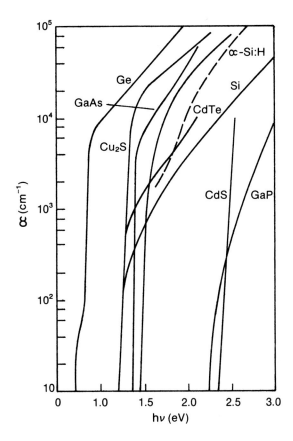

FIGURE 12. Optical absorption coefficients as a function of photon energy for several solar cell materials.

recombine before arriving at the junction are lost to the photovoltaic effect and cannot take part in the generation of J_L (Eq. 12). Diffusion is the mechanism with which the minority carriers move to the depletion region. Therefore, minority carrier diffusion length is a most important material parameter. Electrons and holes have characteristic diffusion lengths (L_n and L_p, respectively) that are material dependent. In general, the diffusion length should be of the order of $1/\alpha$, as discussed in the previous section.

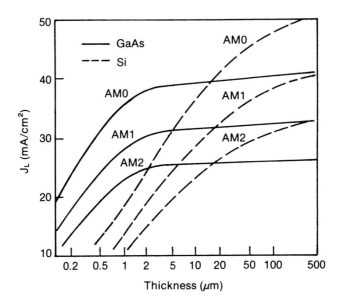

FIGURE 13. Short-circuit current densities (ideal) as a function of solar cell thickness, comparing Si (indirect bandgap) and GaAs (direct bandgap) for various air mass conditions (from Hovel (67)).

The diffusion length can be expressed in terms of basic material parameters (39):

$$L_{n,p} = \left[(kT/q) \mu_{n,p} \tau_{n,p} \right]^{1/2} . \tag{16}$$

Thus, the diffusion length should be expected to depend, as the dependent variables in Eq. 16 do, on

- impurity concentration (40,41)
- crystallinity (single, poly, amorphous) (42)
- crystal orientation (43)
- defect concentrations (39,44)
- stoichiometry (45)

An example of the effect of impurity concentration on the hole and electron diffusion lengths is presented in the GaAs data of Fig. 14. In general, diffusion lengths are very limited at high

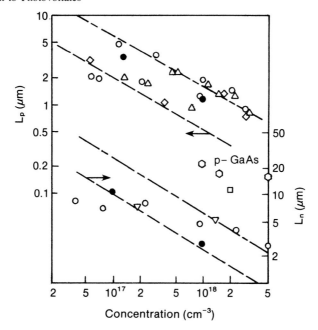

FIGURE 14. Effect of impurity concentration on hole (L_p) and electron (L_n) diffusion lengths in GaAs. (From Marfaing (40)).*

doping levels. An important factor in the fabrication of thin polycrystalline solar cells is the effect of the grain boundaries on minority carrier diffusion length. These boundaries are con-sidered regions of high recombination and sinks for minority carriers (33). One obvious solution is to make grain sizes much larger than the minority carrier diffusion lengths, minimizing the recombination problem. Calculations and data illustrating the effect of grain size on open cell performance have been re-ported in the literature for GaAs (46), Si (47–48), Cu_2S (49), and $CuInSe_2$ (50,51).

D. Minority Carrier Lifetime

The minority carrier lifetime is a major material property that is fundamental in determining the effectiveness of a semiconductor as a photovoltaic candidate. A number of fundamental processes exist that determine the recombination minority carrier lifetime:

- Radiative recombination
- Auger band-to-band recombination
- Electron/hole recombination centers
- Auger recombination through recombination centers
- Defect recombination

The exact identification of the dominant-mechanism is difficult, but electron/hole recombination and Auger recombination are probably the most likely. The development of reliable measurement techniques for minority carrier lifetime determination in solar cells remains a research problem. Although the determination of lifetimes in single-crystal Si that are in the microsecond range has been accomplished (52-53), the accurate measurement of τ in direct bandgap, small grain polycrystalline and amorphous semiconductors, which can have lifetime 10^{-9} sec or less, is much more difficult. An estimation of τ_n in p-type CuInSe$_2$ as a function of grain size is shown in Fig. 15 (50). Its effect on the open-circuit voltage of the CdS/CuInSe$_2$ thin-film polycrystalline solar cell (2 μm grain size) is shown in Fig. 16 (51).

E. Doping

Impurity concentration levels have profound effects on absorption, diffusion length, energy bandgap (54), and minority carrier lifetimes discussed previously. One must be able to dope (either extrinsically or stoichiometrically) absorber and window layers to acceptable levels in order to reduce series resistance and to control whether the photovoltaic activity takes place on the n- or p-side of the junction (55). For example, the Cu$_2$S layer in the conventional CdS/Cu$_2$S solar cell is degenerate

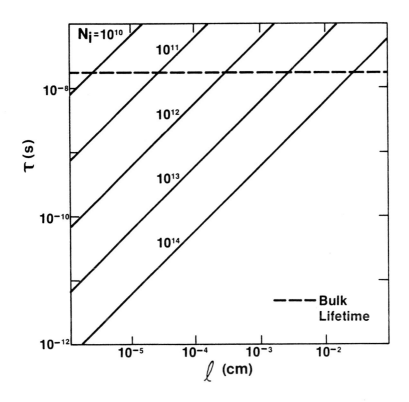

FIGURE 15. The dependence of minority carrier lifetime on grain size in p-type CuInSe$_2$ (from Kazmerski (50)).

($\sim 10^{20}$/cm^3), while the CdS is of moderate concentration (10^{16}/cm^3). This means that the photovoltaic action takes place almost entirely on the CdS - side of the junction. Lowering the concentration would move the active region into the Cu$_2$S (55).

A more subtle consequence of doping is the ability to fabricate homojunctions. This has been a major problem with some materials. Zn$_2$P$_3$, for example, possesses an excellent E$_g$ (1.6 eV), a direct bandgap, high absorption coefficient, and a high minority carrier diffusion length (56); it has, however, been produced only p-type to-date. Thus, it is presently restricted to Schottky

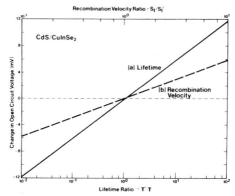

FIGURE 16. *Change in open-circuit voltage upon the (a) minority carrier lifetime, and (b) recombination velocity ratios in the CdS/CuInSe$_2$ heterojunction (AM1 conditions) (from Kazmerski (51)).*

barrier and heterojunction applications.

F. Surface Recombination

In photovoltaic devices, surfaces with high recombination velocities are in competition with the junction (depletion) regions for the collection of photogenerated carriers. This high surface recombination velocity is due to the presence of *surface states* that result from dangling bonds, native oxides, metal/organic precipitates, chemical residues, and similar effects. In an illuminated solar cell, the number of photogenerated carriers is highest at the surface and decreases exponentially into the material (see Eq. 7). Thus the surface recombination velocity is a critical parameter, especially in direct bandgap materials like GaAs in which most carriers are generated close to the surface. The surface recombination velocity for Si and GaAs lies in the 10^5-10^6 cm/s range. Etching has been used to lower this to 10^2cm/s in Si. It is desirable to passivate photovoltaic materials surfaces in order to keep the surface recombination velocity low. The effect of surface recombination velocity on the photogenerated current is shown for Cu$_2$S in Fig. 17 (57) and the open-circuit voltage in Fig. 16 (51). The composite effect of

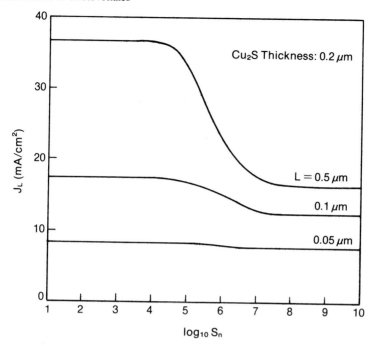

Dependence of the photogenerated current on sur-face recombination velocity for a 0.2 μm thick Cu$_2$S/CdS solar cell. Effect of various diffusion lengths are indicated (from Coutts (57)).

diffusion length is also in evidence.

G. Density of States in Gap

High density of states within the bandgap result in extremely poor diffusion lengths. It is also difficult to dope such mate-rials effectively, form necessary barriers, and obtain reasonable minority carrier lifetimes. Density of state problems are es-pecially critical for *amorphous materials*. Such densities have been reduced by alloying and complexing (58, 59). Fabricating amorphous Si in a hydrogen discharge greatly reduced the density of localized gap states (60). Recently, a glow discharge Si:H:F alloy has been shown to reduce such gap states even higher and therefore holds some promise as a photovoltaic material (61).

V. DEVICE REFINEMENTS

To conclude this discussion of photovoltaics, some examples
of device and materials engineering will be cited to illustrate
the advancement of solar cell technology, and the single-crystal
silicon cell in particular. From the 5-6% devices of the 1950's
to the near-20% (25,62) devices of today, several major developments have been made. In 1972, the *violet cell* was reported by
Comsat Laboratories (63). This silicon cell incorporated a
heavily-doped shallow junction that provided enhanced blue response and less material utilization. The higher resistance emitter necessitated the development of a more sophisticated top
grid. The cell also had an improved ARC and an advanced encapsulation. Some 10-15% improvement in cell output resulted. At this
same time, NASA Lewis Research Center developed the *back-surface-
field (BSF) cell* (64). This design prevented photogenerated minority carriers from recombining at the back ohmic contact surface by placing a low-high junction adjacent to the back contact.
This $p-p^+$ blocking back contact shielded the carriers from the
recombination surface with its electric field. The *black cell,*
reported in 1974, was a refinement of the violet cell (65). The
Si surface was textured to trap more incident radiation. Finally,
a thin metal reflector was used at the back contact to provide
the *back-surface-reflector cell* (66). This thin layer reflected
photons back through the cell, providing multiple passes of the
radiation and improved collection efficiency.

VI. SUMMARY

The purpose of this chapter is to present a *brief* introduction to photovoltaics and to provide a *basis* for understanding
the major properties and requirements of those materials which
are used in fabricating solar cells. Readers are directed toward
several *in-depth* treatments of this subject area given in

references 15, 19, 20, 27, 40, 60, 67-70. The meeting records of the IEEE Photovltaic Specialist Conferences (e.g., ref. 62) provide excellent technical records of all solar cell technologies.

ACKNOWLEDGMENTS

The author expresses his appreciation to Dr. Sigurd Wagner, who helped in the review of this chapter. He also sincerely thanks Ms. Betsy Fay-Saxon, who helped in preparing the manuscript. Finally, the editorial suggestions of Ken Zweibel and the creative help of Ms. Susan Sczepanski and Mr. J. Miller, who prepared the figures, are gratefully acknowledged.

REFERENCES

1. Bequerel, E., *Compt. Rend. 9,* 561 (1839).

2. Adams, W.G., and Day, R.E., *Phil. Mag. 1,* 295 (1877).

3. Reynolds, D.C., Leies, G., Antes, L.L., and Marburger, R.E., *Phys. Rev. 96,* 533 (1954).

4. Shirland, F.A., A.R.L. Tech. Report 60-293 (Harshaw Chemical Co.; 1960).

5. Shirland, F.A. and Heitonen, J.R., Proc. 5th IEEE Photovoltaics Spec. Conf. Sect. IIC-3. IEEE, New York 1965; Also, Shirland, F.A., and Augustine, F., ibid, IIC-4.

6. Shiozawa, L.R., Sullivan, G.A., Augustine, F., Proc. 7th IEEE Photovoltaics Spec. Conf., pp. 39-46. IEEE, New York (1968)

7. Chapin, D.M., Fullerand, C.S., and Pearson, G.L., *J. Appl. Phys. 25,* 676 (1954).

8. See, for example, Wolf, M., Proc. 25th Power Sources, Symp. pp. 120-124, (1972).

9. See, for example, Hovel, H.J., "Solar Cells", pp. 191-203. Academic Press, New York, 1975.

10. Lindmayer, J., Proc. 12th IEEE Photovoltaics Spec. Conf. pp. 82-85. IEEE, New York, 1976.

11. Ciszek, T.F., Schwuttke, G.H., Yang, K.H., *J. Cryst. Growth* 46, 527 (1979).

12. Vojdani, S., Sharifnai, A., and Doroudian, M., *Electron Lett.* 9, 128 (1973)

13. See, for example, Kaldis, E., (Ed.), "Current Topics in Materials Science, Vol. 1", North-Holland, Amsterdam, 1978.

14. Shirland, F.A., "*Adv. Energy Conv.* 6, 201 (1966).

15. See, for example, Hill, R., in "Active and Passive Thin Film Devices" (Coutts, T.J., Ed.) pp. 526-587. Academic Press, New York, 1979. Also, Harrison, R., Jenkins, G., and Hill, R. Solar Cells 1, 55 (1979)

16. Bachmann, K.J., Sinclair, W.R., Thiel, F.A., Schreiber, H., Schmidt, P.H., Spencer, E.G., Buehler, E., and Feldman, W.L., Proc. 13th IEEE Photovoltaics Spec. Conf. pp. 524-527, IEEE, New York, 1978.

17. Chamberlain, R.R., and Skarman, J.S., *J. Electrochem. Soc.* 113, 86 (1966), *Solid-State Electron.* 9, 819, (1966).

18. J.L. Vossen and W. Kern, "Thin Film Processes" Academic Press, New York, 1979.

19. Backus, C.E. (Ed.), "Solar Cells". IEEE, New York, 1976.

20. Neville, R.C., "Solar Energy Conversion: The Solar Cell", Elsevier Scientific, Amsterdam, 1978.

21. Brobst, D.A., and Pratt, W.P., "United States Mineral Resources", Paper 820. United States Geological Publ. Office, Washington, D.C., 1973.

22. Costogue, E.N., "Silicon Material Supply Outlook Study", Proc. DOE Semiannual Review of Photovoltaics, Pinehurst, N.C., Novl. 7, 1979.

23. Maycock, P.D., Proc. 13th IEEE Photovoltaics Spec. Conf., pp. 5-8. IEEE, New York, 1978. Also, Costello, D., and

Posner, D., Solar Cells *1*, 37 (1979).

24. DeMeo, E.A., Bos, P.B., "Perspectives on Utility Central Station Applications", EPRI Rep. ER-589-SR, 1978.

25. Frank, R.I., and Kaplow, R., *Appl. Phys. Lett. 34*, 65 (1979).

26. Loferski, J.J., *Proc. IEEE 51*, 677 (1963).

27. Sze, S.M., "Physics of Semicondutor Devices", pp. 640-653. Wiley, New York, 1969.

28. Sah, C.T., *IEEE Trans. Electron Dev. ED-9*, 94 (1962).

29. Ireland, P.J., Wagner, S., Kazmerski, L.L., and Hulstrom, R.L., *Science, 204,* 611 (1978).

30. Pankove, J.I., "Optical Processes in Semiconductors", pp. 36-38. Prentice-Hall, New Jersey, 1971.

31. Brandhorst, H.W., "Interim Solar Cell Testing Procedures for Terrestrial Applications". NASA TID-26871, July, 1975.

32. Kazmerski, L.L., and Ireland, P.J., Proc. DOE Adv. Mat. Rev. Mtg. -CdS Based Solar Cells, Dept. of Energy, Washington, 1978.

33. Card, H.C. and Yang, E.S., IEEE Trans. Electron Dev. *ED-24,* 397 (1977).

34. Kazmerski, L.L., Solid-State Electron. *21,* 1545 (1978).

35. Rothwarf, A., Proc. 12th IEEE Photovoltaics Spec. Conf., p. 488, IEEE, New York, 1977.

36. Kazmerski, L.L., in "Ternary Compounds 1977", (Holah, G., Ed.), pp. 217-228. Instit. of Physics, Great Britain, 1977.

37. Loferski, J.J., Shewchun, J., Roessler, B., Beaulieu, R., Piekoszewski, J., Gorska, M. and Chapman, G., Proc. 13th IEEE Photovoltaics Spec. Conf. pp. 190-194. IEEE, New York, 1978.

38. Bube, R.H., "Photoconductivity in Solids", pp. 197-254. John Wiley, New York, 1960.

39. Moll, J.L., "Physics of Semiconductors", pp. 110-121. McGraw-Hill, New York, 1964.

40. Marfaing, Y., "Solar Energy Conversion and Applications" p. 671. C.N.R.S., Paris, 1978.

41. Iles, P.A., and Soclof, S.I., Proc. 10th IEEE Photovoltaic Spec. Conf., p. 19. IEEE, New York (1973).

42. Matare, "Defect Electronics in Semiconductors", Wiley-Interscience, New York, 1971.

43. Wang, S., "Solid-State Electronics" pp. 275-308. McGraw-Hill, New York, 1966.

44. Fraas, L.M., *J. Appl. Phys. 49,* 871 (1979). Also, in "Polycrystalline and Amorphous Thin Films and Devices" (L.L. Kazmerski, Ed.) Chapter 5. Academic Press, New York, 1980.

45. Armantrout, G.A., Yee, J.H., Fischer-Colbrie, E., Leong, J., Miller, D.E., Hsieh, E.J., Vindelov, K.E., and Brown, T.G., Proc. 13th IEEE Photovoltaics Spec. Conf. pp. 383-392. IEEE, New York, 1978.

46. Soclof, S., and Iles, P.A., Extended Abst., Electrochem. Soc. Fall Meeting, New York, p. 618 (1974). Also, Blakeslee, A.E., and Vernon, S.M., Solar Cells *1,* 81 (1979).

47. Hovel, H.J., "Solar Cells", pp. 103-109. Academic Press, New York, 1975.

48. Ghosh, A., SERI/DOE Quarterly Reports 1 and 2 (EXXON), January-June, 1979.

49. Rothwarf, A., and Barnett, A.M., *IEEE Trans. Electron Dev. 24,* 381 (1977).

50. Kazmerski, L.L., Sheldon, P., and Ireland, P.J., *Thin Solid Films 58,* 95 (1979).

51. Kazmerski, L.L., Annual AVS Symposium, San Francisco (1978).

52. Wolf, M., Energy Conv. *11,* 63 (1971).

53. Ross, B., and Madiagan, J.R., *Phys. Rev.* 108, 1428 (1957).

54. Hovel, H.J., "Solar Cells" pp. 8-110. Academic Press, New York, 1975.

55. Rothwarf, A., and Boer, K.W., *Prog. in Solid-State Chem. 10,* 71 (1975).

56. Catalano, A., Dalal, V., Devaney, W.E., Gagen, E.A., Hall, R.B., Masi, J.V., Meakin, J.D., Warfield, G., Wyeth, N.C.,

and Barnett, A.M., Proc. 13th IEEE Photovoltaics Spec. Conf. p. 228. IEEE, New York, 1978.

57. Hill, R., "Active and Passive Thin Film Devices" (Coutts, T.M., Ed.) p. 561. Academic Press, London, 1979.

58. Spear, W.E., and LeComber, P.G., *Solid-State Comm. 17,* 1193 (1975).

59. Spear, W.E., and LeComber, P.G., Phil. Mag. *33,* 935 (1976).

60. See, for example, Carlson, D.E., in "Polycrystalline and Amorphous Thin Films and Devices", (L.L. Kazmerski, Ed.), Chapter 7, Academic Press, New York, 1980.

61. Ovshinsky, S. *New Scientist,* p. 674, November (1978).

62. See, for example, "Proc. 14th IEEE Photovoltaics Conf., San Diego", IEEE, New York, 1980.

63. Lindmayer, J., and Allison, J.F., Proc. 9th IEEE Photovoltaics Spec. Conf., p. 83. IEEE, New York, 1977.

64. Mandelkorn, J., and Lamneck, J., Proc. 9th IEEE Photovoltaics Spec. Conf. p. 66. IEEE, New York, 1972.

65. Hyenas, J.G., Allison, J.F., Arndt, R.A., and Meulenberg, A., Proc. Int. Conf. on Photovoltaic Power Generation, p. 487 (1974).

66. Scott-Monck, J., Gay, C., Stella, P., and Uno, F., "High Efficiency Solar Panel (HESP)", p. 487. AFAPL-TR-77-36, July, 1977.

67. Hovel, H.J., "Solar Cells" p. 94. Academic Press, New York, 1975.

68. Pulfrey, L., "Photovoltaic Power Generation". Van Nostrand-Reinhold, New York, 1978.

69. Fonash, S., "Solid-State Energy Conversion". Academic Press, New York, 1980.

70. Fahrenbruch, A., and Bube, R.H., "Photovoltaic Solar Energy Conversion". Academic Press, New York, 1980.

APPENDIX

PHOTOVOLTAIC MATERIALS, DEVICES AND PARAMETERS

Photovoltaic research has expanded rapidly over the past decade. This activity has been directed both toward the development of new materials and various device types. This appendix provides a listing of solar cells according to device configuration (i.e., homojunction, heterojunction, Schottky barrier, metal-oxide-semiconductor and semiconductor-insulator-semiconductor). Parameters including crystallinity (S=single crystal, P=polycrystalline, R=ribbon and A=amorphous); energy gap, E_g; open-circuit voltage, V_{oc}; short-circuit current, J_{sc}; fill-factor, FF; conversion efficiency, η; illumination conditions (either in air mass, AM; or in mW/cm^2); and, antireflective coating, ARC. Although this listing is extensive, it is not all-inclusive. It is hoped that it does provide a flavor of the diversity and amount of R&D activity that has existed in this semiconductor device area.

Lawrence Kazmerski

Homojunctions

Device	Type	E_g (eV)	V_{oc} (V)	J_{sc} $\left(\frac{mA}{cm^2}\right)$	FF	η(%)	ARC	Illumination	Ref.	Comments
Si (n/p)	S	1.11	0.59	46.0	0.76	15.5	Ta_2O_5	AM0	1,2	CNR cell (Comsat)
Si (n/p)	S	1.11		42.9		~18	Ta_2O_5	100	3,2	Violet cell
Si (n/p)	S	1.11	0.61			15.2	Ta_2O_5	AM0	4	(100)Si; $n^+/p/p^+$
Si (p/n)	P	1.11	0.636	24.5	0.79	12.6	Al_2O_3	97.5	5	(111)Si; $p^+/p/n/n^+$ (epi)
Si (n/p)	S	1.11	0.53	29.7	0.68	10.7	none	AM1	6,7	Inversion layer
Si (n/p)	S	1.11	0.58	25.0	0.69	10	$SiO_2;Si_3N_4$; $TiO_2;SiO_2$;	AM1	8	$n^+/p/n^+$ (two junction)
Si (p/n)	S	1.11	0.60	37.1	0.775	17.2	ZrO_2	100.3	9	$n^+/n/p^+$
Si (p/n)	S	1.11	0.52	22.4	0.72	8.2	none	81	10	Corona cell
Si (n/p)	S	1.11				8	none	33	11	Series array device
Si (n/p)	R	1.11	0.572	27.4	0.754	11.8	none	AM1	12,13	EFG ribbon
Si (n/p)	R	1.11	0.560	26.0	0.69	10.0	SiO_2	99	14	$n^+/n/p$(graded)/p-ribbon
Si (n/p)	R	1.11				>7			15	RTR
Si (n/p)	R	1.11	0.51	16.2	0.66	5.8	none	93.8	16	Twinned Si
Si (n/p)	R	1.11	0.55	35	0.753	10.7		AM0	17,18	Web-dendrite
Si (n/p)	C	1.11	0.551	37.8	0.755	11.6	TiO_x	AM0	19-21	(100) p-Silso;10^3μm grain
Si (n/p)	C	1.11	0.57	25.1		10		AM1	22	2-3 mm grain size
Si (n/p)	P	1.11				7-8			23-26	Epi on metallurgical-grade
Si (p/n)	P	1.11	0.51	21	0.70	7.5	SiO_2	100	27,28	Si on ceramic
Si (n/p)	P	1.11	0.4	12	0.66	3.1	none	AM1	10	Corona discharge
Si (n/p)	P	1.11	0.32	16.7	0.06	3.2	SiO	100	29	$n^+/p/p^+$,CVD on Al_2O_3
Si (p/n)	P	1.11	0.22	10	0.6	1.9	none	75	30	Vacuum evaporation
Si (p/n)	S	1.11	0.77		0.77	19.0		300 suns	31	EMV (Concentrator)
Si (n/p)	C	1.11	0.6	38.7		16.0		100	32	Semicrystalline
Si (n/p)	R	1.11	0.525	24	0.54	6.9		AM1	33	n^+ p RAD
Si (p/n)	P	1.11	0.56	27	0.746	9.75	SnO_2	AM1	34	Metallurgical grade Si sub.
Si (p/n)	P	1.11			0.6	2.1		AM0	35	Evaporated
Si (p/n)	R	1.11	0.44			1.9		AM1	36	6.5 cm^2 area
Si (n/p)	P	1.11	0.56	26	0.69	10	SiO_2	99	37	Polysilicon EFG substrate
Si (n/p)	P	1.11	0.56	14	0.73	4.3		AM0	38	Highly doped p-region
Si (n/p)	P	1.11	0.52	19.2	0.75	7.5		AM1	39	30 cm^2 area
Si (n/p)	P	1.11	0.25	15.0	0.56	2.15		AM1	40	Evaporated
Si (n/p)	S	1.11	0.60	26.5	0.792	12.6		AM1	41	15 μm epitaxial n
Si (n/p)	P	1.11	0.57	24.8	0.734	10.3		AM1	41	15 μm epitaxial n on poly

Device	Type	E_g (eV)	V_{oc} (V)	J_{sc} ($\frac{mA}{cm^2}$)	FF	η (%)	ARC	Illumination	Ref.	Comments
Si (n/p)	S	1.11	0.76	31.1	0.807	19.1	Si_3N_4	732 suns	42	Vertical junction
Si (p/n)	S	1.11	0.59	25.6	0.81	15.9	SiO/MgF_2	AM1	43	BSF; p^+-n-n^+
$GaAs$ (n/p)	S	1.4	0.97			20.5		AM1	44	CVD; $n^+/p^+/p/n^+$
$GaAs$ (p/n)	S	1.4				13-14	Anodic oxide	98	45	Radiation resistant
$GaAs$ (n/p)	S	1.4	1.00	21.0	0.82	16.7	Anodic oxide	AM0	46	
$GaAs$ (n/p)	S	1.4	0.79	20.4	0.73	12.0	Anodic oxide	AM1	47	Laser Annealed;$n^+/p/p^+$
$GaAs$ (n/p)	S	1.4	0.92	23	0.76	16.0	Anodic oxide	100	48	MBE;$n^+/p/p^+$
GaP (n/p)	S	2.25	1.15	3.9		3	none	100	49	
$Ga_{1-x}Al_xSb$ (p/n)	S		0.44	16.4	0.43	3.1	none	AM1	50	x=0.36 (best result)
InP (p/n)	S	1.34	0.74	10	0.63	6.7		70	51	LPE
InP (p/n)	S	1.34	0.85	20	0.747	17.2		AM2	52	
$CuInS_2$ (n/p)	P	1.54	0.41	18.2	0.49	3.6	none	100	53,54	1.5 μm grain size
$CuInSe_2$ (n/p)	P	1.02	0.32	19.4	0.5	3.0	none	100	53	2.0 μm grain size
$CdGeP_2$ (n/p)	S	1.72	0.6						55,56	77K; (112)-crystal
$CdSiAs_2$ (n/p)	S	1.55	0.6				none		57	Incandescent lamp
$CdSnP_2$ (p/n)	S	1.17	0.4						58	(112)-crystal
$ZnSiAs_2$ (p/n)	S	1.75	0.6						59	
$CdTe$ (p/n)	S	1.44	0.75	9.8	0.63	6.0	none	77.2	60,61	Electrodeposited
$CdTe$ (n/p)	P	1.44	0.33	9	0.25	1.1	none	AM1	62	

Heterojunctions

Device	Type	E_g (eV)	V_{oc} (V)	J_{sc} ($\frac{mA}{cm^2}$)	FF	η (%)	ARC	Illumination	Ref.	Comments
Cu_xS(p)/CdS(n)	S	1.2/2.41				8		100	63-66	Dry process
Cu_xS(p)/CdS(n)	P	1.2/2.41	0.51	25	0.71	5.5	SiO	100	67-72	Wet process(Clevite proc.)
Cu_xS(p)/CdS(n)	P	1.2/2.41	0.42	19.6		8.6	SnO_x	107.3	73	Spray pyrolysis
Cu_2S(p)/CdS(n)	P	1.2/2.41	0.52	21.8	0.71	4.92	SiO_x	87.9	74	1 cm^2 area; textured
Cu_2S(p)/CdS(n)	P	1.2/2.41	0.48	20.0	0.63	9.15			75	4 cm^2 area; textured cell
Cu_2S(p)/$(Cd,Zn)S$(n)	P	1.2/2.5	0.64	12.7	0.623	6.0	none	80.9	76	Planar cell
Cu_2S(p)/$(Cd,Zn)S$(n)	P	1.2/2.5	0.58	21		6.29		AM1	77-78	Textured cell

Device	Type	E_g (eV)	V_{oc} (V)	J_{sc} ($\frac{mA}{cm^2}$)	FF	η (%)	ARC	Illumination	Ref.	Comments
$InP(p)/CdS(n)$	S	1.34/2.41	0.62	15	0.71	12.5	SiO	53	79	$0.23 cm^2$ area
$InP(p)/CdS(n)$	S	1.34/2.41	0.79	18.7	0.75	15.0	SiO_x	AM2	80,83	(111)-InP
$InP(p)/CdS(n)$	S	1.34/2.41	0.807	18.6	0.74	14.4	SiO_x	77	84	$13.4 mm^2$ area
$InP(p)/CdS(n)$	P	1.34/2.41	0.46	13.5	0.68	5.7	SiO_x	AM2	80,81,85	CVD growth
$InP(p)/CdS(n)$	P	1.34/2.41	0.51	8.0	0.51	2.1	none	100	86	Two-source evaporation
$InP(p)/CdS(n)$	P	1.34/2.41	0.4	21.0	0.31	2.8	SiO	93	87	Carbon substrate
$InP(p)/CdS(n)$	P	1.34/2.41	0.37	18.0	0.30	2.0	SiO	AM1	88	$0.41 cm^2$ area
$Cu_{1.85}Se(p)/InP(n)$	P	1.2/1.34	0.33	5.2		0.7	none	140	89	
$Cu_xSe(p)/InP(n)$	P	1.2/1.34	0.34	11	0.47	1.7	SiO	AM1	88	
$CdSe(n)/CdS(n)$	S	1.74/2.41	0.3						90	CdSe vacuum evaporated
$CdSe(n)/ZnTe(p)$	S	1.74/2.26	0.68			~1		87	91-92	(111)-ZnTe
$CdSe(n)/ZnTe(p)$	S	1.74/2.26	0.80			7		AM2	93	(110)-ZnTe
$CdSe(n)/ZnTe(p)$	P	1.74/2.26	0.61			~1		87	92	
$Cd(S,Se)(n)/ZnTe(p)$	S	2.3/2.26	0.9						94	
$CdS(n)/CdTe(n)/CdTe(p)$	S	2.41/1.44/1.44	0.67	20.4	0.60	10.5	none	68	95,96	
$CdS(n)/CdTe(n)/CdTe(p)$	P	2.41/1.44/1.44	0.69	31.1		8.1	In_2O_3	140	97	$OnIn_2O_3$; $0.36 cm^2$ area
$Cu_xTe(p)/CdTe(i)/CdTe(n)$	S					5			98-100	
$CdS(n)/CdTe(p)$	P					~1				
$CdS(n)/CdTe(p)$	S	2.41/1.44	0.63	16.1	0.658	7.9	ITO and Glycerol	85	101	Evaporated CdS
$CdS(n)/CdTe(p)$	S	2.41/1.44	0.53	17.2		6.6	ITO	85	91,102	Sprayed CdS
$CdS(n)/CdTe(p)$	P	2.41/1.44	0.52	15	0.45	6	none	50	103	CVD CdTe
$CdS(n)/CdTe(p)$	S	2.41/1.44	0.53	16.4	0.63	6.5		85	102	
$CdS(n)/CdTe(p)$	S	2.41/1.44	0.67	17.9	0.60	10.5		68	95	$0.57 cm^2$ area
$CdSe(n)/CdTe(p)$	S	1.74/1.44	0.63	1.43	0.32	0.3	ITO	87	104	
$(Cd,Zn)S(n)/CdTe(p)$	S	2.5/1.44	0.79	12.7	0.64	7.8	ITO	99	105	

Device	Type	E_g (eV)	V_{oc} (V)	J_{sc} ($\frac{mA}{cm^2}$)	FF	η(%)	ARC	Illumination	Ref.	Comments
CdTe(n)/(Cd,Hg)Te(p)	S		0.235	40	0.44	4.0	none	100	106	
ZnO(n)/CdTe(p)	S	3.3/1.44	0.37	18	0.53	4.2	none	85	105	
ZnSe(n)/CdTe(p)	P	2.67/1.44	0.48	1.55	0.36	0.3	ITO	87	104,105	
ZnTe(p)/CdTe(n)	S	2.26/1.44	0.60	0.32	0.37		none	87	104	
ZnSe(n)/ZnTe(p)	P	2.67/2.26	0.69	0.06	0.52		none	87	104	
ZnTe(p)/(Zn,Hg)Te(n)	S	2.26	0.28	38		2.3	none	140	106	
$Cu_xSe(p)/CdSe(n)$	P	1.2/1.74	0.36	7.5	0.40	2		56	107,108	
$Cu_xTe(p)/CdTe(n)$	S	1.08/1.44	0.75			7.5	none	AM1	109,110	
$Cu_xTe(p)/CdTe(n)$	P	1.08/1.44	0.56	12.2		6	none	70	110	η=5% in sun
$Cu_xTe(p)/CdTe(n)$	P	1.08/1.44	0.59	15.9		6.55	yes	100	99,111-113	Flash evaporated Cu_xTe
$Cu_xSe(p)/ZnSe(n)$	S	1.2/2.67	1.1				none		114	Sunlight
$CdS(n)/CuInSe_2(p)$	S	2.41/1.02	0.49	38	0.60	12	SiO	92	115-118	$(112)-CuInSe_2$
$CdS/(n)/CuInSe_2(p)$	P	2.41/1.02	0.49	25.1	0.54	6.6		100	53,119	Dual source evaporation
$CuInSe_2(p)/CdS(n)$	P/S	1.02/2.41	0.26	31.1	0.61	5.6	none	88	120	Dual source evaporation
$CuInSe_2(p)/CdS(n)$	P	1.02/2.41	0.32	25.2	0.54	3.63	none	100	53	Frontwall cell
$CdS(n)/CuInSe_2(p)$	P	2.41/1.02	0.34	31	0.54	5.7	none	100	122	3-source deposition
$CdS(n)/CuInSe_2(p)$	P	2.41/1.02	0.39	33	0.58	7.5	none	100	123	Annealed
$CuInSe_2(p)/CdS(n)$	S	1.02/2.41	0.34	31	0.44	4.7	none	100	124	MBE growth of $CuInSe_2$
$CdS(n)/CuInS_2(p)$	S	2.41/1.54	0.42			0.8			121	
$CdS(n)/CuInS_2(p)$	P	2.41/1.54	0.51	12.5	0.51	3.25	none	100	53,86	1.5 μm grain size
$CdS(n)/CuInTe_2(p)$	P	2.41/0.96				0			53	

Device	Type	E_g (eV)	V_{oc} (V)	$J_{sc}(\frac{mA}{cm^2})$	FF	η (%)	ARC	Illumination	Ref.	Comments
$CdS(n)/CuGaInTe_2(p)$	P		0.65	29	0.69	13.0	none	100	125	$0.2cm^2$; sintered
$CdS(n)/CuInSeTe_2(p)$	P		0.53	28	0.68	10.1	none	100	125	$0.2cm^2$; sintered
$Cu_2Se(p)/AgInTe_2(n)$	S	1.24	0.12						126	
$CdS/(n)/CuGaSe_2(p)$	S	1.68	0.5	13.1	0.55	5	none	71	127	
$AlAs(n)GaAs(p)$	S	2.15/1.43	0.78	28	0.82	18.5	AlAs-oxide	AM1.3	128	(110)-GaAs
$AlAs(n)/GaAs(p)$	P	2.15/1.43	0.65				AM1		128	Graphite substrate
$AlAs(p)/GaAs(p)/GaAs(n)$	S	2.15/1.43	1.0	23	0.8	13.5	SiO_x	AM0	129	
$Ga_{0.5}Al_{0.5}As(n)/GaAs(p)$	S	1.8/1.43	0.88	16.9	0.77	13.5		AM2	130	
$GaAlAs(n)/GaAs(p)$	S	1.43	0.88	27.7	0.76	13.6	Si_3N_4	AM0	131	
$GaAlAs(p)/GaAs(p)/GaAs(n)$	S	1.43	1.015	33.1	0.745	18.5	TiO_2	AM0	132,133	
$GaAlAs(p)/GaAs(p)/GaAs(n)$	S	1.43	0.976	27.8	0.76	21.9	TiO_2	93.9	133	(100)-GaAs
$GaP(n)/GaAs(p)$	S	2.25/1.43	0.49			7			134	(100)-GaAs
$GaP(p)/GaAs(n)$	S	2.25/1.43	0.5			8			134	
$CdS(n)/GaAs(n)$	S	2.41/1.43	0.43	1.7					135	
$Cu_{1.8}Se(p)/GaAs$	P	1.2/1.43	0.54	15	0.52	4.26	Krylon	100	136	
$ZnSe(n)/GaAs(p)$	S	2.67/1.43	0.9	30		8-9	none		137	(111) and (110)-GaAs
$AlGaAs/GaAs$	S		3.96	4.5	0.625	13.85	none	81.95	138	Monolithic Series Array
$InGaP(p)/GaAs(n)$	S		0.96	17.3	0.82	14.0		97	139	Abrupt
$Al_{0.92}Ga_{0.08}As(p)/Al_{0.14}Ga_{0.86}As(n)$	S		1.32		0.84	21.7		1000 suns	140	Concentrator cell
$AlGaAs(p)/GaAs(n)$	S		1.14		0.757	19.0	Si_3N_4	700 suns	141	AM1:V_{oc}=1.01V
$AlGaAs(p)/GaAs(n)$	S		1.07		0.822	24.7		178 suns	142	50°C

Schottky Barriers

Device	Type	E_g (eV)	V_{oc} (V)	J_{sc} ($\frac{mA}{cm^2}$)	FF	η (%)	ARC	Illumination	Ref.	Comments
Al/Si(n)	S	1.11	0.38	30		8.0	none	AM1	143	Only V_{oc}, J_{sc}
Au/Si(p)	s	1.11	0.29	25	0.67	4.8	none	100	144,147	(111)Si
Au/Si(p)	S	1.11	0.37	32	0.59	12	none	52	148,144	(100)Si
Cu/Cr/Si(p)	S	1.11	0.54	25.4	0.62	10.6	SiO	80-100	149	(100)Si
Cu/Cr/Si(p)	R	1.11	0.54	20.1	0.67	7.3	SiO	100	150	
Cr/Cu/Cr/Si(p)	P	1.11	0.43	22.8	0.61	6.4	SiO	92	151	Silso cast
Pt/Si(n)	S	1.11	0.28	19.5	0.58	3.9	In_2O_3	82	152	(111)Si
		1.11	0.41	29.2	0.60	8.8	In_2O_3	82		
Ti/Si(p)	S	1.11	0.55	33	0.65	11.7	TiO_2	AM1	153	(110)Si
Ti/Si(p)	S	1.11	0.50	32	0.5	8	TiO_x	AM1	154,155	(111)Si
Ti/Si(p)	P	1.11	0.38	24	0.48	3.9	TiO_2^x	AM1	156	
Y/Si(p)	P	1.11	0.49	15.7	0.64	5.4	SiO	92	151	Silso cast
Cr/Cu/Cr/Si(p)	S	1.11	0.425	23.5	0.60	6.5	SiO	92	151	(111)Si
Y/Si(p)	S	1.11	0.484	17.8	0.64	6.0	SiO	92	151	
Ag/GaAs(n)	S	1.43	0.765	13.7			none	AM1	157	
Al/GaAs(n)	S	1.43	0.42	14			none	AM1	157	
Au/GaAs(n)	S	1.43	0.452	27.1		8.5	none	100	158-159	(100)-GaAs
Au/GaAs(n)	S	1.43	0.67(0.83)		0.72(0.83)	13-17	Yes	100	158-160	(100)-GaAs
Au/GaAs(n)	S	1.43	0.741	16	0.761	14	none	100	160,161	(111)-GaAs
Au/GaAs(n)	P	1.43	0.705	25.3	0.785	14	Yes	100	162,157	500μm grain size
Au/GaAs(n)	P	1.43	0.79	26.1	0.78	16.2	Yes	100	157	mm grains
Au/GaAsP/(n)	S	1.43				20	Ta_2O_5	100	158	(100)-GaAsP
Au/GaAsP/(n)	S	1.92	1.0	18		10	none	AM0	163	
Au/GaAsP(n)	S	1.92			0.796	22	none	125	163	
Au/GaAsP(n)	S	1.69				12	Ta_2O_5	93.7	164	
Cu/GaAs(n)	S	1.43	0.70	15	0.59	5.1	none	100	157	
Pt/GaAs(n)	P	1.43	0.65	13.3			SiO	100	136	
Au/GaAs(n)	P	1.43	0.49	20.6	0.54	5.45	none	100	165	3.2mm diameter cell; passivated grain boundaries
Ag/CdS(n)	S	2.41	0.45	0.5			none	AM2	166	
Cu/CdS(n)	S	2.41	0.63	8.5			none	AM2	166	Possible pn junction
Ni/CdS(n)	S	2.41	0.33	0.4			none	AM2	166	
Au/CdSe(n)	P	1.74	0.55	5		1.5	none	80	167	
Au/CdTe(n)	S	1.44	0.56	5.8			none	AM0	168,169	

Device	Type	E_g (eV)	V_{oc} (V)	J_{sc} ($\frac{mA}{cm^2}$)	FF	η(%)	ARC	Illumination	Ref.	Comments
Au/CdTe(n)	P	1.44				1	none	50	170	
Pt/CdTe(n)	S	1.44	0.62				none	100	169	
Cu/Cu₂O(p)	S	1.96	0.35	5	0.32	0.56	none	AM1	171	
Cu/Cu₂O(p)	S	1.96	0.30	7.5	0.39	0.8	Epoxy	AM1	171	
Pb/Cu₂O(p)	S	1.96	0.23	0.6				AM1	171	
Al/Cu₂O(p)	S	1.96	0.35	4.0				AM1	171	
Ag/Cu₂O(p)	S	1.96	0.04	1.3				AM1	171	
Sn/Cu₂O(p)	S	1.96	0.45	1.5				AM1	171	
Zn/Cu₂O(p)	S	1.96	0.13	0.6				AM1	171	
Cd/GaSe(p)	S	2.0	0.45	0.9					172	
In/GaSe(p)	S	2.0				>1	none		173	Natural sunlight
Au/Ge$_x$Se$_{1-x}$	A	2.2	0.065	0.1		0.017	none	10	174	(111)-Ge substrate
Au/InSe	S	1.2	0.28	7.5	0.37	1.4	none	36	175	
Bi/InSe	S	1.17				1.5	none	AM2	173	
Au/MoSe₂(n)	S	1.1	0.50	0.6			Ta₂O₅	51	176	
Al/NSe₂(p)	S	1.35	0.504	9.7	0.62	5.3	Ta₂O₅	57	175	
Mg/Zn₃P₂(p)	S	1.35	0.5			6.08	SiO	AM1	177	L_n=6μm
Cu/CdGeP₂(n)	S	1.72	0.6						178	
Cu/CdSiP₂(n)	S	2.1	0.3						179	(112)-CdSiP₂
Cu/CdSnP₂(n)	S	1.17	0.4						58	
In/ZnGeP₂(p)	S	2.0	0.05						180	
Cu/ZnSiAs₂(n)	S	1.75	0.35						59	
Pt/ZnSiAs₂(n)	S	1.75	0.60						59	
Cu/ZnSiP₂(n)	S	2.1	0.3						181	(112)-ZnSiP₂
Au/CdInTe₄(n)	S	0.9	0.2					AM1	182	
MIS										
Al/SiO₂/Si(p)	S	1.11	0.47	26.5		8	SiO₂	AM1	183	20-40Å SiO₂
Al/SiO₂/Si(p)	S	1.11	0.52	33.3	0.75	13	SiO₂	AM1	184	(111)-Si, Inversion
Al/SiOₓ/Si(p)	S	1.11	0.618	32	0.60				186-188	
Al/SiO/Si(p)	S	1.11				~10			185	MISIM
Al/TiOₓ/Si(p)	P	1.11	0.53						184	Inversion
Al/SiOₓ/Si(p)	P	1.11	0.51	15					185	Silso-cast
Au/SiOₓ/Si(n)	S	1.11	0.55		0.72	9	none	100	189	(111)-Si
Al/Si₃N₄/Si(p)	S	1.11	0.48	~30	0.70	10.2		AM1	190	
Al/SiOₓ/Si(p)	S	1.11	0.523	17.7	0.76	8.6	ZnS	AM1	191	0.83cm² area

SIS

Device	Type	E_g (eV)	V_{oc} (V)	J_{sc} ($\frac{mA}{cm^2}$)	FF	η(%)	ARC	Illumination	Ref.	Comments
$Cr/SiO_2/Si(p)$	S	1.11	0.58	29	0.72	12	SiO	AM1	192	1.5cm² area
$Cr/SiO_2/Si(p)$	S	1.11	0.60	26	0.75	11.7	SiO	AM1	192	1.6cm² area
$Ti/SiO_2/Si(p)$	S	1.11	0.54	25	0.72	9.8	SiO	AM1	192	1.5cm² area
$Al.Mg/SiO_x/Si(p)$	S	1.11	0.642	35.6	0.77	17.6	SiO	AM1	193	Active area parameter; grating type cell
$Cr/SiO_x/Si(p)$	S	1.11	0.610	30.0	0.73	13.4	SiO	AM1	194	2.0cm² active area
$Ti/SiO_x/Si(p)$	S	1.11	0.55	33.0	0.65	11.7	TiO_x	AM1	195	2.6cm² active area
$Al/SiO_x/Si(p)$	S	1.11	0.583	30.0	0.6-0.7			AM1	196	BSF
$Be/SiO_x/Si(p)$	S	1.11	0.54	24.6	0.61	9.5	none	85	197	
$Al/SiO_x/Si(n)$	S	1.11	0.62					AM1	198	Oxide-charge-induced BSF

SIS

Device	Type	E_g (eV)	V_{oc} (V)	J_{sc} ($\frac{mA}{cm^2}$)	FF	η(%)	ARC	Illumination	Ref.	Comments
$In_2O_3(n)-Si(p)$	P-S	3.6-1.11	0.16	0.73	0.53	12	none		199,200	(111)-Si
$In_2O_3(n)-Si(n)$	P-S	3.6-1.11	0.23	0.68	0.70	11	none		201,200	(100)-Si
$ITO(n)-Si(p)$	S	3.6-1.11	0.51	32	0.65		none	92	202,203	η vs Sn content
$ITO(n)-Si(n)$	S	3.6-1.11	0.50	32	0.23	1.6	none	100	204,205	Spray pyrolysis ITO
$ITO(n)-Si(p)$	P	3.6-1.11	0.28	25		1.5	none	100	156	
$SnO_2(n)-Si(n)$	S	3.5-1.11	0.521	29	0.64	9.9	none	AM1	206,207	
$SnO_2(n)-Si(n)$	S	3.5-1.11	0.463	26	0.60	7.2	none	AM1	207-210	
$ITO(n)-Si(p)$	P	3.6-1.11	0.48	29	0.65	9	none	AM1	211	Ion-beam ITO
$SnO_2(n)-Si(n)$	S	3.5-1.11	0.55	27.5	0.69	10.5	none	AM1	212	Sprayed ITO; 1cm² area
$SnO_2(n)-Si(n)$	S	3.5-1.11	0.47	28.9	0.66	9.1	none	AM1	212	E-beam ITO; 4cm² area
$SnO_2(n)-Si(n)$	S	3.5-1.11	0.49	28	0.71	9.8	none	AM1	212	E-beam; 1cm² area
$ITO(n)-Si(p)$	S	3.6-1.11	0.46	37	0.76	13	none	AM1	213	
$ITO(n)-Si(n)$	P	3.6-1.11	0.50	36	0.46	9.1	none	AM1	214	Silso substrate; active
$SnO_2(n)-Si(n)$	S	3.5-1.11	0.615	29.1	0.685	12.26		100	215	3.8cm² area; sprayed SnO_2
	P		0.56	26.6	0.68	10.1		100	215	1cm² area
$ITO(n)-CuInSe_2(p)$	S	3.6-1.02		30		8.9	none	100	216	ITO: E-beam deposited
	P	3.6-1.02				3.0-4.5	none	100		
$ITO(n)-GaAs(p)$	S	1.4				5	none	AM2	217	Sputtered ITO
$In_2O_3(n)-GaAs\ (p)$	S	3.6-1.4	0.084	6.2	0.40			AM0	200	
$SnO_2(n)-GaAs(n)$	S	3.5-1.4	0.33	11.4	0.43	1.2		AM0	218	
$In_2O_3(n)-InP(n)$	S	3.6-1.43	0.18	0.6	0.35		none	AM0	200	
$ITO(n)-InP(n)$	S	3.6-1.43	0.76	21.55	0.65	14.4	MgF_2	AM2	219	
$SnO_2(n)-Ge_{1-x}Se_x$ (p)	A	3.5				3×10^{-5}	SnO_2	7.8	220	

Device	Type	E_g (eV)	V_{oc} (V)	J_{sc} $(\frac{mA}{cm^2})$	FF	η(%)	ARC	Illumination	Ref.	Comments
$SnO_2(n)$-$In_{1-x}Se_x(p)$	A	3.5-1.74	0.1	10^{-2}		0.03	SnO_2	7.8	221	
$ITO(n)$-$CdTe(n)$	S	3.6-1.44	0.82	14.5	0.55	8.0	none	85	222	
$ITO(n)$-$CuInGaSeTe(p)$	P		0.72		0.55	12.3	none	100	125	Sintered
$ITO(n)$-$CuInSeTe(p)$	P		0.51	30	0.54	8.3		100	125	Sintered
Amorphous Devices										
$Si(p)/Si(i)/Si(n)$	A	1.55	0.58	10.5	0.40	2.4	ITO	AM1	223,224	Glass covered
$Si(n)/Si(p)$	A/S	1.55/1.11	0.40					15	225	On single crystal Si
Al/Si	A	1.55	0.036	1.43		0.03	none	100	226	Sputtered Si
$ITO/Si(n)$	A	3.6/1.55	0.43	10	0.28	1.2		100	227	Glow discharge a-Si
ITO/Si	A					0.02	none	100	226	Sputtered Si
Mo/Si	A	1.55				3×10^{-4}	none	100	226	Sputtered Si
$Pt/Si(i)/Si(n)$	A	1.55	0.8	7.8	0.38	5.5	ZrO_2,	65	223	Sputtered Si
$Pt/SiO_2/Si(p)/Si(i)/Si(n)$	A		0.803	12		5.5	Si_3N_4	100	223	$0.02cm^2$ area
			0.77	7		3.3	none	100	228	$1.6cm^2$ area
$ITO/Si(p^+)/Si(i)/Si(n^+)$	A		0.560	12		4.5	Yes	80	229	Stainless steel substrate
$Pt/Si(i)/Si(n^+)/Si(i)/Si(p^+)$	A		0.50	15		4.3	none	100	230	Stainless steel substrate
			0.67	9.3		3.4	none	100	231	Stainless steel substrate
$Ni/TiO_x/Si(i)/Si(n^+)$	A		0.63	7.5		4.8	Yes	60	232	Stainless steel substrate

REFERENCES

(1) Arndt, R.A., Allison, J.F., Haynos, J.G., and Muelenberg,
     A., Proc. 11th IEEE Photovoltaic Spec. Conf., pp. 40-43.
     IEEE, New York, 1975.

(2) Brandhorst, H.W., Japan. J. Appl. Phys. 16, 399 (1977).

(3) Lindmayer, J., and Allison, J.F. COMSAT Tech. Rev. 3, 1
     (1973).

(4) Payne, P.A., and Oliver, R.L., Proc. 12th IEEE Photovoltaic
     Spec. Conf., pp. 595-599. IEEE, New York, 1976.

(5) D'Aiello, R.V., Robinson, P.H., and Kressel, H., Appl.
     Phys. Lett 28, 231 (1976).

(6) Norman, C.E., and Thomas, R.E., Proc. 12th IEEE Photovolt-
     aic Spec. Conf., pp. 993-996. IEEE, New York, 1976.

(7) Salter, G.C., and Thomas, R.E., Solid-State Electron 20,
     95 (1977).

(8) Chiang, S.Y., Carbajal, B.G., and Wakefield, G.F., Proc.
     Intern. Photovoltaic Solar Energy Conf., Luxembourg,
     pp. 104-112. D. Riedel, Holland, 1977.

(9) Bae, M.S., and D'Aiello, R.V., Appl. Phys. Lett. 31, 285
     (1977).

(10) Wiehner, R., and Charlson, E.J., J. Electron Mat. 5, 513
     (1976).

(11) Warner, R.M., Murray, E.M., and Smith, W.K., Appl. Phys.
     Lett. 31, 838 (1977)

(12) Ravi, K.V., J. Cryst. Growth 39, 1 (1977).

(13) Serreze, H.B., Swartz, J.C., Entine, G., and Ravi, K.V.,
     Mat. Sci. Bull. 9, 1421 (1974).

(14) Kressel, H., D'Aiello, R.V., Levin, E.R., Robinson, P.H.,
     and McFarlane, S.H., J. Cryst. Growth 39, 23 (1977).

(15) Lesk, I.A., Baghadi, A., Gurtler, R.W., Ellis, R.J.,
     Wise, J.A., and Coleman, M.G., Proc. 12th IEEE Photovolt-
     aic Spec. Conf., pp. 173-181. IEEE, New York, 1976.

(16) Schwuttke, G.H., Phys. Stat. Sol. (a) 43, 43 (1977).

(17) Seidensticker, R.G., *J. Cryst. Growth 39*, 17 (1977).

(18) Davis, J.R., Rai-Choudhary, P., Blais, P.D., Hopkins, R.H., and McCormick, J.R., Proc. 12th IEEE Photovoltaic Spec. Conf., pp. 106-111. IEEE, New York, 1976.

(19) Fischer, H., and Pschunder, W., Proc. 12th IEEE Photovoltaic Spec. Conf., pp. 86-92, IEEE, New York, 1976.

(20) Fischer, H., and Pschunder, W., *IEEE Trans. Electron Dev. ED-24*, 438 (1977).

(21) Fischer, H., in "Photovoltaic Solar Energy" (Strub, A., Ed.) pp. 52-75. D. Reidel, Holland, 1977.

(22) Lindmayer, J., Proc. 12th IEEE Photovoltaic Spec. Conf., pp. 82-85. IEEE New York, 1976.

(23) Chu, T.L., *J. Cryst. Growth 39,* 45 (1977).

(24) Chu, T.L., Chu, S.S., Duh, K.Y. and Mollenkopf, *J. Appl. Phys. 48*, 3576 (1977).

(25) Chu, T.L., *Appl. Phys. Lett. 29,* 675 (1976).

(26) Chu, T.L., and Singh, K.N., *Solid-State Electron. 19*, 837 (1976).

(27) Zook, J.D., Heaps, J.D., and Maciolek, R.B., Proc. Intern. Electron Dev. Mtg., Washington, IEDM, New York, 1977.

(28) Heaps, J.D., Maciolek, R.B., Zook, J.D., and Scott, M.W., Proc. 12th IEEE Photovoltaic Spec. Conf., pp. 147-150. IEEE, New York, 1976.

(29) Saitoh, T., Warabisako, T., Itoh, H., Nakamura, N., Tamura, H., Minegana, S., and Tokuyama, T., *Japan. J. Appl. Phys. 16,* 413 (1977). Also, IEEE Trans. Electron Dev. *ED-24,* 446 (1977).

(30) Feldman, C., Blum, N.A., Charles, H.K., and Satkiewicz, F.G., *J. Electron. Mat. 7*, 309 (1978).

(31) Frank, R.I., and Kaplow, R., *Appl. Phys. Lett. 34*, 65 (1979).

(32) Lindmayer, J., Proc. 13th IEEE Photovoltaic Spec. Conf., pp. 1096-1100. IEEE, New York, 1978.

(33) Fabre, E., Baudet, Y., and Ebeid, S.M., Proc. 13th IEEE Photovoltaic Spec. Conf., pp. 1101-1105. IEEE, New York, 1978.

(34) Chu, T.L., Chu, S.S., Pauleau, M.Y., and Stokes, E.D.,
     Proc. Photovoltaics Adv. R&D Annual Rev. Mtg., Denver,
     CO., pp. 423-440. SERI, Golden, CO., 1979.

(35) Feldman, C., Blum, N.A., and Satkiewicz, F.G., Proc.
     Polycrystalline Silicon Contractors Rev. Mtg., Pasadena,
     pp. 59-82. SERI, Golden, CO., 1979.

(36) Chu, T.L., *J. Vac. Sci. Technol. 12,* 912 (1975).

(37) D'Aiello, R.V., Kressel, H., and Robinson, P.H., *Appl.
     Phys. Lett. 28,* (1976).

(38) Chu, T.L., Chu, S.S., Duh, K.Y., and Yoo, H.I., Proc. 12th
     IEEE Photovoltaic Spec. Conf. IEEE, New York, 1976.

(39) Chu, T.L., "Thin Films of Silicon on Low-Cost Substrates".
     ERDA, Washington, DC; September, 1977.

(40) Feldman, C., Blum, N.A., and Satkiewicz, F.G., Proc.
     Polycrystalline Silicon Contractors Rev. Mtg., Pasadena,
     pp. 59-82. SERI, Golden, CO., 1979.

(41) D'Aiello, R.V., Proc. Polycrystalline Silicon Contractors
     Rev. Mtg., Pasadena, pp. 233-242. SERI, Golden, CO., 1979.

(42) Ekstedt, T.W., Mahan, J.E., Frank, R.I., and Kaplow, R.,
     Appl. Phys. Lett. 33, 422 (1978).

(43) Fossum, J.G., and Burgess, E.L., *Appl. Phys. Lett. 33,*
     228 (1978).

(44) Fan, J.C.C., Bozler, C.O., and Chapman, R.L., *Appl. Phys.
     Lett. 32,* 390 (1978).

(45) Hovel, H.J., and Woodall, J.M., Proc. 12th IEEE Photovolt-
     aic Spec. Conf., pp. 945-947. IEEE, New York; 1976.

(46) Fan, J.C.C., Chapman, R.L., Bozler, C.O., and Drevinsky,
     P.J., *Appl. Phys. Lett. 36,* 53 (1980).

(47) Fan, J.C.C., Chapman, R.L., Donnelly, J.P., Turner, G.W.,
     and Bozler, C.O., *Appl. Phys. Lett. 34,* 780 (1979).

(48) Fan, J.C.C., Calawa, A.R., Chapman, R.L., and Turner, G.W.,
     Appl. Phys. Lett. 35, 804 (1979).

(49) Grimmeiss, H.G., Kischio, W., and Koelmans, H., *Solid-State
     Electron. 5,* 155 (1962).

(50)   Nguyen van Mau, A., Bougnot, G., Muoy, H.Y., and
       Moussalli, G.M., in "Photovoltaic Solar Energy" (Strub, A.,
       Ed.), pp. 405-414. D. Riedel, Holland, 1977.

(51)   Galavanov, V.V., Kundukhov, R.M., and Nasledov, D.N.,
       Sov. Phys.-Solid State 8, 2723 (1967).

(52)   Shay, J.L., Wagner, S., Bettini, M., Bachmann, K.J.,
       Buehler, E., and Kasper, H.M., Proc. 11th IEEE Photovoltaic
       Spec. Conf., pp. 507-507. IEEE, New York, 1975.

(53)   Kazmerski, L.L., in "Ternary Compounds 1977"(Holah,G.D.,
       Ed.), pp. 217-228.   Institute of Phys. Conf. Series 35.
       Instit. of Phys., London, 1977.

(54)   Kazmerski, L.L., and Sanborn, G.A., *J. Appl. Phys. 48*,
       3178 (1977).

(55)   Borshchevskii, A.S., Lebedov, A.A., Mal'tseva, I.A.,
       Ovezov, K., Rud, Y.V., and Undalov, Y.K., *Sov. Phys.-
       Semicond. 9*, 1278 (1975).

(56)   Borchevskii, A.S., Dagina, N.E., Lebedev, A.A., Ovezov,
       K., Polushina, I.K., and Rud, Y.V., *Sov. Phys.-Semicond.
       10*, 934 (1976).

(57)   Dovletmuradov, C., Ovezov, K., Prochukhan, V.D., Rud, Y.V.,
       and Serginov, M., *Sov. Phys.-Techn. Phys. Lett. 1*, 382
       (1975).

(58)   Medvekin, G.A., Ovezov, K., Rud, Y.V., and Sokolova, V.I.,
       Sov. Phys.-Semicond. 10, 1239 (1976).

(59)   Rud, Y.V., and Ovezov, K., *Sov. Phys.-Semicond. 10*, 561
       (1976).

(60)   Naumov, G.P., and Nikolaelva, O.V., *Sov. Phys.-Solid State
       3*, 2718 (1961).

(61)   Vodakov, Y.A., Lomakina, G.A., Naumov, G.P., and
       Maslakovets, Y.P., *Sov. Phys.-Solid State 2*, 1 and 11
       (1960).

(62)   Rod, R., "Improved Semiconductor for Photovoltaic Solar
       Cells", Final Report (49-18)-2457. ERDA, Washington, DC,
       Nov. 1977.

(63)   teVelde, T.S., *Energy Conv. 14,* 111 (1975).

(64)   teVelde, T.S., and Dieleman, J., *Philips Res. Rep.*
       28, 573 (1973).

(65)   Reynolds, D.C., Seies, G., Antes, L.L., Marburger, R.E.,
       Phys. Rev. 96, 533 (1954).

(66)   Das, S.R., Nath, P., Banerjee, A., and Chopra, K.L.,
       Solid-State Comm. 21, 49 (1977).

(67)   Barnett, A.M., Meakin, J.D., and Rothwarf, A., Proc. 12th
       IEEE Photovoltaic Spec. Conf., pp. 544-546. IEEE, New
       York, 1976.

(68)   Boer, K.W., *Phys. Stat. Sol. (a) 40,* 355,(1977).

(69)   Boer, K.W., Proc. 11th IEEE Photovoltaic Spec. Conf.,
       pp. 514-515. IEEE, New York, 1975.

(70)   Palz, W., Besson, J., Duy, T.N., and Vedel, J., Proc.
       10th IEEE Photovoltaic Spec. Conf., pp. 69-76. IEEE, New
       York, 1973.

(71)   Bogus, K., and Mattes, S., Proc. 9th IEEE Photovoltaic Spec.
       Conf., pp. 106-110. IEEE, New York, 1972.

(72)   Burton, L.C., Hench, T., Storti, G., and Haacke, G.,
       J. Electrochem. Soc. 123, 1741 (1976).

(73)   Jordan, J.F., Proc. 11th IEEE Photovoltaic Spec. Conf.
       pp. 508-513. IEEE New York, 1975.

(74)   Barnett, A.M., Bragagnolo, J.A., Hall, R.B., Phillips,
       J.E., and Meakin, J.D., Proc. 13th IEEE Photovoltaic Spec.
       Conf., pp. 419-420. IEEE, New York, 1978.

(75)   Rothwarf, A., Meakin, J.D., and Barnett, A.M., "Polycryst.
       and Amorphous Thin Films and Devices". Academic Press, 1980

(76)   J.D. Meakin, in "Photovoltaics Adv. R&D Annual Rev. Mtg",
       Denver (Deb, S., Ed.) pp. 167-194. SERI, Golden, CO, 1979.

(77)   Meakin, J.D., Proc. CdS/Cu$_2$S and CdS/Cu-Ternary Compound
       Rev. Mtg., Denver, pp. 79-110. SERI, Golden, CO, 1979.

(78)   Hall, R.B., Birkmire, R.W., Eser, E., Hench, T.L., and
       Meakin, J.D., Proc. 14th IEEE Photovoltaic Spec. Conf.

IEEE, New York, 1980.

(79) Wagner, S., Shay, J.L., Bachmann, K.J., and Buehler, E., *Appl. Phys. Lett.* *26*, 229 (1975).

(80) Wagner, S., Shay, J.L., Bachmann, K.J., Buehler, E., and Bettini, M., *J. Cryst. Growth.* *39*, 128 (1977).

(81) Shay, J.L., Bettini, M., Wagner, S., Bachmann, K.J., and Buehler, E., Proc. 12th IEEE Photovoltaic Spec. Conf. pp. 540-543. IEEE, New York, 1976.

(82) Shay, J.L., Wagner, S., Bettini, M., Bachmann, K.J., and Buehler, E., *IEEE Trans. Electron Dev.* *ED-24*, 483 (1977).

(83) Ito, K., and Ohsawa, T., *Japan. J. Appl. Phys.* *14*, 1259 (1975).

(84) Yoshikawa, A., and Sakai, Y., *Solid-State Electron.* *20*, 133 (1977).

(85) Bettini, M., Bachmann, K.J., Buehler, E., Shay, J.L., and Wagner, S., *J. Appl. Phys.* *48*, 1603 (1977).

(86) Kazmerski, L.L., White, F.R., Ayyagari, M.S., Juang, Y.J., and Patterson, R.P., *J. Vac. Sci. Technol.* *14*, 65 (1977).

(87) Bachmann, K.J., Buehler, E., Shay, J.L., Wagner, S., and Bettini, M., *J. Electrochem. Soc.* *123*, 1509 (1976).

(88) Saitoh, T., Matsubara, S., and Minagawa, S., *Japan. J. Appl. Phys.* *16*, 133 (1977). Also, *Japan. J. Appl. Phys.* *16*, 807 (1977).

(89) Fischer, H., Thesis, Tech. Univ. Braunschweig (1970).

(90) Kandilarov, B., and Andreytchin, R., *Phys. Stat. Sol.* *8*, 897 (1965).

(91) Fahrenbruch, A.L., Buch, F., Mitchell, K.W., and Bube, R.H. Proc. 12th IEEE Photovoltaic Spec. Conf., pp. 529-533. IEEE, New York, 1976.

(92) Buch, F., Fahrenbruch, A.L., and Bube, R.H., *Appl. Phys. Lett.* *28*, 593 (1976).

(93) Gashin, P.A., and Simashkevich, A.V., *Phys. Stat. Sol. (a)* *19*, 615 (1973).

(93)   Gashin, P.A., and Simashkevich, A.V., *Phys. Stat. Sol. (a)*
       19, 615 (1973).

(94)   Fedotov, Y.A., Supalov, V.A., Manuilova, T.P., Vanyukov,
       A.V., and Kondaurov, N.M., *Sov. Phys.-Semicond.* 5, 1396
       (1972).

(95)   Yamaguchi, K., Matsumoto, H., Nakayama, N., and Ikegami,
       S., *Japan. J. Appl. Phys. 15*, 1575 (1976).

(96)   Yamaguchi, K., Nakayama, N., Matsumoto, H., and Ikegami,
       S., *Japan. J. Appl. Phys. 16*, 1203 (1977).

(97)   Nakayama, N., Matsumoto, H., Yamaguchi, Y., Ikegami, S.,
       and Hioki, Y., *Japan. J. Appl. Phys. 15*, 2281 (1976).

(98)   Bernard, J., Lancon, R., Paparoditis, C., and Rodot, M.,
       Rev. Phys. Appl. 1, 211 (1966).

(99)   Lebrun, J., *Rev. Phys. Appl. 1*, 204 (1966).

(100) Rodot, M., *Rev. Phys. Appl. 12*, 411 (1977).

(101) Mitchell, K.W., Fahrenbruch, A.L., and Bube, R.H., *Solid-
       State Electron. 20*, 559 (1977). Also, *J. Appl. Phys. 48*,
       4365 (1977).

(102) Ma, Y.Y., Fahrenbruch, A.L., and Bube, R.H., *Appl. Phys.
       Lett. 30*, 423 (1977).

(103) Bonnet, D., and Rabenhorst, H., Proc. 9th IEEE Photovoltaic
       Spec. Conf., pp. 129-132. IEEE, New York, 1972.

(104) Buch, F., Fahrenbruch, A.L., and Bube, R.H., *Appl. Phys.
       Lett. 28*, 593 (1976).

(105) Mitchell, K.W., Fahrenbruch, A.L., and Bube, R.H., *J. Appl.
       Phys. 48*, 4365 (1977).

(106) Cohen-Solal, G., Svob, L., Marfaing, Y., Janik, E., and
       Castro, E., Proc. 9th IEEE Photovoltaic Spec. Conf.,
       pp. 28-36. IEEE, New York, 1972.

(107) Komashchenko, V.N., and Fedorus, G.A., *Ukr. Fiz. Zh. 13*,
       688 (1968).

(108) Komashchenko, V.N., Kinev, S., Marchenko, A.I., and
       Fedorus, G.A., *Ukr. Fiz. Zh. 13*, 2086 (1968).

(109)   Cusano, D.A., *Solid-State Electron*. *6*, 217 (1963).

(110)   Cusano, D.A., *Rev. Phys. Appl*. *1*, 195 (1966).

(111)   Justi, E.W., Schneider, G., and Seredynski, J., *Energy Conv*. *13*, 53 (1973).

(112)   Guillien, M., Leitz, P., Marshal, G., and Palz, W., in "Solar Cells", pp. 207-214. Gordon and Breach, London, 1971.

(113)   Lebrun, J., and Bessoneau, G., in "Solar Cells", pp. 201-206. Gordon and Breach, London, 1971.

(114)   Lozykowski, H., *Czech. J. Phys*. *B13*, 164 (1963).

(115)   Shay, J.L., Wagner, S., and Kasper, H.M., *Appl. Phys. Lett*. *27*, 89 (1975).

(116)   Wagner, S., Shay, J.L., and Kasper, H.M., *J. Phys.-Coll*. *36*, 101 (1975).

(117)   Wagner, S., Shay, J.L., Migliorato, P., and Kasper, H.M., *Appl. Phys. Lett*. *25*, 434 (1974).

(118)   Shay, J.L., Wagner, S., Bachmann, K.J., Buehler, E., and Kasper, H.M., Proc. 11th IEEE Photovoltaic Spec. Conf., pp. 503-507. IEEE, New York, 1975.

(119)   Kazmerski, L.L., White, F.R., and Morgan, G.K., *Appl. Phys. Lett*. *29*, 268 (1976).

(120)   Kokubun, Y., and Wada, M., *Japan. J. Appl. Phys*. *16*, 879 (1977).

(121)   Wagner, S., *J. Cryst. Growth* *39*, 151 (1977).

(122)   Mickelsen, R.A., and Chen, W.S., *Appl. Phys. Lett*. *36*, 371 (1980).

(123)   Mickelsen, R.A., Personal Communication.

(124)   Grindle, S.P., Clark, A.H., Rezaie-Serej, S., Falconer, E., McNeily, J., and Kazmerski, L.L., *J. Appl. Phys*. *51*, 1980.

(125)   Loferski, J.J., and Shewchun, J., in Proc. Photovoltaics Adv. R&D Annual Mtg., Sept. 1979, pp. 182-183. SERI, Golden, CO., 1979.

(126)   Tell, B., and Bridenbaugh, P., *J. Appl. Phys.* *48*, 2477
        (1977).

(127)   Tell, B., Bridenbaugh, P.M., and Kasper, H., *J. Appl.
        Phys.* *47*, 619 (1977).

(128)   Johnston, W.D., and Callahan, W.M., *Appl. Phys. Lett.* *28*,
        150 (1976); Also, *J. Cryst. Growth* *39*, 117 (1977).

(129)   Huber, D., and Bogus, K., Proc. 10th IEEE Photovoltaic
        Spec. Conf., pp. 100-102. IEEE, New York, 1974.

(130)   Konagai, M., and Takahashi, K., Proc. Electrochem. Soc.
        Spring Mtg., Washington, pp. 554-555. Electrochem.
        Soc., New Jersey, 1976.

(131)   Hutchby, J.A., Sahai, R., and Harris, J.S., Proc. Intern.
        Electron. Dev. Mtg., Washington, pp. 91-94. IEDM, New
        York, 1975.

(132)   Hovel, H.J., and Woodall, J.M., Proc. 12th IEEE Photo-
        voltaic Spec. Conf., pp. 945-947. IEEE, New York, 1976.

(133)   Woodall, J.M., and Hovel, H.J., *Appl. Phys. Lett.* *30*, 492
        (1977).

(134)   Purohit, R.K., *Phys. Stat. Sol.* *24*, K57 (1967).

(135)   Yoshikawa, A., and Sakai, Y., *Japan. J. Appl. Phys.* *14*,
        1547 (1975).

(136)   Vohl, P., Perkins, D.M., Ellis, S.G., Addiss, R.R.,
        Hui, W., and Noel, G., *IEEE Trans. Electron Dev.* *ED-24*,
        26 (1967).

(137)   Balch, J.W., and Anderson, W.W., *Phys. Stat. Sol. (a)* *9*,
        567 (1972).

(138)   Borden, P.G., *Appl. Phys. Lett.* *35*, 553 (1979).

(139)   Olsen, G.H., Ettenberg, M., and D'Aiello, R.V., *Appl. Phys.
        Lett.* *33*, 606 (1978).

(140)   Moon, R.L., James, L.W., VanderPlas, H.A., and Nelson,
        N.J., *Appl. Phys. Lett.* *33*, 196 (1978).

(141)   Nelson, N.J., Johnson, K.K., Moon, R.L., VanderPlas,
        H.A., and James, L.W., *Appl. Phys. Lett.* *33*, 26 (1978).

(142)   Sahai, R., Edwall, D.D., and Harris, J.S., *Appl. Phys. Lett. 34*, 147 (1979).

(143)   Card, H.C., Yang, E.S., and Panayotatos, P., *Appl. Phys. Lett. 30*, 643 (1977).

(144)   Ponpon, J.P., and Siffert, P., *J. Appl. Phys. 47*, 3248 (1976).

(145)   Ponpon, J.P., Stuck, R., and Siffert, P., Proc. 12th IEEE Photovoltaic Spec. Conf., pp. 900-903.   IEEE, New York, 1976.

(146)   Lillington, D.R., and Townsend, W.G., *Appl. Phys. Lett. 28*, 97 (1976).

(147)   Childs, R., Fortuna, J., Geneczko, J., and Fonash, S.J., Proc. 12th IEEE Photovoltaic Spec. Conf., pp. 862-867. IEEE, New York, 1976.

(148)   Kipperman, A.H.M., and Omar, M.H., *Appl. Phys. Lett. 28*, 620 (1976).

(149)   Vernon, S.M., and Anderson, W.A., *Appl. Phys. Lett. 26*, 707 (1975).

(150)   Delahoy, A.E., Anderson, W.A., and Kim, J.K., in "Photovoltaic Solar Energy" (Strub, A., Ed.), pp. 308-318. D. Reidel, Holland, 1977.

(151)   Munz, P., and Bucher, E., Proc. 13th IEEE Photovoltaic Spec. Conf., pp. 761-766. IEEE, New York, 1978.

(152)   Matsunami, H., Matsumoto, S., and Tanaka, T., Proc. 12th IEEE Photovoltaic Spec. Conf., pp. 917-919. IEEE, New York, 1976.   Also, *Japan. J. Appl. Phys. 16*, 1491 (1977).

(153)   Fabre, E., Michel, J., and Baudet, Y., Proc. 12th IEEE Photovoltaic Spec. Conf., pp. 904-906. IEEE, New York, 1976.

(154)   Fabre, E., *Appl. Phys. Lett. 29*, 607 (1976).

(155)   Peckerar, M., Lin, H.C., and Kocher, Proc. Intern. Electron Dev. Mtg., Washington, pp. 213-216. IEEE, New York, 1975.

(156) Fabre, E., and Baudet, Y., in "Photovoltaic Solar Energy" (Strub, A., Ed.), pp. 178-186. D. Reidel, Holland, 1977.

(157) Stirn, R.J., (unpublished).

(158) Stirm, R.J., and Yeh, Y.C.M., *Appl. Phys. Lett. 27,* 95 (1975).

(159) Stirn, R.J., and Yeh, Y.C.M., Proc. 11th IEEE Photovoltaic Spec. Conf., pp. 437-438. IEEE, New York, 1975.

(160) Stirn, R.J., and Yeh, Y.C.M., IEEE Trans. Electron Dev. *ED-24,* 476 (1977).

(161) Stirn, R.J., and Yeh, Y.C.M., Proc. 12th IEEE Photovoltaic Spec. Conf., pp. 883-892. IEEE, New York, 1976.

(162) Dupuis, R.D., Dapkus, R.D., Yingling, R.D., Moudy, L.A., Johnson, R.E., and Campbell, A.G., Proc. 19th Electronic Mat. Conf., Cornell Univ. IEEE, New York, 1977.

(163) Stirn, R.J., and Yeh, Y.C.M., Proc. 10th IEEE Photovoltaic Spec. Conf., pp. 15-24. IEEE, New York, 1974.

(164) Yeh, Y.C.M., and Stirn, R.J., Ref. 259, pp. 391-397.

(165) Ghandhi, S.K., Borrego, J.M., Reep. D., Hsu, Y.S., and Pando, K.P., *Appl. Phys. Lett. 34,* 699 (1979).

(166) Grinmeiss, H.G., and Memming, R., *J. Appl. Phys. 33,* 217 (1962).

(167) Bonnet, D., in "Photovoltaic Solar Energy" (Strub, A., Ed.), pp. 630-637. R. Reidel, Holland, 1977.

(168) Bell, R.O., Serreze, H.B., and Wald, F.V., Proc. 11th Photovoltaic Spec. Conf., pp. 497-502. IEEE, New York, 1975.

(169) Ponpon, J.P., and Siffert, P., *Rev. Phys. Appl. 12,* 427 (1977).

(170) Lebrun, J., *Rev. Phys. Appl. 1,* 204 (1966).

(171) Trivich, D., Wang, E.Y., Komp, R.J., and Ho, F., Proc. 12th IEEE Photovoltaic Spec. Conf., pp. 875-878. IEEE New York, 1976.

(172) Abdullaev, G.B., Akhundov, M.R., and Akundov, G.A., *Phys. Stat. Sol. 16,* 209 (1966).

(173) Segura, A., Besson, J.M., Chevy, A., and Martin, M.S.,
       Il Nuovo Cim. 38B, 345 (1977).

(174) Matsushita, T., Suzuki, A., Okuda, M., and Nang, T.T.,
       Japan. J. Appl. Phys. 15, 2461 (1976).

(175) Clemen, C., Saldana, X.I., Munz, P., and Bucher, E.,
       Phys. Stat. Sol. (a) 55, (1979).

(176) Clemen, C., Moller, A., Munz, P., Honigschmid, J., and
       Bucher, E., in "Photovoltaic Solar Energy" (Strub, A., Ed.)
       pp. 638-643. D. Reidel, Holland, 1977.

(177) Catalano, A., Dalal, V., Fagen, E.A., Hall, R.B., Masi,
       J.V., Meakin, J.D., Warfield, G., Convers Wyeth, N.,
       and Barnett, A.M., Proc. 13th IEEE Photovoltaic Spec.
       Conf., pp. 288-293. IEEE, New York, 1978.

(178) Borschevskii, A.S., Lebedev, A.A., Maltseva, I.A.,
       Ovezov, K., Rud, Y.V., and Undalov, Y.K., *Sov. Phys.-
       Semicond. 10*, 934 (1976).

(179) Lebedev, A.A., Ovezov, K., and Rud, Y.V., *Sov. Phys.-
       Semicond. 10*, 78 (1976).

(180) Grigoreva, V.S., Lebedev, A.A., Ovezov, K., Prochukhan,
       V.D., Rud, Y.V., and Yakovenko, A.A., *Sov. Phys.-Semicond.
       9*, 1058 (1975).

(181) Lebedev, A.A., Ovezov, K., Prochukhan, V.D., and Rud,
       Y.V., *Sov. Phys.-Techn. Phys. Lett. 1*, 93 (1975).

(182) Bulyarskii, S.V., Koval, L.S., and Radautsan, S.I.,
       Proc. All Union Conf. Ternary Semicond. and Appl.,
       pp. 165-168. Izd. Shtiintza, Kishinev, 1976.

(183) Charlson, E.J., and Lien, J.C., *J. Appl. Phys. 46*, 3982
       (1975).

(184) VanHalen, P., Mertens, R., Van Overstraeten, R., Thomas,
       R.E., and VanMeerbergen, in "Photovoltaic Solar Energy"
       (Strub, A., Ed.) pp. 280-288. D. Reidel, Holland, 1977.
       Also, Proc. 12th IEEE Photovoltaic Spec. Conf., pp. 907-
       912. IEEE, New York, 1976.

(185) Green, M.A., and Godfrey, R.B., in "Photovoltaic Solar Energy" (Strub, A., Ed.), pp. 299-307. D. Reidel, Holland, 1977.

(186) Green, M.A., and Godfrey, R.B., *Appl. Phys. Lett. 29*, 610 (1976).

(187) Pulfrey, D.L., *Solid-State Electron. 20*, 455 (1977).

(188) St. Pierre, J.A., Singh, R., Shenchun, J., and Loferski, J.J., Proc. 12th IEEE Photovoltaic Spec. Conf., pp. 847-853. IEEE, New York, 1976.

(189) Ponpon, J.P., and Siffert, P., *J. Appl. Phys. 47*, 3248 (1976).

(190) Charlson, E.J., and Richardson, W.F., Proc. 13th IEEE Photovoltaic Spec. Conf., pp. 656-660. IEEE, New York, 1978.

(191) Lillington, D.R., and Townsend, W.G., *Appl. Phys. Lett. 31*, 471 (1977).

(192) Anderson, W.A., "Silicon Photovoltaic Devices for Solar Energy Conversion" Final Rep., June 1976-May, 1978. ERDA, Washington, 1978.

(193) Godfrey, R.B., and Green, M.A., *Appl. Phys. Lett. 34*, 790 (1979).

(194) Anderson, W.A., Delahoy, A.E., Kim, J.S., Hyland, S.H., and Dey, S.K., *Appl. Phys. Lett. 33*, 588 (1978).

(195) Fabre, E., Michel, J., and Baudet, Y., Proc. 12th IEEE Photovoltaic Spec. Conf., pp. 904-907. IEEE, New York, 1976.

(196) Tarr, N.G., Pulfrey, D.L., and Iles, P.A., *Appl. Phys. Lett. 35*, 258 (1979).

(197) Maeda, Y., *Appl. Phys. Lett. 33*, 301 (1978).

(198) Neugroschel, A., *IEEE Trans. Electron. Dev. ED-27*, 287 (1980).

(199) Matsunami, H., Oo, K., Ito, H., and Tanaka, T., *Japan. J. Appl. Phys. 14*, 915 (1975).

(200) Hsu, L., and Wang, E.Y., in "Photovoltaic Solar Energy"
       (Strub, A., Ed.) pp. 1100-1108. D. Reidel, Holland, 1977.

(201) Nagatomo, T., and Omoto, O., *Japan J. Appl. Phys. 15*,
       199 (1976).

(202) DuBow, J.B., Burk, D.E., and Sites, J.R., *Appl. Phys.
       Lett. 29*, 494 (1976).

(203) Burk, D.E., DuBow, J.B., and Sites, J.R., Proc. 12th IEEE
       Photovoltaic Spec. Conf., pp. 971-974. IEEE, New York,
       1976.

(204) Manifacier, J.C., Szepessy, L., and Savelli, M., in
       "Photovoltaic Solar Energy", pp. 289-298. D. Reidel,
       Holland, 1977.

(205) Manifacier, J.C., and Szepessy, L., *Appl. Phys. Lett. 31*,
       459 (1977).

(206) Franz, S., Kent, G., and Anderson, R.L., *J. Electron.
       Mat. 6*, 107 (1977).

(207) Perotin, M., Szepessy, L., Manifacier, J.C., Parot, P.,
       Fillard, J.P., and Savelli, M., in "Solar Electricity",
       pp. 481-492. Maison des Congres, Toulouse, 1976.

(208) Kato, H., Yoshida, A., and Arizumi, *Japan. J. Appl. Phys.
       15*, 1819 (1976).

(209) Nash, T.R., and Anderson, R.L., Proc. 12th IEEE Photo-
       voltaic Spec. Conf., pp. 975-977. IEEE, New York, 1976.

(210) Kato, H., Fujimoto, J., Kanda, T., Yoshida, A., and
       Arizumi, T., *Phys. Stat. Sol. (a) 32*, 255 (1975).

(211) Cheek, G., Ellsworth, D., Genis, A., and DuBow, J.B.,
       J. Electron. Mat. 8, (1979).

(212) Ghosh, A., (Ed.), "Heterojunction Single Crystal Silicon
       Photovoltaic Cell", Quarterly Reports, Sept. 1976-
       Nov. 1977. ERDA, Washington, 1977.

(213) DuBow, J.B., (Ed.) "Characteristics of OSCS Solar Cells",
       Quarterly Rept., March 1977. ERDA, Washington, 1977.

(214) Schunck, J.P., and Coche, A., *Appl. Phys. Lett. 35*, 863
       (1979).

(215) Feng, T., Ghosh, A.K., and Fishman, C., *Appl. Phys. Lett.
       35*, 266 (1979).

(216) Kazmerski, L.L., and Sheldon, P., Proc. 13th IEEE Photo-
       voltaic Spec. Conf., pp. 541-544. IEEE, New York, 1978.

(217) Bachmann, K.J., Sinclair, W.R., Thiel, F.A., Schreiber,
       H., Schmidt, P.H., Spencer, E.G., Buehler, E., and
       SreeHarsha, K., Proc. 13th IEEE Photovoltaic Spec. Conf.,
       pp. 524-527. IEEE, New York, 1978.

(218) Wang, E.Y., and Legge, R.N., Proc. 12th IEEE Photovoltaic
       Spec. Conf., pp. 967-970. IEEE, New York, 1976.

(219) SreeHarsha, K., Bachmann, K.J., Schmidt, P.H., Spencer,
       E.G., and Thiel, F.A., *Appl. Phys. Lett. 30*, 645 (1977).

(220) Okuda, M., Nang, T.T., Matsushita, T., and Yokota, S.,
       Japan. J. Appl. Phys. 14, 1597 (1975).

(221) Nang, T.T., Matsushita, T., Okunda, M., and Suzuki, A.,
       Japan. J. Appl. Phys. 16, 253 (1977).

(222) Fahrenbruch, A.L., Aranovich, J., Courrages, F., Yin,
       S.Y., and Bube, R.H., in "Photovoltaic Solar Energy"
       (Strub, A., Ed.), pp. 608-617. D. Reidel, Holland, 1977.

(223) Carlson, D.E., and Wronski, C.R., *Appl. Phys. Lett. 28*,
       671 (1976).

(224) Carlson, D.E., and Wronski, C.R., *J. Electron. Mat. 6*,
       95 (1977).

(225) Busmundrud, O., and Jayadeviah, T.S., *Phys. Stat. Sol.
       (a) 11*, K47 (1972).

(226) Mizrah, T., and Adler, D., *IEEE Trans. Electron. Dev. 24*,
       458 (1977).

(227) Carlson, D.E., *IEEE Trans. Electron. Dev. ED-24*, 449
       (1977).

(228) Carlson, D.E., Proc. DOE/SERI Photovoltaics Adv. R&D
       Annual Rev. Mtg., Denver, pp 113-120. SERI, Golden, Co.

(229)   Hamakawa, Y., Okamota, H., and Nitta, Y., *Appl. Phys. Lett. 35,* 187 (1979).

(230)   See, for example, Stone, J., in Ref. 228, pp. 69-112.

(231)   Spear, W.E., and LeConber, P.G., *Solid-State Comm. 17,* 1193 (1975).

(232)   Wilson, J.I.B., McGill, J., Kinmond, S., *Nature 272,* 153 (1978).

CHAPTER 16

RESEARCH AND DEVICE PROBLEMS IN PHOTOVOLTAICS

Lawrence Kazmerski

Photovoltaics Branch
Solar Energy Research Institute
Golden, Colorado

I. INTRODUCTION

The state-of-the-art in photovoltaics has been advancing quite
rapidly, especially with the increased emphasis put on this technology by the U.S. government (1,2). Because of this, it is expected that some of the problems underscored in this chapter will
be resolved before publication of this book. It is the intent of
this chapter to focus on *some* current photovoltaic research problems, relating technical barriers and probable areas of future
work. Some effort has been made to present those that are considered most critical to the development of photovoltaics into a
viable energy alternative. Therefore, attention is directed toward
the status and problems of a pair of technologies that have the
predicted application potential discussed in the previous chapter
(3): intermediate-efficiency, *thin-film* solar cells (including
amorphous and polycrystalline Si, GaAs, CdS-based devices, and
some emerging materials) and high-efficiency, *concentrator* devices
of both multijunction and multidevice types. Throughout this overview, the problems associated with the major defect in the polycrystalline device--the *grain boundary*--are emphasized, both in terms
of the effects of these regions upon device performance and also

in terms of research into circumventing the limits they place on
efficient photovoltaic conversion. In addition, the importance of
research into the detection of device/materials degradation mechanisms is stressed, since *reliability* and *device lifetime* are of
an importance equal to *efficiency* in the eventual deployment of
solar cells.

II. OVERVIEW OF RESEARCH PROBLEMS AND TECHNOLOGIES

A. *Thin Films*

Research on thin-film solar cells has centered in two important areas: (i) the *improvement of materials and device properties*
in order to optimize solar cell performance; and, (ii) the *identification of new photovoltaic materials* and the demonstration of
device feasibility for these candidate materials. This section
examines these R&D activities, emphasizing the areas of needed
work.

1. Amorphous Silicon. Research on this material really began
in the late 1960s with the published work of Sterling on dis-

TABLE 1. *Status of Several Major Amorphous Si Solar Cells*

Type/Structure[d]	V_{oc}(V)	J_{sc} (mA/cm^2)	ARC	η(%)	Area(cm^2)	Source
Pt-i-n$^+$	0.803	12	yes	5.5[a]	0.02	Ref. 14
Pt-SiO$_2$-p-i-n	0.770	7	no	3.3[a]	1.6	Ref. 17
ITO-p$^+$-i-n$^+$	0.560	12	yes	4.5[b]		Ref. 96
Pt-i-n$^+$	0.500	15	no	4.3[a]		Ref. 19
n$^+$-i-P$^+$-Cr	0.670	9.3	no	3.4[a]		Ref. 13
Ni-TiO$_x$-i-n$^+$	0.630	7.5	yes	4.8[c]	0.02	Ref. 95

a. *100 mW/cm^2*   c. *60 mW/cm^2*

b. *80 mW/cm^2*   d. *stainless steel substrates*

charge-produced amorphous silicon (a-Si) (4,5). It was determined
that the density of energy gap states was several orders of magnitude lower in discharge-produced A-Si than in its evaporated
counterpart (6). The first photovoltaic devices were actually
fabricated from a-Si in 1974 (7). However, in 1975 this discharge-
produced material was shown to actually be an a-Si/hydrogen alloy
(a-Si:H) (8), and the hydrogen (5-50 atomic-%) accounted for the
reduction in gap state density (9). Recent research on discharge-
produced a-Si alloys has included the use of oxygen (10) - a
bridging atom that heals defects, carbon (11), and fluorine (12) -
a highly electronegative element that eliminates unfavorable bonding, to further reduce unwanted gap states.

From the first report of the doping (n- or p-type) of a-Si:H
(13), a variety of device types have been investigated (see Table
1). The theoretical efficiency limit has been calculated to be
~15% for this material (14), but the best device produced to-date
remains the MIS structure reported by Carlson in 1975 (15) for a
small area cell. (Large area devices have been typically <4%.)
A more rapid advancement of the a-Si photovoltaic technology has
been largely impeded by a lack of knowledge of many of the basic
physical, structural, electrical, and optical properties of this
relatively new material, and some seemingly severe materials limitations that must be overcome to provide efficient and reliable
devices.

Foremost among these *materials problems* is that of rather poor minority-carrier transport. Although the hole diffusion length has
recently been *estimated* to be ~ 400Å in *undoped* a-Si:H (16) and
the hole lifetime *appears* to be of the order of 1 ns, little has
been accomplished in either measuring these parameters precisely
or controlling them. Since the hole diffusion length is limited,
the resulting efficiencies of cells are likewise constrained by
small depletion widths of the order of 0.2 μm under AM1 conditions.
The necessity and value of increasing the depletion width by further lowering the space-charge density has recently been demon-

strated in p-i-n a-Si:H cells (17). Thus, the undoped a-Si need
not have a very large hole diffusion length *if* a significant fraction of the incident radiation can be made to be absorbed in the
depletion layer.

A large number of the materials problems associated with a-Si:H
relate to the preparation of the material. For example, a-Si:H can
be prepared by glow discharge (RF or DC), reactive sputtering,
electrodeposition, ion-plating, CVD, evaporation with either
in-situ or post-deposition hydrogenation and ion-implantation.
However, it is difficult to control the Si-H bonding types (18),
and little is known about either the gas-solid interactions during
a-Si:H formation or the growth kinetics of the material. In general, there are a large number of growth parameters that affect in
varying degree the quality of the material produced. In this
relation, control of defects and impurities is difficult (e.g.,
the built-in potential of p-i-n devices is limited to ~1.1V by
the poor quality of the doped material (17), providing V_{oc} <0.85V).
In fact, the roles of both impurities and modifiers (i.e., H, O,
F, etc.) are not well understood. A good deal of work remains to
be done in optimizing the experimental parameters (19).

In order to solve many of the problems with a-Si based solar
cells, the *materials/device characterization* base must be expanded.
In most cases, conventional semiconductor measurement technologies
(e.g., single-crystal Si related) are not sufficient. For example,
no *reliable* methods exist for determining either the density of
gap states or the *exact* hydrogen content of films[1]. At best, it is
difficult to measure the minority-carrier lifetimes and diffusion
lengths, a procedure very necessary to solving some of the transport problems. Theoretical understanding of any of these phenomena
is limited or nonexistent. Finally, for this technology especially, there is a need to establish relationships between the materi-

---

[1]*Presently, secondary ion mass spectroscopy (SIMS) is the
most dependable, straightforward technique.*

als parameters and the resulting photovoltaic properties. Almost
no work exists in this area, and most results await the development of special measurement techniques specifically designed for
this material.

To date, a-Si solar cells have attained respectable open-circuit voltages (20). However, fill-factors have been low, partly
due to the high series resistance (contact and material caused)
of the devices. However, even if the fill-factor were limited to
0.6, a 10% efficient device can be expected if the small drift
component of current can be improved (e.g., by increasing the
width of the depletion region under illumination).

Amorphous solar cells, like all thin-film devices, are controlled in many cases by the numerous interfaces that are inherent to the complex device structures (21). However, little research has been done on detecting impurities and contaminants at
these interfaces (e.g., using surface analysis techniques (22))
and determining their role on interface state densities and device performance. The performance of a-Si:H devices could be improved significantly *if* a wide bandgap, conductive p$^+$-layer could
be developed (17). Thus, high built-in voltages (open-circuit
voltages) and reasonably light transmission (higher short-circuit
currents) could result.

Finally, and equally important, there is a need to determine
the stability and reliability of these devices. No in-depth
studies have been performed to-date, primarily due to the obvious
and justifiable emphasis on developing and demonstrating working
devices. However, as the performance of a-Si devices improve, the
mechanisms of degradation must be uncovered in order to ascertain
15-20 year lifetimes for commercial devices.

In summary, the important areas for future research include:

• Improvement of minority carrier lifetimes and diffusion
  lengths;

• Better understanding of transport phenomena;

• Development and understanding of the density of states;

- Determination of the nature of chemical bonding;
- Understanding of impurities (incorporation, effects);
- Studies of surface states at interfaces;
- Control of deposition parameters, reactions at gas-solid interfaces;
- Development of theoretical base;
- Development of special measurement techniques; and
- Determination of degradation mechanisms/device reliability.

2. *Polycrystalline Silicon*. In the development of an *inexpensive* solar cell for *terrestrial applications,* the use of thin-film Si seems a logical choice due to the large amount of information available on the well-developed single-crystal device, and to the abundance of the element itself. Early work centered on the fabrication of films and devices on inexpensive, non-silicon substrates. Fang *et al.* produced films on aluminum by electron beam evaporation (23). Above 577°C an Si-Al eutectic is formed, and the growth of large-grained films (with columnar grain characteristics under some conditions) has been demonstrated. However, device efficiencies have been insufficient. Chu *et al.* have investigated the fabrication of homojunction devices using CVD on several low-cost substrates including steel (24,25), graphite (25-28), and metallurgical grade polycrystalline silicon (25,28-30). To date, the best thin-film device produced is represented by the J-V characteristics shown in Fig. 1. This polycrystalline solar cell has a respectable area (9 cm^2) and a 9.75% AM1 efficiency. The device configuration is also presented in this figure. However, the potential of the *thin-film polycrystalline Si* device remains high (31-33). Silicon maintains its premier position substantially because of its great abundance. Compared to the amorphous Si cell, thin-film polycrystalline Si promises about a 50% better performance, using a two-pass light analysis for a 5 μm layer (34).

Two questions often arise for the thin-film polycrystalline Si solar cell:

- What is *thin-film?*

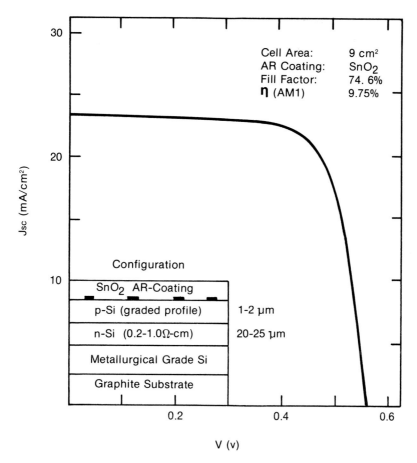

FIGURE 1.   9.75% AM1 polycrystalline Si thin-film solar cell from Chu, et al., ref. 97).

● What is the effect of *grain size*?

The first is difficult to answer either unequivocally or authoritatively. A good portion of advanced Si device R&D is being performed on heterojunction, MIS and SIS structures fabricated on thick (bulk) polycrystalline material (35,36). An example of this is the work of Shewchun *et al.*, who have produced >9.6% (total area, AM1) ITO (thin-film)/Si (polycrystalline) devices using cast material provided by Wacker in Germany (36,37). The substrates in these investigations are typically ~10^3 μm in thickness,

with grain sizes in the 2-10 mm range. The technique used to produce this material (casting followed by wafer slicing) is a *bulk*
rather than a *thin-film* scheme.

The effect of layer *thickness* upon solar cell performance is
twofold: (i) for thinner layers, especially for indirect bandgap
materials such as Si, there is an increasing loss through nonab-
sorption of the incoming radiation; and, (ii) the back contact
becomes critical. If the minority-carrier diffusion length is
comparable or greater than the device thickness, the collection
efficiency is reduced and the dark current increases due to the
excess recombination of both the photogenerated and dark injected
carriers at the back contact. The thickness influence on Si device performance is illustrated in Fig. 2 (38)'. The short-circuit
current density is shown as a function of thickness for an n/p

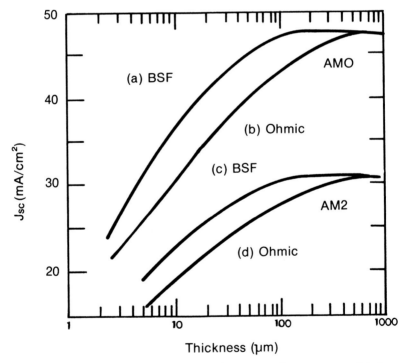

Thickness (μm)

*FIGURE 2. Short-circuit density dependence upon thickness
for an n/p homojunction under AM0 and AM1 conditions (from Hovel,
ref. 38).*

homojunction under AMO and AM2 conditions. For an ohmic back contact, the short-circuit current density is observed to drop for a
thickness less than 500 µm under AM2 conditions. By improving the
back contact, incorporating a back-surface-field (BSF), the thickness influence onset can be seen to begin at ~100 µm. It should be
stressed, however, that the calculations used to generate Fig. 2
consider a *constant* minority-carrier diffusion length. Some recent
work by D'Aiello and Robinson (39) has shown that the actual diffusion length can be much less than expected in their epitaxial
films. As a result, *thinner* layers (15 µm *vs*. 50 µm) actually lead
to somewhat better devices (primarily reflected in higher V_{oc}) for
their material. Defining a thickness-range for "*thin-film* Si solar

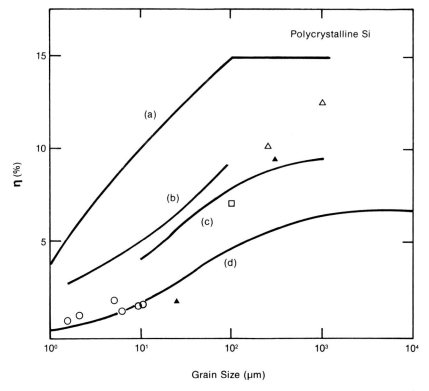

FIGURE 3. *Grain size dependence of efficiency for polycrystalline Si (a) ref. 43; (b) ref. 42; (c) ref. 41; and, (d) ref.
40.*

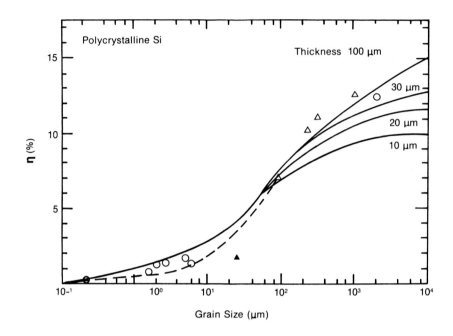

FIGURE 4. *Polycrystalline Si solar cell efficiency dependence on grain size, after calculations of Ghosh, et al.,* ref. *44.*

cells" is difficult and somewhat arbitrary. But for the purpose of
this chapter, cells less than 100 μm in thickness will satisfy the
criterion for silicon.

The limiting effects of *grain size* on polycrystalline Si solar
cell performance have been studied (40-44). Fig. 3 presents the
results of several investigations for a variety of device types.
Recently, Ghosh *et al.* (44) presented a unified theory of grain
size effects. Their calculations of the dependence of efficiency
on grain size are shown in Fig. 4, with some representative data
points included. The calculations for 100-, 30-, 20-, and 10-μm
thick layers are presented. From Fig. 4, it is predicted that *at
least* 150 μm grain sizes are needed for >10% efficient devices.
Using the data of Fig. 4, one can generate a dependence of thickness on grain size for a 10% efficient polycrystalline device.
These data, presented in Fig. 5, indicate that in the case of

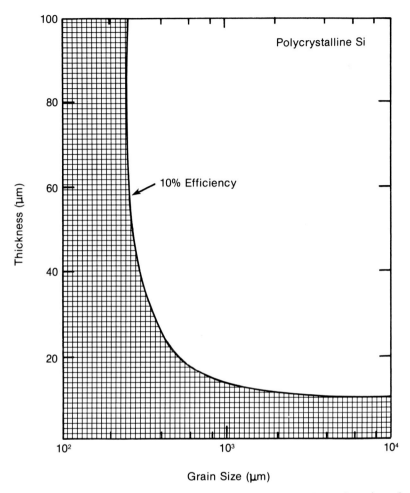

FIGURE 5.   *Relationship between thickness and grain size for a 10% (AM1) efficient solar cell.*

smaller grain sizes, *recombination at the grain boundaries* must be *reduced* if the polycrystalline devices are to realize the 10% efficiency goal. Methods and the status of such attempts at reducing grain boundary recombination (i.e., passivation schemes) are discussed later in this book by D. Ginley (45).

Research in the polycrystalline Si solar cell area presently has three major thrusts:

• materials development,

- device development, and
- basic studies.

The performance of the polycrystalline Si solar cell depends upon the *quality* of the material used in its fabrication. The *first quality factor* is that of impurity content of the starting material. Chu *et al.* build their devices (see Fig. 1) on a deposited metallurgical grade Si layer that has been subjected to a *potentially* inexpensive acid-leaching purification procedure (30). Although the procedure has been shown to be effective from a device fabrication viewpoint, more remains to be learned about the chemistry, cost, and control of the process. Other research programs are concerned with the purification of metallurgical grade Si into a low-cost *solar grade* that can serve as a source for either film or substrate growth (46). The *second quality factor* is that of material crystallinity and reducing losses at grain boundaries. A major portion of present polycrystalline Si research is dictated by finding, evaluating, and characterizing various thin-film growth techniques for producing large grain size films on low-cost substrates. The methods vary from epitaxy (39) (i.e., controlling and maximizing grain size *during* growth) to annealing (30,46) (i.e., post-deposition treatments). For those cases in which grain size limitations seem inevitable, passivation schemes are of interest (45, 47-49). Foremost among the present developments are those that make use of the enhanced diffusion that occurs along the grain boundaries to selectively chemically modify these regions using hydrogen or other chemical species (47-49).

Actually, very little current research is concerned with *thin-film* polycrystalline Si. As previously indicated, most non-single-crystal silicon research is being performed on *bulk* polycrystalline-produced material. Several reasons can be cited. *First*, device quality thin-film Si is difficult to produce on foreign (i.e., non-Si) substrates - especially if low-cost materials are used. (It should be added that the search for *low-cost* substrates for these thin-film devices forms a research area in itself.) *Second*,

there has been more emphasis by research groups in this area on
developing and optimizing the solar cell structures (MIS, SIS,
heterojunction, homojunction) that are compatible with polycrystalline material in order to provide reasonable efficiencies. It is
easier to do this in a controllable and reproducible manner on the
bulk polycrystalline (e.g., cast) material, rather than fighting
the inconsistencies of smaller grain size thin-film Si. This approach is advantageous for optimizing the chemical, structural,
and electrical properties of the insulating layers in the MIS
and SIS devices. Some better indications of the limitations to
polycrystalline solar cell performance are also obtainable. However, these devices are *not* thin-film solar cells and investigators are aware that grain boundaries, oxide growths, diffusion
mechanisms, material quality, *etc.*, in *bulk* polycrystalline semiconductors differ from those encountered in *thin* polycrystalline
films (50).

Because the polycrystalline state presents a multitude of un-
knowns and differences from the crystalline state, a good portion
of present research on solar cells is dedicated to more *basic research*. Most of this basic work is directed toward understanding
the mechanisms that are influenced or controlled by the grain
boundaries. There is an obvious need to develop new measurement
techniques to characterize the structural, compositional, electrical, and optical properties of polycrystalline Si in terms of cell
performance. The electrical transport processes have received much
attention in the recent past (51). However, much remains to be
accomplished in modeling grain boundary effects and correlating
these "models" with direct and comprehensive measurements. It is
possible that such activity can lead to the development of theoretical predictions of the fundamental photovoltaic mechanisms
that control and limit the polycrystalline Si solar cell conversion efficiency.

3. *Thin-Film GaAs.* The cost-practicability of using thin-film
(polycrystalline) GaAs in a nonconcentrating solar cell array is

still somewhat doubtful. This is primarily due to the availability
of the constituent elements (52), especially the gallium. However,
there is a reasonable R&D ongoing, dealing with both single-crystal and polycrystalline thin films (53).

Basically, two fabrication approaches are being pursued in the
fabrication of thin-film GaAs solar cells. The *first* involves the
nonepitaxial growth of GaAs on foreign substrates. These films typically have small grain sizes, with grain diameters approximately
the size of the film thickness. Although the current collection
and resulting short-circuit currents in such devices are reasonable, the overall device performance is limited by low open-circuit voltages and relatively high series resistances (poor fill-
factors). Therefore, there is a need to develop reliable grain
boundary passivation schemes for these films and devices. The
second approach involves the *epitaxial* growth of GaAs on suitable
substrates. This fosters the production of either single-crystal
or large grain size films, in order to minimize or avoid the com-
plicating effects imposed by the boundary regions. In some cases,
this second approach involves the recrystallization of thin layers
(e.g., Ge) deposited on metal substrates to provide the suitable
surface on which the epitaxial growth of the GaAs is initiated
(54). In other cases, the recrystallization of the GaAs layer itself is accomplished using laser techniques (54-56).

The trend in poly-GaAs research is toward the production of
thinner, device-quality films. This direction dictates that progress in materials technology must be emphasized. There is some
evidence that the quality of the material within the grain is at
least as important as the grain boundary in limiting device performance (53). Thus, techniques must be developed not only to
minimize grain boundary effects but also to ensure the growth of
defect-free, low-impurity GaAs grains. Research on the nucleation
and growth of GaAs films is needed to provide insight into these
problems. This can lead to the development of processing techniques that promote grain growth and provide grain sizes with later-

al dimensions exceeding the film thickness.

Although progress has been made at grain boundary passivation,
this area still needs attention. Ghandhi *et al.* have reported success with the selective anodization of boundary regions (57,58).
This method leaves an oxide "cap" over the grain boundary, and reductions in dark reverse-current densities by six orders of magnitude have been observed shown in Fig. 6 (58). This passivation has
resulted in significant improvements in open-circuit voltages
($\Delta V_{OC} \sim 0.3$ V) in these thin-film devices by eliminating grain
boundary shorting effects.

Beyond the materials problems, increased research on device
processing of these thin polycrystalline cells is needed. Techniques for reliably and reproducibly fabricating p-n junctions remain problem areas - for these as well as other thin-film layers.
High-temperature processing must be avoided to minimize material
degradation and enhanced grain boundary diffusion (59). MIS structures appear promising, but control of the oxide properties are
critical (60). As for any of these thin devices, degradation
mechanisms must be identified in order to eventually ensure operating stability (61).

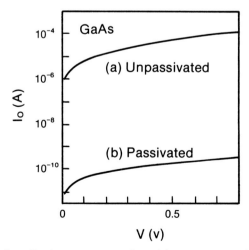

*FIGURE 6.   Dark reverse saturation current in polycrystalline
GaAs solar cells (from Pande, et al., ref. 58).*

4. CdS-Based Devices. Of the thin-film technologies, the $Cu_2S/$
CdS technology is the most advanced. From the discovery of the
photovoltaic effect in CdS by Reynolds (62) in 1954, the device
has advanced to exhibit efficiencies in excess of 9% (63). Recent
efforts in the CdS area have been investigating the use of Cu-
ternaries (64-67) and ternary alloys (67-68) in place of the CdS
in order to improve lattice match and material control. These materials are really in the initial development stages and will not
be emphasized herein. The substitution of Cd(Zn)S for the CdS has
also been instituted, since it provides a better lattice match
with the Cu_2S and a larger bandgap window material (69). Readers
are referred to several reviews of the CdS technologies for in-
depth analyses (63,70,71).

For these CdS-based solar cells, two device geometries are
used:

- *Frontwall configuration,* in which the sunlight impinges on
  the lower bandgap semiconductor (e.g., Cu_2S, $CuInSe_2$,
  $Cu_xIn_yGa_{1-y}Se_zTe_{1-z}$)
- *Backwall configuration,* in which the illumination is through
  the wide bandgap material (i.e., the CdS or Cd(Zn)S).

These device types are represented in cross section in Fig. 7. The
benefits and limitations of each of these structures have been
thoroughly evaluated (71,72), but the major application of each is
tied to economics (for the Cu_2S cell (63)) or junction formation

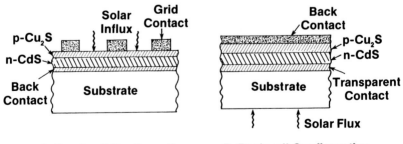

A. Frontwall Configuration   B. Backwall Configuration

*FIGURE 7. Backwall and Frontwall solar cell configurations
(courtesy of S.K. Deb).*

and degradation (for the Cu-ternary devices (66,74)).

For the Cu_2S/CdS (or $Cd(Zn)S$) thin-film solar cell, two device types have been developed. Their categorization results from the qualitative description of the junction geometry:

- textured cells, and

- planar cells.

The *textured cell* is formed conventionally by etching the CdS layer and subsequently dipping it in a Cu_2Cl_2 solution to form the Cu_2S (71). The resultant junction is highly structured and involves deep intrusions of the Cu_2S into the CdS grain boundaries.

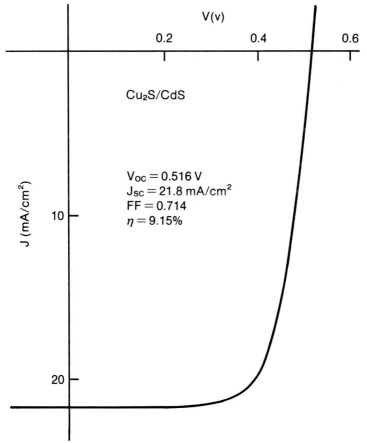

FIGURE 8.   Light J-V characteristics of 9.15% efficient Cu_2S/CdS solar cell (from Barnett, et al.,   ref. 63).

Thus, the junction is *textured*. The best device of this type has
a 9.15% AM1 efficiency, and its light J-V characteristics are
shown in Fig. 8. Following an extensive loss detection and minimization scheme, it has been concluded that the near 10% efficiency range is about maximum for this device type, primarily due to
losses of nonuniform carrier generation/collection and shading at
the grain boundary intrusions. Therefore, the *planar configuration*
has evolved. In this, the junction is formed by a solid-state reaction between CdS and a deposited Cu_2Cl_2 film (71). Thus, the grain
boundary penetration problems are minimized. Planar $Cu_2S/Cd(Zn)S$

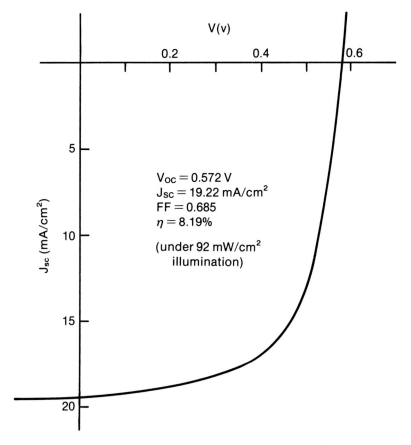

FIGURE 9.   Light J-V characteristics of $Cu_2S/Cd(Zn)S$ solar
cell (from Meakin, ref. 71).

cells have reached the mid-8% (AM1) range and a typical characteristic is shown in Fig. 9. Although these cells have shown good
open-circuit voltages (~0.55V), the major problem has been due to
ineffective light trapping. In order to overcome this, two approaches are being studied. The *first* involves texturing of the
front surface or the substrate surface in order to minimize reflections. The *second* approach is concerned with the development of
more effective antireflection coatings. Primary among these is
the development of two-layer ARCs, of which TiO_x/SiO_2 is presently used, and ZnS (600Å)/MgF (1000Å) is a potential candidate. The
planar $Cu_2S/Cd(Zn)S$ solar cell is expected to reach efficiencies
in excess of 11%.

Major among the problems of the CdS-based solar cells is device stability and reliability. Many of the degradation mechanisms
have been identified and solutions have either been implemented
or proposed (33,94). More basic research is needed, however.
Table 2 summarizes some of the leading problems, their effects
and the solutions. However, more R&D activity is needed in this
area. Degradation must be minimized if these devices are to realize
greater than 15-year lifetimes. Some systematic and reliable (!)
accelerated life-testing procedures must be developed. It is obvious that *encapsulation* is critical for these thin-film solar
cells. The extent and cost acceptability of this encapsulation
must be evaluated. The encapsulation scheme might be ameliorated
by the discovery that the initial cell efficiency of the $Cu_2S/$
Cd(Zn)S device can nearly be restored by a suitable heat-treat-
ment (71,94).

Finally, work is needed on scaling-up the processes that produce these devices. Large-area devices, with comparable efficiencies to research devices of smaller area are necessary. Obviously,
if the technology is to be viable, cost, reliability, and performance limits must be adhered to.

In summary, the major R&D areas that need emphasis for these
devices are:

- identification and solution of degradation mechanisms;
- development of reliable lifetime test procedures;
- development of suitable encapsulants;
- improvement in device performance, reproducibility;
- development of low-cost, large-scale production procedures;
- improvement of device geometries, ARCs, film structures; and
- more basic research, and more research activities in this area

Table 2. Examples of Degradation Problems in Cu_2S/CdS Solar Cells.

Cause of Degradation	Parameter(s) Affected	Solution(s)
1. Exposure to high humidity	J_{sc}, η_{coll}	Recovery: Heat _in-vacuo_ at 180°C, Hermetic Encapsulation
2. Oxidation (e.g., exposure to air above 60°C)	J_{sc}, R_{series}	Hermetic Encapsulation
3. Photo-induced phase change of Cu_2S (e.g., heating above 60°C under illumination)	J_{sc}, (V_{oc}), R_{series}	Provide Cu by evaporation (e.g., 100Å on Cu_2S layer)
4. Delamination by temperature cycling (e.g.,-150 to +60°C)	J_{sc}, R_{series}	Minimum effect on performance, Temperature control

5. *Other Materials*. A number of other materials and materials systems are being investigated for solar cell application. These materials, sometimes called *emerging materials,* possess many of the requirements for a photovoltaic component but are still in their infancies from the device development viewpoint. These include InP (73-75), CdTe (76-78), Cu_2O (79,80), Zn_3P_2 (81), Cu_2Se (82), $ZnSiAs_2$ (83), $CdSiAs_2$ (84), BAs (85), and polyacetylene-

CH_x (86). Research tasks with these devices include:

(1) Theoretical modeling and calculations in order to evaluate the potential (both ideal and attainable) conversion efficiency of the pertinent material and the mechanisms that *limit* the realization of the optimum performance.

(2) The growth and characterization of thin-polycrystalline films (device quality) of these materials, emphasizing the utilization of low-cost substrates.

(3) Understanding of the relationships between the polycrystalline and single-crystal properties of the candidates.

(4) The preparation of devices: pn junctions, MIS, Schottky barriers, heterojunctions. Device performance must be demonstrated.

(5) The understanding and control of material doping.

(6) Development or implementation of experimental procedures to evaluate the basic material and device parameters, most of which have not been measured or reported in the literature.

(7) Investigation of the stability of the device and material parameters under typical environmental and operating conditions.

B. *Concentrator Devices*

The need and economic outlook for photovoltaic concentrators are discussed in the previous chapter. Overall, the concentrator can reduce the impact of cell cost on the system cost, thus providing a viable means of meeting critical cost goals. The cells also exhibit improved solar conversion efficiency with concentration. At constant temperature, the cell efficiency (neglecting series resistance effects) increases with illumination level. Concentrators reduce the requirements for semiconductor materials, so that elements that are apparently less abundant can be utilized. The concentrator system also has application in hybrid photovoltaic/thermal (or cogeneration) systems, since there is a significant demand for *thermal energy* in the 60-100°C range.

Conventional, single-junction solar cells are limited to even
theoretical efficiencies in the 26-28% range (87). The best concentrator cell of this type has been a GaAs device with a 24.7%
efficiency under 178 suns (88). If such devices are to exceed efficiencies of 30% under very high (>1000 suns) concentration, alternate - *unconventional* - device configurations must be developed.
Three such approaches have the potential to satisfy the very high
efficiency constraint and are currently receiving a good deal of
attention in the research community. These are:

- *Multidevice, beam-splitting photovoltaic convertors.* In
  this technique, more than one cell is used in a complex
  spectrum splitting optical system. In this case, the total
  solar aperture is the same for all the devices, but each
  device responds to a different part of the solar spectrum.
  The different portions of the spectrum are provided by
  beam splitting the incident radiation using dichroic filters. Efficiencies (subtracting losses in the optical system) to 28.5% have been demonstrated using a GaAs and Si
  solar cell (89).

- *Monolithic multijunction or cascade solar cells.* This
  technique uses a complex planar technology - similar to
  integrated circuit procedures but with more layers and
  more critical technological constraints for fabrication.
  This device can provide energy conversion efficiencies in
  excess of 30%, even for a two junction device (69). Higher
  efficiencies are predicted for even more complex structures,utilizing three or more junctions. Such a prediction
  is shown in Fig. 10 (69). To date, some progress has been
  made in the demonstration of two junction devices with
  high open-circuit voltages (90), but technical problems
  still limit device performance.

Schematic representations of these two concepts are shown in Fig.
11. A third approach is the:

- *Edge multiple vertical junction device.* This technology

utilizes a vertical groove (presently in an Si wafer) that
is doped to provide the vertical junction. The groove is
approximately 10^{-3} cm across, and 10^{-2} cm deep. Grooves
are separated by ~10^{-2} cm. Efficiencies above 20% have
been demonstrated (91). Major problems with this device
are concerned with the electrical and thermal connections
to the cells. Some modeling problems and performance am-
biguities exist, and calculated attainable efficiencies
are somewhat in doubt.

The *beam-splitting approach* has several advantages over other
high-efficiency approaches. Since individual cells are used, existing technologies are sufficient to fabricate them. No complex
junction coupling is needed, and the cells can be operated independently - at *different temperatures*. Thus, different cells (i.e.,
different materials) can be incorporated into such a system by
proper temperature control for each, optimizing the performance

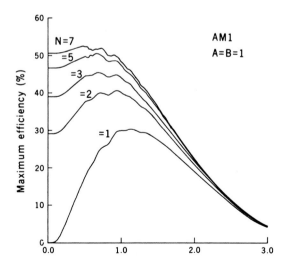

FIGURE 10.   *Multilayer cell efficiency for one to seven
absorber layers (from Ireland, et al., ref. 69).*

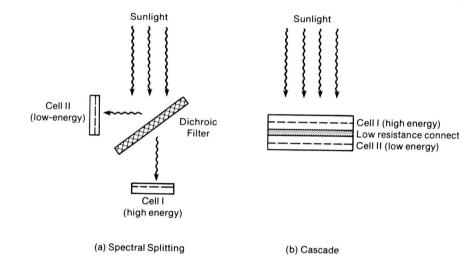

FIGURE 11.   Beam-splitting and multijunction cell concepts.

independently. However, the multidevice approach also has some
severe disadvantages, primarily due to the increased component
population: the more cells, the more cost, the more independent
reliability problems, etc. There is the high cost of using additional substrates. More research remains on optimizing individual cells to different spectral portions, and demonstrating a
greater than two-cell system. Major problems, however, are associated with the optical system, which presently contains a high
loss factor. This is especially true for the optical losses due
to the nonideal dichroic filter.

The *multijunction or cascade concentrator cell* has some con-
trasting advantages compared to the beam-splitting concept. It is
fabricated on a single substrate, and it is a two-terminal device - no matter how many junctions are stacked. This eliminates
both materials and connection costs, as well as the reliability
problems associated with such interconnections. Obviously, the
dichroic filter loss problem is avoided. However, the cell - even
for a two-junction device - is quite complex. From a fabrication
view, many layers (minimum of seven) are required, and stringent

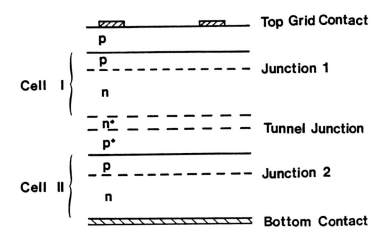

FIGURE 12.   Two-junction cascade solar cells.

material and growth parameters are needed in order to optimize
the device performance. For example, the semiconductors used in
the makeup of such a device - a two-junction version of which is
shown in Fig. 12 - must be nearly lattice matched to maintain
crystal integrity and avoid interface losses. The means to such
lattice matching is to control the composition of III-V binary
alloys. However, as shown in Fig. 13, this not only changes the
lattice parameter, but also the bandgap of the material. The
bandgaps must be precisely adjusted to match effectively the entire solar spectrum. Thus, critical material and composition
choices must be made in the device design, and these must
be strictly adhered to and controlled during device fabrication.   In addition, the layer thicknesses must be precisely
controlled to provide current continuity (i.e., avoid
current losses) in the structure.   Since many thin layers
are used in such a device, growth temperatures should
be as low as possible to minimize interdiffusion problems.

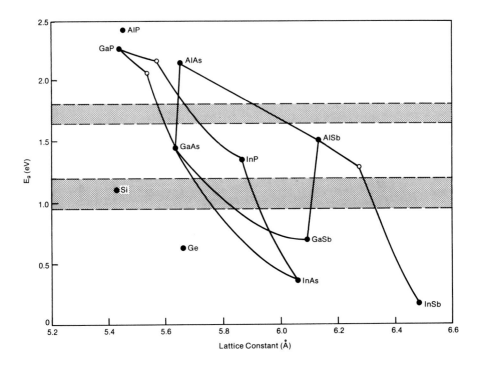

FIGURE 13.   Energy gaps of various semiconductors, indicating lattice matches.

A critical requirement for the device shown in Fig. 12 is the low series resistance connection needed between the junctions to avoid unacceptahle voltage losses.   One approach is to provide a tunnel junction.   However, very high diping levels are needed to accomplish this, and it is quite difficult to achieve such concentrations in the AlGaAs system (92).   Other methods presently under consideration involve bonding two cells together to form the stack (93).   However, only preliminary concepts and few results have been reported with such techniques.

The technological and research problems associated with these very high efficiency concentrator approaches are both varied and shared among the techniques.   Major among these are:

- The *assessment and demonstration* of various III-V semiconductor systems for efficient concentrator solar cells. Methods include LEP, CVD, MBE, VPE, or combinations of such growth techniques.
- Development of the essential *interconnect technologies* - especially the tunnel junction approach.
- Since AlGaAs cells can degrade, *encapsulants* must be developed. *Antireflection coatings* are needed to minimize reflection losses.
- *Contacting* the materials is critical. This is especially true for the high bandgap semiconductors proposed for these devices.
- The *GaAs* (or other *substrates*) must be high quality. However, low-cost methods for providing such substrates are needed.
- As with all solar cell technologies, the *degradation mechanisms* that limit cell reliability must be identified and minimized. Sunlight concentration conditions are an added factor for these devices.
- *Systems-wise,* the necessary concentrator optics (low-loss), tracking, control, thermal cooling, power conditioning, wiring, *etc.,* remain to be developed.

The concentrator solar converter has great potential as a long-range alternative in the photovoltaic area. However, it also represents an area that is in need of a good deal of R&D activity if it is to succeed.

REFERENCES

1. *National Photovoltaic Program Plan,* U.S. Department of Energy (DOE/ET-0035 (78)), Division of Solar Technology, March, 1978.

2. *Solar Photovoltaic Energy Research, Development and Demonstration Act of 1978,* U.S. Public Law 95-590.

3. Kazmerski, L.L., *Introduction to Photovoltaics: Physics, Materials and Technology* (previous chapter).

4. Sterling, H.F., and Swann, R.C.G., *Solid-State Electron. 8.* 653 (1965)

5. Chittick, R.C., Alexander, J.H., and Sterling, H.F., *J. Electrochem. Soc. 116,* 77 (1969).

6. Spear, W.E., and LeComber, P.G., *J. Non-Crystal Solids 8,* 727 (1972).

7. Carlson, D.E., U.S. Patent 4,065,521 (1977).

8. Triska, A., Dennison, D., and Fritzche, H., *Bull.* APS *20,* 392 (1975).

9. Connell,G.A.N., and Pawlik, J.R., *Phys. Rev. B. 13,*787 (1976).

10. Knights, J.C., Street, R.A., and Lucovsky, G., 8th Int. Conf. on Amorphous and Liquid Semicond., in press (1979).

11. Anderson, D.A., and Spear, W.E., Phil. Mag. *35,* 1 (1977).

12. Ovshinsky, S.R., and Madan, A., Nature *276,* 482 (1978).

13. Spear, W.E. and LeComber, P.G., *Solid-State Comm. 17,* 1193 (1975).

14. Carlson, D.E., and Wronski, C.R., *Appl. Phys. Lett. 28,*671 (1976).

15. Carlson, D.E., in "Polycrystalline and Amorphous Thin Films and Devices", (Kazmerski, L.L., Ed.), Chapter 6. Academic Press, New York, 1980.

16. Staebler, D.L., 8th Int. Conf. on Amorphous and Liquid Semicond., in-press (1979).

17. Carlson, D.E., Proc. DOE/SERI Photovoltaics Adv. R&D Annual Rev. Meeting, Denver, pp. 113-120. SERI (TP-311-428),

Golden, CO, 1979.

18. See, for example, Bosnell, J.R., in "Active and Passive Thin Film Devices" (Coutts, T.J., Ed.) pp 245-320. Academic Press, London, 1978.

19. See, for example, Stone, J., Proc. DOE/SERI Photovoltaic Adv. R&D Annual Rev. Meeting, Denver, pp 69-112. SERI (TP-311-428), Golden, CO 1979.

20. Carlson, D.E., in "Polycrystalline and Amorphous Thin Films and Devices". (Kazmerski, L.L., Ed.). Academic Press, New York, 1980.

21. Kazmerski, L.L., in Proc. DOE/NBS Workshop on Stability of (Thin Film) Solar Cells and Materials, pp 128-143. NBS Spec. Publ. 400-58, Dept. of Commerce, Washington, DC 1979.

22. Kazmerski, L.L., *Interface 2,* 6 (1979).

23. Fang, P.H., Ephrath, L., and Nowak, W.B., *Appl. Phys. Lett. 25,* 283 (1974).

24. Chu, T.L., Mollenkopf, H.C., and Chu, S.S., *J. Electrochem. Soc. 12,* 1681 (1975).

25. Chu, T.L., Lien, J.C., Mollenkopf, H.C., Chu, S.S., and Heizer, K.W., *Solar Energy 17,* 229 (1975).

26. Chu, T.L., Mollenkopf, H.C., and Chu, S.S., *J. Electrochem. Soc. 123,* 106 (1976).

27. Chu, T.L., *J. Cryst. Growth 39,* 45 (1977).

28. Chu, T.L., Chy, S.S., Kelm, R.W., and Wakefield, G.W., *J. Electrochem. Soc. 125,* 595 (1978).

29. Chu, T.L., and Singh, K.N., *Solid-State Electron. 19,* 837 (1976).

30. Chu, T.L., van der Leeden, G.A., and Yoo, H.I., *J. Electrochem. Soc. 125,* 661 (1978).

31. Barnett, A.M., and Rothwarf, A., *IEEE Trans. Electron. Dev. ED-27,* in-press (1980).

32. Barnett, A.M., 2nd European Comm. Photovoltaic Solar Energy Conf., Berlin, pp 328-343. D. Reidel Publ. Co., Berlin, 1979.

33. Barnett, A.M., Proc. 14th IEEE Photovoltaics Spec. Conf.,

IEEE, New York, 1980.

34. Kazmerski, L.L., Proc. 14th IEEE Photovoltaic Spec. Conf.,
    IEEE, New York, 1980.

35. Surek, T., Cheek, G.C., and Ariotedjo, A., in Proc. SERI/DOE
    Advanced R&D Annual Review Meeting, Denver, pp 239-256. SERI
    TP-311-428, Golden, CO, 1979.

36. Shewchun, J., Proc. Polycrystalline Si Contractors Meeting,
    Pasadena. SERI, Golden, CO, Dec. 1979.

37. Burke, D., Shewchun, J., Spitzer, M., Loferski, J.J., Scholz,
    F., Singh, R., and Kukulka, J., Proc. 14th IEEE Photovol-
    taics Spec. Conf., San Diego (IEEE, New York; 1980).

38. Hovel, H., "Solar Cells", pg 95. Academic Press, New York,
    1975.

39. D'Aiello, R.V., in Proc. Polycrystalline Si Contractors Rev.
    Meeting, Pasadena, pp 233-242. SERI, Golden, CO, Dec. 1979.

40. Card, H.C., and Yang, E.S., *IEEE Trans. Electron Dev. ED-24*,
    397,(1977).

41. Soclof, S., and Iles, P.A., Electrochem. Soc. Fall Meeting,
    New York, p 618 (1974). Also, Proc. 11th IEEE Photovoltaic
    Spec. Conf. p 56. IEEE, New York, 1975.

42. Lanza, C., and Hovel, H.J., *IEEE Trans. Electron Dev. ED-24*,
    392 (1977).

43. Hilborn, R.B., and Lin, T., Proc. Nat. Workshop on Low-Cost
    Polycrystalline Silicon Solar Cells, pp 246-254. ERDA,
    Washington, DC, 1976.

44. Ghosh, A.K., Fishman, C., and Fang, T., *J. Appl. Phys. 50*,
    (1979).

45. Ginley, D., this book.

46. Fogarassy, E., Stuck, R., Muller, J.C., Grob, A., Siffert,
    P., Salles, Y., and Diguet, D., *Solar Cells, 1*, 23 (1979).

47. Seager, C.H., and Ginley, D., *App. Phys. Lett. 34*, 337 (1979).

48. Seager, C.H., and Pike, G.E., *App. Phys. Lett. 35*, 709 (1979).

49. Matare, H.F., "Defect Electronics in Semiconductors". Wiky-
    Interscience, New York, 1971.

50. Kazmerski, L.L., *Thin Solid Films 57*, 99 (1979).

51. Kazmerski, L.L., in "Polycrystalline and Amorphous Thin Films and Devices", (Kazmerski, L.L., Ed.), Chapter 3. Academic Press, New York, 1980.

52. Brobst, D.A., and Pratt, W.P., "United States Mineral Resources", Paper 820. United States Geological Publ. Office, Washington, DC, 1973.

53. See, for example, Proc. Photovoltaics Adv. R&D Review Meeting, Denver, SERI, Golden, CO, 1979. Also, Proc. 14th IEEE Photovoltaics Spec. Conf.,San Diego. IEEE, New York, 1980.

54. Fan, J.C.C., Proc. Photovoltaics Adv. R&D Review Meeting, Denver, pp 317-323. SERI, Golden, CO,1979.

55. Fan, J.C.C., Calawa, A.R., Chapman, R.L., and Turner, G.W., *Appl. Phys. Lett.* 35, 804 (1979).

56. Fan, J.C.C., Bozler, C.O., and Chapman, R.L., *Appl. Phys. Lett.* 32, 390 (1978).

57. Ghandhi, S.K., Borrego, J.M.,Reep, D., Hsu, Y.S., and Pande, K.P., *Appl. Phys. Lett.* 34, 669 (1979).

58. Pande, K.P., Hsu, Y.S., Borrego, J.M., and Ghandhi, S.K., *Appl. Phys. Lett.* 33, 717 (1978).

59. Kazmerski, L.L., and Ireland, P.J., *J. Vac. Sci. Technol.* 17, in-press, (1980).

60. Kazmerski, L.L., Ireland, P.J., Chu, S.S., and Lee, Y.T., *J. Vac. Sci. Technol.* 17, (in press) (1980).

61. See, for example, Proc. NBS/DOE Workshop on Stability of (Thin Film) Solar Cells and Materials, NBS Spec. Publ. 400-58. U.S. Dept. of Commerce, Washington, 1979.

62. Reynolds, D.C., Leies, G., Antes, L.L., and Marburger, R.E., *Phys. Rev.* 96. 533 (1954).

63. Barnett, A.M., Bragagnolo, J.A., Hall, R.B., Phillips, J.E., and Meakin, J.D., Proc. 13th IEEE Photovoltaic Spec. Conf. pp 419-420. IEEE, New York, 1978.

64. Wagner, S., Shay, J.L., Migliorato, P., and Kasper, H.M., *Appl. Phys. Lett.* 25, 434 (1974).

65. Kazmerski, L.L., White, F.R., and Morgan, G.K., *Appl. Phys. Lett. 29*, 268 (1976).

66. Kazmerski, L.L., in "Ternary Compounds 1977" (G. Holah, Ed.) pp 217-225. Instit. of Phys., London, 1977.

67. Loferski, J.J., Shewchun, J., Roessler, B., Beaulieu, R., Piekoszewski, J., Gorska, M., and Chapman, G., Proc. 13th IEEE Photovoltaic Spec. Conf. pp 190-193. IEEE, New York, 1978.

68. Shewchun, J., Garside, B.K., Polk, D., Loferski, J.J., Burke, D., and Beaulieu, R., Proc. 14th IEEE Photovoltaic Spec. Conf., San Diego. IEEE, New York, 1980.

69. Ireland, P.J., Wagner, S., Kazmerski, L.L., and Hulstrom, R.L., *Science 204*, 611 (1979).

70. Stanley, A.G., "Cadmium Sulphide" (Vol. 1 and 2). JPL Doc. 78-77. Jet Propulsion Laboratory, Pasadena, CA, 1978.

71. Meakin, J.D., Proc. CdS/Cu_2S and CdS/Cu-Ternary Compound Program, Denver, pp 79-110. SERI, Golden, CO 1979.

72. Kazmerski, L.L., Ireland, P.J., White, F.R., and Cooper, R., Proc. 13th IEEE Photovoltaic Spec. Conf., pp 184-189. IEEE, New York, 1978.

73. Chu, T.L., Chu, S.S., Lin, C.L., Tzeng, Y.C., Kuper, A.B., and Chang, C.T., Proc. 14th Photovoltaic Spec. Conf San Diego. IEEE, New York, 1980.

74. Bachmann, K.J., Sinclair, W.R., Thiel, F.A., Schreiber, H., Schmidt, P.H. Spencer, E.G., Buehler, E., and Feldman, W.L., Proc. 13th IEEE Photovoltaics Spec. Conf., pp 524-527. IEEE, New York, 1978.

75. Shay, J.L., Bettini, M., Wagner, S., Bachmann, K.J., and Buehler, E., Proc. 12th IEEE Photovoltaics Spec. Conf. IEEE New York, 1976.

76. Lincot, D., Mimilya-Arroyo, J., Triboulet, R., Marfaing, Y., Cohen-Solal, G., and Barbe, M., Proc. 2nd E.C. Photovoltaic Solar Energy Conf. pp 424-431. D. Reidel Publ. Co., Holland, 1979.

77. Bube, R.H., Buch, F., Fahrenbruch, A.L., Ma, Y.Y., and Mitchell, K.W., *IEEE Trans. Electron Dev. ED-24*, 487 (1977).

78. Nakayama, N. Matsumoto, H., Yamaguchi, K., Ikegami, S. and Hioki, Y., *Jap. J. Appl. Phys. 15*, 2281 (1976).

79. Olsen, L.C., Bohara, R.C., and Urie, M.W., *Appl. Phys.Lett. 34*, 47 (1979).

80. Wang, E.Y., Trivich, D., Komp, R.J., Huang, T., and Brinker, D., Proc. 14th IEEE Photovoltaic Spec. Conf. San Diego, CA. IEEE, New York, 1980.

81. Catalano, A., Bhushan, M., and Convers-Wyeth, N., Proc. 14th IEEE Photovoltaics Spec. Conf. San Diego. IEEE, New York. 1980.

82. Lozykowski, H., Czech, *J. Phys. B13*, 164 (1963). Also, see ref. 91, pp 387-416.

83. Rud, Y.V., and Ovezov, K., *Sov. Phys. Semicond. 10*, 561 (1976).

84. Dovletmuradov, C., Ovezov, K., Prochukhan, V.D., Rud, Y.V., and Serginov, M., *Sov. Phys. Tech. Phys. Lett. 1*, 382 (1975).

85. See ref. 91, pp 387-416.

86. Heeger, A.J., and MacDiarmid, A.G., Proc. DOE/SERI Photovoltaics Adv. R&D Annual Rev. Meeting, pp 417-422. SERI, Golden, CO, 1979.

87. See discussion in previous chapter. Also, Loferski, J.J., *Proc. IEEE 51*, 677 (1958).

88. Sahai, R., Edwall, D.D., and Harris, J.S., *Appl.Phys. Lett. 34*, 147 (1979).

89. Moon, R.L., James, L.W., Vanderplas, H., Yep, T.O., Chai, Y.G., and Antypas, T.A., Proc. 13th IEEE Photovoltaics Spec. Conf. pp 859-867. IEEE, New York, 1978.

90. Bedair, S.M., Phatak, S.B., and Hauser, J.R., Proc. 14th IEEE Photovoltaics Spec. Conf., San Diego, CA. IEEE, New York, 1980. Also, Appl. Phys. Lett. *34*, 38 (1979).

91. See, for example, Mitchell, K.W., Proc. DOE/SERI Photovoltaics Adv. R&D Annual Meeting, pp 329-352. SERI,Denver,CO, 1979.

92. See ref. 89, p 861.

93. See, for example, Mitchell, K.W., Proc. DOE/SERI Advanced R&D Annual Meeting, Denver, pp 329-352. SERI, Golden, CO, 1979.

94. See, for example, Proc. NBS/DOE Workshop on Stability of (Thin Film) Solar Cells and Materials, NBS Spec. Publ. 400-58, (Sawyer, D.E., and Schafft, A.A., Eds.). U.S. Dept. of Commerce, Washington, DC, 1979.

95. Wilson, J.I.B., McGill, J., and Kinmond, S., *Nature* *272*, 153 (1978).

96. Hamakawa, Y., Okamota, H., and Nitta, Y., *App. Phys. Lett.* *35*, 187 (1979).

CHAPTER 17

HETEROJUNCTIONS FOR THIN FILM SOLAR CELLS[1]

Richard H. Bube

Department of Materials Science and Engineering
Stanford University
Stanford, California

I.   INTRODUCTION

There are five basic types of semiconductor junction
systems that may be used for the photovoltaic conversion of
solar energy.

Homojunctions consist of p-n junctions of a single semiconductor material. The best known example is the standard
p-n junction silicon solar cell. In general homojunctions
exhibit a high junction efficiency, but particularly in direct
bandgap materials with a high absorption coefficient, there
are appreciable losses due to front surface recombination,
which reduce the overall solar efficiency of the device by
decreasing the collected current.

Heterojunctions consist of p-n junctions of two different
semiconductors: a small bandgap semiconductor in which optical

[1]Research carried out at Stanford University and described in
this paper was supported by the Division of Materials Science,
Office of Basic Energy Sciences, Department of Energy; by the
Solar Energy Research Institute; and by subcontract from Rockwell
International.

SOLAR MATERIALS SCIENCE

585

absorption takes place, and a large bandgap semiconductor that
acts as a window for the junction. The active portion of the
solar spectrum is that lying between the bandgaps of the two
materials. In the special case that the two semiconductors have
a good lattice constant match so that the density of interface
states is a minimum, heterojunctions may exhibit the optimum
properties of a homojunction constructed from the small bandgap
material without the problems of front surface recombination loss.
If the interface cannot be effectively ignored, however, the
properties of the junction may be deleteriously affected; normally
this takes the form of a reduction in open-circuit voltage
without any necessary large decrease in short-circuit current.

Buried homojunctions consist of p-n homojunctions with an
additional heteroface junction with a large bandgap semiconductor
that acts as a window material for the p-n junction. The high
recombination loss front surface of the p-n homojunction is
replaced by the interface with the large bandgap semiconductor
which may provide improved properties. Buried homojunctions
combine several positive aspects of the corresponding homojunctions and heterojunctions.

Schottky barriers consist of metal-semiconductor junctions
in which a blocking contact is formed. Performance is usually
limited by large thermionic emission currents that reduce the
open-circuit voltage.

*Metal-insulator-semiconductor (MIS) and semiconductor-
insulator-semiconductor (SIS) junctions* are equivalent to the
corresponding Schottky barrier and heterojunction systems,
respectively, with the addition of a thin layer of insulator,
usually an oxide, at the interface to reduce the forward currents
that reduce the open-circuit voltage.

In this chapter we consider the basic properties of
heterojunctions and heteroface junctions, and provide examples of

heterojunctions with good lattice match (CdS/InP), heterojunctions
with poor lattice match (CdS/CdTe, ZnO/CdTe), and buried
homojunctions (ITO/CdTe, ITO/InP).

II. CHARACTERIZING A HETEROJUNCTION SOLAR CELL

In order to obtain an overview of the various properties
that are significant for determining the performance of a typical
heterojunction solar cell, it is useful to consider the types of
measurements and considerations that enter the characterization
of an experimental heterojunction cell.

Bulk properties of the component materials. General
information is needed on the basic electronic properties of the
materials involved: purity, carrier type and density, carrier
mobility, absorption constant as a function of photon energy,
minority carrier lifetime, minority carrier diffusion length.
This information can be obtained by analysis using the electron
microprobe or mass spectrography, and by measurements of Hall
effect, thermoelectric power, optical transmission, electron
beam induced currents (EBIC), Schottky barrier capacitance.
The optically absorbing semiconductor is usually chosen to be a
p-type material because of the longer diffusion lengths for
electrons, and to have a bandgap of the order of 1.4 eV, the
optimum for match with the solar spectrum.

In order to construct an Anderson abrupt junction model
of the heterojunction (1), it is useful also to know the
electron affinities of the materials to be used in forming the
heterojunction. This knowledge allows a choice of materials that
will not have an obvious spike in the carrier transport band that
would limit collection of photogenerated current, and a calculation of the diffusion potential of the junction (2). Since this
junction model assumes no interface states, its predictions must

be considered to have only heuristic purpose.

Crystallographic structure and orientation of component materials. This information is of particular significance for thin film solar cells made from polycrystalline materials. Techniques of X-ray diffraction, scanning electron microscopy (SEM), and transmission electron microscopy (TEM) can be used to obtain data on grain size and film morphology.

Effects of surface treatment on material properties. Knowledge of the bulk properties and the surface properties of the isolated component materials is not sufficient to determine what will happen when a junction is formed. Indeed, the very process used to form the junction may have a major effect on the properties of the junction. As preliminary background for this kind of effect, it is important to know the effect of various kinds of surface treatments on the materials to be used in junction formation. Such surface treatments would include heat treatment in an oxidizing atmosphere, a neutral atmosphere, vacuum, and a reducing atmosphere. Questions to be answered include: Is a surface layer formed? What is it? How will it affect junction properties if such a layer is formed during junction formation? Structural information can be obtained from X-ray diffraction and SEM, chemical information from Auger spectroscopy, and electronic information from measurements of surface conductivity, Hall effect, thermoelectric power, photovoltage, and luminescence. The formation of Schottky barriers on various surfaces can be a valuable diagnostic approach.

Contact resistivities. The contact resistivities of all contacts should be measured in both dark and light. If the expected total series resistance of the cell is estimated by adding all contact and bulk resistances, measurements should be in agreement if the cell is understood.

Dark current vs voltage measurements. Basic information

about the electrical properties of the junction can be obtained
by comparing measured values of current vs voltage with standard
forms for idealized situations:

$$J = J_o \{\exp(qV/AkT) - 1\} \tag{1}$$

$$J = J_o \{\exp(\alpha V) - 1\} \tag{2}$$

A relation such as that given in Eq. (1) corresponds to the
ideal junction behavior in which the diode current is controlled
by diffusion ($A = 1$) or recombination ($A = 2$), while Eq. (2) is
the junction behavior if tunneling currents dominate. These
measurements result in the evaluation of A or α, J_o, and the
series resistance R_s, and the shunt resistance R_p. These
measurements should then be repeated as a function of temperature
and the approximate model for junction transport can be developed.

Light current vs voltage measurements. These standard
measurements enable one to obtain values for the principal
parameters defining solar cell performance: short-circuit current
J_{sc}, open-circuit voltage, V_{oc}, fill factor, and solar efficiency.
These measurements should be repeated as a function of temperature. From the extrapolated value of a V_{oc} vs T plot, one obtains
a measure of the diffusion potential from the value of V_{oc}
corresponding to $T = 0^{\circ}K$.

Light diode parameters. At a particular temperature, a
comparison of the dark J vs V data with the variation of J_{sc} vs
V_{oc} obtained by varying the light intensity, indicates whether
or not the junction transport phenomena are light dependent.
This measurement allows the determination of light values for
A or α, J_o, R_s or R_p, if these are different from the values in
the dark.

Spectral dependence of quantum efficiency. The quantum
efficiency is measured as a function of photon energy. In

addition to the direct data obtained in this way, the variation
of quantum efficiency can often be a simple diagnostic for the
type of junction involved. In an ideal heterojunction the
quantum efficiency will simply be high and constant between the
bandgaps of the two materials. If the minority carrier diffusion
length is small in the absorbing material, the quantum
efficiency decreases toward lower photon energies for which the
absorption coefficient is also lower. If there is appreciable
carrier loss through recombination at interface states, the
quantum efficiency increases under applied reverse bias in a way
that is independent of photon energy. If, instead of a
heterojunction, a particular device is actually a buried
homojunction, the quantum efficiency will decrease toward
higher photon energies due to front interface recombination
on the homojunction. We give examples of these phenomena later
in this chapter.

Reverse breakdown voltage. The junction voltage for
reverse breakdown is measured in dark and light, and as a function of temperature, for comparison with diode models.

Light intensity dependence of cell parameters. Values of
J_{sc}, V_{oc}, fill factor, and solar efficiency are measured over a
wide range (many orders of magnitude) of light intensity. Non-
linear variation of J_{sc} with light intensity indicates a light-
sensitive collection process. Non-linear variation of V_{oc}
with $\ln(I_L/I_o)$ indicates changes in A or α with light intensity.
Comparison of the variation of the fill factor with light
intensity with that of an ideal cell indicates series resistance,
shunt resistance, and voltage-dependent collection phenomena (3).

Junction capacitance measurements. The junction properties
are explored in dark and light by measuring the junction
capacitance as a function of applied voltage. If a simple
junction is approximated, three important pieces of information

are obtained from a $1/C^2$ vs V plot: the carrier density from the slope in reverse bias, the zero-bias depletion layer width from the capacitance for zero bias, and a measure of the diffusion potential from the intercept on the voltage axis.

Electron beam induced current (EBIC). These measurements perform at least two significant functions. By measuring the EBIC signal as a function of distance away from the junction, the minority carrier diffusion length can be measured. By comparing the location of the maximum EBIC signal with the location of the metallurgical junction, it is possible to distinguish between heterojunction and buried homojunction systems.

Photoluminescence and electroluminescence. As indicated above, photoluminescence can be used to indicate the nature of semiconductor surfaces and changes with processing. A particular use is to measure photoluminescence of the p-type material in a junction before and after deposition of the large bandgap n-type material. Changes in luminescence emission may indicate the type of surface effects produced by the deposition method used. Similarly forward bias may be applied and electroluminescence measured, if present, as an indication of the type of defects important for recombination under forward bias conditions.

Chemical composition across the junction. A variety of surface techniques may be used to determine the variation of chemical composition of the material across the metallurgical junction, thus revealing diffusion effects, new compound formation etc.

Physical and structural information on the interface. Detailed information of a physical and structural type can be obtained by applying techniques of TEM to junction interfaces. This is often a rather difficult enterprise since the preparation of suitable samples may be a long and involved task.

III.   TYPICAL HETEROJUNCTION BAND DIAGRAMS

Two extreme cases for typical heterojunctions exist: a
highly conducting p-type optical absorbing material with a
less conducting n-type window material, or a highly conducting
n-type window material on a less conducting p-type absorber.
In the former case, the depletion layer is essentially all in
the window material, no appreciable drift field exists in the
p-type material to aid in collection of carriers, and modulation
of localized charge near the interface in the window material
can modulate the barrier properties. In the latter case, the
depletion layer is essentially all in the p-type absorber, the
depletion field aids in the collection of carriers, and the
n-type window material acts primarily as a unaffected contact
and junction former.

Examples of these two types of junctions are given in
FIGURES 1 and 2, which give approximate band diagrams in the
abrupt junction model. FIGURE 1 shows the Cu_2S/CdS system in
which degenerate p-type Cu_2S is formed, either by displacement
of Cd by Cu in solution or by chemical reaction of CuCl on the
surface of CdS, on n-type single crystal or thin film CdS with
a lower electron density. Cu_2S has a bandgap of about 1.2 eV,
while CdS has a bandgap of 2.4 eV. Lattice mismatch causes the
existence of interface states. Diffusion of Cu into the CdS
causes a broadened depletion layer region in the CdS. If
positive charge is captured in localized states in the CdS near
the junction as the result of photoexcitation, the potential
profile at the junction can be appreciably altered. The condition labeled *enhanced* in FIGURE 1 corresponds to the situation
where positive charge is localized in the CdS near the interface
and the depletion layer is narrowed; the condition labeled
quenched corresponds to the situation where no positive charge is

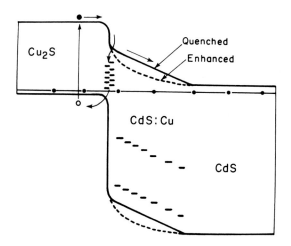

FIGURE 1. *Energy band diagram for* Cu$_2$S/CdS *heterojunction.*

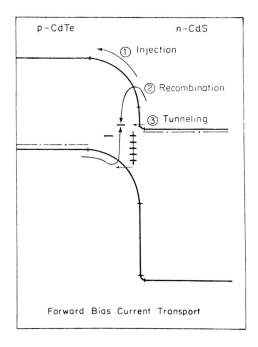

FIGURE 2. *Energy band diagram for* CdS/CdTe *heterojunction.*

localized and the depletion layer is widened (4,5).

FIGURE 2 shows the band diagram for an ideal CdS/CdTe heterojunction, for which the electron density in the CdS is normally larger than the hole density in the CdTe. The bandgap of CdTe is 1.4 eV. This diagram also illustrates the three principal modes of forward bias transport: (1) injection or diffusion, (2) recombination through imperfections or interface states, and (3) tunneling through imperfections or interface states. CdS and CdTe have about 9% lattice mismatch and a relatively high density of interface states is predicted. A tunneling mode of diode transport seems to occur in many hetero-junction systems, with the result that the value of J_o is larger than would be predicted from either diffusion or recombination models.

IV.   CdS/CdTe HETEROJUNCTIONS

Solar cells made from the component materials n-type CdS and p-type CdTe have been made by a variety of processes, involving the deposition on single crystal CdTe of thin film CdS by vacuum evaporation, spray pyrolysis, or chemical vapor deposition. Solar efficiencies up to 12% have been reported for such cells.

CdS/CdTe prepared by vacuum evaporation of CdS have been reported with a V_{oc} = 0.63 V, J_{sc} = 16.1 mA/cm^2 and solar efficiency of 8% (6). The spectral response of the quantum efficiency for this cell is shown in FIGURE 3. It is evident that the cell is a genuine heterojunction. The quantum efficiency decreases toward lower energies because of a small electron diffusion length of 0.4 μm in the CdTe, and reverse bias enhancement of the quantum efficiency is consistent with interface recombination loss of photoexcited carriers.

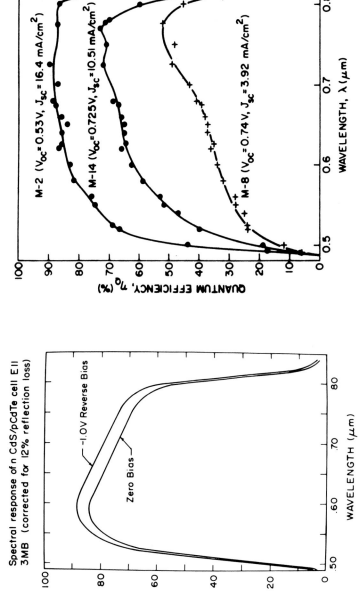

FIGURE 4. Spectral dependence of the quantum efficiency for several CdS/CdTe cells prepared by spray pyrolysis deposition of CdS on single crystal CdTe (7).

FIGURE 3. Spectral dependence of the quantum efficiency for a CdS/CdTe cell prepared by vacuum evaporation of CdS on single crystal CdTe (6).

Other CdS/CdTe cells were prepared by deposition of CdS
by spray pyrolysis onto similar CdTe single crystal substrates.
Variables were the temperature of the substrate during deposition
and the temperature of the heat treatment after deposition (7).
The spectral response for the quantum efficiency of three of the
cells are given in FIGURE 4.   The cell with the highest
efficiency has an essentially heterojunction spectral response,
with V_{OC} = 0.53 V, J_{SC} = 16.4 mA/cm^2 and a solar efficiency of
6.5% without correction for reflection losses.   Cells with
reduced quantum efficiency are also obtained with values of
open-circuit voltage as high as 0.74 V; these cells show a
spectral response more like that expected for a buried homojunction.   Survey of a variety of cells prepared with different
values of the temperature variables indicates a more or less
monotonic decrease in short-circuit current if the cells are
ordered according to an increase in open-circuit voltage.   The
increase in open-circuit voltage with a decrease in short-circuit
current is also evidence for the development of a buried homojunction.

CdS/CdTe heterojunctions prepared by epitaxial vapor growth
of CdS on CdTe in H_2 show V_{OC} = 0.67 V, J_{SC} = 18.4 mA/cm^2, and
a solar efficiency of 12% (8).   These junctions, however, are
reported to be buried homojunctions, presumably because of the
diffusion of indium donors from the CdS into the CdTe in the
process of preparation.   There is some evidence that simple
heat treatment of p-type CdTe in H_2 is sufficient to convert the
surface to n-type (9).   Similar results are indicated for a
CdS/CdTe junction prepared by screen printing techniques with
V_{OC} = 0.69 V, J_{SC} = 31.1 mA/cm^2 (for 140 mW/cm^2 radiation), and
solar efficiency of 8.1%; EBIC measurements indicate that the
method of preparation has produced a buried CdTe homojunction
with the electrical junction about 1 μm inside the CdTe away from

the metallurgical junction.

V.   CdS/InP HETEROJUNCTIONS

Some of the difficulties with the CdS/CdTe system may be
attributed to the lack of lattice constant match between these
two materials; still it should be noted that this did not prevent the realization of high quantum efficiencies, although it
did reduce the open-circuit voltage in the heterojunction configurations. A major improvement should be expected in the
CdS/InP system, for there exists an excellent lattice match
between CdS and InP. The lattice constant of zincblende InP
is 5.869A, and the corresponding parameter ($2^{\frac{1}{2}}a$) of wurtzite CdS
is 5.850A, yielding only 0.32% lattice mismatch between the (111)
plane of InP and the basal plane of hexagonal CdS. In addition,
the tetrahedral atomic distance in InP is 2.533A and in CdS is
2.532A, indicating that the orientation of the InP is not
critical.

The first cells were prepared by vacuum evaporation of CdS
on single crystal InP and showed an efficiency of 12.5%, thus
apparently vindicating the above reasoning (10). Heat treatment of this type of cell increased the open-circuit voltage
and gave a solar efficiency of 14%. Further improvement was made
by going to a chemical vapor deposition of CdS on InP using an
open-tube H_2S/H_2 flow system (11-13). Values were obtained as
follows: V_{oc} = 0.79 V, J_{sc} = 18.7 mA/cm^2 and solar efficiency of
15.0%. Although there was no doubt that these were indeed
heterojunctions, capacitance data indicated an abrupt junction
with a diffusion voltage close to the value of 1.25 V characteristic of an InP homojunction. Thus near-to-optimum heterojunction
performance was being achieved; calculation of the solar
efficiency assuming properties of an InP homojunction indicates

an efficiency of 17.2%, only slightly larger than that realized.

Some of the effect of additional states at the interface is seen by attempts to prepare CdS/InP junctions on polycrystalline InP films. The striking result is that almost the same short-circuit current can be obtained, but the open-circuit voltage is reduced due to increased forward leakage currents. Cells were obtained by chemical vapor deposition of CdS onto poly-crystalline InP (on p^+-GaAs on carbon) with V_{oc} = 0.46 V, J_{sc} = 13.5 mA/cm^2 and a solar efficiency of 5.7%. FIGURE 5 compares CdS/InP cell properties for cells prepared on single crystal InP to those prepared on polycrytalline InP (14,15). Only a relatively small increase in quantum efficiency results from use of single crystal InP, indicating that recombination loss at the interface is small even for these randomly oriented InP grains. FIGURE 5 also shows some response for wavelengths longer than the absorption edge of InP, which might be inter-preted as indicating band tails in the polycrystalline InP grains; the open-circuit voltage increases as the degree of band-tailing decreases. FIGURE 6 shows the reason that the open-circuit voltage is so much lower in the single crystal InP cells; the diode current is about 100 times larger for the polycrystal-line InP.

CdS/InP junction cells have also been prepared by the vacuum evaporation of CdS onto single crystal p-type InP homoepitaxial layers grown by a metalorganic chemical vapor deposition method on p^+ single crystal InP substrates (16). The best cell had an efficiency of 11.9% without an antireflection coating, a typical heterojunction spectral response as shown in FIGURE 7, and photovoltaic parameters summarized in Table I.

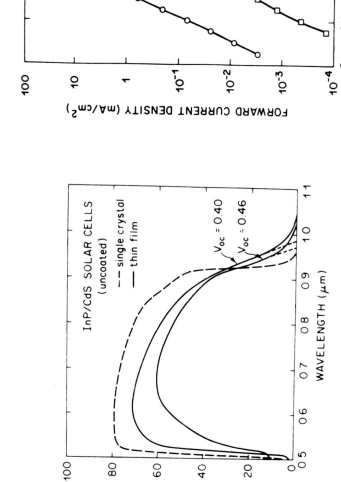

FIGURE 6. Forward bias current-voltage curves for CdS/InP junctions using single crystal InP and using polycrystalline thin film InP (15).

FIGURE 5. Spectral response of the quantum efficiency for CdS/InP cells on single crystal InP and on polycrystalline thin film InP (15).

TABLE I. *Photovoltaic Parameters for CdS/InP and ITO/InP Heterojunctions (16)*

Parameter[a]	CdS/InP	ITO/InP
V_{oc}, V	0.68	0.69
J_{sc}, mA/cm^2	25.5	23.4
Fill factor	0.62	0.65
Solar efficiency, %	12.4	12.4
Diode factor A	2.99	2.35
J_o, A/cm^2	1.3×10^{-6}	1.6×10^{-7}

[a] *Illumination by solar simulator at 85 mW/cm^2.*

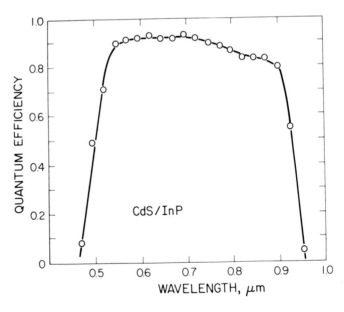

FIGURE 7. *Spectral response of the quantum efficiency for a CdS/InP junction prepared by vacuum evaporation of CdS on single crystal homoepitaxial InP layer deposited by chemical vapor deposition on p^+-InP substrate (16).*

VI.   OXIDE/SEMICONDUCTOR PHOTOVOLTAIC JUNCTIONS

Several different oxides exist that are high conductivity, large bandgap n-type materials, which therefore appear to be ideal candidates for the window material on a heterojunction. Other oxides, as mentioned in Section I, play an important role in MIS and SIS devices.

Compared to the CdS/CdTe heterojunction, for example, a degenerate indium-tin oxide (ITO)/CdTe heterojunction would have a bandgap window 55% larger because of the larger bandgap of the ITO (2.98 eV when non-degenerate but effectively up to over 3.4 eV when strongly degenerate) (17), and a diffusion potential 40% larger as estimated from electron affinities. Almost exactly the same kind of expectation would hold also for a ZnO/CdTe heterojunction.

VII.   ZnO/CdTe HETEROJUNCTIONS

ZnO/CdTe heterojunction solar cells may be prepared by deposition of ZnO by spray pyrolysis on single crystal CdTe substrates (18). Once again, as in the case of the deposition of CdS/CdTe junctions by this process, the critical variables are the temperature of the substrate during deposition and during post-deposition heat treatment in hydrogen.

FIGURE 8 shows the development of the photovoltaic proper-ties of the ZnO/CdTe junction with variations in the substrate temperature during deposition(19). The parameters are clearly very sensitive to the substrate temperature, reaching a maximum solar efficiency of 9.2% for a substrate temperature of 460°C, but being less than half of this at 430°C and showing definite decrease at 470°C. Examination of FIGURE 8 shows that the critical photovoltaic parameter is, as usual, the open-circuit

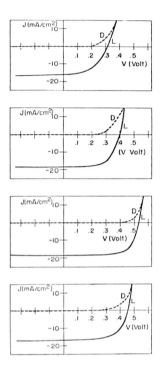

FIGURE 8.   Light current-voltage curves at different
deposition substrate temperatures for ZnO/CdTe heterojunctions
prepared by spray pyrolysis deposition of ZnO.   Substrate
temperatures from top to bottom are 430°, 445°, 460°, 470°C.

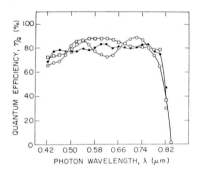

FIGURE 9.   Spectral response of quantum efficiency for
ZnO/CdTe heterojunctions prepared by spray pyrolysis of ZnO.

voltage, which varies rapidly with substrate temperature. The short-circuit current, on the other hand, varies very little with changing substrate temperature. ZnO/CdTe heterojunctions have a lattice mismatch of some 27%, but it is evident from these results and from the spectral response of quantum efficiency data given in FIGURE 9 that high quantum efficiencies are realizable even with very non-ideal heterojunctions. The specific interface conditions that allow high open-circuit voltage, however, are strongly dependent on the preparation conditions.

VIII. MIS AND SIS DEVICES

In a variety of Schottky barrier and heterojunction configurations, at least a potential gain can be achieved by inserting a suitable high bandgap insulator film, frequently an oxide, between the components of the junction. Such an insulating film often serves to increase the open-circuit voltage of the device by means of one of three possible mechanisms: an increase in the effective barrier at the semiconductor surface, an increase in the junction diode factor, or a decrease in J_o without reducing the short-circuit current. Both Si and GaAs form oxides simply by being exposed to air at room temperature; in the case of Si, the formation of SiO_x is a well-integrated aspect of Si technology.

Examples of three MIS devices prepared with Si and three different metals are given in Table II. Cells prepared to date have an efficiency between 8 and 9%, as compared to an estimated efficiency of 13 to 16% maximum for such devices. The effects of an interface oxide layer have also been found critical in improving Schottky barrier performance on GaAs (20,21). FIGURE 10 shows the dramatic effect on the open-circuit voltage of various stages of oxidation of an n-GaAs surface before applica-

TABLE II. MIS Devices on Silicon

Silicon	SiO_2 Thickness,Å	Metal	J_{sc}, mÅ/cm^2	V_{oc}, volts	Solar Eff., %	Ref.
2 ohm-cm p-type	15	Cr	26	0.50	8.1	(22)
3-15 ohm-cm p-type	20-40	Al	26.5	0.45	8	(23)
1-10 ohm-cm n-type	10-40	Au	22	0.55	9	(24)

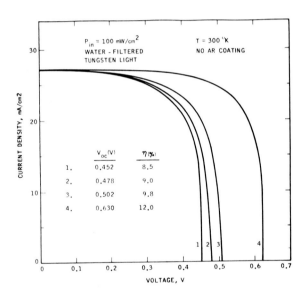

FIGURE 10. Light current vs voltage curves for Au/GaAs Schottky barriers for various treatments of the GaAs surface before application of the metal barrier contact. (1) "Clean" interface, (2) exposed to air at 300°K for 4 hr, (3) exposed to air at 300°K for 94 hr, and (4) exposed to air at 403°K for 70 hr (20).

tion of a Au barrier metal contact. In this case, the magnitude of J_o was found to actually increase with oxide thickness, but the increase in the diode factor caused by the oxide layer more than compensated for this effect and yielded a net increase in open-circuit voltage. Solar efficiencies up to 15% have been reported. In this system, the use of a suitable antiflection coating is essential because of the average 45% reflectivity of the Au/GaAs surface.

Energy band diagrams for MIS and SIS devices are shown in FIGURES 11 and 12 respectively. Increases in barrier height and diode factor appear adequate to explain many of the observed effects to date, but there is evidence that control of the transport currents may also be an important factor, particularly in ITO/Si junctions (25-28). The argument for this behavior goes as follows. If the thickness of the oxide layer is greater than 30A, tunneling currents through the oxide (minority electron flow from the p-type Si in a typical ITO/p-Si junction) is very small, and the ITO and Si are essentially in thermal equilibrium, corresponding to a capacitor with an oxide dielectric. If the oxide layer is less than 10A thick, then the oxide plays little role and a standard Schottky barrier results. If, however, the oxide layer thickness is between 10A and 30A, the tunneling current is large enough to cause the semiconductors to depart from thermal equilibrium and tunnel SIS diodes are formed. Over a certain range of bias voltage, such non-equilibrium SIS devices have a diode current that is controlled by the generation-recombination mechanism in the bulk, and the tunnel current simply provides an Ohmic contact, thus giving a decrease in the magnitude of the J_o of the device. Such an SIS model predicts a possible efficiency of 20% for ITO/Si junctions, and possibly even higher efficiencies for other semiconductor systems with more nearly optimum bandgap.

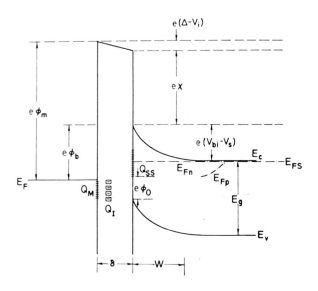

FIGURE 11.   *Energy band diagram for an MIS junction.*

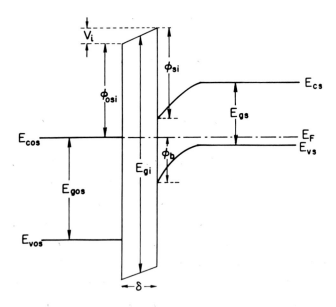

FIGURE 12.   *Energy band diagram for an SIS junction.*

IX. SPUTTERED ITO/CdTe JUNCTIONS

Junctions prepared by sputtering indium-tin oxide (ITO) onto single crystals of CdTe and InP are considered in the final two sections of this paper. As we shall see, both represent some unexpected results with opportunities for detective work using the methods of characterization summarized in Section II. It is found that the method of deposition, in this case rf sputtering at 2.5 KV, 180W, determines the characteristics of the resulting junction, as well as the properties of the two constituent materials.

Typical of the best ITO/CdTe junctions made by this process, without heat treatment, are the photovoltaic parameters: V_{oc} = 0.82 - 0.85 V, J_{sc} = 12 - 15 mA/cm^2, and solar efficiencies of about 8% (29). The large open-circuit voltage is surprising, since values as large as 0.8 V are more normally characteristic of a CdTe homojunction.

The spectral response of two ITO/CdTe cells with quite different electrical junction properties is given in FIGURE 13. A gradual decrease in quantum efficiency is observed with increasing photon energy from a maximum at approximately the bandgap of CdTe. That this decrease cannot be attributed to optical transmission through the ITO film is illustrated by the ITO transmission curve included in FIGURE 13. The shape of the response is, however, consistent with the hypothesis that a buried homojunction has been formed in the CdTe as a result of the sputtering process.

A variety of diagnostic techniques were used in order to evaluate this hypothesis beyond the circumstantial evidence of large open-circuit voltage and shape of the spectral response.

Upon room temperature aging or moderate heat treatment, it was observed that the open-circuit voltage rapidly decreased,

and that the spectral response changed to a form more consistent
with a heterojunction. It thus appears that annealing removes the
buried homojunction and causes the formation of an actual
heterojunction structure, with decreased cell performance.

In order to simulate the effect of sputtering deposition of
ITO on CdTe without actually depositing ITO, sputter etching of
CdTe surfaces was investigated. Surface photovoltage measurements
showed that the band bending at the surface of the CdTe changed
from a downward bending characteristic of a depletion layer on
a p-type surface for the unsputtered surface to an upward bending
characteristic of a depletion layer on an n-type surface with
sputter etching. Changes in the photoluminescence emission of a
CdTe surface of the same type were observed both as the result
of sputter etching and thermal annealing.

An especially useful technique was the preparation of
In/CdTe junctions on CdTe surfaces given various pre-deposition

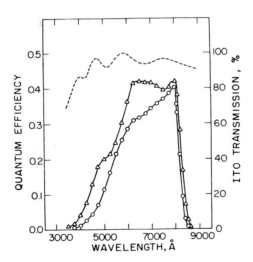

*FIGURE 13. Spectral response of the quantum efficiency for
ITO/CdTe junctions prepared by rf sputtering of ITO onto single
crystal CdTe. Dashed curve shows transmission of ITO film.*

treatments. FIGURE 14 compares the spectral response of the
quantum efficiency for an In/CdTe junction prepared by vacuum
evaporation of In onto chemically etched CdTe (a "normal"
Schottky barrier) with In/CdTe junctions prepared by vacuum
evaporation of In onto sputter etched CdTe surfaces. The latter
all show a sharp decrease in quantum efficiency with increasing
photon energy, becoming more pronounced with higher sputter
etching voltages. The same type of behavior was found for a
In/CdTe junction prepared by sputter deposition of In onto
chemically etched CdTe, as was found for vacuum evaporation of
In onto sputter etched CdTe. The open-circuit voltage for the
In/CdTe cell made on a sputter-etched surface (In contact to

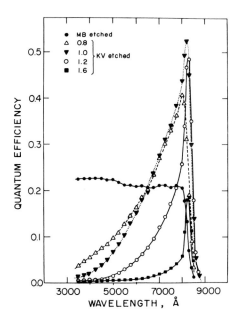

*FIGURE 14. Spectral response of an In/CdTe Schottky barrier
produced by vacuum evaporation of In onto methanol-bromine etched
p-type CdTe. Also spectral response curves for In/CdTe junctions
produced by vacuum evaporation of In onto sputter etched CdTe
surfaces using the indicated sputtering voltages.*

buried CdTe homojunction) was much larger than that for a cell
made by vacuum evaporation of In onto a chemically etched surface;
the 0°K extrapolated value of open-circuit voltage was 1.25 eV
for the sputter-etched surface, compared to 0.85 eV for the
chemically etched surface.

EBIC measurements were also carried out on non-heated
ITO/CdTe junctions. These showed that the maximum current was
observed inside the CdTe, about 1 µm away from the metallurgical
interface.

All of these evidences lead to the conclusion that when an
ITO/CdTe junction is formed by rf sputtering of the ITO, a surface
layer of the p-type CdTe is converted to n-type as a consequence
of thermally induced changes in the defect structure. The result
is a buried homojunction with the band diagram given in FIGURE 15.

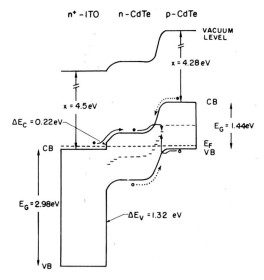

*FIGURE 15. Proposed energy band diagram for an ITO/CdTe
buried homojunction produced by rf sputtering of ITO onto the
surface of a p-type CdTe crystal, showing the n-type layer formed
by thermal effects during sputtering. The diagram is drawn in
the simple form of an abrupt junction.*

X. SPUTTERED ITO/InP JUNCTIONS

Table I compares photovoltaic cell parameters for cells of CdS/InP and ITO/InP, both made on the same homoepitaxial single crystal layers of InP and both exhibiting almost identical properties. The high performance of the CdS/InP cells was attributed to the good lattice match between CdS and InP. What was the reason for the good performance of the ITO/InP cells prepared by sputter deposition of ITO, in view of the fact that ITO and InP have very poor lattice match? Actually even better performance has been reported for similar cells of ITO/InP with solar efficiency of 14.4% (30). A direct clue is obtained by comparing the spectral response of quantum efficiency for the ITO/InP cells of Table I, as given in FIGURE 16, with the spectral response for the CdS/InP cells given in FIGURE 7. It immediately appears that the ITO/InP cells are not heterojunctions, but

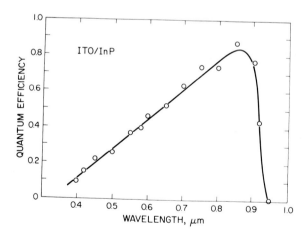

FIGURE 16. Spectral response of the quantum efficiency for the ITO/InP cell of Table I, prepared by rf sputtering of ITO onto homoepitaxial single crystal layer of p-type InP (16).

buried homojunctions, and that the effects described in the previous section for ITO/CdTe junctions also hold for ITO/InP junctions.

There are other ways, however, by which a buried homojunction might be formed. Ion probe analysis has been used to investigate the diffusion of tin impurity from the ITO into the InP during sputtering (31,32). Appreciable Sn diffusion occurs, which would dope the InP n-type, if the InP temperature during sputtering is $250^{\circ}C$, but little or none if the InP temperature is $25^{\circ}C$. Experiments indicate that thermal damage to the InP during sputter deposition of ITO is sufficient to cause conversion of the InP surface to n-type even under situations where diffusion of a donor impurity is not possible (33).

FIGURE 17 shows the spectral response of quantum efficiency for a sputtered ITO/InP cell before and after various degrees of heat treatment. Before heat treatment the spectral response has a maximum near the bandedge of InP and a decrease in quantum efficiency with increasing photon energy, such as is characteristic of a homojunction. With heat treatment there is a change in the spectral response to a heterojunction-like form, moderate heating causing a relatively large increase in quantum efficiency and then further heating causing a decrease.

EBIC measurements on ITO/InP cells made by sputtering showed that the maximum EBIC signal occurred about 1 μm away from the metallurgical junction, inside the InP. After heat treatment, however, the EBIC maximum shifted to about the position of the metallurgical junction.

If sputter deposition of ITO on InP produces a buried homojunction, then similar results should be obtained for a variety of oxides if they are all deposited on InP by the same sputtering process. It has been shown that sputtered In_2O_3, SnO_2, ZnO and CdO all give substantially the same type of junction with InP as

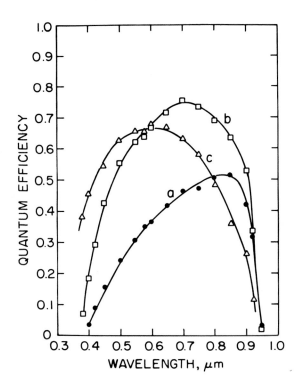

FIGURE 17. Spectral response of an ITO/InP cell produced by rf sputtering of ITO onto single crystal p-type InP. (a) Initial, (b) after heat treatment to reach optimum performance, (c) after excessive heat treatment.

does ITO. All have a spectral response of quantum efficiency that decreases rapidly for higher energy photons, all have an open-circuit voltage between 0.53 and 0.67 V, and all (except the SnO_2 cell which had series resistance problems) have a short-circuit current between 18 and 24 mA/cm^2, yielding an efficiency between 7 and 11%. The fact that such similar results are obtained for ITO and ZnO or CdO, in view of the fact that Zn and Cd are acceptors if incorporated in InP, indicates that the conductivity conversion is due to thermal effects of sputtering and not to impurity diffusion.

To further investigate surface properties of InP important for junction formation, Au/InP junctions were prepared in three different ways: (a) vacuum evaporation of Au onto chemically-etched InP, (b) vacuum evaporation of Au onto sputter-etched InP, and (c) sputtering of Au onto chemically-etched InP. Once again sputter etching was used to simulate the sputtering damage involved in oxide deposition. Au/InP junctions prepared by process (a) exhibited the characteristics of a genuine Schottky barrier with a diffusion potential (measured from the 0°K extrapolated value of open-circuit voltage) of 0.80 V. Au/InP junctions prepared by either process (b) or (c) were quite similar to one another, and quite different from the Schottky barriers produced by process (a); they exhibited a diffusion potential of 1.2 ± 0.1 V and a spectral response of quantum efficiency that was homo-junction-like. Heat treatment caused the characteristics of cells produced by processes (b) and (c) to change in the direction of becoming actual Schottky barriers.

The n-type electronic character of the sputter-etched InP surface was confirmed by direct surface Hall effect measurements. Similarly the effect on the photoluminescence of the InP surface of sputter etching was the same as that caused by sputter deposition of ITO.

An abrupt junction diagram of the n^{+}-ITO/n-InP/p-InP buried homojunction produced by rf sputtering of ITO onto InP, as suggested by the above evidence, is given in FIGURE 18. The ITO is degenerate, with the Fermi level lying about 0.2 eV above the bottom of the conduction band. The n-type InP layer has an electron density of 10^{16} cm^{-3}, assumed to be constant over the whole n-type layer. This damaged n-type layer plays a dominant role in both oxide/InP and Au/InP sputtered junctions, causing them to exhibit properties that depart from those of a simple

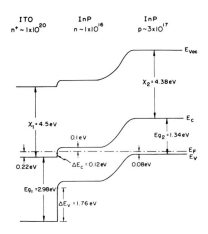

FIGURE 18.   Energy band diagram for an n^{+}-ITO/n-InP/p-InP buried homojunction produced by rf sputtering of the ITO.

heterojunction or a Schottky barrier, respectively, and approximate those of a buried homojunction structure in both types of
junction.

ACKNOWLEDGMENTS

   I would like to take this opportunity to acknowledge my in-
debtedness to my students and colleagues who have worked in the
photovoltaic area with me in recent years: Julio Aranovich, Fran-
cis G. Courreges, Leon B. Fabick, Alan L. Fahrenbruch, William G.
Haines, Y. Yale Ma, Kim W. Mitchell, and Ming-Jong Tsai.   I would
also like to thank the international scholars who have been involved in this work: Professor A. N'Diaye of the University of
Dakar, Professor M. Kasuga of Yamanashi University, Professor T.
Suda of the Institute of Vocational Training in Kanagawa,
A. Ortiz of the Centro de Investigacion de Materiales in Mexico
City, Professor F. Guastavino of Montpellier Univer-

sity, and Dolores Golmayo of IFM in Madrid.

REFERENCES

1. Anderson, R. L., *IBM J. Res. Dev. 4,* 283 (1960).
2. See, for example, Fahrenbruch, A. L., and Aranovich, J.,
   in "Topics in Applied Physics. Vol. 31: Solar Energy
   Conversion. Solid-State Physics Aspects" (B. O. Seraphin,
   ed.), p. 257. Springer-Verlag, Berlin, (1979).
3. Mitchell, K. W., Fahrenbruch, A. L., and Bube, R. H.,
   Solid-State Electronics 20, 559 (1977).
4. Gill, W. D., and Bube, R. H., *J. Appl. Phys. 41,* 3731 (1970).
5. Fahrenbruch, A. L., and Bube, R. H., *J. Appl. Phys. 45,* 1264
   (1974).
6. Mitchell, K. W., Fahrenbruch, A. L., and Bube, R. H., *J.
   Appl. Phys. 48,* 4365 (1977).
7. Ma, Y. Y., Fahrenbruch, A. L., and Bube, R. H., *Appl. Phys.
   Lett. 30,* 423 (1977).
8. Yamaguchi, K., Matsumoto, H., Nakayama, N., and Ikegami, S.,
   Japan. J. Appl. Phys. 15, 1575 (1976); *16,* 1203 (1977).
9. Werthen, J., unpublished result.
10. Wagner, S., Shay, J. L., Bachmann, K. J., and Buehler, E.,
    Appl. Phys. Lett. 26, 229 (1975).
11. Shay, J. L., Wagner, S., Bettini, M., Bachmann, K. J., and
    Buehler, E., *IEEE Trans. Electron Dev. ED-24,* 483 (1977).
12. Bettini, M., Bachmann, K. J., Buehler, E., Shay, J. L.,
    and Wagner, S., *J. Appl. Phys. 48,* 1603 (1977).
13. Bettini, M., Bachmann, K. J., and Shay, J. L., *J. Appl. Phys.
    49,* 865 (1978).
14. Bachmann, K. J., Buehler, E., Shay, J. L. and Wagner, S.,
    Appl. Phys. Lett. 29, 121 (1976).

15. Shay, J. L., Wagner, S., Bettini, M., Bachmann, K. J., and Buehler, E., *IEEE Trans. Electron Dev. ED-24,* 483 (1977).

16. Manasevit, H. M., Hess, K. L., Dapkus, P. D., Ruth, R. P., Yang, J. J., Campbell, A. G., Johnson, R. E., Moudy, L. A., Bube, R. H., Fabick, L. B., Fahrenbruch, A. L., and Tsai, M. J., *13th IEEE Photovoltaic Spec. Conf.,* p. 165 (1978).

17. Haines, W. G., and Bube, R. H., *J. Appl. Phys. 49,* 304 (1978).

18. Aranovich, J., Ortiz, A. and Bube, R. H., *J. Vac. Sci. Technol. 16,* 994 (1979).

19. Aranovich, J., Golmayo, D., Fahrenbruch, A. L., and Bube, R. H., to be published.

20. Stirn, R. J., and Yeh, Y. C. M., *Appl. Phys. Lett. 27,* 95 (1975).

21. Stirn, R. J., and Yeh, Y. C. M., *IEEE Trans. Electron Dev. ED-24,* 476 (1977).

22. Anderson, W. A., Delahoy, A. E., and Milano, R. A., *J. Appl. Phys. 45,* 3913 (1974).

23. Charlson, E. J., and Lien, J. C., *J. Appl. Phys. 46,* 3982 (1975).

24. Pompon, J. P., and Siffert, P., *J. Appl. Phys. 47,* 3248 (1976).

25. Shewchun, J., Dubow, J., Myszkowski, A., and Singh, R., *J. Appl. Phys. 49,* 855 (1978).

26. Shewchun, J., Dubow, J., Wilmsen, C. W., Singh, R., Burk, D., and Wager, J. F., *J. Appl. Phys. 50,* 2832 (1979).

27. Shewchun, J., Burk, D., Singh, R., Spitzer, M., and Dubow, J., *J. Appl. Phys. 50,* 6524 (1979).

28. Ghosh, A. K., Fishman, C., and Feng, T., *J. Appl. Phys. 49,* 3490 (1978).

29. Courreges, F. G., Fahrenbruch, A. L., and Bube, R. H., *J. Appl. Phys.* , to be published (1980).

30. Sree Harsha, K. S., Bachmann, K. J., Schmidt, P. H., Spencer, E. G., and Thiel, F. A., *Appl. Phys. Lett. 30*, 645 (1977).

31. Bachmann, K. J., Bitner, T., Thiel, F. A., Sinclair, W. R., Schreiber, Jr., H., and Schmidt, P. H., *Solar Energy Materials*, in press.

32. Bachmann, K. J., Schreiber, Jr., H., Sinclair, W. R., Schmidt, P. H., Thiel, F. A., Spencer, E. G., Pasteur, G., Feldmann, W. L., and Sree Harsha, K., *J. Appl. Phys. 50*, 3441 (1979).

33. Tsai, M. J., Fahrenbruch, A. L., and Bube, R. H., to be published.

CHAPTER 18

THE OPTIMIZATION OF SOLAR CONVERSION DEVICES[*]

D.S. Ginley
M.A. Butler
C.H. Seager

Sandia Laboratories[†]
Albuquerque, New Mexico

I. INTRODUCTION

For the materials scientist, the potential of solar energy conversion offers an array of opportunities and challenges. The
optimization of materials to achieve optimum device efficiencies
comes into play in almost any conceivable device. The use of
many simple chemical concepts can have considerable utility when
applied to many of these problems. In context, this paper will
discuss the direct and indirect application of simple chemical
potential and electronegativity arguments to two diverse systems
that have considerable potential for solar energy conversion.

First, photoelectrochemical cells (PEC) have shown potential
for the direct conversion of solar energy into useful chemical
products and/or electricity. It will be shown how the key criteria of
these cells can be delineated. Particular attention will be directed toward understanding the biasing requirements of these cells,
and to this end an electronegativity model will be presented that can
be used to quantitatively predict biasing requirements. In this
context, the importance of chemisorbed ions from the electrolyte
on the Helmholtz layer will be discussed. We will also demonstrate how the electromigration of ions in the electrode near-
surface region can substantially affect biasing requirements and
long-term stability of such photoelectrodes.

SOLAR MATERIALS SCIENCE

619

Second, thin-film polycrystalline semiconductors, especially
silicon, show tremendous potential as photovoltaic materials in
flatplate solar collectors. The economics for their deposition
are generally favorable; however, conversion efficiencies are at
present well below the level needed to make the devices economic.

One of the main causes of the reduced efficiencies, compared
to single-crystal materials, is the deleterious effects of the
grain boundaries. Grain boundaries act as recombination centers
for minority carriers as well as shunts across the p-n junction.
We will present our current picture of the energy levels in
silicon grain boundaries, and a brief discussion of the double-
depletion layer, thermal-emission model of conduction over the
potential barrier in the boundary. We will also discuss our
recent results demonstrating that various chemical agents, when
appropriately introduced into silicon grain boundaries, are
capable of modifying grain boundary potential barriers. The
prospects for increasing polycrystalline silicon solar cell
performance by these treatments will be reviewed.

II.   PHOTOELECTROCHEMICAL CELLS (PEC)

Introduction

The energy crisis of the seventies has stimulated research
in energy-related areas, particularly those useful for utilization of solar energy. One of the many research fields which show
promise for solar energy conversion is photoelectrochemistry.
A photoelectrochemical device is one in which a semiconducting
electrode is illuminated in a liquid cell and drives electrochemical reactions at both electrodes. These cells may be of two
types; one directed primarily toward the production of electricity
(wet photovoltaic cell) and one making chemical products through a
chemical change at the electrode or electrolyte. One of the more
attractive chemical reactions is the decomposition of water to
form H_2 and O_2 (photoelectrolysis).

An important aspect of applying photoelectrochemical devices
to solar energy conversion is defining the materials properties
of the light-sensitive electrode necessary to optimize its performance. In order to do this, we must delve into the details
of the mechanisms involved in the photoelectrochemical process.
As will be seen, we have not reached the point where the key
criteria for selecting an optimum electrode can be delineated,
though the problem of finding the material has not been solved.

Although photoeffects at electrodes in electrochemical cells
were first observed by Becquerel (1), it wasn't until Fujishima,
Honda and Kikuchi demonstrated (2) the decomposition of water
at a TiO_2 electrode (in 1969) that this field really commenced
to grow. Since then a number of reviews in the field have been
published (3-5).

The typical photoelectrochemical cell appears as shown in
Fig. 1. Here we illustrate a photoelectrolysis cell for decomposing water into hydrogen and oxygen. Oxidation occurs at the
photoanode and reduction at the metallic or graphite cathode.

PHOTOELECTROLYSIS CELL

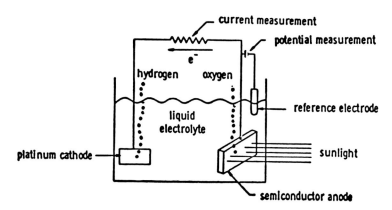

*FIGURE 1.   A photoelectrochemical cell for the decomposition
of water into hydrogen and oxygen. The cell consists of a basic
aqueous electrolyte, a photoactive semiconducting anode and a
platinum or graphite cathode. Experimentally, potential is
measured between the anode and a reference electrode and current
between the anode and the cathode.*

The easiest way to describe the operation of a photoelectrochemical cell is to examine its energy level diagram. The simplest device consists of a semiconducting electrode, a metallic electrode and a "simple" electrolyte as shown in Fig. 2.

The energy in the electrolyte at which electrons must be provided to drive the electrochemical reaction is known as the redox potential and is usually referenced to the normal hydrogen or energy position at which the conduction and valence bands for n- and p-type semiconductors, respectively, intercept the solid/electrolyte interface is known as the flatband potential V_{fb}. This is because V_{fb} is determined from the changing properties of the interface (capacitance, photocurrent, etc.) as the bands are made flat.

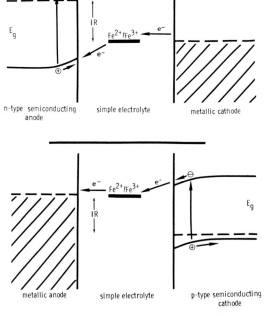

FIGURE 2. *Energy level diagram for simple electrochemical devices which produce electricity but no chemical products. The use of both n- and p-type semiconducting electrodes is illustrated. A simple electrolyte is one in which only a single electrochemical reaction can take place.*

The semiconductor can be used as a light sensitive anode or cathode depending on whether it is n- or p-type, respectively. This is determined by the need for a region depleted of majority carriers at the semiconductor surface. In the depletion region there exists an electric field which is necessary to separate spatially the optically-excited electron in the conduction band from the hole in the valence band. Thus, when illuminated with photons of energy greater than the band gap of the semiconductor an electron is excited into the conduction band and the electron and hole are separated by the electric field in the depletion region before they can recombine. The majority carrier then flows through the electrical load to the metallic electrode and drives an electrochemical reaction. The minority carriers flow to the semiconductor surface driving another electrochemical reaction.

In Fig. 2, we represent the single electrochemical reaction by the reversible ferric-ferrous couple ($F_{aq}^{3+} + e \rightleftarrows Fe_{aq}^{2+}$). In such an electrolyte the reaction is driven one way at the anode ($Fe_{aq}^{2+} \rightleftarrows Fe_{aq}^{3+} + e^-$) and the opposite direction at the cathode ($Fe_{aq}^{3+} + e^- \rightleftarrows Fe_{aq}^{2+}$). Thus, in such a system there is no net chemical change and the power produced must be extracted via the electrical load. Such cells are commonly called wet photovoltaic cells in analogy with the corresponding solid state devices.

Another type of PEC results in the production of a chemical product. Its energy level diagram is illustrated in Fig. 3. The operation of the device is the same as we have just discussed, except that two irreversible electrochemical couples are driven, with one taking place at the anode and the other at the cathode. This results in a net chemical change in the electrolyte. In the figure we illustrate this by the reactions for the decomposition of water to hydrogen and oxygen. At the cathode, the reaction is essentially ($2H^+ + 2e^- \rightleftarrows H_2$) and at the anode the reaction is ($2OH^- \rightleftarrows 2H^+ + O_2 + 4e^-$). Note that the net amount of energy stored is the difference between the two couples.

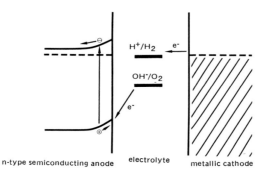

FIGURE 3. *Energy level diagram for electrochemical device which produces a chemical product. The electrolyte is illustrated for the decomposition of water in H_2 and O_2. Only an n-type semiconductor is shown, although a p-type semiconductor could be used as illustrated in Fig. 2. Normally, chemical producing cells are operated under short-circuit conditions to maximize the amount of the chemical product.*

Other types of energy storing reactions exist as well. For example, the production of methanol from CO_2 (6) and of ammonia from H_2O and N_2 (7) are attracting considerable interest.

From examining these energy level diagrams, it is possible to discern the basic criteria that a useful semiconducting electrode must satisfy. The first important aspect is efficient conversion of photons to excited electrons and their efficient utilization in the electrochemical processes. In general, in these cells the generation and separation of carriers in the semiconductor is the rate-limiting step rather than the chemical kinetics at the interface (8A). Thus, the problem becomes the same basic one as is faced in solid-state photovoltaic devices and the same factors are important. In fact, the Schottky barrier formalism has been applied successfully to the semiconductor/electrolyte interface (8). Since only photons of energy larger than the band gap of the semiconductor can be used, the band gap must be chosen to optimize the conversion efficiency. Another factor is the optical absorption depth compared to the depletion layer thickness. Since we need the electric field in the depletion layer to separate the electron-hole pairs and since most semiconductors have short diffusion lengths, it is important to absorb most of the light in the depletion layer region. The optical absorption depth will be significantly different if the gap is direct rather than indirect. The depletion layer thickness depends on doping level and dielectric constant of the semiconductor.

The second most important property of the semiconducting electrode is the location of the energy bands. Aside from determining the band gap, the energy band positions relative to the energy levels in the electrolyte also determine the maximum open-circuit voltage for the wet photovoltaic cells and the biasing requirements if any, for the chemical producing cells. This can be seen by examining Figs. 2 and 3. For the wet photovoltaic cells shown in Fig. 2, under open-circuit conditions the Fermi level in the metal electrode will equal the redox potential in the electrolyte. Under maximum illumination, the bands in the semiconductor will approach flatband condition. Thus, the maximum open-circuit voltage for the cell will be the difference between the redox potential in the electrolyte and the intercept of the conduction band with the interface (the electron affinity of the semiconductor). While the redox potential is known for most couples, the same cannot be said of semiconductor electron affinities. This means that either this information must be determined experimentally in each case or a model must be constructed to predict semiconductor electron affinities.

Similar arguments apply to the chemical producing cell shown in Fig. 3. Since we would like to operate the cell under short-circuit conditions and maximize the production of chemical products, the conduction band must intercept the interface so that a depletion region exists at short-circuit conditions. Thus, for an n-type semiconductor the electron affinity must be smaller than the cathodic redox potential in the electrolyte as measured from the vacuum level. Conversely, for a p-type

semiconductor the valence band should intercept the interface
below the anodic redox potential in the electrolyte.

The third and perhaps most crucial condition that the semiconducting electrodes must satisfy is stability under the rather
rigorous conditions in which they are operated. They must not
only be stable against chemical dissolution in the electrolyte
but also against electrochemical corrosion and photocorrosion.
Most of these effects are not well understood so that choosing
stable materials or modifying their properties or the electrolyte to induce stability is more art than science. Some
progress has recently been made in defining the conditions for
stability of semiconductors in contact with electrolytes (9,10).
The decomposition reactions are merely additional redox couples
in the electrolyte. Thus, the relative postions of these couples
with respect to the semiconductor band edges will determine the
thermodynamic stability of the semiconductor in that particular
electrolyte. However, some semiconductors which are thermodynamically unstable may be effectively stabilized if the kinetics of
the decomposition reaction are slow enough. There is also the
problem of competition between the decomposition reaction and
other possible reactions both beneficial and detrimental.
Stability is a complex question which is quite difficult to
answer even on an individual basis.

These three criteria: (1) Quantum Efficiency, (2) Potential
Behavior, and (3) Stability, define the characteristics desirable
in a semiconducting photoelectrode. The rest of this discussion
will deal with a model we have developed to describe potential
behavior in detail. We will show that the electron affinity of
a semiconductor can be calculated from the atomic electronegativities of its constituent atoms. It will be demonstrated that
the effects of chemisorbed ions are large and can be measured and
taken into account. Finally, it will be demonstrated that the
electromigration of ions into and out of the electrode surface
can affect the flatband potential of the electrode signficantly.

Potential Behavior

Similar to the case of the solid state device, the overall
conversion efficiency in a photoelectrochemical cell is a function
primarily of the short-circuit current, open-circuit voltage and
the fill factor. The short-circuit current and fill factor are
determined by the number of photons absorbed minus the losses in
the system. These are primarily a function of the intrinsic
properties of the semiconductor, band gap, doping level, carrier
lifetimes and number of traps and recombination centers. The
open-circuit voltage for a wet photovoltaic cell or effective
biasing potential for chemical producing cells, however, depends
on the relative positions of the energy levels in both the semiconductor and the electrolyte as shown in Fig. 4.

The energy levels in the electrolyte (redox potential) are
generally known as a function of electrolyte composition (11) or
can easily be determined. However, the energy levels in the
semiconductor are more difficult to obtain. For semiconducting

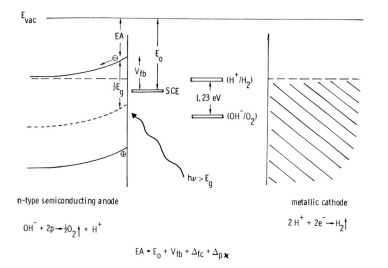

$$EA = E_0 + V_{fb} + \Delta_{fc} + \Delta_{px}$$

FIGURE 4. *Energy level diagram for a photoelectrolysis cell
illustrating the relationship between the electron affinity (EA)
and the flatband potential (V_{fb}). Energy levels are shown for
zero external bias.*

electrodes, the usual procedure is to measure some property
which depends on applied potential and extrapolate to the flatband condition.

The semiconductor/electrolyte interface can be represented
as two capacitances in series; one in the electrolyte near the
surface (Helmholtz layer) C_H and the other capacitor formed in
the semiconductor by the depletion layer C_{sc} (12). Since these
two capacitors are in series and $C_{sc} << C_H$, the net capacitance
is ~ C_{sc}. The width of the depletion layer depends on the
applied potential as in a normal Schottky barrier. Since the
depletion layer width goes to zero at the flatband potential, the
capacitance of the junction goes to infinity. The analysis of
this type of behavior for solid-state devices has shown that C^{-2}
is linear in applied potential (Mott-Schottky plot). This
approach can be applied to the semiconductor/electrolyte junction.[3] The flatband potential V_{fb} is the position of the Fermi
level in the semiconductor when the depletion layer goes to zero
width (no band bending) measured relative to whatever reference
electrode is being employed.

With no depletion layer and therefore no electric field to
separate the photogenerated electron-hole pair, the photocurrent
should also go to zero at this potential. Differences between
the onset of photocurrents and flatband potential determined from
capacitance data indicates the existence of recombination centers
or surface states in the semiconductor gap (13). In many cases
these are absent and the photocurrent onset can be used to determine the flatband potential.

Determination of the flatband potential is of crucial importance since it determines the maximum open-circuit voltage of a
wet photovoltaic cell and the bias requirements of a chemical
producing PEC. The potential difference between the semiconductor band edge and the redox potential at the counter electrode is
the maximum obtainable band bending.

It is particularly useful to be able to predict V_{fb} for a potential semiconductor to evaluate its promise as a photoelectrode. To do this effectively, V_{fb} must be related back to some fundamental property of the semiconductor. This can be done fairly easily, as we have previously demonstrated (14,17) by virtue of the fact that the electron affinity (EA) of the semiconductor is directly related to the flatband potential. This can be seen by referring to Fig. 4 for a typical photoelectrolysis cell. Here we can see that (14)

$$EA = E_o + V_{fb} + \Delta_{fc} + \Delta_{px} \tag{1}$$

where E_o is a constant relating the reference electrode, normally the standard calomel electrode (SCE), to the vacuum level, Δ_{fc} corrects for the difference between the doped Fermi level and the bottom of the conduction band and Δ_{px} is the potential drop across the Helmholtz layer due to ions adsorbed on the semiconductor surface.

The problem then is to evaluate each of these factors so as to be able to predict V_{fb} for any semiconductor/electrolyte combination. While there is some uncertainty, E_o is taken to be 4.75 eV for the SCE at 23°C (15). Δ_{fc} can be evaluated from Seebeck coefficient measurements (16) and for the highly-doped semiconductors commonly employed in PECs is typically 0.1 eV, a small correction. In our future discussions, we will ignore this term. The most important factors are EA and Δ_{px}.

A detailed understanding of adsorption of ions at the semiconductor/electrolyte interface is not possible at the present time. However, a number of things are known. If the flatband potential of the semiconductor shifts with concentration of ions in the electrolyte, then some species related to the ions in solution is specifically adsorbing on the semiconductor surface. For metal oxides in simple acids and bases, the flatband potential shifts by 59 mV/pH unit due to specific adsorption of H^+ and OH^-. These ideas have also been applied to non-oxides such as CdS (17A)

and GaP (17B). The net surface charge depends on the relative
electrochemical potentials for the two adsorbing species in the
adsorbed state and in the solution. Since changing concentration
in the solution changes the electrochemical potentials, it is
possible to vary the relative coverages of the two adsorbing
species and thus the net surface charge. Some concentration of
ions in the solution exists at which the coverages by the two
oppositely charged species are the same and thus the net surface
charge is zero (point of zero zeta potential, PZZP). At this
point, the potential drop due to adsorbed species is zero
($\Delta_{px} = 0$). Fortunately, techniques exist for determining this
concentration of ions for any semiconductor/electrolyte combination (14,17). Thus, for metal oxides we may write:

$$\Delta pH = (59 \text{ mV})(pH_{PZZP} - pH) \tag{2}$$

In general, it is possible to determine the PZZP of a semiconductor having only its undoped powder, and thus, it is not required
to actually fabricate electrodes with all the ensuing difficulties.

The final factor we need to know in order to determine the
flatband potential is the semiconductor electron affinity. This
quantity is difficult to determine experimentally and impossible
to calculate from first principles. However, some electron
affinities have been calculated using atomic electronegativities
(14,17-19). Mulliken defines the electronegativity χ of an atom
as the arithmetic average of the energy to add and subtract a
single electron. Thus, for neutral atoms, we have:

$$\chi_{atomic} = \frac{1}{2} (A + I_1) \tag{3}$$

where A is the atomic electron affinity and I_1 the first ionization potential. Both the electron affinities (30) and ionization
potentials (21) are available for most atoms. For an intrinsic
semiconducting solid, the corresponding energies are the bulk
electron affinity EA and the energy at the valence band EA + E_g.
Thus, in the bulk we have:

$$\chi_{bulk} = EA \frac{1}{2} E_g \tag{4}$$

The problem then is to relate the bulk electronegativity to the atom electronegativities of the constituent atoms. Nethercot has postulated (18) that the bulk electronegativity is the geometric mean of the atom electronegativities of the constituent atoms. This hypothesis seems to work for a large number of compounds. For example, for TiO_2 the electron affinity is:

$$EA(TiO_2) = \chi(TiO_2) - \frac{1}{2} E_g(TiO_2)$$
$$= [\chi(Ti) \, \chi^2(O)]^{1/3} - \frac{1}{2} E_g(TiO_2) \tag{5}$$

$$EA(TiO_2) = 4.33 \text{ eV} \tag{6}$$

If we evaluate the correction term Δ_{px}, V_{fb} can be predicted by using Eq. 1. As we have discussed, depending on the concentration of the potential determining ions in the electrolyte and the intrinsic pK_a of the semiconductor of interest with respect to those ions, differing numbers of anions and cations will be absorbed on the surface. There will be a unique concentration of ions in solution at which equal numbers of anions and cations are absorbed on the surface and Δ_{px} is equal to zero, the PZZP. For TiO_2 at this point

$$V_{fb} = EA - E_o$$
$$= 4.33 - 4.75 = -.42 \text{ V vs SCE} \tag{7}$$

The PZZP can be measured by a variety of different techniques (14, 17) one of which is shown in Fig. 5. The dashed curves are a series of differential potentiometric titrations of TiO_2 with NaOH solution where the small peaks are at the PZZP and scale inversely with the amount of TiO_2 added. The solid lines are theoretical curves for the titration of pure H_2O and two buffered H_2O solutions. The peaks in the data points occur because at the the PZZP oxide powder no longer acts as a buffer and Δ_{px}/Δ_{ml}

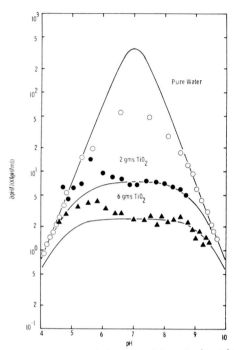

FIGURE 5. *Differential potentiometric titrations plotted
as ∂pH/∂χ(ΔpH/Δml) vs pH for 3 solutions:* ○○○ *600 ml
distilled-deionized water, initial pH adjusted with 0.1 M KOH,*
●●● *600 ml distilled-deionized water 0.001 M in KNO₃ with 2 gm
of suspended TiO₂ (99.995%) initial pH adjusted with 0.1 M KOH,*
▲▲▲ *600 ml distilled-deionized water 0.001 M in KNO₃ with
6 gm of suspended TiO₂ (99.995%). The solid lines represent
theoretical fits for solutions with various concentrations of
buffer. All solutions were stirred under a constant purge of
argon and all titrations were performed with 0.1 M HNO₃.*

increases correspondingly. The PZZP can also be determined by

adding TiO_2 powder to an aqueous solution of known pH. The pH of

the solution will drift upon addition of the powder toward the

PZZP. At the PZZP, the pH of the solution is independent of the

amount of added TiO_2. The PZZP of TiO_2 occurs at a pH of 5.8.

Thus, by using Eqs. 2 and 5 and the PZZP, we can calculate V_{fb}

for any pH.

The technique illustrated for TiO_2 works well for the rest

of the transition metal oxides, and it can be applied to non-

oxides, both n- and p-type. The model works quite well for n-CdS

as is normally employed in wet photovoltaic cells such as n-CdS/
1 M Na_2S, 1 M NaOH/C. Here though, the potential   determining
ions are not OH^- and H^+ as in the metal oxides but are HS^- and
H^+. Thus, the PZZP must be determined with respect to these

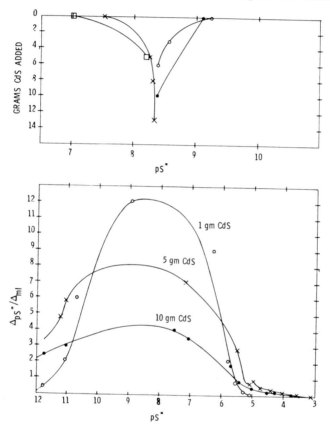

FIGURE 6. Determination of the $S^=$ concentration at the PZZP
of the CdS anode. The top half of the figure represents data
taken by the pX drift technique. In particular, we plot grams of
CdS added to a 600 ml sulfide solution of known initial $pS^=$ and
0.1 M KOH vs. measured $pS^=$. Points on continuous lines represent successive additions to the same solution. The lower half
of the figure consists of three differential potentiometric
titrations of a suspended amount of CdS, 1, 5, 10 gm in 600 ml
0.1M KOH with Na_2S solution. The rate of change of $pS^=$ with
added titrant $\partial pS^=/\partial S$ is plotted vs. $pS^=$. It is of note
that the same peak for the PZZP is observed in both curves and
that $\partial pS^=/\partial x$ scale inversely with the amount of added CdS.

ions as shown for both the drift and differential potentiometric techniques in Fig. 6. Once this is done, the predicted flatband potential can be compared to the experimental value obtained by extrapolating the known dependence of V_{fb} on the concentrations of the potential determining ions as shown in Fig. 7. As can be seen, the agreement is quite good.

The method is also applicable to p-type semiconductors but here Eq. 1 must be rewritten as:

$$EA = V_{fb} + E_o + \Delta_{fc} \Delta_{px} - E_g \qquad (8)$$

where the factor of the band gap takes into account the position of the Fermi level in p-type semiconductors.

The use of Eq. 8 can be illustrated for p-GaP, a promising photocathode in photoelectrolysis cells. We can rewrite Eq. 8 to solve for V_{fb} as:

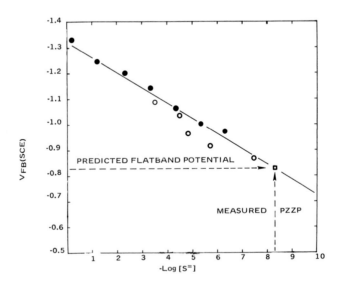

FIGURE 7. Predicted vs. experiment ial for n-Cds. The experimentally measured flatband (V_{fb}) for a number of different sulfide solutions is plotted vs. the measured sulfide ion concentration as $-log[S^=]$. The solid line has a slope of 59 mV per $S^=$ unit. The open circle is the calculated V_{fb} at the measured PZZP.

$$V_{fb} = EA + E_g - E_o - \Delta_{fc} - \Delta_{px} \tag{9}$$

calculating EA

$$EA = \chi_{GaP} - \frac{1}{2} E_g$$

$$EA = (\chi_{Ga} \chi_P)^{1/2} - \frac{1}{2} E_g$$

$$= 4.20 - \frac{1}{2}(2.22)$$

$$= 3.09 \tag{10}$$

At the PZZP $\Delta_{px} = 0$ and neglecting Δ_{fc}

$$V_{fb} = EA + E_g - E_o$$

$$= 3.09 + 2.22 - 4.75$$

$$= .56 \tag{11}$$

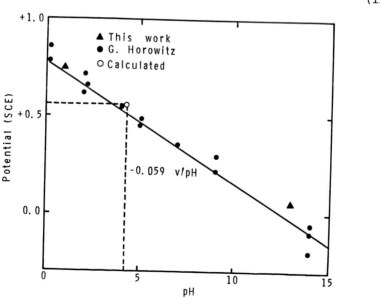

FIGURE 8.   *Predicted vs. experimental flatband potential for p-GaP. Plotted is the variation of the experimentally measured flatband potential with pH. The solid line has a slope of 59 mV per pH unit. The open circle is the calculated V_{fb} at the measured PZZP. [G. Horowitz, J. Appl. Phys. 49, 3571 (1978)]*

In photoelectrolysis cells the experimentally determined PZZP occurs at a pH 4.25.

Fig. 8 shows the dependence of V_{fb} on pH with the position of the predicted V_{fb} shown by the dashed lines. As can be seen, the point falls on the experimental line as expected.

By using the methods discussed to calculate the electron affinity, it is now possible to predict the flatband potential for any semiconductor/electrolyte combination. Correcting all the measurements to the respective PZZPs allows a direct comparison of the calculated electron affinity with the measured flatband potentials. In Fig. 9, we show such a comparison for a number of different semiconductors. As can be seen, the agreement is quite good.

This knowledge and the knowledge of the energy level structure in the electrolyte enables us to predict the maximum

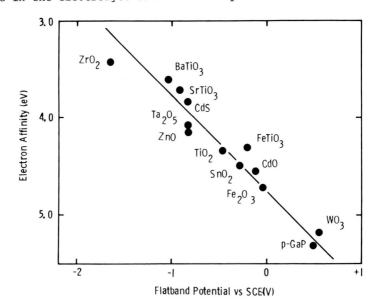

FIGURE 9. *The electron affinity calculated from atomic electronegativities vs. the measured flatband potentials of several semiconductors corrected to their respective PZZPs. The solid line is that expected from Eq. 1. The data are from Refs. 13, 14 and 17. The p-GaP data point refers to EA + E_g rather than the electron affinity.*

open-circuit voltage for a wet photovoltaic cell or the bias requirements for a chemical producing cell.

Another important aspect of the electronegativity model is that it provides an understanding of the role atomic properties play in determining the behavior of a semiconducting electrode. For semiconducting anodes an important requirement is as small an electron affinity as possible (22) to maximize the open-circuit voltage or minimize the applied bias. Since oxygen is a very electronegative atom, the model suggests that metal oxides with small oxygen content would be the most useful in this respect.

Another quantity which appears to correlate with electron affinity for metal oxides is band gap. The argument has been

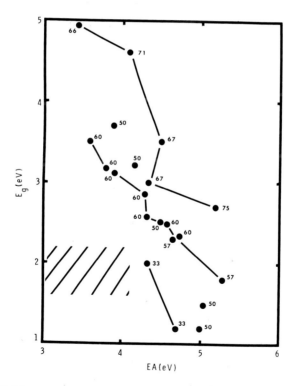

FIGURE 10. Measured band gap versus the electron affinity calculated using the atomic electronegativity model. The numbers are the atomic percent oxygen. The cross-hatched area is the optimum region for photoelectrolysis.

made that since the valence band is formed by O(2p) levels, which
will be approximately the same for all oxides, decreasing electron
affinity corresponds to increasing band gap and vice versa (23).
Both of these arguments are illustrated in Fig. 10. One observes
that there is indeed a correlation between band gap and electron
affinity, if the data are restricted to compounds with about the
same atomic percent oxygen. The figure suggests that compounds
with small metal valences and therefore low oxygen content such
as Cu_2O would be the most fruitful to explore. However, there is
some question as to the stability of this class of compounds.
The atomic electronegativity model has been used to explain the
difference between Fe_2O_3 and $YFeO_3$ where the electron affinity
is lowered by replacing Fe by Y (24).

Electromigration of Ions

   Stability in semiconductors for electrodes in photoelectrochemical cells is perhaps the crucial criterion these materials
must satisfy. Generally, stability is thought of primarily in
the thermodynamic sense, comparing the redox couples of interest
to the decomposition potentials; and a number of recent models
have been published discussing this in detail (9,10). Kinetics
can sometimes dominate and stabilize a thermodynamically unstable
electrode. While these factors are anticipated to be the most
important, recently a number of other mechanisms (25-28) have
been delineated that appear to be quite important with respect
to determining the photoresponse and flatband. Foremost among
these is ion migration in the surface and near-surface region of
many semiconductors. The high electric fields (kV/cm) at the
electrode surface caused by the potential drop across the very
narrow depletion region are sufficient to cause the electromigration of ions in this region (25). We will discuss our observations for the transition metal oxides though analogous arguments
apply to main group semiconductors.

In general, there are two main types of ion migration. If
an electrode is aged at an anodic bias then positive interstitial atoms tend to be removed from the surface region (25). Since
these ions are generally thought to be responsible for the doping
in the transition metal oxides, their removal alters substantially the doping profile in the surface region of the semiconductor. At cathodic bias the reverse process occurs and small
positive ions are injected into the surface region (26). By far
the most important of these is hydrogen, and we feel that the
amount of hydrogen in the near-surface and surface region is
important not only in determining the photoresponse of the
electrode and flatband potential but even its catalytic ability
(29,30). We will illustrate these results for TiO_2 and $SrTiO_3$
and show how the electronegativity model can be applied here to
gain valuable information about the electrodes.

To evaluate the importance of the incorporated hydrogen on
photoelectrochemical properties of TiO_2 anodes in a photoelectrolysis cell, a number of electrochemical experiments have been
carried out on illuminated anodes. In general, these experiments
entail aging lightly-doped TiO_2 anodes in 1 M NaOH in various
potential and current regimes. Fig. 11 and Table 1 enumerate
the results of such an experiment on one rutile anode. Scan 1
is that for the virgin, lightly-hydrogen reduced anode and its
photoresponse is that expected giving an effective gap of
approximately 3.0 eV. Its measured flatband of -0.865 V vs. SCE
is also that normally found (14).

As the sample begins to age cathodically, one sees the
gradual hydrogenation of the surface layer of the electrode and
the formation of a deep blue spot in the region exposed to the
electrolyte. Current controlled aging was employed so as to
to avoid the incorporation of too much hydrogen and subsequent
shattering of the sample (Fig. 12) which occurs at current densities between 1 - 10 mA in 1 to 3 days. As the hydrogen doping

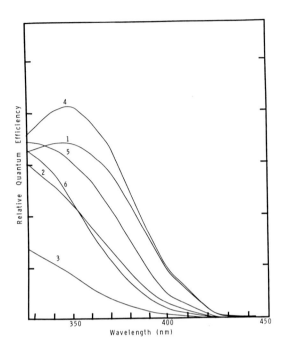

FIGURE 11. *Successive agings of a lightly-hydrogen reduced rutile anode in 1 M NaOH: 1) virgin sample, 2) 100 μA 14 hr. cathodic, 3) 100 μA 20 hr. cathodic, 4) 0.0 V vs. SCE 20 hr., 5) +1.0 V vs. SCE 23 hr., and 6) +5.0 V vs. SCE 4 days.*

TABLE 1. *Dependence of V_{fb} on Aging Conditions for TIO_2*

Experiment	Conditions	V_{fb} (vs SCE)
1	Virgin	-0.865
2	Aged Cathodically 100 μA 14 hrs.	-0.735
3	Aged Cathodically 100 μA add. 20 hrs	-0.750
4	Aged 0.0 V vs SCE 20 hrs.	-0.817
5	Aged Anodically +1.0 V vs SCE 23 hrs.	-0.765
6	Aged Anodically +5.0 V vs SCE 4 days	-0.70

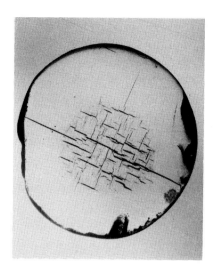

*FIGURE 12.   Photograph 5x magnification of an electrochemi-
cally deuterium-doped rutile wafer.   Doping was accomplished
by current controlled cathodic aging an undoped TiO_2 wafer in
LiOD, D_2O for 3 days at 10 mA.   This resulted in a shattering
of the sample in the region exposed to the electrolyte.*

proceeds, the flatband starts moving to positive values vs SCE
and then levels off. This shift is generated by three effects.
The introduction of hydrogen into the lattice increases the bulk
electronegativity of the oxide so as to move V_{fb} to more positive
values. We can understand this by looking at the electronegativity model. From Eq. 6 we know that EA for TiO_2 is 4.33 eV.
If we return to Eq. 1 and solve for the the electron affinity
change needed to account for the shift in V_{fb} for treatment two,
we find a required EA of 4.46 eV. If this is assumed to be due
exclusively to incorporated hydrogen, which is born out by depth
profiling experiments, then we can solve for the percent hydrogen.
This can be done by returning to our hypothesis that the bulk
electronegativity of a compound is equal to the goemetric mean of
the electronegativities of the constituent atoms. So for a same
unknown amount of hydrogen in the surface region we can write:

$$\chi_{TiO_2H_x} = [\chi_{Ti}\chi_O^2 \chi_H^x]^{1/3+x} \tag{12}$$

Since we know the apparent electron affinity of 4.46 eV, we can solve for x.

$$EA_{observed} = [\chi_{Ti}\chi_O^2 \chi_H^x]^{1/3+x} - 1/2 \; E_{g_{TiO_2}} \tag{13}$$

If this is done we find that if one out of five TiO_2 moities in the near surface were hydrogenated, the appropriate shift in V_{fb} would result. However, two other factors must be evaluated: the change in the band gap of the oxide, which is small based upon photoresponse measurements and the change in the point of zero zeta potential of the surface which may be significant if the relative pK_a of the surface is substantially altered.

The relative lack of change in V_{fb} from scan two to scan three most likely results from the near saturation of the available surface sites with hydrogen. The substantial loss of quantum efficiency for the cathodically aged sample seems to indicate that the hydrogenated surface region acts as if it has many recombination centers.

Treatment four is the reverse aging of the sample under illumination at 0.0 V vs SCE. Most of the hydrogen is removed from the surface layer in this process, but V_{fb} shows that perhaps 20-30% of the hydrogen remains. This is further substantiated by the continued presence of the blue spot. However, as the photoresponse curve number four shows, the quantum efficiency is totally restored. This indicates that the remaining hydrogen is located on a lattice site that is normally not photoactive in the photoelectrolysis process.

Continued aging in the anodic direction at +1.0 and +5.0 V vs SCE, scans five and six respectively, result in the gradual diffusion of Ti^{3+} interstitials out of the surface layer as has recently been discussed by Butler [25]. The removal of the electropositive Ti^{3+} and concentration of cathodically incorporated hydrogen at the surface results in a more positive V_{fb} and the reduced number of donors results in the observed decrease in Φ, (the quantum efficiency) and the observed red shift of the photoresponse due to increased depletion layer width. The blue spot remains indicating the continued presence of hydrogen in the surface.

Similar results are obtained for $SrTiO_3$. These are illustrated in Table 2 and Fig. 13. Here the cathodic and anodic aging experiments were done on two separate nearly identical samples. For cathodic aging the partial reversible incorporation of hydrogen, approximately 1 per 3 titaniums, is observed substantially altering V_{fb} and the magnitude (Φ) but not the shape of the photoresponse curve. Anodically, there appear to be mobile interstitials similar to TiO_2 which when removed from the surface layer result in a diminished and red-shifted photoresponse. Here we can clearly see that V_{fb} has not been shifted by the removal of the interstitials. This is the anticipated result because while the ions removed substantially affect the doping profile the number is well below an atomic percent and would not be

TABLE 2.   *Dependence of* V_{fb} *on Aging Conditions for* $SrTiO_3$

Sample	Light	Conditions	V_{fb} vs SCE
1		Virgin	−1.11
	on	+5 V vs SCE − PC 21 hrs.	−1.15
	on	+1.0 V vs SCE − PC 4 days	−1.15
	off	−1.0 V vs SCE − PC 2 days	−1.16
	off	300°C in air − 16 hrs.	−1.09
2		Virgin	−1.17
	on	.2 mA − cc 25 hrs.	− .95
	on	.2 mA − cc 21 hrs.	− .96
	on	0.0 V vs SCE − PC 27 hrs.	−1.06

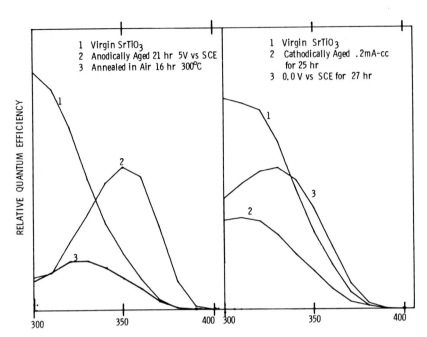

FIGURE 13. *Electrochemical aging experiments on two lightly reduced SrTiO$_3$ wafers. The left-hand portion is for anodic aging, the right for cathodic aging. All experiments were performed in 1 M NaOH.*

expected to change the electron affinity. When the samples are allowed to sit at room temperature, the new ion profiles are stable but at temperatures around 250°C donors are redistributed homogeneously.

Thus, large changes in V_{fb} and Φ in these electrodes under both anodic and cathodic bias indicate that the electromigration and the electrochemical doping of the surface layers in oxide photoanodes may be crucial in evaluating electrode stability and performance. The use of an appropriate electrochemical dopant may result in the advantageous modification of the properties of an electrode.

In conclusion, we have demonstrated how a single concept, electronegativity, can give us quantitative information about the biasing requirements of photoelectrodes and how the same concept

can be employed to give us valuable information about aged
electrodes.

III.   GRAIN BOUNDARY EFFECTS IN POLYSILICON

Introduction

The application of solid state photovoltaic devices to convert solar energy to electrical power progressed rapidly in the
early 60's primarily because of the need for space power systems.
Interest in this field has grown dramatically in the last few
years as attention has been turned towards terrestrial applications.   The pressing need to find alternatives to conventional
power sources ensures continued importance to solar cells as
possible large-scale electrical energy sources.

In the race to find practical devices, one of the attractive
candidates has always been silicon solar cells.   The element is
abundant and the device technology for single-crystal material
well established.   However, while single-crystal silicon solar
cells are reasonably efficient, they are expensive to fabricate
and their feasibility has yet to be demonstrated.   On the other
hand, solar cells constructed from thin films of polycrystalline
silicon (polysilicon) are economically feasible to produce but
suffer from considerably reduced efficiencies (3).   The United
States Department of Energy has estimated that at 10% efficiency
and current thin-film fabrication techniques, polysilicon solar
cells would become a viable energy alternative.

The major reason for the low efficiencies in thin-film polysilicon devices is the effects of grain boundaries.   Due to the
growth conditions most current thin-film devices have relatively
small grain sizes, on the order of 2 to 20 microns.   It is well
known that in efficient single-crystal silicon solar cells minority carriers collected from distances much greater than 100
microns contribute significantly to device currents.   Thus for

polysilicon devices to be adequately efficient, carriers will
have to traverse many grain boundaries. It is known that grain
boundaries possess an attractive potential for photogenerated
minority carriers; more importantly, there exists a high density
of recombination centers that the carriers are exposed to once
they are trapped at the "core" of the boundary. An equally
important effect is the apparent shunting action of grain boundaries which intersect the p-n junction, this causes poor fill
factors and open-circuit voltages (32). These two effects serve
to substantially reduce device efficiencies. Grain boundaries
affect minority carrier lifetimes in semiconductors other than
silicon as well. In any polycrystalline solar cell, especially
thin-film devices, grain boundaries must be seriously considered
as one of the major sources of efficiency loss.

There are two main approaches to alleviating grain boundary
problems. The first is to regrow the material in an attempt to
increase grain size and concurrently reduce the number of boundaries. This may be accomplished by a variety of techniques such
as: laser regrowth, float zone refining, melting and controlled
solidification. Most of these processes are very energy intensive and do not eliminate but merely reduce grain boundary
problems. The second method is to treat the grain boundaries
directly, usually with some variety of chemical agent to render
them electrically inactive (passivated). Historically, the
former method has been the one most often employed because passivating techniques have been unknown until lately. Recently, we
have demonstrated (38) how monatomic hydrogen can indeed passivate
grain boundaries in polycrystalline silicon, and it is this work
which will be discussed here. We will consider the ramifications
of these and other recent results with respect to the improvement
of solar cell performance.

Electrical Nature of the Boundary

To understand the nature of the passivation process it is
necessary to examine closely the electrical nature of the grain
boundary. Grain-boundary potential barriers are formed in semiconductors when the grain boundary region has a lower chemical
potential for majority carriers than the grains. The resultant
influx of electrons or holes into the barrier region creates a
space charge which repels the further flow of majority carriers.

In silicon it has been demonstrated (32) that the Fermi level
in the grain boundary region lies close to the middle of the for-
bidden gap in both n- and p-type material. Since the chemical
potential of the doped grains is not at midgap, significant barriers are observed in most samples of n- and p-type silicon. One
can picture the formation of the potential barriers as is illustrated in Fig. 14. Part A illustrates the neutral grain boundary
region before it is joined to the adjacent (neutral) n-type grains.
This picture is, of course, difficult to duplicate physically
but it serves to illuminate the nature of the charge transfer
process. The boundary width, w, is assumed to be of the order
of ~50 Å, a number which is known to be representative of
the disordered region near the grain boundaries of a material-
like silicon (33). The location of the Fermi level in the
boundary E_{FB} is a function of the nature of the defect states
present in the boundary. These are thought to be primarily due
to unsaturated silicon-silicon bonds, "dangling bonds," and to
trapped impurity atoms, such as oxygen (34). While many bent
bond situations exist, they do not appear to be electrically
active as is born out by the electrical inactivity of twinned
boundaries in germanium and silicon (34).

Fig. 14(b) illustrates the two grains and the grain boundary
region after they have been joined and equilibrium has been established. The accumulated negative charge near x = 0 causes the
energy bands to be bent upwards by an amount ϕ_B and the Fermi

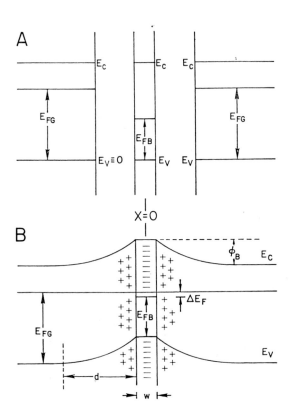

FIGURE 14. Energy band diagram for two semiconductor grains
and their boundary region. (a) This part shows the two grains
and the region of their boundary schematically as they would be
if they were three isolated materials. (b) This part shows the
band bending in the grains and the increased Fermi level in the
boundary which occurs when the materials in (a) are joined at
zero applied bias voltage.

level in the grain boundary region to shift by an amount ΔE_F.

Although the width of the depletion region d, has been drawn

here as comparable to the GB region width, in general W << d.

The width d varies with the doping level in the grains similar

to the dependence found for Schottky barriers.

For wide depletion layers thermionic emission of majority

carriers over the potential barrier is expected to be the domi-

nant conduction mechanism (35,36). Pike and Seager (6) have

recently shown that the grain-boundary resistance in silicon can indeed be modeled by assuming that conduction is dominated by the thermal excitation of carriers over the barrier. Utilizing this model they have developed a deconvolution scheme whereby the energy density of grain boundary states may be obtained from current-voltage measurements. These densities of states have been confirmed recently by the first direct measurements of charge emission from silicon grain boundary defect states (37). This has been accomplished by monitoring the recovery of the nonequilibrium grain-boundary barrier capacitance. The use of the deconvolution scheme for the I-V data and of the capacitance data allows a direct determination of the density of states in an individual grain boundary. The measurement of the density of states serves as a specific probe of the effectiveness of any passivation procedure. The qualitative effectiveness of grain boundary passivation can thus be obtained by a simple four probe determination of the grain boundary conductance. A simple apparatus for these determinations is illustrated in Fig. 15. Current

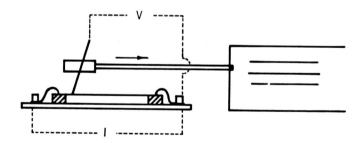

Grain Boundary Conductance Apparatus

FIGURE 15. *The four-probe grain boundary conductance apparatus is illustrated diagramatically. A constant current is applied to the sample (2 mm x 2 mm x 2 cm) through ohmic contacts (the crosshatched areas). A potential-versus-distance plot is obtained by having a movable potential probe traverse the surface at a fixed rate and monitoring the potential referenced to the low-current contact.*

is passed through a polysilicon sample, 2 mm x 2 mm x 20 mm, that has been contacted at both ends. The surface potential is measured with a traveling point contact. The potential drop for the largest barrier is kept to less than kT. If one plots potential versus distance, the vertical potential drops correspond directly to grain boundary impedances as is shown in Fig. 16.

Grain Boundary Passivation

Since "dangling bonds" or impurities are responsible for the grain boundary defect states, it should be possible to introduce

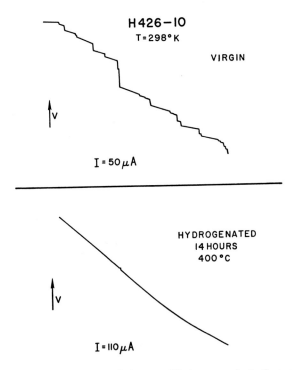

FIGURE 16. A typical potential vs. distance plot for a virgin (top) and hydrogen plasma-treated (bottom) p-type Honeywell silicon on ceramic polysilicon sample. The voltage scales are the same, and approximately 60 mV is dropped across the sample in each case. Treatment conditions were 14 hrs at 400°C in a 500 millitorr hydrogen plasma.

a foreign chemical species into the grain boundary to remove or
neutralize them. In effect by reacting the defects with a reagent
of the appropriate electronegativity, their electronic trapping
levels can be displaced to well below the Fermi level. Similarly,
it might be anticipated that other reagents of appropriate electronegativities could increase the density of states above the Fermi
level. In fact, monatomic hydrogen plasmas have been demonstrated (38,39) to be highly effective at removing grain-boundary
in-gap sites in silicon. Heller and Miller (40) have also recently
demonstrated that Ru serves a similar function in thin-film GaAs.

A typical apparatus is illustrated in Fig. 17. Here all gas
treatments are conducted inside an evacuated quartz vessel, inserted in a horizontal tube furnace. Temperature is monitored with
a platinum/platinum-rhodium thermocouple located near the sample
wafers which are on a tantalum platform. Plasmas were excited by
means of a circumferential metal band which was connected to a
tesla coil (voltage typically 1-5 kV). Plasma composition could

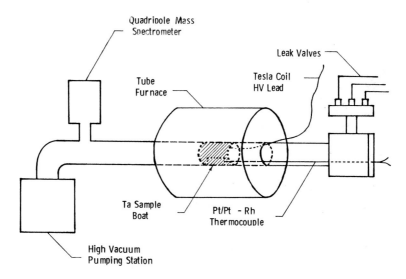

FIGURE 17.   *Diagramatic illustration of the hydrogen passi-*
vation apparatus. The sample resides in the Ta boat in the cross-
hatched region, which is the region of highest plasma density.

be monitored with the attached quadrupole mass spectrometer.
Fig. 16 illustrates two potential vs distance traces for a slice
of Honeywell p-type silicon on ceramic with approximately 100
micron grains. The upper portion of the figure is for the virgin
slice, the vertical steps are across individual grain boundaries.
The lower curve is for the same sample, exposed to a 500 millitorr hydrogen plasma for 14 hours at 400°C (note the increased
current). Significant increases in grain-boundary conductance
are seen and the overall sample conductance approaches that of
the bulk. Having established that the passivation process is
quite effective at altering grain boundary conductances, we have
begun to perform experiments to understand and optimize the
process. The rest of the discussion will deal with some of the
observations we have obtained to date.

Fig. 18 illustrates the effect of a series of treatments on
the zero-bias conductance, G_o, of an individual grain boundary in
a sample of neutron transmutation doped polysilicon. Treatments
a, c, f, i and k are hydrogen plasma treatments of various dura-
tions at temperatures between 330 and 400°C and pressures of 60-
100 millitorr. In each case substantial increases in G_o are
observed. Treatments b, d and g are vacuum anneals carried out
at 1×10^{-6} torr or less at 620°C. At this temperature a complete return to virgin conductance levels is observed. Treatment
e was performed under identical conditions to the hydrogen plasma
treatments except for the absence of the electrical discharge.
The conductance actually decreases somewhat which may indicate
the absorption of impurities in the grain boundary. From this
observation, it is clear that the atomic hydrogen generated in
the plasma is essential to the process effectiveness. Treatment
h is at very low hydrogen pressure (10^{-2} torr); this is apparently below the optimum pressure range for passivation (~100-
500 millitorr). Treatment j is an anneal in a low-partial
pressure of oxygen 5×10^{-4} torr at 625°C. This illustrates

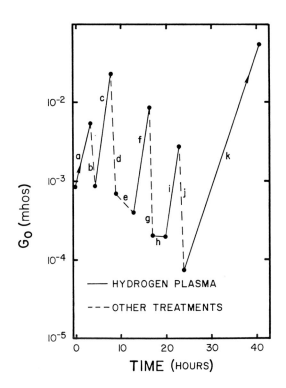

FIGURE 18.   The zero-bias conductance at 300°K of one grain boundary potential barrier in a $10^{16}-P/cm^3$ neutron transmutation (NTD) doped polycrystaline silicon sample after a series of gas treatments.

that the incorporation of impurity atoms can reduce the grain boundary conductance to values significantly below virgin levels. This effect can be enhanced by higher pressures and has been observed for O_2, SF_6 and N_2.

The zero-bias conductance of two different grain-boundary potential barriers in neutron-transmutation doped polycrystalline silicon are illustrated in Fig. 19. The solid curves show the virgin conductance values, the dashed curves the results after hydrogenation at 400°C for ≈5 hrs. In addition to marked increases in barrier conductance, hydrogenation weakens the temperature variation of the barrier conductance; in the case of the

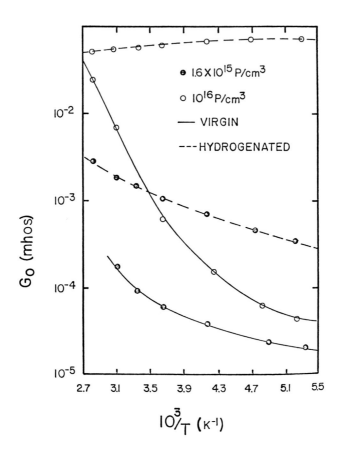

FIGURE 19. *The zero-bias conductance of two different grain
boundary potential barriers in neutron transmutation doped polycrystalline silicon. The solid curves show the virgin conductance values, the dashed curves the results after hydrogenation
at ~400°C, 500 microns H_2, for 5 hr.*

10^{16} P/cm^3 sample the observed temperature dependence loses

its activated character and takes on an opposite, but weak,

variation, characteristic of the conversion of the grain boundary

to a simple scattering center for majority carriers. A closer

examination of the temperature dependence of the conductance of

a virgin and treated sample is illustrated in Fig. 20. The

activation energy for conductance in the virgin state is approxi-

mately 0.62 eV while that in the hydrogenated sample is 0.035 eV.

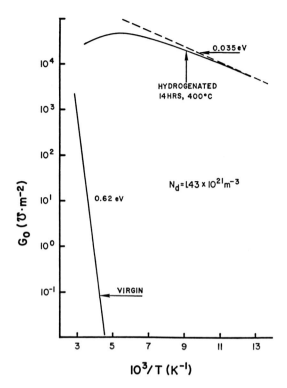

FIGURE 20. Activation energy determinations for a barrier before and after hydrogen plasma treatment in a sample of neutron transmutation doped (10^{16}P/cm^3) polysilicon.

This is indicative of a substantial reduction in the number of trap states in the boundary and consequent reduction of barrier height and ease of emission over the barriers. This is further illustrated in Fig. 21, where for the same sample we plot directly the effective density of grain boundary states $N_t{}^*$ versus energy (in eV) with the zero of energy being the valence band maximum.

The data are obtained by deconvoluting the current-voltage data according to the Pike-Seager deconvolution scheme (36). A substantial reduction in the density of states is observed as expected from the temperature dependence of G_o. Thus, the hydrogen plasma treatment seems to drastically reduce the density of

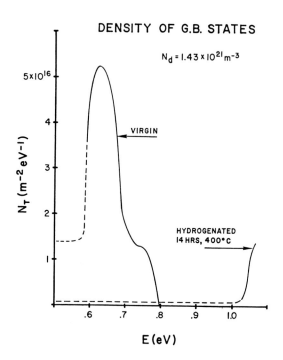

FIGURE 21. The effective density of grain boundary states (N*$_t$) for a virgin and hydrogen plasma treated boundary vs. energy (in eV with zero being the valence band maximum (same sample as in Fig. 20). The data are obtained by deconvoluting the current-voltage data according to the Pike-Seager deconvolution scheme.[36]

states in the grain boundary region, giving rise to significant reductions in the double depletion layers and consequent increases in the conductance.

In the course of optimizing the treatment it has been observed that the condition of the surface is crucial for optimum hydrogen penetration. This is illustrated in Fig. 22. Here we show a 9 Ωcm float-zone refined n-type polysilicon sample before (top) and after (bottom) hydrogen plasma treatment at 500 microns, 400°C for four hours. The middle portion of the sample had 25 microns etched off with an HF/CrO$_3$/H$_2$O etch, and both ends were left as they were after diamond-wheel cutting. The middle section

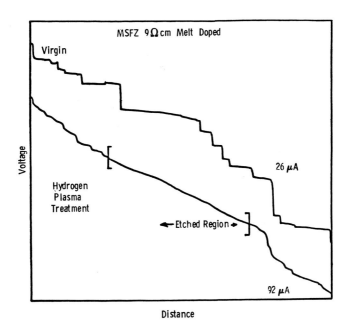

FIGURE 22.   The potential vs. distance plots for a virgin
and hydrogen plasma treated sample of 9 ohm-cm float-zone refined
n-type polysilicon.   The middle portion of the sample (brackets
had 25 microns etched off with an HF/CrO$_3$/H$_2$O etch.

is extensively passivated but both of the saw damaged ends are
only lightly passivated.   The definition between etched and
unetched areas after treatment is sharp.   This may indicate that
grain boundary diffusion and not bulk diffusion is the main
mechanism for hydrogen introduction into the boundaries.

Fig. 23 also illustrates that the passivation process in the
melt-doped polysilicon is as effective as in the neutron trans-
mutation-doped polysilicon.   This is an indication that any
grain boundary dopant segregation present is not important in
determining the electronic properties of the barriers in the
melt-doped materials.   This is further borne out by the I-V
properties of single melt-doped boundaries which can be modeled
quite well by thermal emission theory.

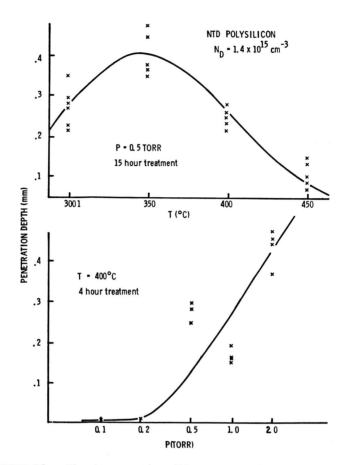

FIGURE 23. The top portion illustrates the effective penetration depth (mm) of the passivation process vs. temperature
at a fixed pressure of 500 microns H_2. The x's represent an
accumulation of data from different boundaries. The bottom
portion illustrates the effective penetration depth (mm) vs. H_2
pressure at a fixed temperature of 400°C. Above 2 torr it
becomes difficult to sustain a plasma.

In order to optimize the process a means to evaluate the
effective penetration depth of the treatment has been developed.
A series of etches are performed with the HF/CrO$_3$ etch while monitoring the grain-boundary conductance. The current flow in the
grain boundary is modeled as a sum of flow through two regions,
the passivated region and the essententially untreated region.

By ratioing the slope of the grain boundary conductance to that
of the bulk, one can calculate the thickness of the conductive
region. Fig. 23 illustrates a plot of depth of penetration of
passivation (in millimeters) versus temperature on the top and
pressure on the bottom. Optimum temperature is around 350°C and
optimum pressure is betwen 1 and 2 torr. Above 2 torr it is
difficult to sustain a plasma.

It is clear that under relatively mild conditions, passivation can be observed to considerable depths. Thus penetration is
very much in excess of the anticipated thickness of useful thin-
film (10 to 100 microns) devices. In thin films treatment times
of 10-15 minutes might well be sufficient. As Fig. 16 demonstrates, grain boundary impedances can be virtually eliminated (on
a relatively thick, 400 microns, sample). Having established the
viability of the passivation process for bulk and thin-film polycrystalline silicon, we turned our attention to some exploratory
experiments on prototype solar cells. The utlization of solar
cells allows a direct probing of minority carrier lifetimes and a
tentative evaluation of the ability of the process to increase
efficiencies. Fig. 24 illustrates typical I-V curves for two
cells before and after treatment. Overall efficiencies are low
because the neutron transmutation doped (NTD) polysilicon employed
suffers from a very low bulk lifetime. The upper set of curves
are for a boron diffused solar cell fabricated from NTD (phosphorous) doped polysilicon (200 micron grains). The dashed curves
represent the light and dark traces for the virgin cell; the
solid lines are the curves for the plasma treated cell. The
efficiency increase (AM1) of a factor of   26 is due to two distinct factors. One is an increase in the short-circuit current
ostensibly due to improved transport of minority carriers across
grain boundaries. The other is a substantial improvement in the
open-circuit voltage due to an elimination of the shunting nature
of the grain boundaries. The lower half of Fig. 24 shows three

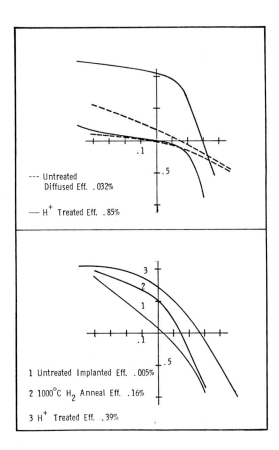

FIGURE 24. *The upper portion of the figure is the light and dark I-V curves for a prototype solar cell at n-type neutron transmutation doped polysilicon with a boron diffused p-layer before (dashed curves) and after (solid curve) hydrogen plasma treatment. The efficiencies are AM1. The current is plotted vertically in mA and potential horizontally in volts. The lower portion is a similar cell with a boron ion implanted surface layer. The curves illustrated are the illuminated I-V curves only for the virgin cell, the cell after a 1000°C thermal anneal in hydrogen and after hydrogen plasma treatment. The efficiencies are AM1.*

illuminated I-V curves for a boron ion implanted bulk NTD phosphorous doped polysilicon solar cell. The 1000°C H_2 anneal is an attempt to remove some of the ion implantation defects. Curve 3 again shows that the hydrogen plasma treated cell increases in efficiency as a consequence of improvements in both the short-circuit current and open-circuit voltage in the cell.

The lifetime of the minority carriers and the spectral response of some of the ion implanted cells has been probed more directly as shown in Fig. 25. The lower curve and number refer to the virgin sample, the upper set are after hydrogen plasma treatment. The lifetimes were determined by monitoring the short-circuit current decay after a 5 nsec x-ray pulse. The data indicate that there is substantial increase in bulk minority carrier lifetime after plasma treatment. This is substantiated by an examination of the spectral response curves for the cell.

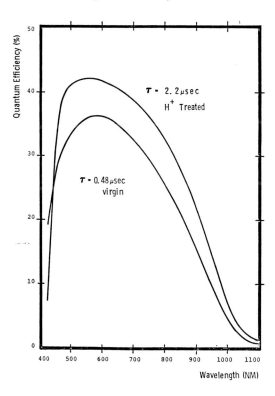

FIGURE 25. Spectral response curves (quantum efficiency vs. wavelength) for an n-type neutron transmutation doped polysilicon with a boron p-type ion implanted junction before and after hydrogen plasma treatment. The lifetimes are from current vs. time data after the cell receives a 5 nanosecond x-ray pulse.

There is a large increase in the quantum efficiency of the long
wavelength tail where the photons are adsorbed well below the
junction.

In conclusion, we have demonstrated that by modifying the
chemical potential of the grain boundary defective regions we can
substantially alter their electronic properties in the polycrystalline material. It has been demonstrated that boundaries can
be passivated by relatively mild treatments in hydrogen plasma
and that grain boundary potential barriers can be increased by
treatment with O, N and F. The hydrogen passivation process is
very effective on thin films and experiments on prototype solar
cells indicate that the process has considerable potential for
increasing efficiencies in thin-film solar cells.

ACKNOWLEDGMENTS

*This work was supported by the U. S. Department of Energy
under Contract DE-AC04-76-DP00789.
†A U. S. Department of Energy facility.

REFERENCES

1. Becquerel, E., *C.R. Acad. Sci. 9,* 561 (1839).

2. Fujishima, A., Honda, K., and Kikuchi, S., *J. Chem. Soc. Japan 72,* 108 (1969).

3. Gerisher, H., in "Physical Chemistry: An Advanced Treatise" (H. Eyring, D. Henderson, W. Jost, eds.), Vol. 9A. Academic Press, New York, (1970).

4. Nozik, A.J., *Ann. Rev. Phys. Chem. 29,* 189 (1978).

5. Butler, M.A., and Ginley, D.S., *J. Mat. Science,* in press.

6. Inoue, T., Fujishima, A., Konishi, S., and Honda, K., *Nature 277,* 637 (1979).

7. Nozik, A.J., 2nd International Conference on Photochemical Conversion and Storage of Solar Energy, Cambridge, England, Aug. 10, 1978.

8. (A) Butler, M.A., *J. App. Phys. 48,* 1914 (1977); (B) Kennedy, J.H., and Frese, K.W., *J. Electrochem. Soc. 125,* 709 (1978); (C) Wilson, R.H., *J. Appl. Phys. 48,* 4292 (1977).

9. Gerischer, H., *J. Electroanal. Chem. 82,* 133 (1977).

10. Bard, A.J., and Wrighton, M.S., *J. Electrochem. Soc. 124,* 1706 (1977).

11. Dobos, D., "Electrochemical Data", Elsevier, New York, (1975).

12. Vijh, A.K., "Electrochemistry of Metals and Semiconductors", Dekker, New York, (1973).

13. Ginley, D.S., and Butler, M.A., Electrochem. Soc. Mtg., Boston, MA, May 6-11, 1979.

14. Butler, M.A., and Ginley, D.S., *J. Electrochem. Soc. 125,* 228 (1978).

15. Lohmann, F., *Z. Naturforsch. Tiel A. 22,* 843 (1967).

16. Hannay, N.B., "Semiconductors", Reinhold, New York, (1959).

17. (A) Ginley, D.S., and Butler, M.A., *J. Electrochem. Soc. 125,* 1968 (1978); (B) Butler, M.A., and Ginley, D.S., *J. Electrochem. Soc.,* submitted.

18. Nethercot, A.H., *Phys. Rev. Lett. 33,* 1088 (1974).

19. Poole, R.T., Williams, D.R., Riley, J.D., Jenkins, J.G., Liesegang, J., and Leckey, R.C.G., *Chem. Phys. Lett. 36,* 401 (1975).

20. Hotop, H., and Lineberger, W.C., *J. Phys. Chem. Ref. Data 4,* 539 (1975).

21. White, F.A., "Mass Spectrometry in Science and Technology", Wiley, New York, (1968).

22. Mavroides, J.G., Tchernev, D.I., Kafalas, J.A., and Kolesar, D.F., *Mat. Res. Bull. 10,* 1023 (1976).

23. Kung, H.H., Jarrett, H.S., Sleight, A.W., and Ferretti, A., *J. Appl. Phys. 48,* 2463 (1977).

24. Butler, M.A., Ginley, D.S., and Eibschultz, M., *J. Appl. Phys.* *48*, 3070 (1977).

25. Butler, M.A., *J. Electrochem. Soc.* *126*, 338 (1979).

26. Ginley, D.S., and Knotek, M.L., *J. Electrochem. Soc.*, Dec. 1979.

27. Heller, A., Schwartz, A.P., Vadimsky, R.G., Menezes, S., and Miller, B., *J. Electrochem. Soc.* *125*, 1156 (1978).

28. Gerischer, H., and Gobrecht, J., *Ber. Bensenges Phys. Chem.* *82*, 520 (1978).

29. Knotek, M.L., Proceedings of the 151st Electrochem. Soc. Meeting, 77-6, 234 (1977).

30. Wilson, R.H., 153rd Electrochem. Soc. Meeting, Seattle, WA, Abs. 415 (1978).

31. Chu, T.L., Liu, J.C., Mollenkopf, H.C., Chu, S.C., and Heizer, K.W., *Solar Energy 17*, 229 (1975).

32. Seager, C.H., and Castner, T.G., *J. Appl. Phys.* *49*, 3879 (1978).

33. Matare, H.F., "Defect Electronics in Semiconductors", Wiley-Interscience, New York, (1971).

34. Koper, A.B., Proceedings of the 13th IEEE Photovoltaic Specialists Conference, 1978, IEEE, New York, (1979).

35. Taylor, W.E., Odell, N.H., and Fan, H.Y., *Phys. Rev.* *88*, 867 (1952).

36. Pike, G.E., and Seager, C.H., *J. Appl. Phys.* *50*, 3414 (1979).

37. Seager, C.H., Pike, G.E., and Ginley, D.S., *Phys. Rev. Lett.* *43*, 532 (1979).

38. Seager, C.H., and Ginley, D.S., *Appl. Phys. Lett.* *34*, 337 (1979).

39. Campbell, D.R., et al., *Bull. Am. Phys. Soc. 24*, Abstract TK, 435 (1979).

40. Heller, A., private communication.

CHAPTER 19

INTRODUCTION TO BASIC ASPECTS OF PLASMA-DEPOSITED AMORPHOUS
SEMICONDUCTOR ALLOYS IN PHOTOVOLTAIC CONVERSION

Richard W. Griffith[1]

Department of Energy and Environment
Brookhaven National Laboratory
Upton, New York

I. INTRODUCTION

The genre of amorphous semiconductor materials currently en-
joys a renaissance owing to recent technological applications in
areas that span photovoltaic conversion, microelectronics, and a
variety of photoelectronic semiconductor devices. An especially
promising class of such materials is that of the hydrogenated
amorphous silicon alloys (a-Si:H). The role of $\sim$ 5-20% alloyed
hydrogen in a-Si:H films is twofold: i) bonded hydrogen passivates electrically-active defects such as dangling bonds and weak
molecular bonds, and ii) the chemical bonding of the alloy *per se*
creates a new electronic material. The resulting improvement in
optoelectronic quality relative to pure a-Si permits systematic
electronic doping of the material (1-3) and the fabrication of
developmental solar cells (4).

The demonstration of photovoltaic feasibility was preceded
by the evolution of several seminal concepts in noncrystalline
semiconductor theory. These developments include: i) localiza-
tion of electronic states in (substitutionally) disordered systems (5,6); ii) the proof that local chemical bonding, i.e.,
short-range structural order alone, is sufficient to predict the

[1]*Work performed under the auspices of the U.S. Department of
Energy.*

existence of the fundamental semiconductor bandgap (7); and iii)
the formulation of the mobility edge concept, together with the
idea that overlapping tails of states (albeit disorder-induced)
pin the Fermi level in the bandgap (8,9). Since the first successful fabrication of solar cells in 1976, a veritable explosion
in experimental information has resulted from renewed interest in
the characterization of a-Si:H materials. In particular, the
following features have been elucidated: silicon-hydrogen bonding configurations as a function of deposition conditions (10-
12); the nature of defect states responsible for the recombination of photo-generated carriers (13-15); modifications in optoelectronic properties of films via impurity incorporation (16,
17); the growth of potentially interesting related alloys such as
a-Si:(F,H) (18); and the influence of gap states upon diode characteristics and device performance (19-23).

In this Chapter we shall selectively review progress that has
been made in advancing an understanding of the materials properties of plasma-deposited a-Si:H alloys. Therefore, an emphasis
will be placed upon the basic aspects of such materials, but with
a focus toward photovoltaic conversion. In developing the subject, our philosophy is to take an integrated approach in which
are interwoven discussions of: i) the plasma chemistry responsible for film growth, ii) the crucial effects of plasma processing conditions upon both chemical bonds in the plasma and in
synthesized alloys, and iii) optoelectronic film properties and
device characteristics. Accordingly, in Sect. II we begin by
discussing the nature of the plasma, proceeding from preliminary
concepts of the rf glow discharge (Sect. II.A) to a more detailed
description of the plasma chemistry in the silane glow discharge
(Sect. II.B). Experience with plasma etching is briefly reviewed
as a paradigm for plasma deposition (Sect. II.B.1). We then discuss emitting reactive species that are identified in plasma
deposition using optical emission spectroscopy (Sect. II.B.2),
and end with a summary of electron-impact processes that are

pertinent to the silane glow discharge (Sect. II.*B.3*).

In Sect. III we discuss the characterization of plasma-
deposited silicon-hydrogen alloys. The introduction to chemical
bonding in Sect. III.*A* consists of discussions on short-range
structural order (Sect. III.*A.1*), the role of hydrogen in the
alloy (Sect. III.*A.2*), and the influence of residual gap states
upon diode characteristics (Sect. III.*A.3*). In Sect. III.*B* we
explore the following dilemma: From measurements of quenched
photoluminescence efficiency (and of degraded device performance), electronically doped alloys unfortunately exhibit reduced
minority-carrier transport. As a consequence, highly conductive,
efficiently doped alloys are relegated the menial role of effecting Ohmic contacts. Finally, in Sect. III.*C* we discuss the effects of nitrogen and oxygen impurities with regard to: i) the
optoelectronic properties of a-Si:H alloys; and ii) the photovoltaic conversion efficiencies of diagnostic devices that are
fabricated using such materials. It follows that an ambiguity
arises in the definition of the nominally intrinsic silicon-
hydrogen alloy a-Si:H. Indeed, the usual alloy should more accurately be regarded as the alloy a-Si:(H,O,N,C,...). The
Chapter is concluded by an expanded discussion of anomalous behavior in photoconductivity (Sect. III.*C.2*). Such behavior involves new recombination processes in alloys prepared under
certain low impurity conditions.

II. NATURE OF THE PLASMA

Many deposition techniques now exist for the growth of thin-
film a-Si:H materials: rf or dc glow-discharge decomposition of
silane (SiH_4); rf reactive sputtering of a silicon target in an
atmosphere of argon and hydrogen; low-temperature chemical vapor
deposition; e-beam evaporation of silicon into atomic hydrogen;
electrodeposition; etc. So far, the most efficient solar cells
have been fabricated from a-Si:H alloys grown via the glow-

discharge (plasma deposition) technique. Therefore, we will discuss the deposition chemistry specifically in the context of processes characteristic of plasma deposition. Moreover, we will
mainly confine our remarks to plasma processes that occur for
the rf glow-discharge decomposition of silane in a capacitively-
coupled system. In this case, alloys are typically produced
under the following operating conditions: chamber pressure p =
= 0.1-1.0 Torr, substrate temperature T_s = 225-400°C, and rf
power density = 0.01-0.3 W cm^{-2}.

Glow discharges ignited in hydride or halide gases create a
dilute plasma state that consists of electrons and many different, mostly neutral, radical species. For example, the silane
glow discharge (Sect. II.B) is a metaphorical "zoo" of reactive
species that contains SiH, SiH$^+$, SiH$_2$, SiH$_3$, SiH$_3^+$, Si$_2$H$_6$, H, H$_2$,
etc. The nature and concentrations of such species vary sensi-
tively with the processing conditions. Now it is possible that
a particular composition of radical species in the plasma (and
in the "sheath" regions) may be correlated with specific bonding patterns in films, and thence with distinct optoelectronic
and microstructural properties of such films. Thus, the plasma
processing conditions play a key role in determining film properties. Of course, the specific correspondences between molecular bonds of reactive species in the plasma and chemical bonds
in as-deposited films may not be especially simple. Indeed,
complex electron-impact processes create reactive species in the
plasma, while subsequent gas-phase (homogeneous) and *surface*
(heterogeneous) reactions promote film nucleation and growth on
substrates. As a consequence, for example, it may be naive to
expect that concentrations of the radical SiH$_2$ (detected in the
plasma by molecular absorption techniques) could be correlated
with dihydride chemical bonding =SiH$_2$ (detected in films by
vibrational spectroscopy).

We again emphasize that the basic plasma chemistry is importantly influenced by the variation of the processing parameters

in deposition. Moreover, these parameters constitute a rather large multidimensional space: substrate temperature (T_s); rf power (P); chamber pressure (p); concentrations of hydrides or halides in carrier gases; flow rates; electrode biases; impurity levels; etc. Therefore, a detailed understanding of the relationship between the plasma processing conditions and the plasma chemistry would greatly facilitate the optimization of a-Si:H materials intended for photovoltaic conversion as well as for a host of other photoelectronic applications.

A. Preliminary Features of the RF Glow Discharge

Historically, it was Faraday who first discovered the (dc) glow discharge by initiating luminous zones in air with a 1000 V power source at the Royal Institution (1831-1835). The striking feature of the discharge, namely the visible luminosity, mainly results from the de-excitation of emitting molecular and atomic species contained in the plasma. However, as we indicated previously, a plasma chemistry of considerable complexity belies the beauty of the glow discharge.

The rf glow discharge in a gas is sustained by inelastic electron-impact processes that are initiated by electrons which have acquired sufficient energy from rf fields as a result of successive elastic collisions with gas molecules. In principle, ion-molecule reactions also play an important role in the kinetics of the discharge. An electron loses an infinitesimal fraction ($\sim m_e/M$) of its kinetic energy in elastic collisions, but such collisions provide a random component to the electron velocity that allows the particle to absorb net power. As we mentioned above, when a glow discharge is ignited in a molecular gas such as silane, the resulting plasma "zoo" contains a wide variety of species: electrons, excited neutrals (free radicals and gas molecules), ions, and photons. All of these species participate in the plasma chemistry in varying degree.

Equations 1-6 summarize general aspects of typical rf glow-discharge plasmas:

$$n_e \sim 10^8 - 10^{12} \text{ cm}^{-3} \qquad (\sim \text{ cation density}) \tag{1}$$

$$<E> \sim 0.5 - 5 \text{ eV} \qquad (T_e \sim 10^4 - 10^5 \text{ K}) \tag{2}$$

$$E_e = \frac{e^2}{2m} \left(\frac{F_o^2}{\omega_c^2 + \omega^2} \right) = \frac{1}{2} m \omega^2 A^2 \tag{3a}$$

$$\sim (F_o/p)^2 \qquad (\omega_c >> \omega) \tag{3b}$$

$$\text{Ions} \sim 10^{-4} \cdot \text{Neutrals} \tag{4}$$

$$V_{dc} \simeq - \left(\frac{3}{2\epsilon} \frac{\omega_p^2}{\omega^2} \right) \frac{E_e}{e} \tag{5}$$

$$T_{gas} << T_e \quad , \qquad T_{gas} \sim T_s = 300 - 650 \text{ K} \tag{6}$$

Equations 1 and 2 list ranges of values for the electron density n_e in the plasma, and for the average electron energy $<E>$, or equivalently, the electron temperature ($\frac{3}{2} kT_e = <E>$). The electron energy distribution, which is only approximately Maxwellian, has an extended tail that, as will be discussed later, is responsible for high-lying excited and ionized states of atoms and molecules in the plasma. It is interesting that the solar corona is another example of a plasma for which the electron density is lower ($n_e \sim 10^7 \text{ cm}^{-3}$) but for which, as anticipated, the average energy is higher ($<E> \sim 100 \text{ eV}$).

Equation 3a expresses the average electron energy E_e acquired between successive elastic collisions in terms of the rf electric field amplitude F_o and the collision frequency ω_c. Since ω_c is proportional to the gas pressure p, the quantity E_e scales with

the ratio (F_o/p), as shown in Eqn. 3b. This simple result also
suggests that the electron temperature, hence the rates of
electron-impact processes, should approximately depend upon the
ratio P/p^2, where P is the rf power. The electron component of
the plasma oscillates in space with rf frequency ω and amplitude
$A = \mu_e F_o/\omega$ owing to the high electron mobility $\mu_e = e/m\omega(\omega_c^2+\omega^2)^{\frac{1}{2}}$
compared to that of the ions $(\mu_i/\mu_e \sim m_e/M)$. Indeed, neglecting
ambipolar diffusion, all electrons that are within a distance A
of a surface are removed from the plasma in each half cycle,
leaving a positive space-charge region of width $\sim A$ near the surface. (This region is often referred to as a "dark space", or
"sheath".) It is for this reason that ungrounded electrodes in a
capacitively-coupled plasma deposition system acquire a negative
potential relative to the plasma. Equation 3a also shows that
the electron in the plasma behaves like a harmonic oscillator
with force constant $m\omega^2$, driving frequency ω, and amplitude A.

The dilute plasma condition expressed by Eqn. 4 is easily
understood on the basis of available low-lying excited states
relative to high ionization potentials for atomic and molecular
species in the plasma. For example, consider the case of the
molecule SiH in the silane glow discharge. In the plasma it is
more probable that electron-impact processes lead to a low-lying
excited state of the neutral molecule, e.g., the state $A^2\Delta - X^2\Pi$
(3 eV above the ground state), than that ionization of the SiH
molecule (ionization potential of 8 eV) is effected by the tail
of the electron energy distribution. The influence of ions and
electrons upon film growth and film properties, however, may be
rather greater than suggested by cursory deduction from Eqn. 4.

Equation 5 gives a relation for the negative dc bias V_{dc} that
develops between either electrode of a symmetric system and the
plasma owing to oscillation of the electron component (see, e.g.,
Eqn. 3a). V_{dc} is expressed in terms of the "potential drop"
E_e/e, multiplied by a dimensionless factor that contains the
plasma frequency $\omega_p^2 = n_e e^2/m\varepsilon_o$ and the relative permittivity ε of

the plasma ($\varepsilon \gtrsim 1$). For a typical plasma, the condition that rf radiation is absorbed, namely ($\omega^2 + \omega_c^2$) > ω_p^2, is indeed satisfied. (Otherwise, radiation would be reflected, as with radio waves that bounce off the ionosphere.) Equation 5, which must be regarded as a crude approximation, was derived under the following assumptions: i) diffusion by electrons to chamber surfaces could be neglected; and ii) the cation space charge density is constant in space and equal to n_e. Precisely because of the existence of the space charge region, the flux of positive ions onto an ungrounded electrode (cathode) is high relative to the concentration of cations in the discharge expressed in Eqn. 4. Moreover, the bias accelerates cations onto the cathode with mean energy $\sim e|V_{dc}|$ that greatly exceeds thermal energies for neutrals. Therefore, it is possible that properties of alloys grown on the cathode are influenced by energetic cation bombardment that, e.g., alters existing bonding arrangements, incorporates new bonding patterns, and influences the rates of gas-surface reactions. A flux of electrons and anions may also influence the properties of films grown on the grounded electrode (anode).

For a numerical example using Eqns. 3 and 5, we assume typical processing conditions such that $n_e = 10^9$ cm^{-3}, $\omega = 13.56$ MHz, $\omega_c = 5 \times 10^9$ s^{-1}, and $F_o = 10$ V cm^{-1}. Then we obtain the representative values $E_e = 0.035$ eV, $A = 0.26$ cm (a measure of the "dark space" width), and $V_{dc} = -9.3$ V (the potential drop across the dark space).

Equation 6 summarizes a fundamental feature of the rf plasma: the decoupling of the two characteristic temperatures of the plasma, i.e., the decoupling of the gas temperature T_{gas} and the electron temperature T_e. We previously cited the origin of this circumstance. Namely, in elastic collisions very little kinetic energy is transferred from the energetic electrons to the molecules. Moreover, the inelastic collisions of electron-impact processes predominantly lead to *internal* excited states of product species, i.e., to excited electronic states of atoms, and to excited electronic, vibrational, and rotational states of molecules. The net result is that the kinetic energy (translational energy) of molecules, hence T_{gas}, is not significantly increased in the plasma which contains the coexisting hot electron component.

B. *Plasma Chemistry of the Silane Glow Discharge*

In order to achieve an understanding of the basic plasma kinetics and surface reactions encountered in plasma deposition, a variety of complementary techniques must be employed to study reactive radical species. Such techniques include optical emission spectroscopy, molecular absorption spectroscopy, mass spectrometry, and novel laser techniques with good spatial resolution for investigating the gas-solid interface of nascent films. We will begin this Section with a brief discussion of plasma etching, a dry etching process of silicon (or SiO_2, Si_3N_4, etc.) for which complementary plasma techniques have already given considerable insight. The specific interest in the chemistry of plasma *deposition*, however, is more acute than in the case of plasma etching. The ultimate objective in deposition is to find correspondences between chemical bonds in the plasma and chemical bonds in the condensed state, i.e., in films with desired optoelectronic properties. In the case of etching, material is ablated, and chemical bonds simply vaporize into the reactor effluent. Later in the Section we explore optical emission spectroscopy as a technique for the identification of emitting reactive species in the plasma deposition of a-Si:H alloys. The Section ends with a partial list of possible electron-impact processes that occur in the silane glow discharge.

1. Plasma Etching Paradigm. The etch rate of silicon in a CF_4 plasma is known to increase with the addition of oxygen to the plasma. This problem was solved in general terms using both optical emission spectroscopy (OES) and mass spectrometric

techniques (24-26). Since the first technique may be less familiar to the reader, we note here that the de-excitation of atoms
and molecules in the plasma gives rise to photons that (in the
visible part of the spectrum) constitute the luminosity of the
"glow discharge". Spectral analysis of the emission from the
discharge (using a monochromator) allows the identification, in
principle, of the subset of *emitting* reactive species in the
plasma. Using OES, the high etch rate for $\sim$ 15-25% O_2 in the
CF_4 gas stream was correlated with a high emission intensity
from both atomic F and molecular CO, and with suppressed emission from atomic O. In addition, mass spectrometry of the reactor effluent, downstream of the reaction chamber, showed that
the stable end products of the etching are SiF_4, CO, CO_2, and
COF_2. It was therefore concluded that: i) the oxidation of
carbon-bearing species releases atomic F; and ii) F is responsible for etching the silicon and produces only one stable
species, namely SiF_4. Since it is known that CF_4 does not
directly react with atomic O, the radical CF_3 is suspected as a
reaction intermediate that is oxidized. Subsequent reactions
then lead to free atomic F. The recombination of fluorine with
adsorbed carbon may be strongly suppressed owing to the presence
of atomic O. This would be an additional mechanism that aids
the etching.

It is also of interest that in plasma etching the etch rate
is suppressed by the presence of impurity H_2O ($\lesssim$ 20%). Using
OES in a diagnostic mode, this effect can be correlated with
quenched emission from atomic F, and with enhanced emission
from atomic hydrogen that is produced via the decomposition of
H_2O in the plasma (25).

2. *Toward the Identification of Species in Plasma Deposition
of a-Si:H Alloys Using Optical Emission Spectroscopy.* Figure 1 illustrates part of an optical emission spectrum for a
silane glow discharge with 600 ppm added N_2 impurity (17). Since

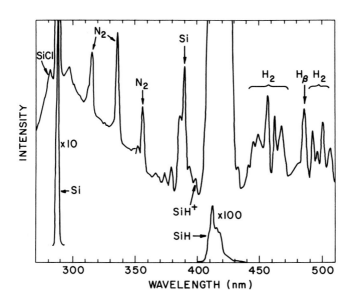

FIGURE 1. *Optical emission spectrum of plasma with 600 ppm*
N_2 *added. (50 W, 250 mTorr, T_s = 270°C.)*

optical emission spectroscopy (OES) does not perturb the discharge, this technique is routinely used by our group to monitor
the plasma during film deposition. The particular deposition
conditions used here were: P = 50 W rf power, p = 250 mTorr,
100 scc min^{-1} silane flow rate, and substrate temperature T_s =
= 270°C. The following emitting reactive species were detected
in the plasma: Si, SiH, SiH$^+$, H_2, H, and the impurities SiCl and
N_2. Moreover, when oxygen is intentionally added to the silane
gas stream, or when a sizable airleak occurs in the deposition
system, weak emission from the molecule SiO is also observed
(27). Figure 2 shows an emission spectrum for an oxygenated discharge (1000 ppm added O_2) scanned at higher resolution and sensitivity.

The silicon multiplets of Fig. 1, Si(288 nm) and Si(391 nm),
or those of Fig. 2, arise from distinct electronic transitions

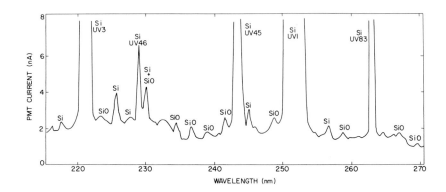

FIGURE 2. Optical emission spectrum of plasma with 1000 ppm
O_2 *added. (15 W, 100 mTorr, T_s = 270°C.)*

of atomic silicon. Some of these Si lines, such as Si UV1 (251-
253 nm) in Fig. 2, consist of several components owing to fine
structure effects (spin-orbit coupling). By contrast, the prominent SiH band (413-428 nm), together with weaker bands between
386-396 nm (not explicitly identified in Fig. 1), correspond to
the *same* electronic transition. This transition, $A^2\Delta$-$x^2\Pi$, is due
to de-excitation from levels that lie ∿ 3 eV above the electronic
ground state. Each band, however, comprises distinct vibronic
bands (i.e., specific vibrational transitions with unresolved
rotational structure). The SiH band (413-428 nm) contributes to
the rather striking deep violet appearance of the silane glow
discharge. Bands from the electronic transition $A^2\Delta$-$x^2\Pi$ are also
observed in sunspots and weakly in the solar disc. As we mentioned in Sect. II.*A*, the weak emission in Fig. 1 from the cation
SiH^+ (399 nm) can easily be explained: The probability is higher
for exciting low-lying levels of the neutral molecule SiH than
for ionizing this species with an ionization potential of 8 eV.
Another possible source of SiH^+ is the dissociative ionization of
SiH_4.

Emission from molecular hydrogen, H_2 (450-630 nm), is also evident in Fig. 1. Moreover, the continuum in Fig. 1 below approximately 400 nm is due to H_2 emission. The extent of the tail for the electron energy distribution can be appreciated by noting that initial energies for the emitting state in the H_2 spectra lie as high as 14 eV above the ground state. Emission from H_β (486 nm) in the Balmer series signals the presence of reactive hydrogen in the plasma. The dark red emission of H_α (656 nm) is also observed (not shown), while under some processing conditions H_γ (434 nm) can be detected. Besides hydrogen impurity in silane tanks, sources of H_2 in the plasma include: i) subsequent gas-phase reactions of primary products of the silane decomposition; and ii) surface reactions at the gas-solid interface of the growing a-Si:H alloy for which the strong bond of H_2 is energetically favored (4.5 eV).

The observed emitting species Si and SiH could arise directly from the decomposition of excited (and unstable) silane molecules SiH_4^*, or via the decomposition of SiH_2 and SiH_3. Unfortunately, these last species have no known emission spectra. Therefore, using the OES technique alone, it is not possible to study such species in the silane glow discharge. On the other hand, the importance of SiH_2 and SiH_3 to the growth of a-Si:H alloys may be inferred from a mass spectrometric study (28) of the 147 nm photolysis of silane (photon energy $h\nu = 8.4$ eV). In that case the two reactions[2]

$$h\nu + SiH_4 \rightarrow SiH_4^* \rightarrow SiH_2 + 2H \quad , \tag{7}$$

$$\rightarrow SiH_3 + H \tag{8}$$

were calculated to have quantum yields of 0.83 and 0.17, respectively. From the study it was also concluded that wall reactions

---

[2] *In photolysis, the threshold for the formation of SiH_4^* is $h\nu = 7.5$ eV. Electron energies >11.7 eV are required for ionization of SiH_4.*

of SiH_3 and SiH_2SiH_2 radicals led to film formation. For processing conditions that are more appropriate to the plasma deposition of a-Si:H alloys, film growth may instead proceed from complex surface reactions that involve the gas phase species H, SiH,
SiH_2, and SiH_3 (and perhaps certain ions as well). It is clear
that in order to extract the maximal information about the plasma
chemistry in film deposition, OES must be complemented by additional techniques.

Pursuing our analogy with the plasma etching paradigm of
Sect. II.B.1, we now turn to the impurities that were detected in
plasma deposition, namely SiCl(281 nm) and $N_2(C^3\Pi_u-B^3\Pi_g)$ of Fig.
1, and $SiO(A^1\Pi-X^1\Sigma^+)$ of Fig. 2. The high sensitivity of OES for
the first two impurities suggests that this technique may play a
useful diagnostic role in the routine monitoring of the plasma.
The presence of oxygen, however, is more easily detected by infrared transmission spectroscopy of vibrational modes, such as
the Si-O-Si stretch modes, in as-deposited films. (See Sect.
III.)

At higher resolution, the emission from SiCl can be resolved
into at least two components, i.e., vibronic bands that belong
to the common electronic transition $B^{-2}\Delta-X^2\Pi$. The presence of
this molecule in the plasma can be attributed to the decomposition of trace chlorosilane impurities originally contained in
the silane process gas. Analysis by secondary-ion-mass-
spectrometry (SIMS) showed a qualitative correlation between the
concentration of Cl incorporated in a-Si:H alloys and the emission intensity of SiCl in the plasma when using different silane
tanks. However, for 2-600 ppm incorporated chlorine, no obvious
correlation was found between the chlorine concentration and the
electrical properties of a-Si:H alloys, although an inverse relation with photoconductive response is possible. One difficulty
is that the effects of chlorine are masked by the presence of
other impurities in silane tanks. If the chlorine acts to passivate defects by participation in bonds such as $\equiv$SiCl and

=SiHCl, it is conceivable that a small concentration would not have an appreciable effect upon electrical properties. The efficiency for defect passivation by chlorine, compared to that by hydrogen (or fluorine), is presumably influenced by the relatively large covalent radius $\simeq 0.99$ Å for this element.

The molecule N_2 is a ubiquitous impurity of the usual plasma deposition system. The high sensitivity of OES for emission from N_2 (second positive system) allows quantitative calibration of the N_2 content in the plasma. This feature is quite useful since concentrations of N_2 as low as several hundred ppm in the plasma can affect the electrical properties of a-Si:H alloys. As will be amplified in Sect. III, the modification of film properties is related to the ability of incorporated N to introduce shallow defects, or to act as a donor, i.e., an electronic dopant, in the amorphous network.

The small relative emission intensity of SiO with respect to atomic Si is apparent from Fig. 2. Each SiO peak, nestled among the "sequoia trees" that are the Si multiplets, corresponds to a distinct vibrational-rotational transition. The common electronic transition $A^1\Pi-X^1\Sigma^+$ involves a de-excitation from an initial state ~ 5.3 eV above the ground state. Since oxygen exhibits extreme reactivity toward silicon hydrides, a-Si:(H,O) alloys are deposited under the processing conditions of Fig. 2. Because SiO is intrinsically a good emitter, the weak relative emission of SiO in Fig. 2 implies that the observed SiO was not a reactive intermediate in the incorporation of oxygen in the a-Si:(H,O) alloy. Instead, this emitting species must be a by-product of that incorporation. As a consequence, the SiO emission is not expected to be a quantitative measure of oxygen in the plasma that could be used to calibrate oxygen incorporation in films.

3. Electron-Impact Processes. We review typical processes that in principle give rise to the primary species and major

reactive intermediates engaging in the plasma chemistry of a
silane glow discharge. Some of the reactions are proven on the
basis of the preceding results, but the importance of others can
only be speculative at this stage, pending innovative experiments
for their elucidation.

   a. Electron-molecule (-atom) reactions.

Excitation: $e + Si \rightarrow Si^* + e$ (5.0-6.8 eV)

                  $e + SiH \rightarrow SiH^* + e$ (3.0 eV)

                  $e + SiH^+ \rightarrow (SiH^+)^* + e$ (3.1 eV)

                  $e + H \rightarrow H^* + e$ (12-13 eV)

                  $e + H_2 \rightarrow H_2^* + e$ (11.9-14.0 eV)

The energies in parentheses refer to measurements by OES in Ref.
27 and denote the positions of the initial emitting states above
the ground state. Of course, the reactions themselves are more
general, and different levels of excitation would depend upon
the particular plasma processing conditions involved.

Dissociation: $e + SiH_4 \rightarrow SiH_3 + H + e$

                    $\rightarrow SiH_2 + 2H + e$

                    $\rightarrow SiH + 3H + e$

                    $\rightarrow Si + 4H + e$

These equations are the analogs of Eqns. 7 and 8 for photolysis
in which the threshold for absorption and dissociation of SiH_4
is 7.5 eV (165 nm). Of course, the silicon-hydride products in
these reactions can be further dissociated by subsequent
electron-impact processes, e.g., $e + SiH_3 \rightarrow SiH + 2H + e$. One
may naively expect that each successive reaction above requires
an additional increment in electron energy equal to the Si-H

bond energy, i.e., equal to 3.06 eV (29) or perhaps 3.6-4.1 eV
(30). On such a model, Eqn. 3b would imply that higher rf power
and lower silane pressure would favor the production of SiH, Si,
and H_2 (H) primary products in the gas phase.

Ionization: $e + SiH \rightarrow SiH^+ + 2e$ (IP = 8 eV)

$e + SiH_4 \rightarrow SiH_4^+ + 2e$ (IP = 11.7 eV)

Dissociative $e + SiH_4 \rightarrow SiH_2^+ + H_2 + 2e$ (AP = 11.9 eV)
Ionization:
$\rightarrow SiH_3^+ + H + 2e$ (AP = 12.3 eV)

The appearance potential for the cation SiH_2^+ shows that the
cation SiH_4^+ itself is unstable, i.e., the reaction $SiH_4^+ \rightarrow$
$\rightarrow SiH_2^+ + H_2$ is endothermic by only 0.2 eV (30). These ionization processes replenish the plasma with electrons and thereby
sustain the discharge. Such "breeder" reactions compete with
electron attachment, certain electron-cation reactions, and the
loss of electrons from the plasma by diffusion and drift to
walls and electrodes.

Attachment: $e + H \rightarrow H^-$ (affinity = 0.75 eV)

$e + SiH_4 \rightarrow (SiH_4^-)^*$

Dissociative $e + SiH_4 \rightarrow SiH_3^- + H$
Attachment:
$\rightarrow SiH_2^- + H_2$

The dissociative attachment reactions could involve *low energy*
resonances between electrons and gas molecules (31). In this
regard, recall that the average electron energy $<E>$ of Eqn. 2 is
rather low. A steady state could rapidly be attained for the
anion population as a consequence of competing anion destruction
processes such as associative detachment.

b. Electron-cation reactions.

Recombination: $e + SiH_2^+ \rightarrow SiH_2$

Dissociative $e + SiH_2^+ \rightarrow SiH + H$
Recombination:

Other processes will now be listed that are *a priori* of considerable importance to film growth.

c. Surface reactions (at electrodes).

$$SiH_2 \underset{s}{\rightarrow} Si(film) + H_2$$
$$SiH_3 \underset{s}{\rightarrow} SiH(film) + H_2 \quad , etc. \tag{9}$$

$$SiH^+ + SiH_3(film) \underset{s}{\rightarrow} SiH_2^+(film) + SiH_2(film)$$
$$Si^+ + SiH_2(film) \underset{s}{\rightarrow} SiH^+(film) + SiH(film) \tag{10}$$

(cation bombardment of films)

As discussed in Sect. II.*A*, the small relative population of ions in the plasma may be a poor measure of the importance of ions to film growth that proceeds by surface reactions. Indeed, the effects of energetic ion bombardment may be twofold: i) a direct influence on the conversion and incorporation of chemical bonds in films (Eqn. 10); and ii) the stimulated increase in the reaction rate of other gas-surface interactions, e.g., those of Eqn. 9. Finally, we mention that catalytic effects of chamber surfaces can even alter gas-phase reactions.

d. Ion-molecule reactions.

$$SiH_2^+ + SiH_4 \rightarrow SiH_3^+ + SiH_3$$
$$\rightarrow Si_2H_4^+ + H_2 \quad , etc.$$

In contradistinction to the surface reactions involving ions,
these gas-phase reactions are restricted by Eqn. 4 and may not
play a large role in plasma deposition at lower power and pressure.

III. CHARACTERIZATION OF PLASMA-DEPOSITED SILICON-HYDROGEN ALLOYS

Evaporated and sputtered films of pure a-Si exhibit a large
density of gap states $\sim 10^{20}$ cm^{-3} eV^{-1}, a large electron-spin-
resonance (ESR) signal $N_s \sim 10^{19}$-10^{20} spins cm^{-3}, high dark conductivity, poor photoconductivity, and little or no photoluminescence at low temperatures. By contrast, a-Si:H alloys can be
deposited with a low density of gap states ($\sim 10^{15}$-10^{16} cm^{-3} eV^{-1}
near midgap), low spin density $N_s \sim 10^{14}$-10^{15} spins cm^{-3}, high
resistivity, high photoconductivity, and nearly 100% efficient
peak photoluminescence. As a consequence, a-Si:H alloys can be
electronically doped (1-3) and used to fabricate promising
photovoltaic devices (4,23). It is interesting to note that
these alloys are still not completely well-characterized materials. Nonetheless, certain salient materials properties have
been established, including features of the silicon-hydrogen
bonding (Sect. III.*A*) and many optoelectronic properties (Sect.
III.*B*). In addition, in Sect. III.*C* we shall explore the evolving chemical definition of the nominally intrinsic silicon-
hydrogen amorphous alloy a-Si:H. Indeed, it has been found that
the distribution of gap states for "a-Si:H" alloys may sensi-
tively depend upon the degree of incorporation of common atmospheric impurities encountered in plasma deposition (17). In
fact, the more correct designation for the usual a-Si:H alloy
should perhaps be a-Si:(H,O,N,C,...), where concentrations of
the impurities can be typically as high as $\sim$ 0.1-1%.

A. *Introduction to Silicon-Hydrogen Bonding in the Alloy*

1. Short-Range Structural Order. Radial distribution functions from electron diffraction data for evaporated films of pure a-Si show that: i) the first and second coordination shells are nearly identical with those of the ideal diamond lattice; but ii) the third-neighbor coordination does not exist (32). Hence, whereas a-Si lacks the long-range structural order of the perfectly periodic crystal, this noncrystalline phase retains *short-range* structural order. That is, the classic tetrahedral bonding of nearest neighbors depicted in Fig. 3a is approximately preserved for a-Si. In contrast to random network models for a-Si that are consistent with the diffraction data, structural models for the alloy a-Si:(H,O,N,C,...) are quite complex and depend on the degree and kind of alloying involved. (See Fig. 3b.) An important principle to note for both materials, however, is that most fundamental semiconductor concepts follow from short-range structural order alone in a covalently-bonded solid, i.e., from the local chemical bonding. In particular, the existence of characteristic energy bands separated by a "gap" in electron energy is independent of whether the lattice is ordered (crystal), or disordered (amorphous substance). Indeed, this fact was always implicit in the chemists' picture of a semiconductor as a giant molecule in which the conduction band is composed of "anti-bonding" states and the valence band is composed of "bonding" states.

An important theorem due to Weaire and Thorpe (7) proved that the infinite-range correlation of a perfect lattice is not required to predict the existence of a bandgap. In order to appreciate this assertion, we first sketch the usual argument that is employed to predict energy bands for the crystal phase in the tight-binding approximation. In this case, a Wannier expansion of the electron wavefunction that embodies Bloch symmetry is inserted into Schrödinger's equation for the crystal. The

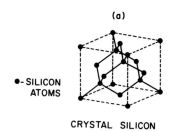

(a)

●- SILICON
ATOMS

CRYSTAL SILICON

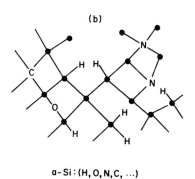

(b)

a- Si : (H, O, N, C, ⋯)

*FIGURE 3.   Comparison of crystal structure with that of amor-
phous "silicon-hydrogen" alloy: (a) unit cell of c-Si; (b) schematic planar representation of the alloy a-Si:(H,O,N,C,...).   In
(b) the concentrations of contaminants in the alloy relative to
hydrogen have been exaggerated for the purpose of illustration.*

retention of nearest-neighbor non-diagonal matrix elements of the
Hamiltonian then gives the elementary energy eigenvalue spectrum
for bands found in textbooks. The bandgap itself emerges from
consideration of the first Bragg reflection at the boundary of
the first Brillouin zone. In other words, the crystal bandgap
is derived by initially assuming an exact Bloch wavefunction, and

then approximating the Hamiltonian with nearest-neighbor (and diagonal) interactions between atomic lattice sites. By contrast, on the model of Ref. 7 one *begins* with a Hamiltonian $\hat{H}$ that incorporates overlapping bonding orbitals between nearest-neighbor atomic sites, and then one solves for the exact energy eigenvalue spectrum. Topological disorder is thereby implied. In the simplest formulation,

$$\hat{H} = V_1 \sum_{n,a,a'} |n,a\rangle\langle n,a'| + V_2 \sum_{(n,a),(m,b)} |n,a\rangle\langle m,b| \qquad (11)$$

$\qquad\qquad$ (atomic sites) $\qquad\qquad\qquad$ (overlapping bonding orbitals)

$$\equiv zV_1\hat{H}_1 + V_2\hat{H}_2 \ . \qquad\qquad\qquad\qquad (12)$$

In Eqn. 11 the state $|n,a\rangle$ represents the orbital a at atomic site n; V_1 is the self-energy, i.e., the interaction energy between orbitals at the same site; and V_2 is the interaction energy for all overlapping orbitals (n,a) and (m,b) that constitute two-center bonds. In Eqn. 12, z is the coordination number. The operators $\hat{H}_{1,2}$ have simple interpretations: $\hat{H}_1$ takes the orbital at any site and averages it over all orbitals at that site, and $\hat{H}_2$ transforms any orbital into the nearest-neighbor orbital that shares the common bond. Making use of these properties, and employing operator algebra manipulations, one can solve for the eigenvalues of $\hat{H}_1$, $\hat{H}_2$ and, hence, of $\hat{H}$. The result is that $\hat{H}$ is a two-band Hamiltonian, and that the bands are separated by an energy gap of width $|zV_1 - 2V_2|$. Of course, $z = 4$ for the sp^3 orbitals of silicon. Thus, the existence of short-range structural order alone is sufficient to guarantee the existence of a bandgap for a topologically disordered semiconductor.

2. Role of Hydrogen in a-Si:H Alloys. The demonstration that hydrogen is indeed bonded in plasma-deposited materials is best provided by measurements of infrared transmission spectroscopy

on as-deposited films. Figure 4 illustrates the characteristic
vibrational modes due to silicon-hydrogen bonding in a film produced at medium-high substrate temperature (T_s = 270°C) and relatively low pressure (p = 250 mTorr). For this alloy with predominant monohydride bonding, the stretch mode of the complex
$\equiv$SiH is readily identified at $\tilde{v}$ = 2000 cm^{-1}. The degenerate wag
and rock modes of this singly hydrogen-bonded complex lie at $\tilde{v}$ =
= 630 cm^{-1}. Calibration of the integrated absorption with analyses by SIMS indicates that $\sim$10% hydrogen is incorporated in this
film. Assuming that the mild absorption at $\tilde{v}$ = 880 cm^{-1} can be
identified with the bend-scissors and twist modes of the dihydride bonding =SiH$_2$, it can be estimated that the proportion
of dihydride modes in this particular alloy is only $\sim$4%. By
comparison, this proportion can be reduced below detectability
for alloys grown at lower rf power (15 W) and pressure (100
mTorr). As first pointed out by Brodsky, et al. (10), low substrate temperature and high pressure are processing conditions
that conspire to give a larger proportion of polyhydride bonding
in alloys. Examples are dihydride and trihydride bonding, and
(SiH$_2$)$_n$ polymer bonds that are associated with electrically-
active defects (11). The non-controversial signature of dihydride or (SiH$_2$)$_n$ polymer bonding is a bend scissors mode at
$\tilde{v}$ = 880-890 cm^{-1}. In the case of (SiH$_2$)$_n$, a wag mode at $\tilde{v}$ =
= 845 cm^{-1} is also observed. Both complexes may exhibit a
stretch mode at $\tilde{v}$ = 2090 cm^{-1}. However, it has recently been
suggested that this 2090 cm^{-1} absorption may require a more complex interpretation (33). Evidence is mounting that alloys with
predominant monohydride bonding, i.e., no 880-890 cm^{-1} modes,
exhibit better optoelectronic response (photoconductivity,
photoluminescence), thereby resulting in improved device characteristics.

By comparing film properties of pure a-Si with those of the
a-Si:H alloy, it is concluded that the role of hydrogen in the
alloy is twofold: i) hydrogen passivates electrically-active

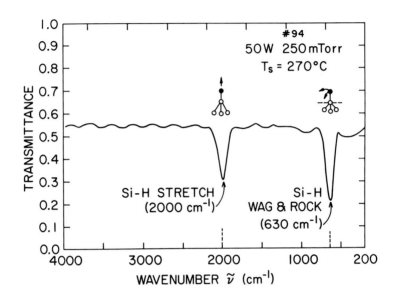

FIGURE 4.   Infrared transmittance spectrum of predominantly monohydride a-Si:H alloy.   Film thickness is 4 μm.

defects in the amorphous network[3]; and ii) the silicon-hydrogen bonding *per se* modifies the distribution of bonding and anti-bonding states of otherwise pure silicon, creating a new electronic material.   To illustrate the first role, the minimal density of spins in a-Si:H films ($N_s \sim 10^{14}$-10^{15} cm^{-3}) is consistent with the idea that hydrogen passivates a whole class of defects with positive correlation energy[4] $U > 0$, namely, the dangling bond states ($\equiv$Si·).   By contrast, we noted before that $\sim 10^{20}$

---

[3]*Although in large concentrations, perhaps >25%, hydrogen incorporation may actually create new defects.   As an example, dangling bonds could be created in voids associated with $(SiH_2)_n$ polymer chains (11).*

[4]*The electron-electron correlation energy U may be thought of as the energy needed to add an extra electron to the defect; it contains the Hubbard energy of Coulombic repulsion.*

paramagnetic states cm^{-3} ($g \simeq 2.0055$) exist in pure a-Si (34,35).
It is therefore reassuring to observe that the spin density of
ultrahigh-vacuum-evaporated a-Si is quenched below $N_s = 10^{17}$ cm^{-3}
when the material is subsequently exposed to a hydrogen plasma at
400-500°C (36). Now in a typical a-Si:H alloy with monohydride
bonding the concentration of hydrogen is ∿5-10%. However, the
concentration of paramagnetic centers in a-Si is only a maximum
of ∿0.2%. Therefore, even if some minimum concentration of hydrogen (∿1%?) were consistent with the best electronic properties
of a-Si:H alloys, hydrogen would still be expected to passivate
other, more numerous, defects. One obvious candidate is the
class of defects with paired electron spins. For such localized
states, an energy-dependent and negative correlation energy
$U(E) < 0$ could imply *a spectrum* of weak molecular bonds in the
amorphous network[5]. Thus the class of spin-paired defects spans
weakly reconstructed Si-Si bonds, divacancies, etc. Incorporated
hydrogen may convert, for example, a reconstructed bond into
≡SiHHSi≡, a broken Si-Si bond with two H atoms inserted.

  The second role of hydrogen is suggested by measurements of
both photoemission spectroscopy (PES) and optical absorption.
For example, using PES it was found that the valence band edge
recedes up to 1 eV relative to a-Si as a function of hydrogenation in sputtered and plasma-deposited a-Si:H (37). That is,
Si 3p states that constitute the leading edge of the a-Si valence
band were chemically modified by the silicon-hydrogen bonding.
Since the leading edge of the Si-H bonding states would be expected[6] to lie below the top of the distribution of Si-Si

---

[5]*However the detailed role of the electron-phonon interaction
in promoting spin-pairing is not well understood at present*

[6]*Because of the small covalent radius of hydrogen, the Si-H
bond strength, 3.06 eV, exceeds the Si-Si bond strength, 1.94 eV
(29).*

bonding states, the valence band should indeed recede and the
energy gap should widen. Using the PES technique (37), states
that arise from the overlap of Si 3s, 3p orbitals and the H 1s
orbital can be identified with the bonding configurations $\equiv$SiH,
$=$SiH$_2$, $-$SiH$_3$, and (SiH$_2$)$_n$. Consistent with the findings of PES,
the optical bandgap also widens with increasing hydrogenation
(38). For the same kind of alloy depicted in Fig. 4 ([H] = 10%),
Fig. 5 shows the determination of the effective optical gap
$E_o \simeq 1.65$ eV from reflection and transmission measurements at
visible and near-infrared wavelengths. E_o was extrapolated from
a linear fit to the quantity $(\alpha h\nu)^{\frac{1}{2}}$, where $\alpha(\nu)$ is the fundamental optical absorption coefficient and $h\nu$ is the photon energy
(39). By comparison, the optical gap for a pure a-Si film, also
prepared at $T_s \simeq 270°$C, lies in the range $E_o = 1.20$-1.35 eV.
Consistent behavior is found when alloys are annealed at

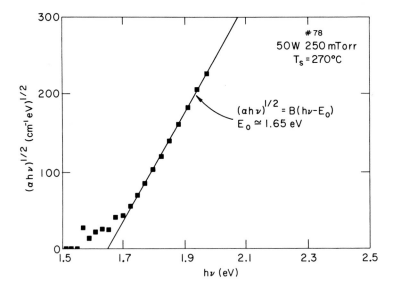

FIGURE 5.   Optical bandgap E_o deduced from plot of $(\alpha h\nu)^{\frac{1}{2}}$
versus photon energy $h\nu$ at room temperature.

450–600°C: hydrogen effuses, N_s increases owing to the regenera-
tion of dangling bond defects, and E_o decreases toward values
characteristic of pure a-Si (40).

It is evident that the density and distribution of gap states
should have a profound effect upon the optoelectronic properties
of the material. As previously discussed, the incorporation of
hydrogen in a-Si:H alloys significantly decreases the density of
gap states relative to the case of pure a-Si. To illustrate this
feature, Fig. 6 shows the density-of-states distribution $n(E)$
for an a-Si:H film deposited at about T_s = 250°C (41). Also
shown is the density of states in the upper half gap for an
a-Si:(F,H) alloy that was plasma deposited with a gas ratio
$[SiF_4]/[H_2]$ = 5 at the nominal substrate temperature T_s = 380°C

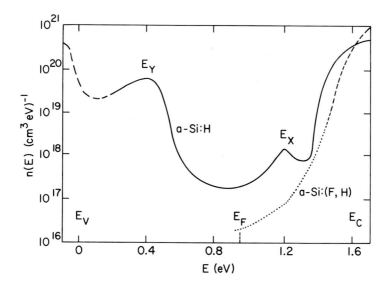

*FIGURE 6.   One-electron density-of-states distribution n(E)
for a-Si:H at $T_s \approx 250°C$ (Ref. 41), and for a-Si:(F,H) at nominal
$T_s = 380°C$ (Ref. 18).*

(18). For both alloys $n(E)$ was deduced from field-effect-conductance measurements. The energies E_C and E_V denote the "mobility edges" of the conduction band and valence band, respectively. These edges are sharp demarcations between free-carrier states and band-tail states that are localized owing to random fluctuations in bond angles and bond lengths. The energies E_Y and E_X refer to peaks in the gap state density that are widely disparate in magnitude. Because of the large difference in magnitude of these peaks, donor-like and acceptor-like states of the *same* defect cannot be invoked to explain their origin. It should be pointed out, however, that the specific presence of two peaks in $n(E)$ invites controversy since the analysis of field-effect-conductance data is subject to many assumptions. Foremost among these limitations is the neglect of surface or interface states that leads to an overestimation of gap state density. Other techniques, in particular deep-level-transient-spectroscopy (42), appear to be promising alternatives for the determination of $n(E)$. A less controversial feature can be abstracted from Fig. 6, namely, that overlapping tails of defect densities pin the Fermi level E_F near the minimum in $n(E)$. On this model (41), donor-like states that are neutral when occupied lie mostly in the lower half of the gap; acceptor-like states that are neutral when unoccupied lie mostly in the upper half gap. (See also Sect. III.*A.3.*) Near the Fermi level, unoccupied donor defects (occupied acceptor defects) then give rise to positive (negative) charge distributions. Without invoking structure such as peaks in $n(E)$, the density of donor-like states may be an order-of-magnitude larger than for the acceptor-like states. The high proportion of hole-trapping states in the lower half of the gap *must fundamentally limit minority-carrier transport* in these materials.

3. Influence of gap states on diode characteristics. The degree to which gap states affect device properties is well

illustrated by the case of the dark forward current $J =$
$= J_o\{\exp(eV_F/nkT) - 1\}$ for Pt/a-Si:H Schottky diodes. (Films
of a-Si:H are "n-type" when not intentionally doped, hence such
diodes are n-type Schottky devices.) Both the diode quality
factor n and the saturation current J_o are found to be sensitive
functions of the alloy preparation conditions (21). In particular, ideal diode rectification characteristics, namely $n \simeq 1$
and small values $J_o \sim 10^{-11}$ A cm^{-2}, are obtained for low deposition pressure p, high substrate temperature T_s, high flow rates
of silane, and post-anneal of the Pt contact. For example, Fig.
7 shows the deviation from diode perfection at $T_s = 300°C$ as a
function of increasing chamber pressure from $p = 60$ mTorr to
$p = 300$ mTorr (21). As discussed previously, the processing
conditions that result in more ideal diode parameters are
consistent with a higher proportion of monohydride bonding $\equiv$SiH
versus polyhydride bonding in the bulk of the a-Si:H alloy.
These results suggest that the recombination current[7] increases
relative to the usual injected current with increasing polyhydride content. Therefore, polyhydride bonding is correlated
with defects in the bulk of the alloy that, in turn, lead to
poor diode characteristics. Now because of band bending, the
Fermi level lies lower in the gap toward E_V as a function of
position nearer to the metal-semiconductor interface. Hence on
Spear's model (41), deep-lying unoccupied defect donors: i) contribute to the net positive charge of the space-charge region;
ii) self-consistently help to pin E_F; and iii) act as recombination centers that affect transport. The insensitivity of the
open-circuit voltage V_{oc} to preparation conditions is taken as

---

[7]For the n-type Schottky diode, recombination in the space-
charge region near the metal-semiconductor interface gives rise
to a current that behaves as $J_r \sim J_{or}\exp\{(\phi_n-\phi_p)/mkT\} \sim$
$\sim J_{or}\exp(eV_F/mkT)$, $m \simeq 2$, compared to the usual injection current $J_i \sim J_{oi}\{\exp(eV_F/kT)-1\}$.

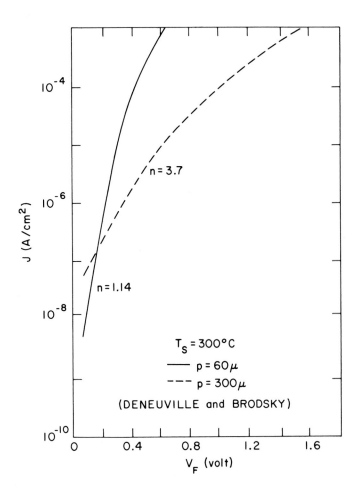

FIGURE 7. Dark current versus forward bias for two dif-
ferent silane pressures, showing deviation from diode perfec-
tion with increasing polyhydride bonding (Ref. 21).

evidence that E_F is pinned by bulk states in the alloy (21). On
the other hand, interface states may also be insensitive to
preparation conditions. Such states with density $\sim 10^{13}$ $cm^{-2} eV^{-1}$
(43) may assist in pinning E_F and in determining the barrier
height ϕ_{Bn}. In either case, bulk states in the alloy strongly

influence recombination rates and therefore affect the transport parameters n and J_o.

Another interesting example of the influence of gap states on device measurements is the following: these states lead to a more complicated interpretation of the differential barrier capacitance C for Schottky diodes. As a consequence, the usual expressions for C in crystalline diode theory cannot naively be applied to the present case. There are two major aspects that must be taken into account for a description of C in the case of amorphous-semiconductor Schottky diodes (19,20): i) the character of the density-of-states distribution $n(E)$ will determine C, and ii) this relationship will be dependent upon the measuring frequency ν of the applied potential. Conversely, with regard to aspect i), we initially note that measurements of C may, in principle, help determine general features of $n(E)$. It is not clear, however, whether detailed structure, such as the peaks E_X and E_Y in Fig. 6, could be proven or disproven by this means. With regard to aspect ii), if states deep in the gap do not have time to respond during the variation of an applied potential, characterized by period $\tau \equiv 1/\nu$, then such states cannot contribute to the space-charge density that determines C. In order to arrive at a relation that crudely quantifies this statement, we proceed as follows (19). Thermal equilibrium is established between the conduction band and an energy level E in the gap with the characteristic thermal response time:

$$\tau_E = (v\sigma_{nE}N_C)^{-1}\exp\{(E_C-E)/kT\} \sim 10^{-13}\cdot\exp\{(E_C-E)/kT\} \text{ s} \quad , \quad (13)$$

since $(v\sigma_{nE}N_C) \sim (10^7 \text{ cm s}^{-1})(10^{-15} \text{ cm}^2)(10^{21} \text{ cm}^{-3}) = 10^{13} \text{ s}^{-1}$.

For states at E to make a contribution to C, we then require that

$$\tau \equiv 1/\nu > \tau_E \quad . \quad\quad\quad\quad (14)$$

For example, for states located near the midgap, $(E_C-E) \simeq$
$\simeq 0.80$ eV, Eqn. 13 implies that $\tau_E \sim 2.3$ s at $T = 300$ K.   Hence,
$\nu < 1$ Hz is required from Eqn. 14 -- tantamount to a dc measurement!

Returning now to the behavior with bias of the differential
barrier capacitance C (per unit junction area) we note that for
n-type c-Si Schottky diodes the presence of *shallow* donors simply implies a constant charge density N_D in the space charge
region.   (Donor states of phosphorous in c-Si lie 45 meV below
the conduction band edge.)   The solution of Poisson's equation
for the barrier profile potential $\Psi_b(x)$ then results in the
usual linear relation between the quantity C^{-2} and the bias V
($V > 0$ for forward bias):

$$C^{-2} = 2(V_o-V)/\varepsilon\varepsilon_o eN_D \quad \text{(Crystal Case), (15)}$$

where V_o is the "built-in voltage" of the barrier[8].   Extrapolation of Eqn. 15 to large forward bias then yields V_o whereas
the slope determines N_D.

By contrast, in the case of the metal/a-Si:H diode, Ψ_b will
be self-consistently determined by the gap state density.   Measurements of C at $\nu \simeq 0$ Hz then probe all states in the gap.   In
order to obtain an *exact* analytic result for C that illustrates
this circumstance, we make a simplifying assumption.   Namely, we
assume that overlapping *exponential* tails of donor/acceptor defect states pin E_F in the gap as a function of position throughout the space-charge region.   On this model, $n(E) = n_D(E) + n_A(E)$,
where $n_D(E) = N_o \exp\{-\beta_o(E-E_o)\}$, for $E > E_o$, and $n_A(E) =$
$= N'_o \exp\{-\beta'_o(E'_o-E)\}$, for $E < E'_o$.   In addition we assume that the
slopes satisfy $\beta_o = \beta'_o$, but this is not an essential assumption

---

[8]We use the convention for the space charge region that $V_o =$
$= \Psi_b(x)+\Psi(x)$ where $\Psi(x)$ is the true potential difference, i.e.,
$\Psi(0) = 0$ and $\Psi_b(0) = V_o$ at the metal-semiconductor interface
$(x=0)$.

for the basic result. As a consequence, the space-charge density may be expressed as:

$$\rho(x) = \rho[\Psi_b] = \rho_o \sinh\{e\beta_o \Psi_b(x)\} \qquad (\nu = 0), \qquad (16)$$

where $\rho_o \equiv en(E_{Fo})\beta_o^{-1} \equiv en(E_{Fo})kT_o$ and E_{Fo} is Fermi level in
the bulk (quasi-neutral region). Solving for C^{-2}, we receive
instead of Eqn. 15:

$$C^{-2} = 2/\varepsilon\varepsilon_o e^2 n(E_{Fo})\{\cosh\{e\beta_o(V_o-V)\} + 1\} \qquad (\nu = 0) . \qquad (17)$$

An immediate consequence of Eqn. 17 is the behavior $C^{-2} \sim$
$\sim \exp\{-e\beta_o(V_o+V_R)\}$ that results for increasing reverse bias
$V_R = -V > 0$. This (exponential) decrease with reverse bias is
obviously completely opposite to the linear increase with V_R
that would be predicted by Eqn. 15.

Figure 8 shows data at $\nu \simeq 0$ Hz (19) that does indeed confirm a rapid decrease in C^{-2} with increasing V_R. On the other
hand, measurements at $\nu = 10$ Hz (20) probe states slightly nearer
to the conduction band: For increasing V_R, the behavior in
Fig. 8 appears more representative of Eqn. 15. Data is also
shown in Fig. 8 for measurements of C^{-2} at $\nu = 100$ Hz on a
Pt/a-Si:H diode (44). At this frequency, states nearer to E_C
presumably dominate the behavior of C^{-2}. However, if Eqn. 15
were literally applied to this case, the physical interpretation
of the extracted parameters N_D and V_o would be somewhat dubious.
We therefore provisionally note that the two straight line seg-
ments of the 100 Hz data yield "densities" $N_{D1} = 5 \times 10^{15}$ cm^{-3}
and $N_{D2} = 1.4 \times 10^{16}$ cm^{-3}. Under 0.02 sun illumination, measurements at $\nu = 100$ Hz on the same n-type Schottky diode also
gave linear behavior for C^{-2} with V_R. The extracted density
parameter $N_L = 3.8 \times 10^{16}$ cm^{-3} exceeds the densities N_{D1}, N_{D2}
obtained in the dark (44). Provided that Eqn. 15 could at least
be used as a relative measure of space charge densities in the
dark and under illumination, it is possible to conclude the
existence of: i) a sizable optically-induced space-charge

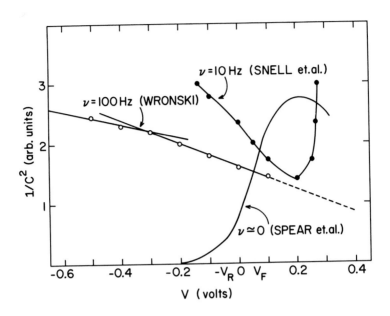

FIGURE 8. C^{-2} plots versus bias V for Schottky diodes measured at indicated frequencies: a) Au/a-Si:H at $\nu \simeq 0$ Hz (Ref. 19); b) Au/a-Si:H at $\nu = 10$ Hz with upturn for increasing V_F due to equivalent circuit assumed for bulk corrections (Ref. 20); and c) Pt/a-Si:H at $\nu = 100$ Hz, bulk corrections unspecified (Ref. 44).

density, and/or ii) a (reversible) photostructural-induced change in $n(E)$ itself (45,46). For measurements of C at $\nu \simeq 0$ Hz that probe deep gap states, it is interesting to speculate about the first possibility. For example, would optically-induced states in the space-charge region of the illuminated Schottky diode share common origins with the optically-induced spin resonances detected for doped and undoped alloys (14,47-49)? In that event, the positive space charge would be enhanced under illumination by contributions from holes that condense into various defect states. Hole polarons may also play a role (50), i.e., free-carrier holes could subsequently self-trap (broad line in ESR centered near $g \simeq 2.01$). Hole polarons may be linked as well to recombination in photoluminescence (15).

Finally, we mention that preliminary studies have been carried out in which solar cell efficiency is correlated with substrate temperature and other processing variables, hence, in principle, with different gap state densities (51).

B. The Dilemma of Efficient Doping but Reduced Minority-Carrier Transport

1. *Electronic Doping and Increased Conductivity.* The first systematic substitutional doping of a-Si:H alloys demonstrated remarkable changes in dark conductivity (1) and in photoconductivity (52,53) as functions of concentrations of the dopant gases phosphine (PH_3) and diborane (B_2H_6). For example, n^+-type films were prepared at $T_s \simeq 250°C$ for which the dark conductivity increased by seven orders of magnitude relative to the undoped state: $\sigma_d(300 \text{ K}) \simeq 10^{-2} \ (\Omega \cdot \text{cm})^{-1}$ for 2000 ppm PH_3 added to the silane gas stream. It was estimated that the substitutional doping efficiency is roughly 15% of the PH_3 concentration in the silane gas stream, and that a maximum of $\sim$200 ppm P atoms can be incorporated as donors in the a-Si:H network (1). To illustrate these remarks, we show our own data in Figs. 9 and 10 for films doped by phosphorous under various processing conditions (54). Figure 9 shows that at room temperature, $\sigma_d \simeq$ $\simeq 10^{-2} \ (\Omega \cdot \text{cm})^{-1}$ was attained with 300 ppm PH_3 dopant. It is probable that the high-temperature activation energy (E_σ) in Fig. 9 pertains to free electron transport. Hence $E_\sigma = (E_C - E_F) =$ $= 0.28$ eV, the distance of the Fermi level from the mobility edge of the conduction band. In the absence of intentional doping $E_\sigma \simeq 0.77$ eV under the same processing conditions. Following Ref. 52 the shallower slope in Fig. 9, namely 0.16 eV, may be due to electron hopping in the donor band, or the explanation may require "polaronic" concepts (55). Figure 10 illustrates classic behavior in photoconductivity for a doped film produced under the indicated processing conditions. The room-temperature

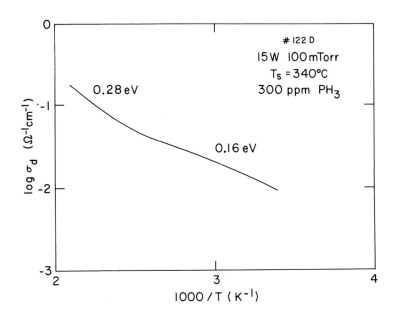

FIGURE 9. Dark conductivity versus reciprocal temperature for phosphorous-doped a-Si:H alloy.

photoconductivity dramatically increased from $\sigma_p = 1.8 \times 10^{-8}$ $(\Omega \cdot cm)^{-1}$ for the undoped material to $\sigma_p = 4.8 \times 10^{-6}$ $(\Omega \cdot cm)^{-1}$ at a flux of 10^{14} γ's $cm^{-2}s^{-1}$ for material doped by 600 ppm PH_3 in the plasma. For this material, the doping reduced the high-temperature activation energy of σ_d from $E_\sigma = 1.05$ eV to $E_\sigma =$ = 0.34 eV: a movement of E_F by ≈ 0.7 eV.[9] Of course, photo-conductivity is a measure of *majority-carrier* transport, in this case by electron carriers. Therefore, high photoconductivity is only one figure-of-merit in selecting candidate materials for photovoltaic conversion. Another, more critical factor is the

---

[9]*On quite general grounds one can understand that as E_F approaches E_C, more and more recombination centers by definition become electron occupied. This situation persists under illumination (quasi-equilibrium), so that recombination is decreased and photoconductivity is increased.*

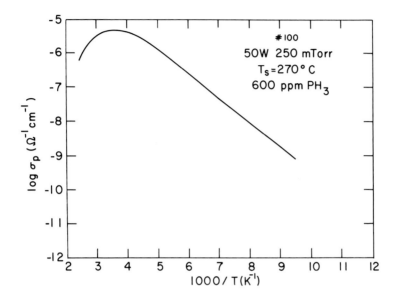

FIGURE 10.   *Photoconductivity versus reciprocal temperature*
for phosphorous-doped a-Si:H alloy.   Flux $f_\lambda = 10^{14}$ γ's $cm^{-2}s^{-1}$
at λ = 600 nm.

quality of minority-carrier transport (Sect. III.*B.2*).

An obstacle to increased doping efficiency of a-Si:H alloys
is, of course, the residual gap states shown in Fig. 6.   In this
regard, it is instructive to compare the relative doping efficiencies of a-Si:H and single crystal silicon.   For example,
in the case of c-Si, $\sigma_d(300 \text{ K}) = 1 \ (\Omega \cdot cm)^{-1}$ is typically attained with only 0.1 ppm P dopant (n-type) or with 0.3 ppm B
dopant (p-type).   Moreover, minority-carrier diffusion lengths
in these doped c-Si materials are very large, approximately L ∿
∿ 100 μm.

Higher *n-type* doping efficiency with phosphorous or arsenic
has been reported for a-Si:(F,H) alloys prepared at nominal substrate temperature T_s = 380°C (18).   For example, $\sigma_d(300 \text{ K})$ ≃
≃ 5 $(\Omega \cdot cm)^{-1}$ and E_σ ≃ 0.05 eV resulted from adding only 500 ppm
PH_3 to the $SiF_4 + H_2$ gas stream.   This doped material would

appear to be degenerate in the sense that E_F penetrated the
bandtail. By contrast, p-type doping with B_2H_6 is no more efficient than found for a-Si:H alloys, i.e., E_σ is limited to
$\simeq$ 0.20 eV. From these results it is possible to conclude that
the density of states for a-Si:(F,H) is lower in the upper half
gap relative to the usual a-Si:H alloy (e.g., as suggested in
Fig. 6), but that the density of states in the lower half gap is
comparable for both materials.

However, it is interesting to note that extremely conduc-
tive, σ_d(300 K) $\sim$ 100 ($\Omega \cdot$cm)$^{-1}$, phosphorous-doped silicon-
hydrogen alloys have been grown at high power density and $T_s =$
= 300°C (56). But from X-ray diffraction studies it was determined that these films had, in fact, a *microcrystalline*
structure. Moreover, very conductive polycrystalline films with
σ_d(300 K) $\sim$ 5 ($\Omega \cdot$cm)$^{-1}$ result from high temperature anneals of
doped a-Si:H alloys (57). From these two studies on a-Si:H
alloys it can be concluded that: i) The degree of incorporation
of P atoms and the efficiency of conversion to substitutional
sites both *increase* with the degree of crystallinity[10]; ii) For
high doping levels of PH$_3$ in the plasma, the crystallization
temperature decreases (to $\simeq$ 510°C at $\simeq$ 2400 ppm); and iii) The
"sign anomaly" for the Hall mobility of doped a-Si:H is removed
for both partially-crystallized (56) and fully-crystallized
films (57). On the basis of these results for a-Si:H, it would
perhaps be enlightening to carry out measurements of Raman
scattering and X-ray diffraction in order to test the amorphic-
ity of doped (and undoped) a-Si:(F,H) alloys. Even the high
mechanical stress found for a-Si:(F,H) films might be correlated
with the possible appearance of microcrystalline structure.

---

[10]*From Ref. 56 as many as $\sim$ 2000 ppm P atoms can be incorporated. From Ref. 57 the conversion to substitutionality was
100% for completely crystallized material.*

2. Quenched Photoluminescence Efficiency. Since a-Si:H alloys can be electronically doped, it might appear possible to
fabricate reasonably efficient p-n homojunctions, and other devices, using such doped materials for the active layers. Unfortunately, this expectation has not yet been fully realized[11],
and intentionally doped materials, such as n^{+}-type a-Si:H alloys, are essentially relegated the role of effecting Ohmic contacts (4,58,59). The reason for this situation is qualitatively
the following. Under illumination a solar cell is a "minority-
carrier device", i.e., the photocurrent is carried by minority
carriers. But doping increases the recombination rate for
minority carriers in a-Si:H alloys by: i) introducing additional defects into the gap; and ii) increasing the occupation
of gap states by majority carriers. As a consequence, minority-
carrier transport is suppressed relative to the case of the undoped a-Si:H material. This statement can be appreciated by
noting that even for undoped a-Si:H (or a-Si:(F,H)), minority-
carrier transport is already a limiting factor for solar cells.
In particular, hole diffusion lengths in some "intrinsic" (n-
type) a-Si:H materials were estimated to be only L_p < 400 Å
(60). Small diffusion lengths $L_p \sim$ 1000 Å appear likely for
even the best materials prepared to date. Again, this feature
is a consequence of the high density of states in the lower
half gap (see Fig. 6). Thus the photocurrent of solar cells
based on a-Si:H is composed essentially of carriers that were
generated in the depletion region alone and that are subsequently swept out by the drift field. On such a model, the
internal spectral response (for the Schottky diode) can be
crudely approximated by the expression $SR(\lambda) \simeq \eta_{\lambda}\{1-\exp(-\alpha_{\lambda}w)\}$,
where w is the width of the space-charge region and $\eta_{\lambda} \leq 1$

---

[11]*For example, in Ref. 4 it was found that the short-circuit
current was reduced if the "intrinsic" layer in a p-i-n cell
were doped.*

denotes the quantum efficiency for carrier creation and separation in that region.

For undoped films, the efficiency of low-temperature "band-
tail" photoluminescence (PL) peaking at the canonical value
$\sim$ 1.3 eV (61,62) has often been taken as a measure of the defect
density that limits minority-carrier transport at room temperature. For this purpose, it must be assumed that deep-lying defect states in the gap increase the non-radiative recombination
rate that directly competes with the radiative recombination
rate, i.e., with the photoluminescence. A connection then has
to be established between the defect states and minority-carrier
recombination processes at T = 300 K. For example, material
produced at low substrate temperatures contains a higher proportion of dangling bond states that is correlated with reduced
PL intensity and hence with a higher non-radiative recombination
rate via paramagnetic defects. Indeed, it was found that the
spin density at T_s = 25°C increased to $N_s \sim 10^{18}$ cm^{-3}, while the
PL intensity was reduced by a factor of 10^{-3}-10^{-2} (63). Such
material, like pure a-Si, presumably exhibits poor minority-
carrier transport.

The effect of doping upon the PL emission spectrum is twofold: i) The overall PL intensity is quenched as a function of
increasing dopant incorporation (apparently independent of the
chemical nature of the dopant); and ii) a low-energy spectral
component emerges at 0.8-1.0 eV that grows relative to the
canonical component and dominates the latter over a wide temperature range. This quenching behavior is consistent with a
qualitative connection of photoluminescence efficiency with
minority-carrier transport. The low-energy component was first
noted in the case of phosphorous doping (64). It may well be
related to defects in c-Si introduced by ion implantation (65)
and to defects in "poor" undoped a-Si:H materials (61). Figure 11 illustrates this circumstance in the case of substitutional doping by nitrogen (17). In this figure PL emission

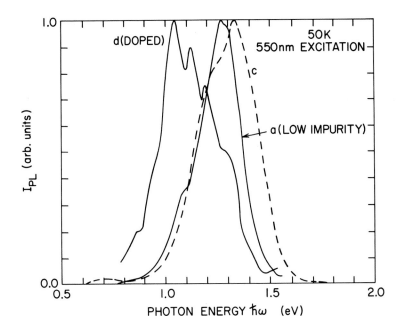

FIGURE 11.   Photoluminescence emission spectra at 50 K for a-Si:H alloys with different impurity levels.   Film d is electronically doped by nitrogen.   (Fringes due to interference effects.)

spectra at 50 K are compared for films prepared at 15 W, 100 mTorr with varying levels of incorporated nitrogen and oxygen impurities. For film a produced under low impurity conditions, the spectrum is dominated by the canonical peak at $\sim$ 1.3 eV. Film c was produced by adding 1200 ppm N_2 and 700 ppm O_2 to the silane gas stream, but from the dark conductivity data, little or no movement of E_F indicates the lack of pronounced doping. The PL spectrum for this film may be slightly shifted to higher energy. Film d was plasma-deposited in the presence of an air leak. The electronic doping of this film by active nitrogen is manifested by: i) the small activation energy of the dark conductivity, $E_\sigma = (E_C - E_F) \simeq 0.5$ eV; and ii) extremely high "classic" photoconductivity that exceeds the photoconductivity

of the P-doped film in Fig. 10 by over an order-of-magnitude at
$T = 300$ K.

The PL spectrum for this particular film is dominated by a
low-energy component at ∿ 1.0 eV. In fact, Fig. 12 shows that
the low-energy component $d(1.0$ eV) dominates the high-energy
component $d(1.2$ eV) for all temperatures ≳ 4.2 K. This temperature dependence is also consistent with behavior found for
the low-energy component in phosphorous- and boron-doped alloys
(64). One effect of doping is therefore to introduce (or acti-
vate) a shallow defect (radiative recombination center) that
traps electrons which subsequently recombine with trapped holes
(perhaps self-trapped holes) to give the low-energy spectral PL

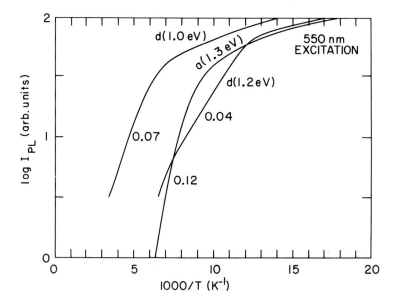

*FIGURE 12.   Temperature dependence of PL spectral components
for doped film d.   Also shown is canonical component of film a
produced under low impurity conditions.   (Arbitrary normaliza-
tion.)*

component. Evidently competitive, non-radiative recombination centers are also introduced that can be associated with para-magnetic centers (63). These centers must be responsible for the overall quenching of the PL intensity, the increased minority-carrier recombination, and ultimately for degraded minority-carrier transport in doped junction layers of devices. Finally, we point out that the degree of the aforementioned effects in nitrogen-doped alloys seems to be a very sensitive function of plasma processing conditions, and hence of the level and efficiency of doping. Preliminary data (54) indicate that doped alloys can be deposited that do not exhibit significant quenching of the PL spectrum or the emergence of a low-energy component at $T = 50$ K even though the Fermi level has shifted by as much as 0.4 eV from the undoped state (Sect. III.*C.1.b*).

C. The Effects of Incorporated Atmospheric Impurities: The Alloy a-Si:(H,O,N,C,...)

1. Modifications in Optoelectronic Behavior. Many sources of common atmospheric impurities (nitrogen and oxygen) are encountered in the plasma deposition of a-Si:H alloys: ubiquitous airleaks in the system; backstreaming from the pumping system; outgassing of adsorbed chamber air and moisture; impurities in process and carrier gases, etc. It is little wonder, then, that the optoelectronic properties of films reported in the literature to date reflect in varying degree the unintentional incorporation of such impurities as nitrogen, oxygen, carbon, and other elements. Indeed, for otherwise fixed processing conditions in a given deposition system, the proverbial irreproducibility of materials properties may well be ascribed to variable sources and concentrations of impurities from one deposition run to the next.

At the time of writing, the effects of impurities upon the photovoltaic quality of a-Si:H alloys have barely been explored. In any definitive study, the solar cell conversion efficiency,

the open-circuit voltage, and the short-circuit current would
all need to be correlated with calibrated concentrations of impurities contained in the alloy layers of a diagnostic device.
Of course, as a function of impurity content, it may also be
useful to test the hypothesis that measurements of low-
temperature photoluminescence could provide a guide to minority-
carrier transport properties that relate to solar cell efficiencies. The systematic effects of nitrogen and oxygen impurities upon both solar cell efficiency and relative photoluminescence efficiency are topics of a current comparative
investigation (66).

Accordingly, we first discuss the documented effects of
atmospheric impurities upon majority-carrier transport (Sect.
III.C.1.a) and then briefly note trends that are emerging with
regard to the effects of such impurities upon solar cell conversion efficiency (Sect. III.C.1.b). In Sect. III.C.2 we explore in more detail the nature of anomalous photoconductivity,
including the universality of such behavior.

a. *Majority-carrier transport*. Rather striking modifications of dark conductivity and especially of photoconductivity
are found as the result of the incorporation of nitrogen and
oxygen impurities in a-Si:H alloys (17). The separate effects
of each impurity have already been studied in the case of reactively sputtered a-Si:H alloys (16,67). The details of preparing materials under low impurity conditions, or with calibrated concentrations of impurities in the plasma, are discussed
elsewhere (17,27,54). We simply note here that the phrase "low
impurity" specifically pertains to minimizing sources of impurities during plasma deposition, hence the resulting alloys
will be produced with a minimal level of incorporated atmospheric impurities *for a given set of processing conditions*.
Varying the processing conditions clearly alters the film deposition rate, the relative concentrations of reactive species

in the plasma due to residual impurities, etc. Therefore actual
concentrations of contaminants in "low-impurity" materials vary
from one processing condition to the next.

Figure 13 shows the effects upon the dark conductivity of
adding 1000 ppm N_2 to the silane gas stream at $P = 50$ W, $p =$
$= 250$ mTorr, and $T_s = 260°C$. As illustrated in Fig. 1, the N_2
impurity in the plasma can easily be detected by optical emission spectroscopy. In Fig. 13 the dark conductivity for the
low-impurity material is (approximately) singly-activated with
slope $E_\sigma = 1.02$ eV. From an extrapolation of the quantity
$(\alpha h\nu)^{\frac{1}{2}}$, the optical gap for the same material is found to be
$E_o \simeq 1.65$ eV. (See also Fig. 5.) Hence, E_F lies below midgap
for this low-impurity alloy produced at medium-high substrate

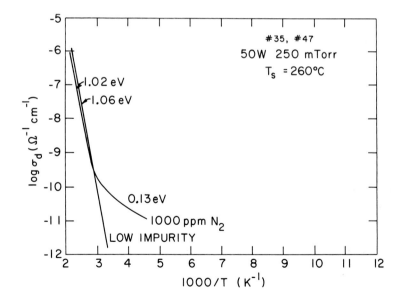

FIGURE 13. Dark conductivity versus reciprocal temperature
for a-Si:H alloy produced under low impurity conditions, and
for alloy grown with 1000 ppm N_2 in the silane gas stream.

temperature. By contrast, the film with the nitrogen impurity
exhibits phonon-assisted hopping transport below about $T \simeq 360$ K.
Depending upon the processing conditions, as little as several
hundred ppm N_2 can introduce shallow defects that engender hopping in dark transport. The relevance of this observation is,
of course, that N_2 is a ubiquitous impurity encountered in
plasma deposition. In fact, despite manufacturers' specifica-
tions, such levels of N_2 contamination are occasionally found in
silane tanks.

Figure 14 illustrates modifications in photoconductivity
versus reciprocal temperature owing to the incorporation of
oxygen and nitrogen impurities in a-Si:H alloys. Alloys that are
deficient in both impurities, or in either impurity, exhibit
anomalous behavior in photoconductivity. In particular, the
phenomenon of thermal quenching (peaks near room temperature or
below) is apparent for Film a (low impurity conditions), and for
Films b (1000 ppm N_2) and c (1000 ppm O_2). Anomalous photoconductivity in these alloys is actually characterized by as many
as three phenomena: i) the thermal quenching; ii) supralinear
dependence upon illumination intensity; and iii) infrared
quenching (68). By contrast, *synergistic effects* of oxygen and
nitrogen impurities restore classic behavior in photoconductivity as illustrated for Film d (540 ppm N_2 and 1000 ppm O_2).
Analyses by SIMS show that the transformation $a \rightarrow d$ in Fig. 14
is effected by a twentyfold increase in the *incorporated* impurity content: from 0.2% O and 20 ppm N to 4% O and 400 ppm N.

The anomalous thermal quenching phenomenon occurs in regions
of temperature for which the excess majority carriers far out-
number the dark carriers. (See Fig. 14.) On the other hand,
the thermal quenching behavior at *high* temperature in the classic photoconductivity of Film d is easily understood by noting
that at this flux, the number of dark carriers now exceeds the
excess carrier density. Under these circumstances, the density
of hole-occupied recombination centers for free-carrier

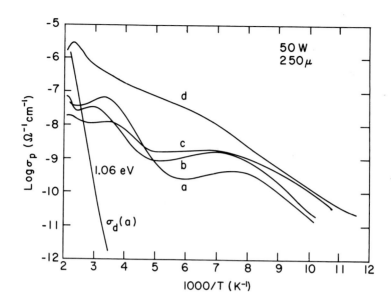

FIGURE 14.   *Photoconductivity versus reciprocal temperature as a function of impurity content in the silane gas stream (50 W, 250 mTorr, T_S = 260°C). Low impurity Film a was deposited with no intentional impurities; Film b with 1000 ppm N_2; Film c with 1000 ppm O_2; and Film d with 540 ppm N_2 and 1000 ppm O_2. Flux f_λ = 10^{14} γ's $cm^{-2}s^{-1}$ at λ = 550 nm.*

electrons is nearly the same as in thermal equilibrium: the quasi-Fermi levels are not displaced far from the dark Fermi level.   Consequently, raising the temperature has the effect of: i) thermally exciting additional holes into the centers; ii) increasing the recombination rate of majority-carrier electrons; and thereby iii) quenching the electron photoconductivity as observed.   As a function of illumination intensity, monomolecular recombination $\sigma_p \sim f_\lambda$ is consistently found in this region of temperature.

It is interesting that the introduction of calibrated air-leaks directly into the plasma deposition chamber exactly confirms the aforementioned modifications of film properties for medium-high substrate temperature, e.g., $T_S \simeq 270°C$.   At rather

higher substrate temperatures, alloys produced in this manner
exhibit more pronounced effects of electronic doping by nitrogen. For example, Fig. 15 shows the photoconductivity and dark
conductivity of an alloy deposited at $T_s \simeq 370°C$ in the presence
of ~ 3000 ppm air. The high-temperature activation energy is
only $E_\sigma \simeq 0.4$ eV, whereas an optical gap of the normal size,
namely $E_o = 1.7$ eV, was found from extrapolation of the quantity
$(\alpha h\nu)^{\frac{1}{2}}$. Consistent with the doping of this material, unusually
high photoconductivity was also attained: $\sigma_p(300 \text{ K}) = 4 \times 10^{-5}$
$(\Omega \cdot cm)^{-1}$ at a flux of 10^{14} γ's $cm^{-2}s^{-1}$. Thus, depending upon
the processing conditions, nitrogen has the ability to introduce
shallow defects (Fig. 13), or to act as a fourfold-coordinated
substitutional donor in the amorphous network (Figs. 15 and 3b).

Even in the absence of an airleak, high substrate temperatures $T_s \simeq 325-375°C$ may be expected to promote impurity incorporation in materials by at least two mechanisms. Firstly,
high T_s significantly increases the outgassing rate of atmospheric (and other) impurities from chamber surfaces into the
plasma where active species are created. Of course, this partial chamber "bakeout" prior to deposition may actually decrease
certain residual impurity levels, notably water vapor that is
one source for oxygen. Secondly, the effect of high T_s is to
heal the amorphous network (provided that sufficient hydrogen
is incorporated), thereby decreasing the gap state density and
permitting more efficient substitutional doping by impurity
nitrogen in particular. Both mechanisms will be enhanced for
materials grown at low deposition rates. The first mechanism
can be illustrated by comparison with an example taken from a
different deposition process: For pure a-Si that is evaporated
at ultrahigh-vacuum with a low deposition rate ($\lesssim 1$ Å s^{-1}),
enough contamination was found to occur at $T_s = 380°C$ that approximately 5×10^{18} spins cm^{-3} were quenched in the *in situ*
ESR signal (35). Both mechanisms appear to be illustrated by
the case of an alloy that was plasma-deposited at a low

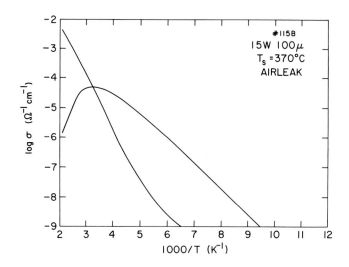

*FIGURE 15.   Dark conductivity and photoconductivity versus
reciprocal temperature for n^+-type, nitrogen-doped film produced in presence of airleak (15 W, 100 mTorr, T_S = 370°C).
Flux $f_\lambda = 10^{14}$ γ's $cm^{-2}s^{-1}$ at λ = 600 nm.*

deposition rate under conditions similar to those shown in Fig.
15, but with no intentional airleak. Depending upon illumination intensity, this material exhibited classic behavior in
photoconductivity versus reciprocal temperature. The dark
activation energy for this alloy, $E_\sigma \simeq 0.77$ eV, and the photoconductivity, $\sigma_p(300\ K) \sim 10^{-6}\ (\Omega\cdot cm)^{-1}$ at 10^{14} γ's $cm^{-2}s^{-1}$, are
more representative of values found in the literature. Of
course at high T_s, both impurity incorporation *and* altered
intrinsic Si-H bonding may be responsible for this behavior.

     In order to illustrate an extreme example of impurity in-
corporation, 1000 ppm O_2 was added to the silane glow discharge
at low pressure, i.e., no premixing of oxygen and silane occured
in the gas manifold. Whereas only weak emission was observed
from molecular SiO in the plasma (Fig. 2), the resulting film
was a true alloy with oxygen, a-Si:(H,O). As shown in the infrared transmission data of Fig. 16, the effects of the oxygen

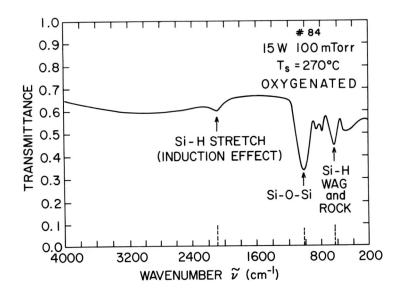

FIGURE 16. Infrared transmission spectrum of oxygenated alloy a-Si:(H,O).

impurity upon the chemical bonding of this alloy are manifested by two features: i) the Si-O-Si asymmetric stretch mode at $\tilde{\nu} \sim$ ~ 1000 cm^{-1} (16); and ii) the "induction effect" in which the monohydride $\equiv$SiH stretch mode at 2000 cm^{-1} has been displaced to ~ 2100 cm^{-1} owing to the backbonding of oxygen in the configuration =SiHO (69). (Films produced at 15 W, 100 mTorr contain nearly 100% monohydride bonding.) Using the integrated absorption of the Si-O-Si stretch mode alone, it was estimated that the a-Si:(H,O) alloy contained $\sim 15\%$ oxygen, whereas calibration with analyses by SIMS indicates $\sim 37\%$ incorporated oxygen. Optical and electrical parameters for this highly oxygenated alloy were typical of a dielectric, e.g., the wider optical gap $E_o = 2.0$ eV and excessively low photoconductivity σ_p(300 K) = $= 6 \times 10^{-11}$ $(\Omega \cdot$cm$)^{-1}$ at a flux of 10^{14} γ's cm^{-2}s^{-1}. For smaller concentrations of incorporated oxygen, the presence of the Si-O-Si bridging configuration suggests that oxygen may passivate various defects, such as weak molecular bonds.   (As depicted in Fig. 3b, the stable complex $\equiv$Si-O-Si$\equiv$ is formed in that case, bound by perhaps $\sim$ 7 eV relative to $\equiv$Si-Si$\equiv$.)   Indeed, several hundred ppm O_2 added to the silane gas stream removes the hopping transport depicted in Fig. 13 and results in dark conductivity identical with the low impurity alloy in that figure.

   b. Minority-carrier transport. As a function of impurity incorporation, preliminary correlations were made between the relative photoluminescence efficiency of alloys and the solar cell efficiency of diagnostic Schottky devices in which such alloys constitute the bulk layers (66). For this purpose, a typical device structure was chosen to be

$$SiO/Pd/a\text{-}Si:(H,O,N)/n^+a\text{-}Si:H/Mo$$

in which the bulk layer is either an a-Si(H,O,N) alloy or a-Si:H material produced under low-impurity conditions. No attempt was initially made to optimize the efficiency of this structure since the relative conversion efficiency was of interest for various alloys.

   Relative photoluminescence efficiency was measured at low temperatures for alloys deposited at several processing conditions[12]. As one example, emission spectra were compared at 50 K for alloys produced under low-impurity conditions versus controlled impurity conditions at P = 15 W, p = 100 mTorr, and T_s = 370°C. The highest photoluminescence efficiency was attained for the material deposited under low-impurity conditions. The relative peak photoluminescence efficiency was found to decrease by only 15% for an alloy prepared by admitting 3000 ppm air directly into the plasma deposition chamber. In addition to the lack of significant spectral quenching, no low-energy

---

[12]*Excitation of photoluminescence was accomplished using a low-power He-Ne laser with a diffuse beam (2 mW, hν = 1.96 eV).*

spectral component was observed at 50 K despite the evidence of
fairly pronounced impurity doping by the airleak as illustrated
in Fig. 15.

By contrast, the solar cell efficiency dramatically decreased owing to the same airleak rate during the growth of the
bulk alloy layer. In particular, the efficiency was reduced by
51% at 100 mW cm^{-2} illumination relative to the solar cell fabricated with the alloy produced under low-impurity conditions.
This decrease is mainly attributed to a reduction in the short-
circuit current J_{sc}. Therefore, solar cell efficiency (and
particularly J_{sc}) is *a more sensitive measure* of impurity incorporation than that provided by the low-temperature photoluminescence efficiency. Now according to Fig. 15, very high
photoconductivity was exhibited by the alloy produced in the
presence of the airleak. In fact, $\sigma_p(300 \text{ K}) \sim 10^{-3}$ $(\Omega \cdot \text{cm})^{-1}$ for
AM1 sunlight implies a small bulk series resistance, perhaps
$R_s = 0.1\text{-}1$ $\Omega \cdot \text{cm}^2$. This improvement is probably masked, however,
by other series resistances. Evidently, minority-carrier re-
combination is increased owing to impurity incorporation so that
better majority-carrier transport is not a guarantee of improved
solar cell efficiency.

The illuminated $J\text{-}V$ characteristics of Fig. 17 confirm these
trends at a lower substrate temperature, namely at $T_s = 270°C$.
Again, the performance of two Schottky devices is compared as a
function of impurity incorporation in the bulk alloy layer. For
Fig. 17a, the alloy was produced under low-impurity processing
conditions. For Fig. 17b, the a-Si:(H,O,N) alloy was deposited
in the presence of 3000 ppm air. The indicated efficiency for
each characteristic is an external efficiency for the active
area. It was measured with a tungsten lamp for which the spectral distribution is shifted to the red compared to AM1 sunlight.
In true sunlight (100 mW cm^{-2}), the current J_{sc} and the efficiency are roughly doubled. Figure 17 shows that the solar
cell efficiency decreased by 43% as a consequence of increased

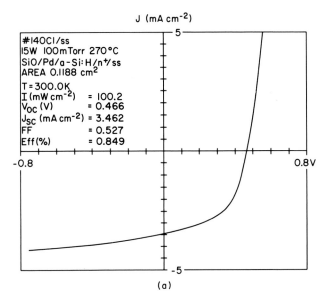

J (mA cm^{-2})

```
#140Cl/ss
15W 100mTorr 270°C
SiO/Pd/a-Si:H/n⁺/ss
AREA 0.1188 cm²

T=300.0K
I (mW cm⁻²)      = 100.2
Voc (V)          = 0.466
Jsc (mA cm⁻²)    = 3.462
FF               = 0.527
Eff(%)           = 0.849
```

-0.8   0.8V

(a)

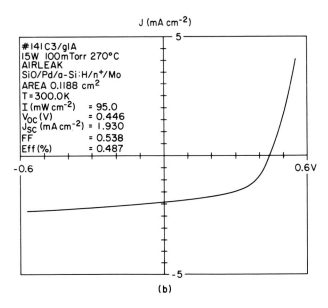

J (mA cm^{-2})

```
#141C3/glA
15W 100mTorr 270°C
AIRLEAK
SiO/Pd/a-Si:H/n⁺/Mo
AREA 0.1188 cm²
T=300.0K
I (mW cm⁻²)      = 95.0
Voc (V)          = 0.446
Jsc (mA cm⁻²)    = 1.930
FF               = 0.538
Eff (%)          = 0.487
```

-0.6   0.6V

(b)

FIGURE 17.   *Effects of impurity incorporation on illuminated J-V characteristics for Schottky solar cell structure
SiO/Pd/a-Si:(H,O,N)/n⁺a-Si:H/substrate. Bulk alloy layer is
produced: a) under low-impurity processing conditions; b) in
the presence of 3000 ppm air.*

impurity incorporation in the bulk alloy layer. It is also of
interest that the more efficient solar cell was fabricated with
an alloy that exhibits an anomalous peak in photoconductivity at
low temperature ($T \simeq 160$ K) because of the low-impurity processing conditions.

    *2. Expanded Discussion of Anomalous Behavior in Photoconduc-
tivity.* In order to explore the nature of anomalous photoconductivity further, we proceed to examine the illumination dependence of the anomalous peaks. Figure 18 shows two anomalous
peaks in photoconductivity versus reciprocal temperature for an
alloy produced at 50 W, 250 mTorr, and $T_s = 270°C$[13]. This particular data was obtained at the flux $f_\lambda = 10^{14}$ γ's cm^{-2}s^{-1} for
wavelength $\lambda = 550$ nm ($h\nu = 2.25$ eV). For the thermal quenching
side of the T_A peak (high-temperature side), Fig. 19 now illustrates the novel phenomenon of supralinearity in photoconductivity as a function of illumination intensity. For reference,
the dashed line in this figure denotes exact monomolecular behavior, i.e., the linear growth $\sigma_p \sim f$. Figure 19 therefore
shows that there is a region of illumination intensity for which
$\sigma_p \sim f^\nu$ where the power ν *exceeds unity.* Supralinear behavior
with a power as large as $\nu \simeq 1.6$ is attained for the thermal
quenching side of the T_A peak. A similar plot for the T_B peak
reveals more moderate supralinear behavior with $\nu \simeq 1.2$. With
increasing monochromatic intensity f_λ, this peak shifts to
higher temperature. Infrared quenching of photoconductivity
is also exhibited in the region of the T_B peak, i.e., this peak
is reduced when a broad spectrum of sub-bandgap radiation is
simultaneously imposed (68). The recombination kinetics tends
toward bimolecularity with $\nu \simeq 0.5$ for phototransport on the
low-temperature side of this anomalous peak (Fig. 18).

[13]*A concentration of 100 ppm N_2 impurity was incidentally
present in the silane gas stream.*

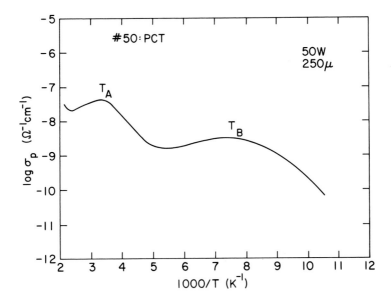

FIGURE 18. *Anomalous peaks with thermal quenching in photo-conductivity versus reciprocal temperature. Flux $f_\lambda = 10^{14}$ γ's cm^{-2}s^{-1} at λ = 550 nm.*

Anomalous behavior in photoconductivity is observed over a wide range of processing conditions (under appropriate impurity controls, as discussed): p = 30-500 mTorr, P = 15-100 W, and a range of substrate temperatures up to at least $T_s \simeq$ 300°C. More-over, this behavior is independent of: i) the type of substrate (7059 glass, quartz, or sapphire); ii) contacts on top, or under-neath, the film in a gap cell geometry; iii) the nature of the Ohmic contact (e.g., Mo versus Ag); and iv) film thickness over the range 0.1-10 μm (true bulk effect). The anomalous peaks are also observed to persist under the following conditions: i) broadband illumination (450-650 nm); ii) dc measurement (no chop-ping of illumination); and iii) heating versus cooling cycles for the measurement (implying no thermally-stimulated currents).

The universality of condensed matter science is again dem-onstrated by the observation of anomalous photoconductivity in a

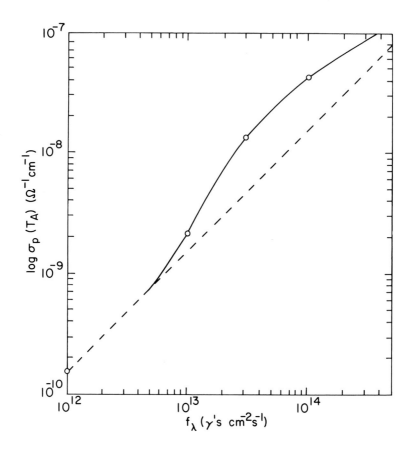

FIGURE 19. Supralinearity in photoconductivity as a function of illumination intensity (λ = 550 nm). Dashed line is monomolecular (linear) growth.

wide variety of materials. The phenomena of thermal quenching, supralinearity, and infrared quenching have been observed for both amorphous and crystalline semiconductors, as well as for certain insulators. For example, Fig. 20 illustrates thermal quenching of photoconductivity (and in some cases, supralinearity) for materials as diverse as: c-CdSe:I:Cu (70), c-Ge:Mn (71), a-ZnSe (72), a-Se (73), and the single-crystal insulator β-AgI (74). Together with related results for a-Si:H alloys,

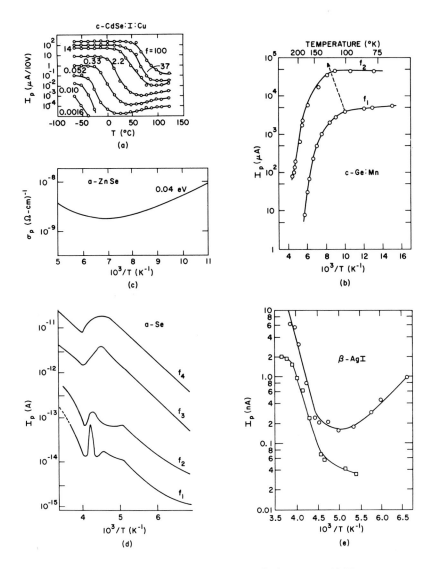

FIGURE 20. Anomalous photoconductivity for different materials: (a) c-CdSe:I:Cu (Ref. 70), (b) c-Ge:Mn (Ref. 71), (c) a-ZnSe (Ref. 72), (d) a-Se (Ref. 73), and (e) β-AgI (Ref. 74).

this universality of behavior does not, of course, suggest precisely the same chemical and physical origins for states that are responsible for the phenomena. For example in the case of

c-Ge (or c-Si), substitutional double-acceptor transition metals
such as Mn, Co, Ni, Fe, and Zn have been indicted as sources of
recombination centers in the crystal bandgap that give rise to
anomalous photoconductivity.

In the case of a-Si:H alloys, there are several open possibilities for such recombination centers, including: i) certain
intrinsic states of the silicon-hydrogen alloy, and ii) centers
due to extrinsic trace elements that are somehow incorporated
during plasma deposition. With regard to concentrations of
transition metals that could conceivably contribute to the second possibility, analyses of our alloys by SIMS reveal the presence of several ppm Cu and several tens ppb Cr. But the elements Fe, Ni, and Zn are all essentially undetectable in the
bulk of the films. Whatever the nature of the centers, a synergism of incorporated nitrogen and oxygen effects the transformation to classic behavior in photoconductivity, e.g., the
transformation $a \rightarrow d$ at T_s = 260°C in Fig. 14. Such transforma-
tions at medium-high T_s may be caused by a chemical modification
of the centers *per se*, and/or by displacements (however slight)
of the Fermi level from the vicinity in the gap of the sensitiz-
ing energy levels. The absence of anomalous photoconductivity
at very high T_s may not, at this stage, be taken as unambiguous
evidence for intrinsic centers due to silicon-hydrogen (or even
silicon-silicon) bonding. As we discussed in Sect. III.*C.1.a*,
even though the effect of high T_s is to reduce the overall density of gap states, chamber impurities (except, perhaps, H_2O)
are less controlled in this case. These questions are currently
being addressed (54).

Despite the fact that the specific chemical bonds responsible
for anomalous photoconductivity have not yet been identified, the
essence of such behavior can be understood on the basis of an
adaptation of a simple two-level model (75). As illustrated in
Fig. 21, it is assumed on the model that there exist two levels
of recombination centers, levels "*1*" and "*2*", located at

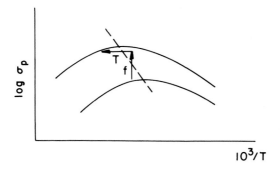

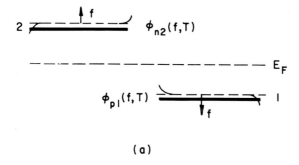

(a)

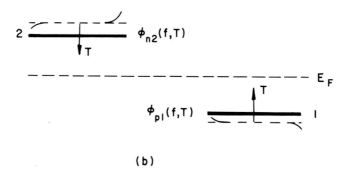

(b)

FIGURE 21.   *Simple two-level model for explaining anomalous photoconductivity (top figure): (a) supralinearity (increasing f); (b) thermal quenching (increasing T).*

$E_2 > E_F > E_1$ with respect to the dark Fermi level. (Presumably these centers help to pin E_F.) For the case of majority-carrier electron phototransport, it is then essential that these levels have *widely disparate* kinetic-capture cross sections for electrons:

$$\sigma_{2n} \gg \sigma_{1n} . \tag{18}$$

Let us denote the excess carrier density for electrons by $\Delta n = f/(\omega_1 + \omega_2)$, where $\omega_i \equiv \tau_i^{-1} \equiv v\sigma_{in} p_{ri}$ $(i = 1,2)$ is the recombination rate per electron by the hole-occupied recombination center of density p_{ri}. For specificity we shall also consider an anomalous peak at low temperature, such as peak T_B of Fig. 18 that lies at $T_B \sim 130$ K. Then, as appropriate for low-temperature photoconductivity $\sigma_p = e\Delta n\mu$, the density Δn pertains to low-mobility electrons that carry the photocurrent by phonon-assisted hopping. Hence mobility μ is thermally activated.

We shall firstly explain the phenomenon of supralinearity with reference to Fig. 21. At a given level of illumination, initially consider a value of σ_p on the thermal quenching side of the anomalous peak (high-temperature side). Recombination in this region is dominated by states *2* with the larger capture cross section for electrons. If we now imagine that the flux f is increased at fixed temperature, then, as depicted in Fig. 21a, the quasi-Fermi level for trapped electrons $\phi_{n2}(f,T)$ and the quasi-Fermi level for trapped holes $\phi_{p1}(f,T)$ depart more and more from the thermal equilibrium position E_F. In this configuration for which $\phi_{n2}(f,T) \gtrsim E_F$, level *2* is being depopulated of holes. Inversely, level *1* is acquiring additional holes. Now in a restricted region of the flux f, $\omega_2 \gg \omega_1$ is still satisfied precisely because of the relatively large capture cross section σ_{2n}. That is, the recombination rate for photocarriers is still determined by states *2* so that $\Delta n \simeq f/\omega_2$. But the recombination rate decreases for increasing f, following the rate of decrease in the number of holes $p_{r2} = (\omega_2/v\sigma_{2n})$. Hence, $\omega_2 \sim 1/f^\alpha$, where α is some power. Therefore in that region of flux (and

temperature) we have $\Delta n \simeq f/\omega_2 \sim f^{1+\alpha}$, i.e., supralinear behavior $\sigma_p = e\Delta n\mu \sim f^{1+\alpha}$ is obtained.

For a large enough increase in flux, a point is attained on
the low-temperature side of the anomalous peak (see Fig. 21).
The capture kinetics on this side proceeds via bimolecular recombination. If this transition to bimolecularity were taken as
evidence for a change in dominant recombination path, then the
recombination rate by states 1 would finally exceed that by
states 2, i.e., $\omega_1 > \omega_2$ and $\Delta n \simeq f/\omega_1$. On this scenario, accurately bimolecular recombination is consistent with $\Delta n \sim p_{r1} \sim$
$\sim (f/v\sigma_{1n})^{\frac{1}{2}}$. Hence, the activation energy of σ_p at sufficiently
low temperatures is equal to the thermal activation energy of the
mobility μ. Depending upon the exact relative sizes of σ_{2n} and
σ_{1n}, and of the densities for states 2 and 1, it is conceivable
instead that states 2 could dominate recombination over the entire temperature range of an anomalous peak. (Of course, at high
enough flux and low enough temperatures, states 1 would finally
have to dominate.) As in the case of c-Si, however, this situation would probably require a different (more positive) charge
state for hole-occupied states 2 versus states 1 so that
$\sigma_{2n}/\sigma_{1n} \sim 10^6$. This condition would not be met if, for example:
i) hole-occupied states 1 were identified with numerous self-
trapped holes below E_F, and ii) hole-occupied states 2 were
identified with some positive, hole-occupied donor defects lying
above E_F. (It can be argued that hole polarons have a smaller
capture cross section for electrons than positive, hole-occupied
donors, but a larger capture cross section than neutral centers.)
Subject to further experimental tests, we shall merely assume for
definiteness (and consistency) that states 1 do indeed dominate
recombination on the low-T side of the peak.

Proceeding now to an explanation of the thermal quenching
phenomenon, we note that an increase in the temperature T at
fixed illumination intensity acts to restore thermal equilibrium:
i) the quasi-Fermi levels for trapped carriers, $\phi_{n2}(f,T)$ and

$\phi_{p1}(f,T)$, retreat toward E_F , and ii) the Fermi-Dirac distributions become more extended. (See Fig. 21b.) The net effect is
to increase the population of holes in level 2, and inversely to
decrease the population of holes in level 1. For a large enough
increase in temperature, $\omega_2 > \omega_1$ is restored and recombination
is again dominated by states 2 that thermally quench the photoconductivity as depicted in Fig. 21. The shift in the maximum of
the anomalous peak to higher temperature with increasing flux is
implicit in our discussion. Namely, the interplay of flux and
temperature influences the relative populations of hole-occupied
states 1 and 2. An increase in the flux shifts holes away from
states 2, tilting the balance in favor of states 1 (low-T side
of peak). A subsequent increase in the temperature repopulates
states 2 with holes at the expense of states 1, restoring the
thermal quenching. As a consequence, the maximum in the peak
moves to higher temperature for a higher flux. It should be
noted that if phototransport were bipolar at low temperatures,
then anomalous photoconductivity can be explained by an exactly
symmetrical argument for hole-carrier recombination in which
$\sigma_{1p} \gg \sigma_{2p}$ is now assumed to be valid.

Finally, as shown in Fig. 10, anomalous behavior disappears
when the dark Fermi level is displaced in the gap by intentional
doping with phosphorous (or boron). Presumably, in the absence
of significant chemical modification of the material by the
dopant itself, the effect of doping has simply been to vitiate
the condition $E_2 > E_F > E_1$ that is necessary for anomalous behavior. In principle, this situation may be contrasted with the
synergistic incorporation of nitrogen and oxygen in alloys. In
the latter case, new chemical bonds (e.g., "bridging" or "backbonding" configurations) could be formed that eliminate either
states 1, 2 or both. Now the classic photoconductivity depicted
for the "airleak" alloy in Fig. 15 certainly was accompanied by
movement of E_F (to within 0.4 eV of E_C). On the other hand,
results have been obtained in which the transformation from

anomalous to classic photoconductivity has been accompanied by little, if any, change in dark conductivity activation. These results could suggest chemical modification of the centers responsible for anomalous behavior. Studies of very lightly doped alloys are currently in progress to test this idea (54).

ACKNOWLEDGMENTS

The Brookhaven data discussed in this Chapter is the result of a group effort, and I gratefully acknowledge in this regard the outstanding contributions by P. E. Vanier (electrical measurements on films), F. J. Kampas (optical studies of the plasma), A. E. Delahoy (device measurements), and M. D. Hirsch (optical measurements on films). I would also like to acknowledge R. Gremme for his technical assistance with film preparation, and L. C. Arns for her very fine typing skills.

REFERENCES

1. Spear, W. E., and Le Comber, P. G., *Sol. Sta. Comm. 17*, 1193 (1975).

2. Chittick, R. C., Alexander, J. H., and Sterling, H. F., *J. Electrochem. Soc. 116*, 77 (1969).

3. Knights, J. C., *Phil. Mag. 34*, 663 (1976).

4. Carlson, D. E., and Wronski, C. R., *Appl. Phys. Lett. 28*, 671 (1976).

5. Anderson, P. W., *Phys. Rev. 109*, 1492 (1958).

6. Ziman, J. M., *J. Phys. C2*, 1230 (1969).

7. Weaire, D., *Phys. Rev. Lett. 26*, 1541 (1971); Weaire, D., and Thorpe, M. F., *Phys. Rev. B4*, 2508 (1971).

8. Mott, N. F., *Adv. Phys. 16*, 49 (1967).

9. Cohen, M. H., Fritzsche, H., and Ovshinsky, S. R., *Phys. Rev. Lett. 22*, 1065 (1969).

10. Brodsky, M. H., *Thin Solid Films 40*, L23 (1977); Brodsky, M. H., Cardona, M., and Cuomo, J. J., *Phys. Rev. B16*, 3556 (1977).

11. Lucovsky, G., Nemanich, R. J., Knights, J. C., *Phys. Rev. B19*, 2064 (1979).

12. Freeman, E. C., and Paul, W., *Phys. Rev. B18,* 8 (1978); Connell, G. A. N., and Pawlik, J. R., *Phys. Rev. B13*, 787 (1976).

13. Spear, W. E., Loveland, R. J., and Al-Sharbaty, A., *J. Non-Crys. Sol. 15*, 410 (1974).

14. Knights, J. C., Biegelsen, D. K., and Solomon, I., *Sol. Sta. Comm. 22*, 133 (1977).

15. Tsang, C., and Street, R. A., *Phil. Mag. B37*, 601 (1978).

16. Paesler, M. A., Anderson, D. A., Freeman, E. C., Moddel, G., and Paul, W., *Phys. Rev. Lett. 41*, 1492 (1978).

17. Griffith, R. W., Kampas, F. J., Vanier, P. E., and Hirsch, M. D., *J. Non-Crys. Solids 35/36, Part I,* 391 (1980).

18. Madan, A., Ovshinsky, S. R., and Benn, E., *Phil. Mag. B40,* 259 (1979).

19. Spear, W. E., Le Comber, P. G., and Snell, A. J., *Phil. Mag. B38*, 303 (1978).

20. Snell, A. J., Mackenzie, K. D., Le Comber, P. G., and Spear, W. E., *Phil. Mag. B40*, 1 (1979).

21. Deneuville, A., and Brodsky, M. H., *J. Appl. Phys. 50*, 1414 (1979).

22. Wronski, C. R., *Sol. Energy Mat. 1*, 287 (1979).

23. Carlson, D. E., *J. Non-Crys. Solids 35/36, Part II*, 707 (1980).

24. Mogab, C. J., Adams, A. C., and Flamm, D. L., *J. Appl. Phys. 49*, 3796 (1978).

25. Harshbarger, W. R., Porter, R. A., Miller, T. A., and Norton, P., *Appl. Spectrosc. 31*, 201 (1977).

26. Winters, H. F., Coburn, J. W., Kay, E., *J. Appl. Phys. 48*, 4973 (1977).

27. Kampas, F. J., and Griffith, R. W., *J. Solar Cells* (to be published).

28. Perkins, G. G. A., Austin, E. R., and Lampe, F. W., *J. Amer. Chem. Soc. 101*, 1109 (1979).

29. Pauling, L., and Pauling, P., "Chemistry", p. 740, W. H. Freeman, San Francisco, (1975).

30. Potzinger, P., and Lampe, F. W., *J. Phys. Chem. 73*, 3912 (1969).

31. Cottrell, T. L., and Walker, I. C., *Trans. Faraday Soc. 63*, 549 (1967).

32. Moss, S. C., and Graczyk, J. F., *Phys. Rev. Lett. 23*, 1167 (1969).

33. Paul, W., Harvard preprint (1979).

34. Brodsky, M. H., Title, R. S., Weiser, K., and Pettit, G. D., *Phys. Rev. B1*, 2632 (1970).

35. Thomas, P. A., Brodsky, M. H., Kaplan, D., and Lépine, D., *Phys. Rev. B18*, 3059 (1978).

36. Kaplan, D., Sol, N., Velasco, G., and Thomas, P. A., *Appl. Phys. Lett. 33*, 440 (1978).

37. von Rödern, B., Ley, L., Cardona, M., and Smith, F. W., *Phil. Mag. B40*, 433 (1979).

38. Griffith, R. W., in "Sharing the Sun! Vol. 6, Photovoltaics and Materials" (K. W. Böer, ed.), p. 205, Amer. Sect. Int. Sol. Energy Soc. (1976).

39. Tauc, J., Grigorovici, R., and Vancu, A., *Phys. Stat. Sol. 15*, 627 (1966).

40. Fritzsche, H., in "Amorphous and Liquid Semiconductors" (W. E. Spear, ed.), p. 3. University of Edinburgh, Edinburgh, (1977).

41. Madan, A., Le Comber, P. G., and Spear, W. E., *J. Non-Crys. Sol. 20*, 239 (1976); Spear, W. E., and Le Comber, P. G., *Phil. Mag. 33*, 935 (1976).

42. Cohen, J. D., *J. Solar Cells*, to be published.

43. Wronski, C. R., and Carlson, D. E., *Sol. Sta. Comm. 23*, 421 (1977).

44. Wronski, C. R., *IEEE Trans. Elec. Devices ED-24*, 351 (1977).

45. Jousse, D., Viktorovitch, P., Vieux-Rochaz, L., and Chenevas-Paule, A., *J. Non-Crys. Solids 35/36, Part II*, 767 (1980).

46. Staebler, D., and Wronski, C. R., *Appl. Phys. Lett. 31*, 292 (1977).

47. Pawlik, J. R., and Paul, W., in "Amorphous and Liquid Semiconductors" (W. E. Spear, ed.), p. 437. University of Edinburgh, Edinburgh, (1977).

48. Stuke, J., *ibid.* p. 406.

49. Biegelsen, D. K., and Knights, J. C., *ibid.* p. 429.

50. Emin, D., in "Electronic and Structural Properties of Amorphous Semiconductors" (P. G. Le Comber and J. Mort, eds.), p. 261. Academic Press, London and New York, (1973).

51. Hanak, J. J., Korsun, V., and Pellicane, J. P., to be published.

52. Anderson, D., and Spear, W. E., *Phil. Mag. 36*, 695 (1977).

53. Rehm, W., Fischer, R., Stuke, J., and Wagner, H., *Phys. Stat. Solidi (b)79*, 539 (1977).

54. Brookhaven Group, to be published.

55. Emin, D., in "Amorphous and Liquid Semiconductors" (W. E. Spear, ed.), p. 249. University of Edinburgh, Edinburgh, (1977).

56. Usui, S., and Kikuchi, M., *J. Non-Crys. Solids 34*, 1 (1979).

57. Reilly, O. J., and Spear, W. E., *Phil. Mag. B38*, 295 (1978).

58. Carlson, D. E., Wronski, C. R., Triano, A. R., and Daniel, R. E., *Proc. 12th IEEE Photo. Spec. Conf.*, p. 893. IEEE, New York, (1976).

59. Wilson, J. I. B., McGill, J., and Robinson, P., *Proc. 13th IEEE Photo. Spec. Conf.*, p. 751. IEEE, New York, (1978).

60. Staebler, D. L., *J. Non-Crys. Solids 35/36, Part I*, 387 (1980).

61. Engemann, D., and Fischer, R., in "Structure and Excita-
     tions of Amorphous Solids" (G. Lucovsky and F. L. Galeener,
     eds.), p. 37. Amer. Inst. Phys., New York, (1976).

62. Street, R. A., *Phil. Mag. B37*, 35 (1978).

63. Street, R. A., Knights, J. C., and Biegelsen, D. K., *Phys.
     Rev. B18*, 1880 (1978).

64. Nashashibi, T. S., Austin, I. G., and Searle, T. M., in
     "Amorphous and Liquid Semiconductors" (W. E. Spear, ed.),
     p. 392. U. of Edinburgh, Edinburgh, (1977).

65. Pankove, J. I., and Wu, C. P., *Appl. Phys. Lett. 35*, 937
     (1979).

66. Delahoy, A. E., Griffith, R. W., and Vanier, P. E., to be
     published.

67. Baixeras, J., Mencaraglia, D., and Andro, P., *Phil. Mag.
     B37*, 403 (1978).

68. Vanier, P. E., and Griffith, R. W., *Bull. Amer. Phys. Soc.
     25*, 330 (1980); and to be published.

69. Lucovsky, G., *Sol. Sta. Comm. 29*, 571 (1979).

70. Bube, R. H., *J. Phys. Chem. Solids 1*, 234 (1957).

71. Newman, R., Woodbury, H. H., and Tyler, W. W., *Phys. Rev.
     102*, 613 (1956).

72. Brodie, D. E., in "Amorphous and Liquid Semiconductors"
     (W. E. Spear, ed.), p. 472. U. of Edinburgh, Edinburgh,
     (1977).

73. Viger, C., Lefrancois, G., and Fleury, G., *J. Non-Crys.
     Solids 33*, 267 (1979).

74. Govindacharyulu, P. A., and Bose, D. N., *Phys. Rev. B19*,
     6532 (1979).

75. Rose, A., *Phys. Rev. 97*, 322 (1955).

PROBLEMS and References

The mysteries of matter have stimulated the great intellectual exploration of our time. There are two reasons why we should share in its excitements. One is for the sheer fun, the esthetic pleasure, call it what you like, of reaching deeper into the unknown. The other is for the understanding to be gained as a result.

——————— C.P. Snow

CHAPTER 20

PROBLEMS AND REFERENCES

L.E. Murr

Department of Metallurgical and Materials Engineering
New Mexico Institute of Mining and Technology
Socorro, New Mexico

(1)   About 1855 James Clerk Maxwell discovered equations which
govern the behavior of electromagnetic oscillations, and
showed that they were applicable to light. In their (scalar)
integral form, Maxwell's equations can be written

$$\int E_s \, ds = - \int\int \mu \, \frac{\partial H_n}{\partial t} \, dS$$

and

$$\int H_s \, ds = \int\int \left(\rho V_n + \sigma E_n + \varepsilon \frac{\partial E_n}{\partial t} \right) dS.$$

The integrals of the tangential components of the electric
and magnetic field intensities, E_s and H_s on the left are
taken around the peripheries of the surfaces over which are
integrated the corresponding normal surface components, E_n
and H_n, respectively on the right; V_n is the velocity of
moving charge; ρ is the density of electric charge; σ is the
conductivity; μ is the permeability; and ε is the permittivity (or dielectric constant). We see that in these two
fundamental equations, the propagation not only of light,
but all electromagnetic radiation is dependent upon materials
properties: σ, μ, and ε.

In homogeneous, nondissipative media (defined by $\rho = 0$
in the second equation above), the velocity of light is

given by $C = 1/\sqrt{\mu\varepsilon}$, and a plane wave incident at a plane
interface between two *different* homogeneous nondissipative
media is partially "reflected" into the first medium (1),
and partially "refracted" into the second medium (2). If
the incident wave angle relative to the interface is denoted
I_1, the reflected angle R_1 and the angle of refraction r_2,
then $I_1 - R_1$ and Snell's law

$$\cos r_2/\cos I_1 = C_2/C_1$$

govern the angle of refraction. It is perhaps obvious that
if a third medium (3) is added, the initially refracted wave
acts as the incident wave upon the interface between medium
(2) and medium (3).

Considering these very basic, historical wave optics
phenomena, discuss the important features which must be considered in the design of "idealized" reflectors, absorbers,
and coatings for photoconversion devices, etc. That is,
using some simple sketches, show the effects of various
overlays of say three different optically "transparent"
materials (which can include air as a "layer"). Show all
the angles of incidence, reflection, and refraction and
write the appropriate equations for the angles at the interfaces. Indicate what portion of the "initial" wave is
finally "transmitted" through the three layers and what
portion is emitted or re-radiated back. Is there a dependence of the transmitted and emitted portion on the "initial"
angle of incidence? If "yes", then show it. Finally, briefly describe the dependence of ε and μ for materials in
general on such features as crystal structure and composition and give some examples, using optically "transparent"
single crystals which could ideally be used to construct a
three-layer regime alluded to above. How is the index of
refraction related to either ε or μ?

REFERENCES

Born, M., "Principles of Optics: Electromagnetic Theory of Propagation, Interference, and Diffraction Light", Pergamon Press, New York, (1959).

Carlson, F.P., "Introduction to Applied Optics for Engineers", Academic Press, New York, (1977).

Glazebrook, R.T., "James Clerk Maxwell in Modern Physics", MacMillan and Co., New York, (1896).

Knittil, Z., "Optics of Thin Films: An Optical Multilayer Theory", Wiley-Interscience, New York, (1976).

Weast, R.C. (ed), "Handbook of Chemistry and Physics", The Chemical Rubber Co. (CRC), (1978).

(2)   Many excellent materials available for solar applications - reflectors, absorbers, conversion devices, storage devices, etc. - are in such short supply, or are otherwise so expensive to produce, that their use is virtually untenable, and this feature is an important motivating feature in solar materials research and development. Compare two typical (common) materials in each category for use as reflecting surfaces, absorbers, and photocells or photovoltaic materials and describe the annual production, production cost, and show a projection of costs likely to occur in the next decade (1980-1990). Discuss some recycling or other production alternatives which may help alleviate some of the critical supply-cost problems, or alternatives which might be pursued with regard to new materials development. You should specifically describe the development of silicon technology in some historical perspective (indicating if possible advances or variations in production costs, material purity and perfection, etc. over the past 2-3 decades) and discuss how this might be transferred to a new material development, or how it may be indicative of current feelings of the futility in seeking alternative material development in the context of photocells or photovoltaic materials. Speculate on breakthroughs which could conceivably alter this attitude.

REFERENCES

Halacy, D.S., "The Coming Age of Solar Energy", Harper and Row,
    New York, (1973).
McVeigh, J., "An Introduction to the Applications of Solar
    Energy", Pergamon Press, New York, (1977).
Merrigan, J.A., "Sunlight to Electricity: Prospects for Solar
    Energy Conversion by Photovoltaics", MIT Press, Cambridge,
    Mass., (1975).
Palz, W., "Solar Electricity: An Economic Approach to Solar
    Energy", Butterworths, London, (1977).
Stein, C. (ed), "Critical Materials Problems in Energy Produc-
    tion", Academic Press, New York, (1976).

(3)   Referring to Fig. 1 of Chapter 2, show in a geometrical
      sketch that the face-centered tetragonal lattice shown is
      equivalent to a body-centered tetragonal lattice unit cell
      in which the side of the unit cell base is $1/\sqrt{2}$ times that
      for the fct cell. Show the (111) plane in the fct unit cell
      and calculate the spacing between (111) planes if $c/a = 1.95$.
      (Make a sketch of this cell to scale.)   If a vacancy were
      formed in every other unit cell center of the bct lattice
      above, what is the density compared to a perfect material?
      Make a two-dimensional sketch of this structure and comment
      upon the microstructure created.   If an edge dislocation
      were associated with every four unit cell basal planes, cal-
      culate the dislocation density.   In other words a disloca-
      tion line terminates on the basal plane corresponding to
      every four unit cells.   Discuss how this compares with
      measured dislocation densities in deformed metals and alloys.
      Finally, show the pseudomorphic, epitaxial growth of a cubic
      deposit upon this tetragonal (fct) substrate material in
      two-dimensions (the substrate is viewed along c and a).

REFERENCES

Barrett, C.S. and Massalski, T.B., "Structure of Metals", 3rd
    Edition, McGraw-Hill Book Co., Inc., New York, (1966).

Buerger, M.J., "Elementary Crystallography", J. Wiley and Sons,
    Inc., New York, (1960).
Guy, A.G., "Introduction to Materials Science", McGraw-Hill Book
    Co., Inc., New York, (1972).
Hirth, J.P. and Lothe, Jens, "Theory of Dislocations", McGraw-Hill
    Book Co., Inc., New York, (1968).
Kuhlmann-Wilsdorf, D., "Workhardening", J.P. Hirth and J.
    Weertman (eds), Gordon and Breach, New York, (1968).
Pearson, W.B., "A Handbook of Lattice Spacings and Structures of
    Metals and Alloys", Pergamon Press, New York, (1958).
Richman, M.H., "An Introduction to the Science of Materials",
    Blaisdell Publishing Co., Walthan, Mass., (1967).
Rigney, R.A., *Scripta Met.*, Vol. 13, p. 353 (1979).

(4) Optical absorptance is related to the complex dielectric
     constant by $\alpha = |k| \varepsilon''$, where $\underset{\sim}{k}$ is the wave vector (having a
     magnitude of $2\pi/\lambda$), and ε'' is the dielectric loss factor.
     Optical properties in general are related to the complex
     index of refraction. In transparent materials the dielec-
     tric constant can be measured optically. Describe how (con-
     sider Prob. 1 in retrospect). Another more general method
     for measuring the dielectric constant of a medium would be
     to utilize a parallel-plate capacitor. For example, a
     coating or deposit of a medium of interest could be grown
     upon a flat metal substrate to a thickness d. A metallic
     layer could then be deposited upon the medium, or another
     flat metal plate could be placed upon the medium forming a
     simple, parallel, flat-plate capacitor. Describe how the
     dielectric constant could be measured.

     In a two-layer dielectric or a Maxwell-Wagner two-layer
     capacitor as shown below, the optical dielectric constant
     is given by

$$\varepsilon'' = \frac{d/\varepsilon_o}{d_1/\varepsilon_1 + d_2/\varepsilon_2} .$$

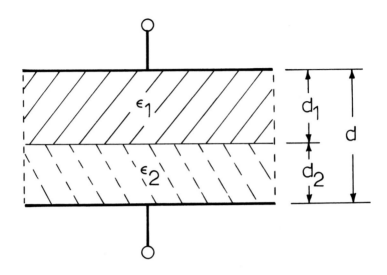

Describe how you would measure this dielectric constant in
an experiment identical to the one above. If it can be
assumed that this model is applicable to a dielectric layer
or a bi-layer upon a metal plate, describe how a flat-plate
absorber can be made more efficient (characterized by an
increase in absorptance). If the absorber layer were composed of two media of thickness d with one component suspended in the other (for example, spheres of one in the
other) how could you approximate the dielectric constant if
the dielectric constants of the two component media were
known? [As a practical example see Niklasson, G.A. and
Granqvist, C.G., J. Appl. Phys., 50(8), 5500, (1979).]
From this analysis, what kinds of materials would seem to
be the best optical absorbers?

REFERENCES

Condon, E.U. and Odishaw, Hugh (eds), "Handbook of Physics",
    McGraw-Hill Book Co., Inc., New York, (1958).
Harrop, P., "Dielectrics", Wiley, New York, (1972).
Kaminow, I.P., "An Introduction to Electrooptic Devices",
    Academic Press, New York, (1974).

Winch, R.P., "Electricity and Magnetism", Prentice-Hall, Inc.,
   New York, (1955); or related text in the general area.

(5)   Figure P. 5 below shows vacuum vapor-deposited tin formed on
       the surface of and ionic crystal.   Such a thin layer can be
       used to monitor the selective effects of particle size dis-
       tributions on the selective optical response of a surface.
       Make a plot of the particle size from the transmission elec-
       tron microscope view in Fig. P. 5(b) and describe the dis-
       tribution.   Comment upon the possibility of selective
       absorption or reflection associated with the particle size
       distribution.   If the particle size distribution were natural
       dust or related environmental particulates, would you expect
       the particles to have settled out or to have been attracted
       by static charge differences (surface accumulation); or to
       be a residue in moisture drops?   Discuss these options.

REFERENCES

Friedlander, S.K., "Smoke, Dust and Haze: Fundamentals of Aerosol
   Behavior", Wiley Interscience, New York, (1977).
Green, H.C. and Lane, W.R., "Particulate Clouds: Dusts, Smokes
   and Mists; Their Physics and Physical Chemistry and Industrial
   and Environmental Aspects", 2nd Edition, Van Nostrand, Prince-
   ton, New Jersey, (1964).
Murr, L.E., "Electron Optical Applications in Materials Science",
   McGraw-Hill Book Co., New York, (1970).
Powell, C.F., Oxley, J.H., and Blocher, Jr., J.M., "Vapor Deposi-
   tion", J. Wiley and Sons, Inc., New York, (1966).

(6)   Figure P. 6 shows two typical aerosol particles attached to
       a float glass protective layer on a flat-plate silver re-
       flector.   Which particle would be the more easily removed?
       Show, using simple surface tension relationships which
       particle has the highest surface tension?   If a detergent
       were used in removing the particles in Fig. P. 6(b), what
       must the detergent do, in principle, in order to aid the
       efficient particle removal?   Show this effect schematically.

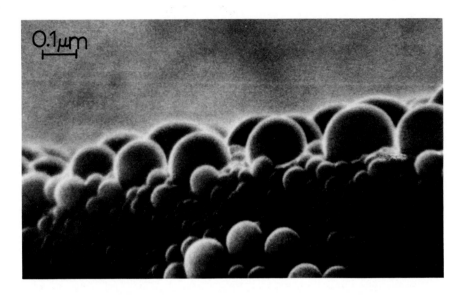

(a)

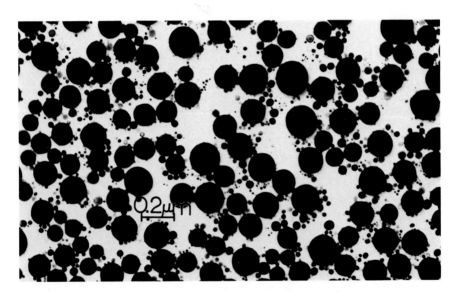

(b)

FIGURE P.5.   (a) SEM view of tin particles.   (b) TEM view of
same particles.

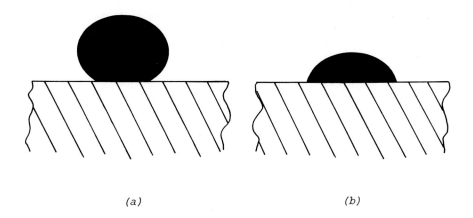

(a)   *(b)*

FIGURE P.6.

Describe how the adsorption or affinity for adsorption and
the adhesion "strength" of particles could be reduced by
changing the protective coating material. If the accumulation of dust or aerosol particles on a reflector/protector
layer develops and is not removed, show the effects on the
reflectivity and discuss the difference which could be expected if the initial silver reflectivity were 0.95 as
compared to 0.90 as a result of differences in, for example,
deposition characteristics. What structural or microstructural features of the silver reflecting layer could account
for these initial reflectivity differences?

REFERENCES

Adamson, A.W., "Physical Chemistry of Surfaces", Wiley-Intersci-
    ence, New York, (1960).
Bikerman, J.J., "Physical Surfaces", Academic Press, New York,
    (1970).
Murr, L.E., *Mater. Sci. Engr.*, Vol. 12, 3277 (1973).
Murr, L.E., "Interfacial Phenomena in Metals and Alloys", Addison-
    Wesley Publishing Co., Reading, Mass., (1975).
Murr, L.E., ASTM Special Technical Publication No. 640, "Adhesion
    Measurement of Thin Films, Thick Films, and Bulk Coatings",
    K.L. Mittal (ed), Philadelphia, Pennsylvania, p. 82, (1978).

Richman, M.H., "An Introduction to the Science of Materials",
    Ginn/Blaisdell Publishing Co., Waltham, Mass., (1967).
Woodruff, D.P., "The Solid-Liquid Interface", Cambridge University
    Press, London, (1973).

(7) Total hemispherical emissivity is expressed by

$$\bar{\varepsilon}_H = \int_0^{\pi/2} (\frac{\varepsilon_{TE} + \varepsilon_{TM}}{2}) \cos\theta \sin\theta \, d\theta \cdot \int_0^\infty L_w dw \bigg/ \int_0^{\pi/2} \cos\theta \sin\theta \, d\theta$$

$$\cdot \int_0^\infty L_w dw$$

show that $\bar{\varepsilon}_H(TE) = \varepsilon/3$ and $\bar{\varepsilon}_H(TM) = \varepsilon (\varepsilon = \bar{\varepsilon}_N)$; and consequently $\bar{\varepsilon}_H = 4\bar{\varepsilon}_N/3$. (TE and TM refer to the electric and
magnetic wave components, respectively; $\varepsilon_{TE} = \varepsilon \cos\theta$,
$\varepsilon_{TM} = \varepsilon/\cos\theta$.) Discuss how the hemispherical emissivity
can be decreased by altering the crystal structure of a
material. Under what conditions would the emissivity of a
metal approach zero?

REFERENCES

Condon, E.U., and Odishaw, H. (eds), "Handbook of Physics",
    p. 4-75 and 6-14, McGraw-Hill Book Co., New York, (1958).
Goodenough, J.B., and Whittingham, M.S. (eds), "Solid-State
    Chemistry of Energy Conversion and Storage", (Adv. in Chem.
    Serier 163, Amer. Chem. Soc.), 1977.
Guy, A.G., "Introduction to Materials Science", p. 371, McGraw-
    Hill Book Co., New York, (1972).
Müller, W.M. (ed), "Energetics in Metallurgical Phenomena",
    Vol. 1, Gordon and Breach, Science Publishers, New York,
    (1965).

(8) Referring to the tin vapor deposit of Fig. P. 5(b) above,
     show that the absorptivity in the U.V. will be essentially
     the same as that for a continuous film having an absorptiv-
     ity of $2/\omega \tau$ if the principles of geometrical optics are
     assumed to apply for the range of larger particle sizes

shown. Would the emissivity of a thick tin film on a tungsten surface increase or decrease the room temperature
emissivity? By how much? If the absorptivity of a black-
chrome electrodeposit is 0.97 and the emissivity is measured
to be 0.12 at 100°C, what must be done in order to maintain
the same $\alpha_s/\bar{\varepsilon}_H$ ratio at 400°C?

REFERENCES

Heavens, O.S., "Optical Properties of Thin Solid Films", Dover,
    New York, (1965).
Kerker, M., "The Scattering of Light and Other Electromagnetic
    Radiation", Academic Press, New York, (1969).
Pramanik, D., Sievers, A.J., and Silsbee, R.H., "The Spectral
    Selectivity of Conducting Micromeshes in Solar Energy
    Materials", in press, (1979).
Seraphin, B.O. (ed), "Solar Energy Conversion-Solid State Physics
    Aspects, Topics in Applied Physics", Vol. 31, Springer-Verlag,
    Berlin, (1979).

(9) Given the following data for a composite absorber coating,
     make a plot of optical extinction coefficient versus wave-
     length and briefly discuss the merits of the composite as
     an optical absorber:

λ (μm)	ε (total relative dielectric constant)	R (reflectivity)
0.3	9	0.30
0.6	16	0.40
1.4	25	0.45
2.6	16	0.36

All data apply at 500 K.

Is it sufficient to consider absorptivity of such a composite independent of other properties? Discuss your answer.
That is, discuss what features are important in good optical
absorbers.

REFERENCES

Craighead, H.G., Bartynski, R., Buhrman, R.A., Wojcik, L., and
    Sievers, A.J., *Sol. Energy Mater.*, 1, 105 (1979).
Fulrath, R.M., and Pask, J.A. (eds), "Ceramic Microstructures,
    1976, with Emphasis on Energy-Related Applications", Proc.
    6th Int. Mater. Symposium, Westview Press, Boulder, Colorado,
    (1977).
Heavens, O.S., "Optical Properties of Thin Solid Films", Dover,
    New York, (1965).
Kerker, M., "The Scattering of Light and Other Electromagnetic
    Radiation", Academic Press, New York, (1962).

(10) Describe an efficient method for fabricating $Au-Al_2O_3$ selec-
     tive absorbers, including the choice of a substrate. Dis-
     cuss how the absorptivity and emissivity can be controlled
     by the adjustment of fabrication parameters. What two
     fabrication parameters can be most effectively controlled to
     produce selective absorptivity and emissivity, and how can
     these be changed? On considering the reference to work of
     Murr and Bitler listed below, is there any evidence of
     intrinsic structural effects on the optical response of thin
     Au films in the solar portion of the spectrum? Discuss your
     answer. (Note the effect of temperature on the grain size
     in evaporated Au.) Recall that in Chapter 2 it was noted
     that reflectivity at $\lambda = 10.6$ μm was influenced by grain
     size in silver films, while in the reference cited above
     there is no evidence of similar microstructural influence.
     Discuss this apparent discrepancy.

REFERENCES

Abeles, F. (ed), "Optical Properties and Electronic Structure of
    Metals and Alloys", North-Holland, Amsterdam, (1966).
Chopra, K.L., "Thin Film Phenomena", McGraw-Hill Book Co., New
    York, (1969).
Craighead, H.G., Bartynski, R., Buhrman, R.A., Wojick, L., and
    Sievers, A.J., *Sol. Energy Mater.*, 1, 105 (1979).
Craighead, H.G., and Buhrman, R.A., *Appl. Phys. Lett.*, 31, 423
    (1977); *J. Vac. Sci. Technol.*, 15, 269 (1978).

Murr, L.E., and Bitler, W.R., *Mater. Res. Bull.*, 2, 787 (1967).
Szilva, W.A., and Murr, L.E., *Phys. Stat. Sol. (a)*, 40, 211
    (1977).

(11) Many simple solar collection schemes can be envisioned as
      being composed of a variety of materials, and of course many
      indeed are. Early collectors and many contemporary designs
      use a variety of metals and alloys which are somewhat in-
      compatible in terms of corrosion. The case in point would
      be copper tubing connected to iron or iron alloy hot-water
      units, which can be prone to galvanic corrosion. In general,
      galvanic corrosion for a single electrode process can be
      characterized by

$$E = E^\circ + \frac{RT}{ZF} \ln K,$$

where E is the electrode potential, E° is the metal electrode potential (standard electrode potential), R is the
gas constant, T is the absolute temperature, F is the Faraday constant [23,600 cal/volt (gram equivalent)], Z is the
metal (ion) charge, and K is the equilibrium constant for
the single electrode process (or activity of the metal involved). In effect, this expresses the theoretical limit of
oxidation of a metal (at equilibrium) in contact with the
standard hydrogen electrode (SHE). Find the activity of
copper and iron separately, and then find the ratio of
activities ($K_{Fe}/K_{Cu} = a_{Fe^{2+}}/a_{Cu^{2+}}$). Discuss the implications of these results. (Hint: E in the equation will be
the hydrogen electrode potential.) What could happen if the
system were aluminum and copper rather than iron and copper?
What would the potential (emf) be if a zinc-copper cell were
created somewhere in the system, i.e., what would the difference be?

REFERENCES

Evans, U.R., "The Corrosion and Oxidation of Metals: Scientific
   Principles and Practical Applications", St. Martin's Press,
   New York, (1960); plus supplement, (1968).
Fontana, M.G., and Greene, N.D., "Corrosion Engineering", McGraw-
   Hill Book Co., New York, (1967).
Guy, A.W., "Introduction to Materials Science", McGraw-Hill Book
   Co., New York, (1972).
Uhlig, H.H., "Corrosion and Corrosion Control: An Introduction to
   Corrosion Science and Engineering", Wiley, New York, (1971).
Weast, R.C. (ed), "Handbook of Chemistry and Physics", The
   Chemical Rubber Co., Cleveland, Ohio, (1978).

(12) A heat transfer tube in a solar furnace is subjected to a
      maximum stress (tensile) of 20 Ksi. The tube alloy has a
      yield stress of 50 Ksi and an ultimate tensile stress (UTS)
      of 110 Ksi. The tube is 20 cm in outside diameter and has
      a 1 cm wall thickness. Because of chloride-ion contamina-
      tion in the heat-transfer fluid, the inside tube wall is
      observed to corrode away at a rate of 0.1 mm/month. If a
      safe system design were to consider a safety factor of 2
      based upon the UTS, when would the tube have to be consid-
      ered for replacement, if the maximum load never changes?
      During a focusing accident, the load is removed on an iden-
      tical tube in a similar furnace and the tube wall is rapidly
      annealed, resulting in a grain size in a circumferential
      section which is a factor 10 greater than the initial grain
      size. What effect would you expect this to have on the
      yield stress in this region when normal operation resumes?
      That is, what change, if any, would occur in the yield
      stress? (Hint: Consider a Hall-Petch relationship applies.)

REFERENCES

Ailor, W.H. (ed), Symposium on the State of the Art in Corrosion
   Testing, "Handbook on Corrosion Testing and Evaluation",
   Wiley, New York, (1971).

Berry, W.E., "Corrosion in Nuclear Applications", Wiley, New
   York, (1971).
Cernica, J.N., "Strength of Materials", Holt, Rinehart and
   Winston, Inc., New York, (1966).
Guy, A.W., "Introduction to Materials Science", McGraw-Hill Book
   Co., New York, (1972).
Tegart, W.M.M., "Elements of Mechanical Metallurgy", The MacMil-
   lan Co., New York, (1966).

(13) Compare the thermal energy (heat) which can be stored on
      melting 1 ft^3 (0.029 m^3) of sodium chloride, sodium sulfate,
      sodium hydroxide, silicon, sodium, and gallium. Which of
      these materials would you choose for a solar home heating
      system? Discuss your choice based upon efficiency, econ-
      omics, and materials availability, compatibility and other
      environmental considerations. Compare the difference in
      energy required in a solar-industrial recycling operation
      for reprocessing iron and aluminum scrap. (Assume the
      metals are melted for pouring into ingots and consider their
      temperature must be raised from room temperature, i.e.,
      20°C.)

REFERENCES

Duffie, J.A. and Beckman, W.A., "Solar Energy Thermal Proces-
   ses", J. Wiley and Sons, New York, (1974).
Telkes, M., *ASHRAE Journal,* September, 40 (1974).
Telkes, M., "Critical Materials Problems in Energy Production",
   C. Stein (ed), Chap. 14, Academic Press, New York, p. 440,
   (1976).
Ubbelohde, A.R., "Melting and Crystal Structure", Clarenden
   Press, Oxford, (1965).
Weast, R.C. (ed), "Handbook of Chemistry and Physics", Chemical
   Rubber Co., Cleveland, Ohio, (1978).

(14) A small industrial photoprocess plant measures 200 ft
      (61.5 m) and 100 ft (30.8 m) (wide). The exterior walls have
      no windows and are 6 inches (15.24 cm) thick. The flat
      ceiling and roof of the single-level building is 14 inches
      (35.56 cm) thick, and has a height of 15 ft (4.6 m). The

walls and ceiling have the same (or uniform) average thermal
conductivity, K = 0.05 (in units of $BTU/ft^2\ hr/°F/ft$). If
the building is solar heated, what is the minimum storage
volume of sodium sulfate decahydrate to offset the heat loss
in a 12 hour period when the collector is not operating if
the average inside temperature is to be maintained at 68°F
and the average outside temperature is 0°F during this
period? What percent of the heat lost can be saved if
another 0.5 ft of insulation is added to the ceiling area?
(Assume the thermal conductivity of the insulation is the
same as noted.) What type of energy resource does this constitute?

REFERENCES

Bayley, F.J., Owen, J.M., and Turner, A.B., "Heat Transfer",
    Barnes and Moble, New York, (1972).
Chapman, A.J., "Heat Transfer", 3rd Edition, MacMillan Publish-
    ing Co., Inc., New York, (1974).
Duffie, J.A., and Beckman, W.A., "Solar Energy Thermal Processes",
    Wiley, New York, (1974).
Ozisik, M., "Basic Heat Transfer", McGraw-Hill Book Co., New York,
    (1977).
Telkes, M., "Critical Materials Problems in Energy Development",
    Chap. 14, C. Stein (ed), Academic Press, New York, (1976).
Young, H.D., "Fundamentals of Mechanics and Heat", McGraw-Hill
    Book Co., New York, (1974).

(15) Consider the thermal decomposition reaction: $CaCO_3$(solid) =
      CaO(solid) + CO_2(gas). Illustrate how this reaction can be
      utilized in the storage of solar heat at high temperature.
      Calculate the standard free energy of dissociation at 937°C.
      Show how the associated entropy change and enthalpy change
      could be calculated for $CaCO_3$.

REFERENCES

Darken, L.S. and Gurry, R.W., "Physical Chemistry of Metals",
    McGraw-Hill Book Co., New York, (1953).

Gaskell, D.R., "Introduction to Metallurgical Thermodynamics",
    McGraw-Hill Book Co., New York, (1973).
Hammes, G.G., "Principles of Chemical Kinetics", Academic Press,
    New York, (1978).
Huang, F.F., "Engineering Thermodynamics", MacMillan Publishing
    Co., Inc., New York, (1976).

(16) Describe the principal contemporary thermochemical reactions
      of interest for solar energy storage and give some examples
      of each. Consider the reversible reaction $CH_4 + H_2O \longrightarrow$
      $CO + 3H_2$. Make a sketch of the complete reactor system and
      describe any special materials problems attendant to this
      system.

REFERENCES

Duffie, J.A., and Beckman, W.A., "Solar Thermal Processes", J.
    Wiley and Sons, New York, (1974).
Guy, A.G., "Introduction to Materials Science", McGraw-Hill Book
    Co., New York, (1972).

(17) A critical feature in semiconductor junction devices, in-
      cluding solar cells, is the metal/semiconductor junction
      response. That is, depending upon the type and chemistry of
      semiconductor and the type of contact metallization, the
      junctions can respond as either rectifying junctions or
      ohmic junctions. Consider the p-n junction device config-
      uration shown below. Draw the energy band diagram for the
      complete device system shown and an equivalent circuit,
      thereby indicating whether the junctions are ohmic or rec-
      tifying. What determines the nature of the junction, i.e.,
      what are the conditions for p- or n-type materials joined
      to a metal which make it ohmic or rectifying? What can you
      say about the metal/metal junctions? (Hint: The p-Si
      and n-Si work functions are 3.6 and 3.7 eV, respectively.)

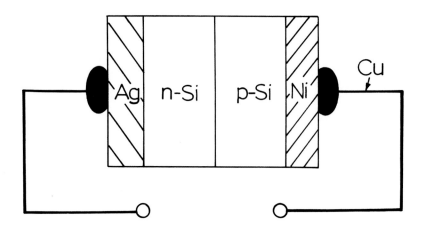

FIGURE P.17.

REFERENCES

Deboo, G.J. and Burrous, C.N., "Integrated Circuits and Semi-
    conductor Devices: Theory and Applications", 2nd Edition,
    McGraw-Hill Book Co., New York, (1977).
Driscoll, F.F. and Coughlin, R.F., "Solid-State Devices and
    Applications", Prentice Hall, Inc., Englewood Cliffs, New
    Jersey, (1975).
Fan, John C.C., "Solar Cells: Plugging Into the Sun", Technology
    Review, August/September, 14 (1978).
Merrigan, J.A., "Sunlight to Electricity: Prospects for Solar
    Energy Conversion by Photovoltaics", MIT Press, Cambridge,
    (1975).
Murr, L.E., "Solid-State Electronics", Marcel Dekker, Inc., New
    York, (1978).
Tickle, "Thin Film Transistors", J. Wiley and Sons, Inc., New
    York, (1969).

(18) A feature of interest in contemporary devices such as sili-

      con solar cells is the so-called grain boundary passivation.

      Some polycrystalline silicon, when specially solidified,

      contains a multitude of boundaries which appear to be twin

      boundaries. How could one unambiguously identify twin

      boundaries in silicon? If passivation is associated with

      segregation of hydrogen along grain boundaries at 500°C, and

      this is induced by high interfacial energies, what can you

say about twin boundaries in silicon with regard to segregation if the twin boundary/grain boundary (high-angle)
energy ratio is the same as that in indium? [Assume the
grain boundary energy in silicon to be 0.3 F_S, where F_S is
the surface energy given by $F_S = 1.2 \ (\gamma_{LV})_m + 0.45 \ (T_m - T)$;
where $(\gamma_{LV})_m$ is the liquid (liquid/vapor) surface free
energy at the melting point, T_m is the melting temperature
(°C), and T is the temperature in the solid state.] Are
there other types of grain boundaries which could have
special energies (and structures)?

REFERENCES

Azaroff, L.V., "Introduction to Solids", McGraw-Hill Book Co.,
    New York, (1960).
Chalmers, B., "The Photovoltaic Generation of Electricity",
    Scientific American, November (1976), p. 34.
Murr, L.E., "Interfacial Phenomena in Metals and Alloys",
    Addison-Wesley Publishing Co., Reading, Mass., (1975).
Seager, C.H. and Ginley, D.S., Appl. Phys. Letters, 34(5), 337
    (1979).

(19) Consider the idealized case of a Cu_2S/CdS single-crystal
      heterojunction. Make a sketch of the associated energy-
      level diagram and discuss how this is characteristic of a
      junction (P/n) device. Show the lattice mismatch associated
      with this junction and make a sketch indicating how this
      mismatch can be accommodated by the formation of interfacial
      dislocations. Describe the mismatch at ZnO/Cd Te junctions
      for the idealized case of single crystal films. (Hint:
      Consult the ASTM X-ray Card Fild and associated references.
      Do not consider epitaxial growth. The junctions are directly
      bonded.)

REFERENCES

Bube, R.H., Chap. 16 in "Critical Materials Problems in Energy
    Production", C. Stein (ed), Academic Press, New York, (1976),
    p. 486.
Bube, R.H., "Photoconductivity of Solids", J. Wiley and Sons,
    New York, (1960). (Reprinted by Krieger, New York, 1978).
Larach, S., "Photoelectronic Materials and Devices", Van Nostrand,
    Princeton, New Jersey, (1965).
Milnes, A.G., and Feucht, D.L., "Heterojunctions and Metal-Semi-
    conductor Junctions", Academic Press, New York, (1972),
Murr, L.E., "Interfacial Phenomena in Metals and Alloys",
    Addison-Wesley, Reading, Mass., (1975).
Seraphin, B.O. (ed), "Solar Energy Conversion", Springer-Verlag
    New York, Inc., New York, (1979).

(20) Dislocations in or near a bonded Cu_2S/CdS heterojunction

      interface can be associated with three different regimes:

      Cu_2S, the interfacial plane, or CdS. What types of Burgers

      vectors would you expect for the dislocations in the Cu_2S

      and CdS? Discuss your response. What effect would you

      expect dislocation density to have? Make a sketch of open

      circuit current versus dislocation density at some constant

      solar flux. (Hint: In addition to the general references

      suggested below you should search the journal and periodical

      literature.)

REFERENCES

Bube, R.H., "Electronic Properties of Crystalline Solids: An
    Introduction of Fundamentals", Academic Press, New York,
    (1974).
Guy, A.G., "Introduction to Materials Science", McGraw-Hill Book
    Co., New York, (1972).
Hirth, J.P., and Lothe, J., "Theory of Dislocations", McGraw-Hill
    Book Co., New York, (1968).
Milnes, A.G., and Feucht, D.L., "Heterojunctions and Metal-Semi-
    conductor for Junctions", Academic Press, New York, (1972).
Murr, L.E., "Interfacial Phenomena in Metals and Alloys",
    Addison-Wesley Publishing Co., Reading, Mass., (1975).
Sharma, B.L., and Purohit, R.K., "Semiconductor Heterojunctions",
    Pergamon Press, New York, (1974).

(21)   Describe the zeta potential and its measurement. How does
       the zeta potential determine the dlotation characteristics
       of inorganic minerals? How does the zeta potential influ-
       ence the operation of the wet photovoltaic cell, and how is
       this concept related to the zeta potential role in mineral
       flotation?

REFERENCES

Adamson, A.W., "Physical Chemistry of Surfaces", Interscience
    Publishers, New York, (1967).
Bockris, J. O'M., and Drazic, D.M., "Electrochemical Science",
    Barnes and Noble, New York, (1976).
Bockris, J. O'M., and Reddy, A.K.N., "Modern Electrochemistry",
    Vols. 1 and 2, Plenum Publishing Corp., New York, (1970)
    (Second Reprint in paperback, 1976).
Gaudin, A.M., "Flotation", McGraw-Hill Book Co., New York, (1957).
Holmes, J.P. (ed), "The Electrochemistry of Semiconductors",
    Academic Press, New York, (1962).
Osipow, L.I., "Surface Chemistry", Reinhold Publishing Corp., New
    York, (1962).

(22)   Grain boundary passivation in polycrystalline silicon occurs
       for diffused hydrogen and magnesium, but not for N, F, or
       atomic oxygen. Show the possible effect of all these ele-
       ments adsorbed along a grain boundary in silicon and their
       passivation or non-passivation role using simple sketches or
       atomic models. Make a model of a simple symmetric grain
       boundary between two (010) silicon grains having a misorien-
       tation of $5°$ and another having a misorientation of $15°$.
       If hydrogen passivation by bond satisfaction occurs, how
       much more hydrogen is contained in the $15°$ boundary than the
       $5°$ boundary?

REFERENCES

Gleiter, H., and Chalmers, B., "High-Angle Grain Boundaries",
    Pergamon Press, New York, (1972).
McLean, O., "Grain Boundaries in Metals", Clarenden Press, Oxford,
    (1957).

Murr, L.E., "Interfacial Phenomena in Metals and Alloys",
    Addison-Wesley Publishing Co., Reading, Mass., (1975).
Read, W.T., "Dislocations in Crystals", McGraw-Hill Book Co.,
    New York, (1953).

(23) Figure P.23 below shows black-chrome surface topography
      after argon ion bombardment. Note that the surface appears
      to be composed of tiny spheres or sphere-like "particles".
      What additional techniques could be used to determine the
      structure of the layer shown in Fig. P.23? Refer to Chap. 3
      and discuss the techniques available for examining the
      chemical composition of the layer. How could you determine
      whether the layer was a composite of Cr or Cr-rich "parti-
      cles" within a Cr_2O_3 matrix?

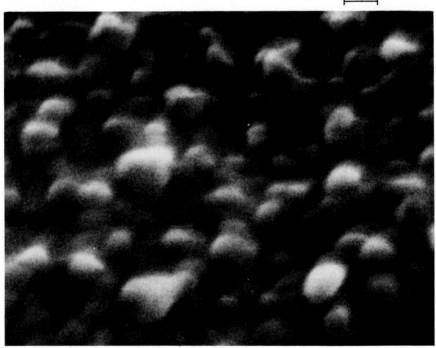

*FIGURE P.23.   Scanning electron micrograph showing ion-etched
black-chrome coating electrodeposited onto nickel.*

REFERENCES

Buck, Otto, Tien, J.K., and Marcus, H.L. (eds), "Electron and
   Positron Spectrocopies in Material Science and Engineering",
   Academic Press, Inc., New York, (1979).
Czanderna, A.W., and Vasofsky, R., *Progress in Surface Science,*
   9, 43 (1979).
Czanderna, A.W. (ed), "Methods of Surface Analysis", Elsevier,
   Amsterdam, (1975).
Murr, L.E., "Electron Optical Applications in Materials Science",
   McGraw-Hill Book Co., New York, (1970).

(24) Is it possible to examine amorphous silicon structure?
   That is, can the atomic structure in amorphous silicon be
   directly observed? How? Describe the technique. What has
   already been learned about the "structure" of amorphous
   materials, including metals and alloys? Is hydrogen passivation effective in amorphous silicon? Why?

REFERENCES

Buck, Otto, Tien, J.K., and Marcus, H.L. (eds), "Electron and
   Positron Spectroscopies in Material Science and Engineering",
   Academic Press, Inc., New York, (1979).
Goldstein, J.I., Hren, J.J., and Joy, D.C. (eds), "Introduction
   to Analytical Electron Microscopy", Plenum Press, New York,
   (1979).
Howie, A., *J. Non-Crystalline Solids,* 31, 41 (1978).
Valdre, U., and Ruedl, E. (eds), "Electron Microscopy in Materials
   Science", Parts I-IV, Commission of the European Communities,
   Luxembourg, (1975).

(25) Discuss the current efforts to understand materials problems
   associated with the development of a viable solar technology.
   Do you envision changes in the traditional approaches to
   materials science and engineering in regard to future solar
   technology developments? If so, describe them and discuss
   their possible consequences.

REFERENCES

Cohen, M. (ed), "Materials Science and Engineering: Its Evolution, Practice, and Prospects", Special Issue of *Materials Science and Engineering,* 37(1), (1979).

Roller, D.H.D. (ed), "Perspectives in the History of Science and Technology", Univ. of Oklahoma Press, Norman, (1971).

PROBLEM
SOLUTIONS
and Discussion

I give you now Professor Twist,
A conscientious scientist.
Trustees exclaimed, "He never bungles!"
And sent him off to distant jungles.
Camped on a tropic riverside,
One day he missed his loving bride.
She had, the guide informed him later,
Been eaten by an alligator.
Professor Twist could not but smile.
"You mean," he said, "a crocodile."

——————— The Purist, by Ogden Nash

CHAPTER 21

PROBLEM SOLUTIONS AND DISCUSSION

L.E. Murr

Department of Metallurgical and Materials Engineering
New Mexico Institute of Mining and Technology
Socorro, New Mexico

(1) As shown below, plane waves incident upon a plane interface
are related by the angles shown: $90° - I_1$, $90° - k_1$,
$90° - r_2$, and

$$\cos r_2/\cos I_1 = \sqrt{\mu_1 \varepsilon_1/\mu_2 \varepsilon_2}$$

since $C_j = 1/\sqrt{\mu_j \varepsilon_j}$; $j = 1, 2, \ldots$

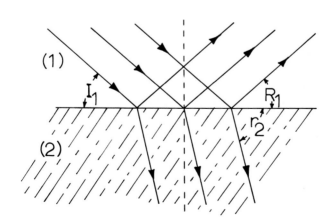

If we add a third medium (3) we have the following:

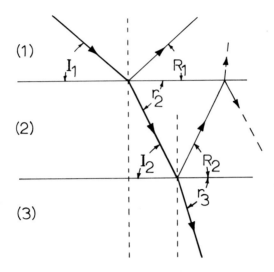

From the three-medium sketch above showing the angles of the
incident, reflected, and refracted wave portions in (1), (2),
and (3), it can be observed that all angles depend upon the
properties of the media, and all are interconnected. Thus,
it is apparent that any transmitted portion and reflected
portions of the wave are dependent upon the angle of incidence in medium (1), I_1. The angles in all media show a
dependence on μ or ε which are dependent upon the particular
kinds of atoms composing the media, the way in which they
are bound, and the crystal structure. Furthermore, the
media may be composites or aggregates which could alter μ
and ε (especially ε) depending upon the aggregate size and
distribution. The effects leading to variations in μ or ε
occur mainly in response to magnetic or electric polarizabil-
ities. Polarization of either magnetic or electric dipoles
and indeed the creation of such dipoles is dependent upon
the atomic structure, the electronic structure, the crystal
structure, and on a larger scale, the microstructure. The
index of refraction, n, is strongly coupled to the dielectric
constant through a relationship of the form $n = \sqrt{\varepsilon_r \mu_r}$ where

ε_r is referred to as the relative dielectric constant and defined by the ratio of permittivity in a particular medium to that of free space, denoted ε_o. Thus $\varepsilon_r = \varepsilon/\varepsilon_o$; and correspondingly $\mu_r = \mu/\mu_o$. For nonmagnetic material, $\mu \cong \mu_o$ and $n = \sqrt{\varepsilon_r}$.

(2)   Two prominent reflector materials would of course be silver and aluminum. Silver is becoming increasingly expensive. For example, the market price of silver has increased nearly by a factor 3 in the period 1976 - 1979. Similarly, aluminum cost has escalated, but aluminum is comparatively cheap. Raw material pricing of aluminum (principally bauxite) has increased more than a factor 4 in the period above. Chromium, another reflecting coating is not even produced in the United States and has not been produced since 1964. Consequently, it is imported. The United States has few large sources of bauxite, so it too is imported. Prominent absorber candidates include black chrome and black nickel, along with other oxide candidates. Most can be considered to be reasonably cheap except when large volumes are required, and for massive solar proliferation this will be the case. For a comparison of metal costs, etc. you might consult "Metal Statistics", American Metal Market, Fairchild Productions Division of Capital Cities Media, Inc., 7 East 12th St., New York for appropriate years, and other related references.

Photovoltaics include of course the common silicon solar cell or metal sulfides, especially sulfide heterojunctions. These are relatively expensive and are at present inefficient. Current costs for silicon cells are roughly $10/watt and this constitutes a size of roughly 0.01 m^2. Historically, silicon technology in America has accelerated steadily since about 1950, and very standard techniques are utilized in producing very pure, defect free, oriented single crystal chips or slices which are routinely $\sim 0.01 \text{ m}^2$ in disc form, grown from boules in continuous growth processes.

Polycrystalline materials are also rapidly grown with a high
degree of quality control. New photovoltaic technology must
face the quality control problem, the production of large-
area arrays, and a host of complex technological and produc-
tion-related problems. The issue is really how long would
it take to develop such a manufacturing capability for a
new photovoltaic scheme? In some cases the projections are
decades. Consequently, a current principal thrust is to
alter current manufacturing processes to improve contemporary device efficiencies. The cost to do this could be far
cheaper than a massive redevelopment program. However, new
concepts such as the improvement of efficiency in so-called
amorphous semiconductors could provide a breakthrough because such devices do not require rigid quality control,
grain boundary passivation (because they do not in principal
have grain boundaries), or related process controls, and
they could be manufactured like sheet glass or plastics in
enormous areas in very little time.

(3) As shown below a single bct cell (heavy lines) can be drawn
within two fct cells, where it can be readily observed that
the a and b unit cell dimensions of the bct cell differ by a
factor $\sqrt{2}$ from those of the fct cell since they are half-
diagonals of the fct cell.

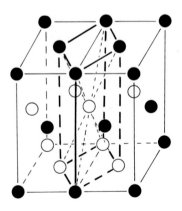

The distance between planes in a tetragonal lattice is given by the general equation for orthogonal unit cells:

$$d = 1 \Big/ \sqrt{(h/a)^2 + (k/b)^2 + (\ell/c)^2}$$

where h, k, and ℓ are the Miller indices for the plane (hkℓ), and a, b, and c are the lattice parameters. Since in a tetragonal cell, a = b, and given c/a = 1.95, d can be calculated for (111) to be a/1.5. If a vacancy were formed in every other cell center in the bct cell above (composed within the fct cells), the number of atoms per unit cell would be reduced by twenty-five percent. The structure created would be a vacancy defect structure and a vacancy sublattice (or superlattice) will occur.

If an edge dislocation intersects, or is associated with, every four unit cell basal planes, we would have one in an area of $4a^2$ and the density would be $(4a^2)^{-1}$. Assuming a value of a of say 4Å, this would give a dislocation density of 1.6×10^{14} cm^{-2}, which is about two orders of magnitude higher than normally measured or expected as a limiting value in deformed metals and alloys.

A deposit layer grown pseudomorphically upon the basal plane of an fct substrate would be cubic because the dimensions in the plane of the overgrowth would be a, and would assume an epitaxial (001) orientation.

(4)   In optical materials where the relative permeability is unity ($\mu = \mu_o$), the index of refraction, n, equals the square root of the relative dielectric constant. Consequently ε can be measured by measuring n.

In a parallel plate capacitor the capacitance is given by $C = C_o K_r$ where $\varepsilon_r \equiv K_r = \varepsilon/\varepsilon_o$. So $K_r = C/C_o$. We need only measure the capacitance of an identical parallel-plate capacitor in air for comparison. The same is true of a two-layer dielectric or a complex dielectric, i.e., the same parallel plate set-up is used, with the same plate spacing, etc. If a flat-plate absorber is increased in thickness the dielectric effect will increase or if the absorber is a bi-layer, the more efficient layer can be increased in thick-ness or occupy a greater percentage of the total layer thickness. For complex mixtures of two different dielec-trics, a logarithmic mixing rule is commonly used in the form $\log K_m = \theta_1 \cdot \log K_1 + \theta_2 \cdot \log K_2$ where K_m is the dielectric constant of the mixture; K_1 and K_2, the dielec-tric constant of the components, and θ_1 and θ_2 are the volume ratios of the respective components.

Since dielectric constants are important in optical ab-sorption, it would seem that materials with high dielectric constants and especially high dielectric loss factors would be the better absorbers or absorber complexes. Consequently, oxide systems would be the best absorbers, including perhaps ferroelectric materials such as $BaTiO_3$ or other materials having highly polarizable crystal structures embedded in materials of lower or different dielectric constants.

(5) A plot of particle size in Fig. P. 5(b) will show a bimodel size distribution with two average particle sizes less than 0.2 μm diameter. This size is below the solar (optical) portion of the spectrum and as a consequence there will be no absorption because the surface will be technically "re-flecting" because even the spaces between the particles are in this range, and the surface will be absorbing only in the U.V. part of the spectrum. Since the size distributions are below 0.2 μm, such particulates in the natural environment could certainly not settle out because only particles greater than about 50 μm readily settle out. Such small particu-lates could be deposited in rain, and could more easily be attached electrostatically to the surface.

(6)   The easiest particles to remove from a surface are those
      least attached.   Consequently particles in Fig. P. 6(a)
      would probably be the easiest to remove.   Removal really
      depends upon the adhesive strength or adhesive energy, and
      this depends upon the balance of interfacial energies in
      the system:

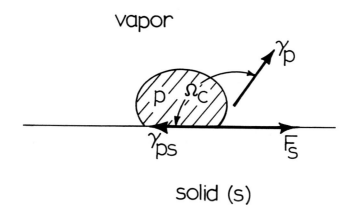

In the notation shown

$$\gamma_{ps} = \gamma_p \cos(180 - \Omega_c) + F_s$$

where γ_{ps} is the particle substrate interfacial free energy,
γ_p is the particle surface free energy (or surface tension)
and F_s is the substrate surface free energy (or surface tension).   We see that when the contact angle, $\Omega_c < 90°$, the
particle tends to "wet" the substrate or spread over the
substrate while for $\Omega_c > 90°$ the particle tends to loose
such a tendency.   If the particles in Fig. P. 6 are attached
to the same substrate, the particle in Fig. P. 6(a) would
have the highest surface tension.   Consequently, a detergent
must increase the particle surface tension or lower the
interfacial free energy in order to reduce the wetting, and

thereby decrease the contact angle. Ideally, a particle
which is completely non-wetting and therefore not technically in contact with the surface is one where $\Omega_c = 180°$,
while complete wetting occurs when $\Omega_c = 0°$. By changing the
substrate, the interfacial free energy can be changed because
F_s will change. Consequently the contact angle can be
altered. A practical example is the application of a wax
layer to prevent water wetting of a surface, e.g., the bead-
ing of water on a freshly polished automobile.

If dust accumulation produces a plateau of optical response, then the effect performance might be characterized
by, say the difference from the initial conditions to the
plateau. Consequently, for a reflectivity plateau, the
better performance would result from a higher initial reflectivity. Reflectivity in certain parts of the spectrum
is highly dependent upon grain size of a reflecting coating
as well as porosity. These could account for reflectivity
variations in silver reflecting coatings. This is at
present a somewhat hypothetical situation in the solar
spectrum.

(7)   Assuming either ε_{TM} or ε_{TE} to be zero, respectively, or
simply integrating the entire equation, using standard
integral tables we find:

$$\frac{\varepsilon}{2} \int_0^{\pi/2} \cos^2 \theta \sin \theta \, d\theta = \varepsilon/6; \quad \frac{\varepsilon}{2} \int_0^{\pi/2} \sin \theta \, d\theta = \varepsilon/2$$

$$\text{and} \quad \int_0^{\pi/2} \cos \theta \sin \theta \, d\theta = \frac{1}{2} \int_0^{\pi/2} \sin 2\theta \, d\theta = 1/2.$$

Since $\overline{\varepsilon}_H = 4\overline{\varepsilon}_N/3$, and $\overline{\varepsilon}_N \, \alpha \rho_{d.c.} \cdot (T)$ it is possible to alter
the emissivity by altering the resistivity. For example, it
is known that resistivity increases with a significant

increase in crystal defects such as point defects and dislocations. There is also a rather abrupt change in resistivity
when an order-disorder transition occurs. Consequently, when
the resistivity approaches zero the emissivity will also
approach zero. This occurs in metals at very low temperature, or ideally where a metal becomes superconducting
$(\sigma \to \infty; \; \rho \to 0)$.

(8)   When geometrical optics apply (as in the visible portion of
the spectrum) Kirchoff's law states that absorptivity =
emissivity; consequently absorptivity of a continuous film =
$2/\omega \, \tau$. Similarly, the absorptivity of a micromesh, which
the vapor deposit in Fig. P. 5(b) might be considered to
represent, is given by $(2/\omega \, \tau)(d/2a)$, where d is the distance
between the particles and a is their diameter. It can be
observed (by direct measurements) in Fig. P. 5(b) that the
distance between the larger particles is approximately twice
the particle diameter. Consequently the factor $(d/2a) = 1$,
and the absorptivity will be essentially the same as a continuous tin film. Since the emissivity is in proportion to
the resistivity, and since the room temperature resistivity
of tungsten is about 5.5 microhm-cm as compared to 11.5
microhm-cm for tin, (c.f. Handbook of Chemistry and Physics),
the emissivity would increase by nearly a factor 2. Since
the emissivity increases with temperature if a black-chrome
coating is to maintain a constant ratio $\alpha_s/\varepsilon_H = 0.97/0.12$ the
absorptivity will have to be increased because the emissiv-
ity at $400°C$ will increase over that at $100°C$.

(9)   Consider

$$R = \left[\frac{(n-1)^2 + k^2}{(n+1)^2 + k^2} \right] ; \quad n = \sqrt{\bar{\varepsilon}}$$

and solve for

$$k = \frac{\sqrt{R(n+1)^2 - (n-1)^2}}{(1-R)} \, .$$

We can then plot values for k versus λ as shown below, where it can be observed that the composite is a somewhat selective absorber in the solar spectrum ($0 < \lambda < 2$ μm), but is characterized by a large thermal broadening.

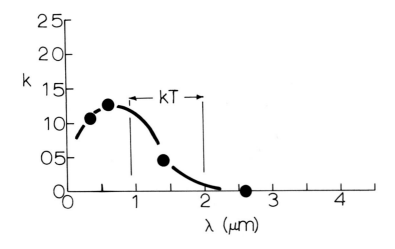

It is well established that variations in high values of absorptivity usually involve variations in emissivity. Both absorptivity and emissivity are important. Ideally one strives to achieve the highest absorptivity and a corresponding low emissivity. For example, $\alpha_s > 0.95$, $\varepsilon_H < 0.1$.

(10) Selective composite (cermet) absorbers can be fabricated by co-evaporation or sputtering of Au and Al_2O_3 in vacuum or inert atmosphere upon a reflecting substrate providing good adhesion for the composite overgrowth. The substrate must be heated to provide for a specific size of Au particles and the evaporation rates can be systematically adjusted to allow for a gradient in the metallic fraction with overgrowth thickness. Absorptivity and emissivity can be most effectively controlled by varying the metal (Au) fill fraction

and/or film thickness. The thickness is simply adjusted by
evaporation rate or sputtering time and the fill fraction
can be varied by varying the ratio of deposition rates for
metal and ceramic, or by varying the temperature and thereby
increasing (with increasing temperature) or decreasing (with
decreasing temperature) the mean particle (metal inclusion)
size.

In the case of vapor-deposited gold films there is no
evidence of variations in grain size having a measurable
effect on optical properties (absorption) in the visible
portion of the spectrum for grain sizes greater than about
0.01 μm. The same was observed for silver. While an effect
of grain size has been observed in silver films at 10.6 μm,
this is more than an order of magnitude greater wavelength
and the electromagnetic interactions would be expected to be
considerably different.

(11) In the single electrode process for iron we have

$$0 = -0.41 + \frac{0.0592}{2} \log a_{Fe^{2+}}$$

where -0.41 is the electrode potential (from the electromotive or electrochemical series) for iron (Fe^{2+}), $2.3\,RT/F =$
0.0592, and the denominator represents the charge. Solving
for $a_{Fe^{2+}}$ we find $\sim 10^{15}$ (an impossibly large activity).
Conversely, solving for copper (Cu^{2+}) we find $a_{Cu^{2+}} \simeq 10^{-11}$
(an impossibly small activity). Consequently the ratio
$a_{Fe^{2+}}/a_{Cu^{2+}} \simeq 10^{26}$, indicating an enormous difference in the
theoretical limit of oxidation. Replacing the iron by
aluminum simply changes the electrode potential and the
corrosion characteristics, but there would still be corrosion of the aluminum provided an oxide were not formed.
Unlike iron, aluminum normally forms a very tenacious protective oxide (Al_2O_3). Consequently there may be a significant reduction in corrosion even though the electrochemical

conditions are possible. If a zinc-copper cell were formed
in the system, the electrode potential (or potential difference) would be - 0.763 - (0.337) volts or - 1.1 volts.
(Data from the electrochemical series.)

(12) The initial and constant maximum load is determined from
$P = \sigma A = 20$ Ksi$[\pi(10^2 - 9^2)]/(2.54)^2 = 185$ K. Then for a
safety factor 2 based upon UTS implies a maximum permissible
stress of 55 Ksi (110 Ksi/2). We set this equal to 183
K$/\pi[10^2 - (9 + 0.01$ t$)^2]/(2.54)^2$, and solving for t in
months we find t = 64 months (or 5.33 years). Considering
the annealed tube wall section, the yield stress can be
approximated by a Hall-Petch relationship, $\sigma_y = \sigma_o + KD^{-1/2}$.
If σ_o and K do not change, then σ_y (the yield stress), would
be reduced since the new grain diameter, D' = 10D.

(13) Compare the heats of fusion and melting points as follows:

Material	Heat of fusion	Melting point	Heat storage
Sodium chloride	124 cal/g	800°C	30,551 BTU/ft^3
Sodium sulfate (anhydrous)	41 cal/g	884°C	12,432 BTU/ft^3
Sodium sulfate decahydrate	60 cal/g	32°C	10,476 BTU/ft^3
Sodium hydroxide	50 cal/g	322°C	11,790 BTU/ft^3
Silicon	337 cal/g	1427°C	88,015 BTU/ft^3
Sodium	27 cal/g	98°C	2,989 BTU/ft^3
Gallium	19 cal/g	29°C	12,716 BTU/ft^3

The heat storage is obtained by converting heat of fusion in
cal/g to BTU/ft^3, considering the density for each material.
It can be observed that the only practical materials for
home solar heat storage are sodium sulfate decahydrate,
sodium, and gallium since their melting points are in the
range of such systems. But sodium melts a bit too high and

stores little heat. It is also more expensive and difficult
to store. Sodium sulfate decahydrate is similar to gallium
in melting point and heat storage but it is roughly a factor
10^3 less expensive. Sodium sulfate decahydrate is therefore
the logical choice. It is plentiful and easily stored in
plastic containers, etc.

The heat energy required to melt aluminum is about the
same as that required to melt iron (since the specific heats
for Al and Fe are 0.250 cal/g and 0.150 cal/g, respectively,
and the heats of fusion are 96 cal/g and 64 cal/g for Al and
Fe, respectively); even though the melting point for Fe is
more than twice that for Al.

(14) For steady-state heat transfer through plane walls and
ceilings

$$Q = - KA \frac{dT}{dx} \text{ (BTU/hr)}$$

where A is the area through which the heat flows, T is the
temperature, x is the wall or ceiling thickness. Since heat
flows from higher to lower temperature we can solve the above
for

$$Q = \frac{KA(T_1 - T_2)}{x_2 - x_1} = \frac{KA\Delta T}{\Delta x}$$

So, for the walls, Q_1 = (0.05)(9000)(68)/0.5 = 61200 BTU/hr.
and for the ceiling and roof, Q_2 = (0.05)(20000)(68)/1.17 =
58120 BTU/hr. Since the heat of fusion of sodium sulfate
decahydrate is 108 BTU/lb, and since its density is 97 lb/ft^3,
the minimum volume of salt storage will be 137 ft^3. If an
extra 6 inches of insulation are added, it can be observed
that the savings will be around 17000 BTU/hr. or 14%. This
is of course an important part of energy conservation, sometimes referred to as an energy resource.

(15) At any temperature

$$\Delta G^\circ(T) = - RT\ell n\ K_{eq}(T)$$

where for the dissociation of $CaCO_3$

$$K_{eq} = \frac{a^{Solid}_{CaO}\ a^{Gas}_{CO_2}}{a^{Solid}_{CaCO_3}} = P_{CO_2};\quad (P\ in\ atmospheres)$$

Therefore we can write

$$\Delta G^\circ(1210^\circ K) = - R(1210^\circ K)\ell n(1.77) = - 1373\ cal/mole$$

Since $\Delta G^\circ(T) = \Delta H^\circ - T\Delta S^\circ$, we could compare the standard free energies at two different temperatures:

$$\Delta G^\circ(T_1) = \Delta H^\circ - T_1\Delta S^\circ$$

$$\Delta G^\circ(T_2) = \Delta H^\circ - T_2\Delta S^\circ$$

So $\Delta S^\circ = - [\Delta G^\circ(T_2) - \Delta G^\circ(T_1)]/(T_2 - T_1)$, etc.

(16) Principal thermochemical systems include catalyzed reactions such as $2SO_3 \longrightarrow 2SO_2 + O_2$ and $CH_4 + H_2O \longrightarrow CO + 3H_2$; thermal decomposition reactions such as $CaCO_3 \longrightarrow CaO + CO_2$ and $Ca(OH)_2 \longrightarrow H_2O$; solution-dissolution reactions such as the mixing of water with acid (H_2SO_4) to generate exothermic reaction heat. In the methane cycle, hydrogen embrittlement must be considered a serious potential materials problem especially because of the high temperatures involved.

(17) For the device in equilibrium as shown, the energy-level diagram is depicted below along with the equivalent-circuit junction components. Note that the first feature of any energy-level system is to draw the Fermi level. This never varies. Biasing causes band bending (conduction and valance bands).

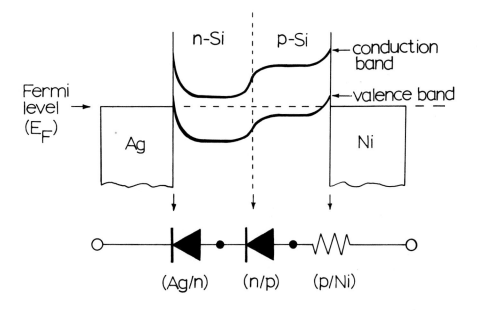

The metal work functions are larger than the semiconductor work functions (4.8 eV and 4.6 eV, respectively for silver and nickel). The rules for contacts are as follows: ($E_{w(m)}$, $E_{ws(n)}$, and $E_{ws(p)}$ refer to the metal, n-type semiconductor, and p-type semiconductor work function, respectively).

$$E_{w(m)} > E_{ws(n)} \blacktriangleright \text{RECTIFYING}$$

$$E_{w(m)} < E_{ws(n)} \blacktriangleright \text{OHMIC}$$

$$E_{w(m)} < E_{ws(p)} \blacktriangleright \text{RECTIFYING}$$

$$E_{w(m)} > E_{ws(p)} \blacktriangleright \text{OHMIC}$$

The metal junctions can also become active as voltage generators because of work function differences, i.e., consider the Peltier effect.

(18) If the boundaries are twin boundaries they would lie predominantly along {111} plane directions. This could be
verified by either transmission electron diffraction for
very small grain sizes or X-ray diffraction for very large
grain sizes. If the twin boundaries have energies given by
γ_{tb}/γ_{gb} = 0.055 (ratio of twin to grain boundary free energy),
then the twin boundary energy is very low since $(\gamma_{LV})_m$ =
730 ergs/cm^2 and at 500°C, $F_S \simeq$ 1.2 (730) + (1410 - 500)(.45)
$\simeq$ 1286 ergs/cm^2. So for γ_{gb}/F_S = 0.3 we have γ_{gb} = 386
ergs/cm^2 and γ_{tb} = 21 ergs/cm^2. This is a very low energy
indicating the twin boundaries are very well ordered, and
certainly not conducive to accepting adsorbed hydrogen
atoms. Other similar coincidence boundaries could have the
same effect since they are by their nature of low energy.
Even coincidence boundaries having energies twice the
coherent (111) twin boundary energy would still be low
energy since the grain boundary (high-angle) energy would
be an order of magnitude larger.

(19) The normal form (crystal structure) for Cu_2S is orthorhombic
(a = 11.9 Å, b = 27.3 Å, c = 13.5 Å) while CdS can have
cubic (sphalerite) (a = 5.8 Å), or wurtzite structure. It
is therefore not possible to develop a coherent interface
through structural accommodations by dislocations except
through epitaxial growth on specific orientations. This is
also the case for ZnO/CdTe junctions since ZnO is hexagonal
(wurtzite structure with a = 3.2 Å and c - 5.2 Å) and CdTe
is cubic (sphalerite structure with a - 6.5 Å). A coherent
interface can, however be created at the P-CdTe/n - CdS
junction interface if CdS is cubic. The misfit here, defined
by 100[a(CdTe) - a(CdS)]/a(CdTe), in percent, will be
100(6.5 - 5.8)/6.5 = 11%. This very large mismatch can be
accommodated by creating dislocations in the (001) plane
interface after every nine {010} planes in the CdS. That is,
an extra-half plane will be an (010) plane in the CdS.

(20)   Since Cu_2S is orthorhombic, dislocations can occur on several
different slip planes and will have a variety of Burgers
vectors. This is a large lattice and dislocations will be
influenced by temperature, stress, etc. If the CdS is cubic
(sphalerite) the slip plane will be {111} and the Burgers
vectors will be of the form $\frac{a}{2}$ <110>. If the CdS has the
wurtzite structure, the slip plane will be (00.1) and slip
directions (Burgers vectors) will have a form <110>. Since
the c/a ratio is > 1.6, slip is restricted to (00.1). Dislocations at the interface or within the matrix can cause
recombination and trapping and as a consequence the greater
the density the lower the device efficiency. A simple plot
could be made of open circuit current versus dislocation
density at some constant solar flux which would show current
decreasing as dislocation density increases. The curve
would probably be different depending upon whether dislocations were being measured in the junction interface or in
the matrix on either side of the junction (in the Cu_2S or
the CdS).

(21)   The zeta potential or electrokinetic potential (ζ), in
millivolts, is related to the electric double layer capacitance associated with a surface: $\zeta = \sigma/C$; where σ is the
charge per unit surface and C is an equivalent capacitance.
Such a double layer exists between any solid surface and an
ionic liquid. It can be measured by several electrokinetic
methods and at the point of zero zeta potential obviously
$\sigma = 0$. This means that the solid surface will be uncharged.
At this point mineral floculation will occur which can promote settling. On the otherhand, by adjusting the magnitude
and sign of ζ (or σ) dispersion of particles can be induced.
The zeta potential influences wet photovoltaic cell biasing
through the effect of hydrogen ion (pH) on the photoelectrodes. In CdS/sulfide electrolyte wet cells there is a

dependence of the point of zero zeta potentia. .PZZP) on
S^{2-}. In general pH changes are zero at the PZZP. Consequently, zeta potentials and their relationship to pH are
somewhat critical in determining wet photocell operation as
well as the flotation features for many minerals.

(22) While it is not known with any certainty how passivation
occurs, the most obvious mechanism involves bond satisfac-
tion. Consequently, hydrogen would be most effective
because of its valence and its small atomic size. While
magnesium is small, its valence would require it to satisfy
two dangling Si bonds simultaneously. This is certainly a
possibility but this mechanism is probably not as effective
as with atomic hydrogen. The sizes of N, F, and O are
somewhat larger and the electronic structure is not conducive to the type of favorable electron sharing provided by
H. However F could be effective because of its valence
although it is roughly four times larger than H. On considering the symmetric tilt boundary shown below, the
boundary dislocations are observed as sites of unsatisfied
Si bonds. Consequently, hydrogen would be expected to
selectively and chemically attach to these sites as shown.
The spacing of dislocations in the boundary, d, and therefore the number per unit length of boundary is related
approximately by $d = |b|/\Theta$, where $|b|$ is the magnitude of
the Burgers vector and Θ is the misorientation (in radians).
Since the hydrogen in the boundary will be associated with
individual dislocations, the difference in concentration
will simply be the ratio of dislocation spacings or
$\Theta_{15}/\Theta_5 = 3$. Consequently, the 15° boundary would have
roughly 3 times the hydrogen as the 5° boundary.

(23) It must be recognized in Fig. P.23 that the topography is exaggerated by the ion etching which is more rapid within the grain boundaries. This effect produces a "particle"-like typography. Careful inspection will indicate that the "particles" are not spheres but have in some cases well-defined facets. By thinning the substrate from the reverse side a thin film could be made of the black chrome. This would allow one to determine that the "particles" are actually contiguous grains. Auger electron or other surface analytical techniques could be employed in determining the composition of individual "particles". This could also be accomplished in the electron microscope (TEM) if instrumental attachments are available. This mode of analysis is referred to as analytical electron microscopy (AEM).

(24) High-resolution (lattice-imaging) transmission electron microscopy techniques are allowing for the direct observation

of amorphous structures, where it has been observed that in
some cases the "amorphous" structure is really a micro-micro
crystalline regime. (See for example the article by M.
Fukarrachi, K. Hashimoto and H. Yoshida, *Scripta Met.*, 13,
807 (1979).) Since amorphous silicon could be considered
a micro-micrograin regime, hydrogen passivation can be
effective by satisfying the multitude of unsatisfied bonds
in the microcrystallite "boundaries". Furthermore, the
large "grain boundary" area in such a micrograin regime
would be conducive to rapid diffusion of hydrogen throughout. Dangling bonds are reduced in hydrogenated amorphous
silicon by roughly a factor 10^4.

(25) Some particular, contemporary insight can be gained especially from part II of the first reference [R.S. Claassen and
A.G. Chynoweth, *Mater. Sci. Engr.*, 37(1), 31 (1979)]. A
very extended historical perspective with regard to metallurgical technology and its very early development can also
be gleaned from "De Re Metallica" (translated by H.C. Hoover
and L.H. Hoover and published by Dover Publications, Inc.,
New York, 1950). This work describes very early metallurgical technology up to the 16th Century. While this may
seem remote in the context of modern times and future needs,
the importance lies in an understanding of the connection
between the needs of a period, the existing social and
economic systems, and the availability of materials and
methods of production, etc.

INDEX

A

Absorbers
  alumina composite, 281–316
  selective solar, 24, 28
Absorption
  curve, ideal, 293
  electronic, 152
  interband, 152
  ionic, 152
  molecular dipole, 151
  optical coefficients for solar cells, 511
  of solar radiation, 152
Absorptivity, spectral, 237
Accelerated life tests, 323
Adhesion of materials, 85, 337
Adsorbate–solid interactions, 108, 109
Aerosols, 218
Alkali nitrate salt, mixtures, 328
Alloys
  high strength development, 329
  high temperature, 327
  hydrogenated amorphous silicon, 665
  plasma-deposited, 666–685
  silicon–hydrogen, 683–727
Alumina composite absorbers, 281–316
Aluminized acrylic film
  reflectance for, 179
Aluminum
  polished, 190
  reflectance properties of, 185
Amorphous solids, 55–60
  extinction coefficient for, 291
  hydrogenated silicon, 553–556, 665, 683–692
  photovoltaic, 493, 494, 517, 533, 551–556
  plasma-deposited, 666–689
  silicon, 552–556, 665
  structure of, 55–60, 685
Anhydrous salts, 377
Antireflection coatings, 577
Aqueous systems, 334

Array
  field, 492
  photovoltaic, 492
a-Silicon (amorphous silicon), see Amorphous solids, silicon
Atom probe, 107
Auger electron spectroscopy (AES), 104, 105
Auto-electrodeposition, 82, 88

B

Back-surface
  field cell, 518
  reflector cell, 518
Band diagrams
  heterojunction, 495, 592–594, 606, 610, 615
  homojunction, 495
  metal-insulator–semiconductor, 495
  metal-semiconductor, 495
  for photoelectrolysis cell, 627
  for simple electrochemical devices, 495, 622, 624
Bandgap
  Fermi level in, 666
  fundamental semiconductor, 666
BET (Brunauer, Emmett, Teller) method, 101
Bidirectional reflectometer, see Reflectometer
Binary phase diagram, 60
Biofouling, 130
Black-chrome
  absorptance, 85
  adhesive strength of, 85
  clusters on substrate, 84
  coating, 278
  degradation, 124
  reflectance, 25, 164
  SEM observations of, 157
  transmittance, 163

Bonding
  monohydride, 693
  polyhydride, 693
Botryoidal growth, 88
Bravais lattices, 56
Burgers vector, *see* Dislocation
Burried junctions (homojunctions), 586

                         C

Calomel electrode, standard, 629
Catalysis, 115
  in thermochemical reactions, 472–482
Catalyst development, for thermochemical
    reactions, 479–482
Catalyzed reactions, 472, 474
Cation-anion, vacancy pair, 62
Cementation, of metal ions, 321, 370
Cermet, 157, *see also* Composite materials
Chemical heat pump system, *see* Heat pump
Chemisorption, 115
Chlorosilane, 678
Chromium black, *see* Black-chrome
Clausius–Mossotti equation, 155
Coatings
  antireflection, 577
  black-chrome, 278
  degradation of, 336
  graded composition, 297–302
  protective, 191
  selective solar, stability of, 334
Collector, *see* Solar Collector
Color center, 54, 85
Composite film selective absorbers, 277–316
  engineering of, 292–316
  microstructure, 280
  performance parameters for, 306
  physical and optical properties of, 279–
    292
  production of, 279, 280
Composite materials
  spectral selectivity of, 255–275, 277–316
Composition depth profile, 300
Compound formation, 60
Concentration, solar, 38
Concentrator devices, photovoltaic, 571–
    575
Concentrators
  point focus, 362
  solar, 171, 336–338
  standards for, 336–338
  thermal systems, 325
Conductance, zero bias, 653

Conducting mesh, 261
Contact resistivities, 588
Conversion processes, solar, 34–47
Coolant chemistry, 333
Copper emissivity, 247
Corrosion, 117–119
  electrochemical relationships for, 351
  erosion, 370
  fatigue, 321, 370
  galvanic, 321, 371
  intergranular, 371
  kinetics, 347, 348
  rate calculations, 352–366
  of receiver tubes, 335
  science, 346–365
  sodium, 329
  of solar materials, 321–365
  stress, 321, 372
  of thermochemical systems, 447–479
Creep, 322
Critical materials, issues identified, 14
Crystal imperfections, *see* Crystal lattice de-
    fects
Crystal lattice defects, 54–88
  electrically active, 665
Crystal structures, 56
  of silicon-hydrogen alloys, 685
Cuprous oxide
  degradation, 128
  solar cell, 128, 129

                         D

Dark mirrors, 263–272
Deformation gradient wavelength, 77
Density of states, 517
  grain boundary, 656
Depth profile, 300
Desiccants, water vapor sorption by, 127
Device physics, 498–506
Dichroic mirror, 41
Dielectric
  coatings, *see* selective solar absorbers
  mismatch, 298
  response, 152
Diffusion length, in solar cells, 512
Digital AC impedance measurements, 356–
    360
Direct bandgap materials, 510, 512
Dislocation, 54
  Burgers vector, 63–65, 71
  cells, 69
  edge, 63–65, 71

line, 62–67
loop, 66, 87
partial, 65, 70
screw, 63–65
total, 64
Dissociative attachment, 681
Dissociative ionization, 681
Drude electron model, 235–237
Dust
  adhesion, 215–219
  cleaning, 222, 223
  deposition, 215–219
    study of, 219, 220
  particles, 214–219
  on solar surfaces, 29, 205–210

E

Effective sunshapes, 183, 184
Effective medium theory, 153–157, 281–290
Electrochemical cell, 495, 496
Electrodeposit, 67
Electrolyte–semiconductor junction, 495, 496, 627
Electromigration of ions, 638–645
Electron-beam deposition, 302
Electron-impact processes, 679
Electron microprobe, 107
Electron microscopy, 66–69, 146
Electron transmission image, 66, 68, 69
Elements, heat of fusion/entropy of fusion, 410–414
Emissivity
  calculated, 301
  changes by sintering, 328
  of composite films, 266–272, 301
  effective, 258
  as a function of
    d.c. resistivity, 249–251
    temperature, 274, 309
  of graded composition films, 301
  of heat mirrors, 258
  hemispherical, 231, 242
  of high temperature selective surfaces, 233, 309
  of metals, 229–254, 328
  model corrections, 245–248
  normal, 230, 241
  spectral, 233
  total, 239–248, 265
Energy band diagrams, *see* Band diagrams
Energy gap, in solar cell (photovoltaic) materials, 507, 509

Entropy of fusion, 377
  of compound (tables), 431, 432
  of elements (tables), 411–414
  periodic system, 430
  of salt hydrates (tables), 379, 380, 390
Epitaxial growth, 83, 564
Erosion, 370
ESCA, 104, 105
Exothermic reactor, 444
Extinction coefficient, 210–213

F

Fatigue
  corrosion, 322, 370
  thermal, 322, 330
F-center, 61
  aggregate, 85, 87
Field emission microscopy (FEM), 103, 111
Field-ion microscopy, 57
  images, 58, 59, 72, 84, 87
Figure of merit, 294, 295
Flat-band potential, 629–645
Fretting, 371
Fusion, heat of (tables for salt hydrates), 379, 380
  for bromides (table), 428
  calculation from entropies of fusion, 386, 387, 416
  calculation from heat of solution, 385, 386
  for chlorides (tables), 418–420
  of compounds (tables), 431, 432
  for elements (tables), 410–414
  estimating, 405–410
  for fluorides (tables), 422, 423
  high values for, 434
  for hydroxides (table), 427
  for iodides (table), 429
  for oxides (tables), 424, 425

G

Gallium arsenide thin film, 526, 563–565
Galvanic attack, 321
Gibbs dividing surface, 76
Glassy metal, 57
Glow discharge, 668
  mass spectrometry (GOMS), 107
  optical spectroscopy (GDOS), 107
  rf features, 669–672
  silane, 673–675
Gold
  black, 157, 158

Gold:
  reflectance properties of, 185
Graded composition coating, 297–302
Grain boundary, 71–79, 119
  dislocations (GBD), 71–76
    extrinsic, 73–75
    intrinsic, 72, 73, 75
  effects in polysilicon, 645–662
  electrical nature of, 647–650
  energy, 71
  large-angle, 71, 72
  ledge, 73, 76
  low-angle, 71
  misorientation, 71
  in molybdenum, 73, 75
  passivation, 561, 650–662
  in photovoltaic materials, 551, 557–563
  structure, 71–76
Grain size effect, 77–79
  on minority carrier lifetime in solar cells, 515
  in polycrystalline silicon, 556–563
  on reflectivity, 79
  on resistivity, 79
  in solar cells, 556–563
  on yield stress, 77

                     H

Hagen–Rubens limit, 238
Hall–Petch law, 77
Heat of fusion, see Fusion, heat of
Heat of solution, 385
Heat mirrors, transparent, 257–263
Heat pump, chemical, 448, 467, 468, 469
Heat storage materials, selection of, 405–
  435
Heat transfer materials, 34
Heliostat, 32, 33
Heterojunction, 494, 495
  CdS/CdTe, 527, 594
  CdS/InP, 597
  ITO/CdTe, 532, 607
  ITO/InP, 532, 611
  tables of properties, 525–533
  theoretical efficiencies, 509
  ZnO/CdTe, 601
High temperature alloys, 327
  carburization, decarburization of, 327, 370
High temperature reactor (HTR), 448
High-temperature selective absorbers,
  302–316
  production of, 302–312
  stability of, 312–316

Homojunction, tables, 525, see also, Photovoltaic device or solar cell
HTR, see High-temperature reactor
Hydrogen passivation, in polysilicon, 650–
  662, 665–727
Hydrogenated amorphous silicon alloys,
  553–556, 665–727

                     I

Imperfections, see Crystal lattice defects
Incident angle properties, 192
Incongruent melting, of salt hydrates, 378
Indirect bandgap materials, 501, 510, 512
Indium–tin–oxide (ITO), 496, 532
  devices (table), 532
Interfaces, 54, 67–84, 97–112, 119, 120
  adhesion at, 337
  characterization of, 101, 104–112
  in photoelectrical devices, 624
  photovoltaic, 591, 592
  polymer/metal (oxide), 125–127, 130–133
  semiconductor/electrolyte, 624, 628
  silver, 121–124
  in solar cells, 591, 592
  solid/liquid (S/L), 97
  solid/solid (S/S), 97
  solid/vapor (S/V) or solid/gas (S/G), 97
  structure of, 67–84
  types of, 97
Interstitials, 54, 61, 62
Ion etching, 104
Ion scattering spectroscopy (ISS), 104, 105
  results on copper oxide films, 133–136
  results for polypropylene, 136–140
Ion boride, 59
ITO (indium tin oxide), 496, 532
  devices (table), 532, 607–615
  energy band diagram for, 610, 615
  junctions (table), 532, 607–615

                     J

Junction capacitance measurements, 590,
  591
Junction device, see Photovoltaic device
Junction device, edge multiple vertical, 572,
  573

                     K

Kirchoff's law, 292, 294
Kramers–Kronig transform, 235

L

Laser optics, 220
Latent heat, 340
Lattice matches, in semiconductors, 576
Ledge
  grain boundary, 73–76
  surface, 81
Leaching, selective, 372
LEED (low-energy electron diffraction), 112, 119
Light diode parameters, 589
Linear polaization method, 354
Line defects, *see* Dislocation
Liquid metal systems, 328

M

Materials
  cover plate, 18
  development framework for, 4
  mirror, 19
  research, 15
  thermal properties of, 18
Maxwell–Garnett theory, 153–157, 281–290
  see also Effective medium theory
Melting of salt hydrates, 377–380
Metal colloids, 85
Metal/dielectric composite, 277–316
Metallized plastics, 189, 190
Metal-plus-dielectric film, 266–269
  plus absorbing film, 269–272
Methanation reaction, 447, 448, 452
M–G theory (Maxwell–Garnett theory), *see* Effective medium theory
Minority carrier lifetime, 515, 516
Mirror materials, 171–194, 257–262
  figure of merit for slab type, 22
  glass, 29
  incident angle properties of, 192
  mechanical properties of, 22
  soiling on, 199–224
  specular reflectance of, 19
MIS (metal-insulator–semiconductor) device, 496, 531, 586, 603–606
  energy band diagram for, 606
Module, photovoltaic, 491, 492
Molten salt systems, 328
  stress corrosion cracking in, 331
Mott-Zener limit, 238

N

Nickel electrodeposit, 67, 68, 72
Nucleation, 81–84

O

Ocean thermal energy conversion, 345
Oil systems, solar, 334
Optical electron spectroscopy (OES), 673–679
Optical measurements, 173–182
Optical tailoring, 162–167
Optoelectronics, properties, 691
  behavior in silicon alloy solar cells, 707–727
Order-disorder phenomena, 55–60
  in silicon-hydrogen bonding, 684, 685
OTEC (Ocean thermal energy conversion), 345
Oxidation of solar materials, 323, 326
Oxide semiconductor photovoltaic junctions, 601

P

Particle distribution on mirror surfaces, 216–219
Passivation
  of amorphous silicon, 665
  grain boundary, 561
  in silicon–hydrogen alloys, 687, 688
Phase boundary, 80–85, 119
Photobiological effects, 47
Photoconductivity
  anomalous, 722–724
  in silicon alloys, 711, 718–727
Photoconversion, 37, 44
  paths, framework for, 37
Photoelectrochemical cells (PEC), 620–645
  energy level diagram for, 622, 624, 627
  potential behavior, 627–645
Photoelectrochemical systems, 46
Photoelectrode, semiconducting, 626
Photoelectrolysis, 46, 620, 630–637
  cell, 621, 627
  optimization, 637
Photoemission spectroscopy (PES), 689
Photoluminescence, in silicon hydrogen alloy solar cells, 703–707
Photolysis, of silane, 677
Photovoltaic converter, 572–575

Photovoltaic device, 489
  amorphous, 493, 494, 517, 533, 551–556,
    665–727
  array, 571
  categories, 493
  concentrators, 571–575
  degradation, 570
  development, 571
  effect of grain size on, 556–563
  efficiency, 509, 560
  energy gap, 507
  heterojunction, 509, 525–533, 585
    band diagrams, 592–594
  hierarchy, 490–493
  homojunction, 494, 499, 525, 585, 586
  introduction to, 489–499
  lattice matches, 576
  materials, 506–509, 525–533
  module, 492
  oxide/semiconductor, 601
  physics of, 498–506
  polycrystalline, 493, 556–563, 645–662
  single crystal, 493
  thin film, 493, 552–571
  wet, 623
Photovoltaic effect, 492
Photovoltaic materials, properties and re-
    quirements, 506–517
Photovoltaic systems, 39
Planar defects, 67–86
Plasma
  chemistry, 680–683
  deposition, 666, 668
  etching paradigm, 673, 674
Plastics, metallized, 189, 190
P-n junction, 492, 585
Point defects, 61, 62
  aggregates of, 54
Point of zero zeta potential see PZZP
Polarization characteristics, 354, 372
Polished aluminum, 190
Polycrystalline silicon, 556–563
Polypropylene/copper oxide interface, 130
Polypropylene degradation, 130–133
Polysilicon, grain boundary effects in, 645–662
Pourbaix diagrams, 341, 350, 372
Protective coatings, 191
Pseudomorphic growth, 83
PZZP (point of zero zeta potential), 630

Q

Quantum efficiency, 589, 590, 611–614
Quantum processes, 36

R

Receiver, central tower, 33, see also Helio-
    stat design, 332
Recombination and trapping
  of electrons, 81
  in solar cells, 514
Reflectance
  of composite, 314, 315
  effect of dust on, 205–210
  hemispherical, 30, 189–205–208
  of mirror materials, 19, 184–191
  specular, 31, 32, 173, 177, 184, 205–208,
    213
  versus wavelength, for black chrome on
    nickel, 25
Reflective coatings, degradation of, 336, 337
Reflectometer, 175, 176, 181
  bidirectional, 31, 175, 176
Reflector, parabolic, 23
Reflector structure, typical, 186
Refraction, index of, 29, 283
Reversible chemical reactions, 439–454
Rutherford back scattering (RBS), 107

S

Salt hydrates, 377–399
  bromides, 390
  calorimetric measurements of, 388, 389
  chlorides, 397
  crystal seeding of, 381–383
  heat removal from, 384
  iodides, 390
  nitrates, 397
  nucleation, 381–383
  properties of (tables), 379, 380, 390, 399
  rate of crystal growth, 383, 384
  sulfates 393
  supercooling of, 381
SCANIR, see Surface analysis of neutral
    and ion impact radiation
Scanning electron microscopy (SEM), 88,
    112, 161
Scattering function, angular, 210
Scattering theory, 210–215
SCE, see Standard calomel electrode
Schottky barrier, 495, 586
  devices, 530, 586, 693–699
  junction, 586
Schottky diode, 693–698
  see also Schottky barrier
Secondary ion mass spectroscopy (SIMS),
    104, 105, 554, 678

Selective absorber, *see* Solar absorber
Semiconductor/electrolyte interface, 624,
    628
Semiconductor surfaces, 116
Silane (SiH$_4$), 667
    dissociation reactions, 680
    glow-discharge decomposition of, 667,
        673
    photolysis, 677
Silicon-hydrogen alloys, 683–692
    characterization of, 683–727
    doping in, 699–702
    effects of impurities on, 707–727
    oxygenated, 714
    photoluminescence, 703–707
    role of hydrogen in, 686–692
    solar cell effciency of, 716
    structure of, 685
Silicon solar cell, *see* Solar cell or Photovol-
    taic device
Silver, reflectance properties, 185
Silvered glass, 187–189
Silver–glass interface, 122, 123
SIMS, *see* Secondary ion mass spectros-
    copy
SIS (Semiconductor–insulator–semi-
    conductor) device, 495, 496, 532,
    586, 603–606
    energy band diagram for, 606
Soiling on solar mirrors, 199–224
    cleaning cycle for, 202, 203
    natural, 200
Solar absorber, surfaces
    engineering of, 292–316
    high temperature, production of, 302–311
    optical tailoring of, 162–167
    optimized, 167
    properties of, 17
    selective composite, 277–316
    spheroid model for, 158–162
    stability at high temperature, 312–316
Solar array, photovoltaic, 492
Solar battery, *see* Photovoltaic device
Solar cell, 40–44, 489, 491
    absorption coefficients, 511
    amorphous, 493, 494, 517, 533, 551–556
    arrays, 490–493
    backwall configuration, 566
    cascade, 572
    concentrators, 571–
        575
    defects in, 512
    degradation, 570
    development, 571

doping, 514
    effect of grain size in, 556–563
    efficiency, 497, 505, 560
    equivalent circuit, 504, 505
    generic, 491
    heterojunction, 509, 525–533, 585
        band diagrams, 592–594
        characterization, 587
        efficiencies, 509
    homojunction, 499, 508, 585
    lattice matches, 576
    minority carrier lifetime, 513
    module, 492
    physics, 498–506
    planar, 567
    reverse breakdown voltage, 590
    Schottky, 717
    Schottky barrier, 496
    silicon, 40–42, 492, 493, 496, 551–571,
        645–662
    silicon–hydrogen alloy, 665–727
    spectral splitting, 43
    stacked, 42
    tables, 525–533
    textured, 567
    thickness effects, 512
    thin-film, 494, 552–571
Solar collector
    cleaning and maintenance, 344
    efficiency, 26, 27
    flat-plate, 26, 171, 342, 343
        efficiency of, 26
    parabolic–cylindrical, 14, 20
    soiling, 199
    types, 16
Solar energy
    collector, 8, 14, 20, 26, 27
    components, functions, and disciplines, 9
    conversion paths, 5
    conversion processes, 6, 34
    cost reduction, 7
    discipline relationships, 12–14
    storage, 338, 377
Solar mirror materials, 171–194
    cleaning cycle for, 202, 203
    effects of soiling on, 199–224
    reflectance properties, 177
Solar photovoltaic device, *see* Photovoltaic
    device
Solar power plant, 446
Solar receiver materials, 325–327
Solar selective absorbers, 24, *see* also Solar
    absorber
Solar selective step function, 165

Solar spectra, 502
Solar storage
  latent heat, 340
  in salt hydrates, 377–399
  thermal, 338
  thermochemical, 36, 341
Solar thermal sytems, 324, 439–483
Solid forms, 102
Soltherm, 452
Space-charge region, 76
Space heating/cooling, 448, 449
Spectral selectivity, of composite materials,
    255–275
Spectroscopy
  Auger electron, 104, 105
  ion scattering, 104, 105
  secondary ion mass, 104, 105
  X-ray photoelectron, 104, 105
Specular reflectance, effect of dust on,
    205–210
Spheroid model, 158–162
Sputtered ITO/CdTe junctions, 607–611
  energy band diagram for, 610
Stacking fault, 67, 70
  energy, 69
  intrinsic, 70
Standard calomel electrode, 629, 631
Steam generator tubes, caustic cracking of,
    330
Stress corrosion cracking, 321
  by molten salts, 331
Substitutional impurity, 62
Sunshapes, 183, 184
Supralinearity, in anomalous photoconduc-
    tivity, 718–727
Surface analysis of neutral and ion impact
    radiation (SCAN 11 R), 107
Surface
  analysis, 101, 104–112
  area, 101
  characterization, 101, 104–112
  clean, 101–103
  composition, 103–106
  corrosion, 117–119
  equilibrium shape of, 58, 107, 108
  figure, 173
  ledge, 81
  properties, measurement of, 112
  purity, see Surface, composition
  reactions, 116
  real, 101–103
  recombination, 516
  science of solar materials, 112–130
  structure, 80, 81, 97–100, 103
  terrace, 81

thermodynamics, 107, 108
topography, 103
types (S/S, S/L, S/V or S/G), 97

T

Tafel extrapolation method, 352, 353
Teflon, aluminized, 31
Thermal decomposition reactions, 473
  entropies of (table), 471
  volume changes for (table), 475
Thermal energy
  entropy considerations, 469
  storage, 377, 405, 439, 443, 450
  transport, 445–447, 451–453
Thermal fatigue, 322
  of receiver tubes, 335
Thermochemical concepts development,
    459
Thermochemical reactions, 459–483
  catalysis of, 479–483
Thermochemical storage, 36, 439–483
  characteristics of, 440–443
  materials corrosion, 477
  thermodynamics of, 459–472
Thin film
  a–Si:H materials, 667
  device, see Photovoltaic device
  solar cells, 552–571
  surfaces, 116
Transmitters, solar, 28
Tungsten, emissivity of, 247
Twins, 68–70

U, V

Unit cells, 56
Vacancies, 54, 61, 62
Vacancy–inst<ial pairs, 61, 62
Voterra defects, 62
Volume defects, 85, 86

W

Water of crystallization, 378
Wet photovoltaic cell, 623
Wind tunnel, 219

X, Y, Z

X-Ray photoelectron spectroscopy (XPS or
    ESCA), 104, 105, 161
XPS, see X-Ray photoeclectron spectros-
    copy
Zeta potential, see PZZP